A CITIZEN'S DISCLOSURE ON UFOS AND ETI

VOLUME FIVE

EVIDENCE OF A TYPE TWO CIVILIZATION IN OUR SOLAR SYSTEM

TERENCE M. TIBANDO

A CITIZEN'S DISCLOSURE ON UFOS AND ETI

VOLUME FIVE

EVIDENCE OF A TYPE TWO CIVILIZATION
IN OUR SOLAR SYSTEM

Copyright Page

In writing this book, I sought out the best possible evidence available on this subject whether that was from numerous UFO and ETI related books, networking with other UFO authors, researchers and first-hand witnesses to UFO sightings, from films and TV documentaries, from internet searches, or just from my personal sightings and contact experiences.

When material is quoted in this book full acknowledgement is given to the author or source of that material as indicated by the extensive bibliography, webliography and videography at the back of the book.

When photographic images are used in this book that are obtained from the internet, usually from Google Images, a full search was made to determine copyright information, the author's name, or address or email address or phone number or copyright mark in order to asked permission to use their photographs. In almost all cases where such images are posted to the internet, there was no satisfactory way to identify the owner of the image even through Google because they left no identification of themselves to be found. When an author's name does appear on an image, written permission was sought or it was not used at all; more often than not, there usually was no reply or response back from the owner.

This lack of due diligence to place a copyright mark or the owner's name is all it would take for that person to claim ownership of a picture, yet the lack of it creates major problems for many people, especially for other authors who lawfully seek their permission to use their photo images.

Because this book is one of six volumes in a series created as public educational material and is of a transitional nature, and I have quoted or referenced the websites from where I obtained the photo images and therefore, I am invoking the Fair Use Doctrine also known as Fair Usage Clause to publish these images in my book.

I will of course give full acknowledgement and credit to the author's and owners of such images in all my future book publications in recognition of their work if they come forward to be identified.

A CITIZEN'S DISCLOSURE ON UFOS AND ETI

VOLUME FIVE

EVIDENCE OF A TYPE TWO CIVILIZATION IN OUR SOLAR SYSTEM

TERENCE M. TIBANDO

"Hggna"

A Cosmic Cousin
Publication

Other Publications by the Author Page

Although, this book is the author's first publication it forms a part of six smaller books or volumes that was originally written as one massive tome of UFO and ETI information entitled: **"A Citizen's Disclosure On UFOs And ETI Visiting The Earth"** which began in March 2009 and was completed in August 2016.

Other books/volumes by the author in this series:

1. Book One (Volume One): **"Global Evidence of the UFO and ETI Presence"**

2. Book Two (Volume Two): **"UFO Disclosure and Covert Programs of Deception"**

3. Book Three (Volume Three): **"Military Intelligence Industrial Complex, USAPs and Covert Black Projects"**

4. Book Four (Volume four): **"In Search of Extraterrestrial Intelligence"**

5. Book Five (Volume Five): **"Evidence of a Type Two ET Civilization in Our Solar System"**

6. Book Six (Volume Six): **"The Rosetta Stone of ETI Contact and Communications"**

Introduction

This book is the fifth volume of "A Citizen's Disclosure on UFOs and ETI" which originally formed part of a 3500 page encyclopedic tome. It provides the latest and best evidence on this subject matter that comes from my 65years of experience dating back to 1953 when I and my mother witnessed flying saucer type craft hover over a Canadian Air Force Base in St. Jeans, Quebec. This was followed by an ET visitation by three ghost-like beings that came into my bedroom a couple of days later.

Over my lifetime, there followed many more UFO sightings and ETI encounters and interactions many in group witness settings as a part of a CSETI field expedition to initiate human contact and communications with these diverse extraterrestrial beings visiting our planet.

This phenomenon is real, regardless of what you have read or have been told by the mainstream media or the science community. There is no transparency in their public disclosure but rather deception, lies and cover-ups designed to betray the public's trust in officialdom. Increasingly, the distrust toward the official position that dispenses their version of knowledge and truth has become suspect which has lead to a public awaking to start investigating truth for themselves.

This fifth volume in a series of six volumes hopes to add to the public's resource for truth and honesty which will lead the reader to do further research and investigation on this subject. The unfettered self-investigation of truth is a God-given right of everyone to understand the world in which they live in.

As you read through this volume and the other volumes in this series, there will be much that is familiar to the senior UFO researcher and much that is new, especially with some alternative and perhaps, controversial perspectives that re-evaluate some long, widely-held traditional beliefs held within the UFO Community.

This book and the other volumes of "A Citizen's Disclosure on UFOs and ETI" in this series connects the dots and assembles the pieces of the UFO and ETI puzzle to form a much larger picture of this phenomenon. The implications and potentialities for humanity derived from understanding this interstellar phenomenon are vast which could positively revolutionize our future for generations to come. The task at hand is for the mainstream science community and our respective governments and its officials is to be honest and straight forward with the general public in what it discloses on this subject.

In writing a book of this complexity and length, errors in content are hopefully minimal or non-existent, and although with careful research and entry of evidence, errors inevitably creep in because new evidence comes to light to prove an alternative reality. This author has endeavoured to minimize such errors as the UFO database is already corrupted and such errors seem to be perpetuated due to poor research or the deliberate ignoring or marginalizing of real facts. With this in mind, corrections have been made where necessary while still maintaining the integrity of the evidence as it was originally reported.

With such material gleaned from my predecessors in understanding the "Big Picture" of this phenomenon, full credit and acknowledgement is given to these Ufologists in the footnotes (author's rants), web links and bibliography of this book.

Please note the colour format in quoting from a source is used to distinguish the web links from video links and bibliography which is indicated in the Bibliography section of this book.

In essence, this book is an interactive book, perhaps the first of its kind, where web links and video links are used to enhance the research value for the student and the seasoned Ufologist and thus, it extends a 3500-page textbook into a 10,000-page tome!

The reader will be able to follow the web link references to see where I have collected the information for this book. Unfortunately, at the time of publication of this book, some of these web links may no longer be active, it is the nature of the internet and it ephemeral web links!

My apologies for any inconvenience this may cause, however, try "googling" a specific word or phrase; this may reveal the same or a similar website!

This book should be considered a reference manual only, leading to further research and investigation for the serious UFO researcher.

As the reader goes through this fifth volume, they will realize the extent to which there has been a suppression and cover -up of this phenomenon by many sectors of society, especially by the military, the intelligence and science communities, as well as the private industrial sector and by our elected government officials. The cover up of the existence UFOs and ETI is interlaced with deceptive and misleading research programs from the above mentioned communities of officialdom.

Back-engineering of the alien spacecraft by WWII Nazi Germany, and by postwar Britain, Canada, and especially the USA has been in full over-drive by those covert arms of the military to discover the science of the alien propulsion system. This reverse engineering also includes the understanding and manipulation of alien DNA and its various hybridization PLF programs. This also includes military engagements and shoot-down programs as well as tracking, monitoring and crash retrieval programs. All the while the news media has been coerced and threatened to no longer function as the public's "Fifth Estate" having been usurped by the military industrial intelligence community. Marginalization and ridicule hold the public's interest in this phenomenon is at bay while the M.I.I.C. leap generations ahead of mainstream science jealously guarding its secrets with threats , even death to those who get to close to the truth.

The depths and convolutions of going down the rabbit hole to discover the truth behind the UFO and ETI phenomenon include deep black projects, SAPs, USAPs, military super weapons, even a secret black space program and a false flag alien invasion agenda! The corrupt and evil machinations of the M.I.I.C. know no bounds and are being played out as a lob-side chess game against an unsuspecting public.

Complicit in this cover and suppression of UFO and ETI knowledge has been the government agency that was supposed to be a public agency, but still remains an agency of the Department of Defense, namely, NASA!

This this volume, we will continue our virtual tour at Mars and then, venture to the outer planets and their moons in our Solar System and beyond. The search for life on other planets is only one of the mission agendas of most space probes sent from Earth, but the real quest is in search of alien technology and marking its location for future manned space missions, when and where possible!

Mars does not disappoint on all levels of exploration as Mars was and may still be home to a very ancient civilization recovery from an ancient global cataclysm whether natural or through warfare! We hope to discover which end brought the destruction of Mars in this book!

As we venture to Jupiter, Saturn, Uranus, Neptune, and to Pluto and beyond, we discover that the Zone of Habitability extends beyond our solar system and the cosmic surprises will be mind-boggling!

This book can help in changing the outcome of that 'chess game' in favour of the common people and the laying of the foundation of a good hopeful future! **Enjoy the ride!**

Terry Tibando --- September 2015

BOOK FIVE (VOLUME FIVE)
EVIDENCE OF A TYPE TWO CIVILIZATION
IN OUR SOLAR SYSTEM

CHAPTER 81

THE MONUMENTS OF MARS - A TYPE TWO CIVILIZATION IN RUINS

Signs of Intelligent Life – Cydonia, Pyramids, and the Face on Mars

The one remaining question that has not been asked or proven thus far is the one that is the proverbial "nail in the coffin" in terms of evidence for life on Mars, "has there ever been intelligent life on Mars similar to human beings of Earth, if so, what evidence is there to support their former existence? A subset to this important question: "Is there any artificial structures left behind on the Martian surface that were constructed eons ago"? "Did a Martian race leave any technology behind on Mars as evidence of their civilization"? "What happened to the planet Mars and the Martian civilization"? "Do Martians still exist today"?

No other question is more disturbing in its implications to humanity than the question of other intelligent life in the universe, particularly a life that may exist so close to our own as to be neighbours. Given our current state of our mindless technological achievements run amok and our atrophied spiritual development in not recognizing the unique age in which we live, we are sadly still a planet divided, in desperate need of unification and peace with itself! Yet, from out of this chaos of global adolescence, we have begun searching for other intelligent life on our closest planetary neighbor, Mars.

There is a visceral, almost genetic realization that life exists on Mars and that somehow we are all connected to it in some way. Only honest and open disclosure from every discovery made by the various space agencies now exploring Mars will reveal that connection.

When **NASA** and the **Russian Space Agency** first sent its space probes to Mars, almost immediately signs of intelligent life in, on and around Mars made itself known. Even before the question of whether or not, "did Mars have an atmosphere conducive to life?" even, before the question of "is there water on Mars?" even, before the question was asked, "is there plant life on Mars?" even before the question arose: "is there animal life on Mars?" Before all these questions were even considered, the evidence for Extraterrestrial intelligent life on Mars had been confirmed.

It was in fact the **Viking 1 spacecraft** orbiting around Mars photographing its surface that it imaged an unusual feature in an area of Mars that borders plains of **Acidalia Planitia** and the **Arabia Terra** highlands known as the **"Cydonia Mensae"** region which sparked a furor of controversy and debate among planetary scientists and generated a lot of public interest. Ufologists and professional debunkers inevitably weighed in on the debate as to whether an unusual Viking photograph was the first indication of intelligent life that may have existed on Mars in the distant past. The photograph in question showed an unusual mesa that NASA Viking

chief scientist **Gerry Soffen** referred to as simply a trick of "light and shadow", an albedo illusion; nothing more than that. But, to the general public it has become known as the **"Face on Mars",** an enigma that screamed out artificiality and high intelligence.
http://en.wikipedia.org/wiki/Cydonia_%28region_of_Mars%29

On July 25, 1976, in one of the images taken by Viking 1, a 2 km (1.2 miles) long Cydonian mesa, situated at 40.75° north latitude and 9.46° west longitude, had the appearance of a ***"humanoid face".*** Could this just be a natural geological formation, an optical illusion, a **pareidolia** seen at different times of the day, similar to geological rock or cliff formations found on Earth like the ***"Old Man of the Mountain",*** the ***"Pedra da Gávea",*** the ***"Old Man of Hoy"*** and the ***"Badlands Guardian"?***
http://en.wikipedia.org/wiki/Cydonia_%28region_of_Mars%29

The Cydonia region and the "Face" on Mars taken by the Viking 1 orbiter was released by NASA/JPL on July 25, 1976. It has sparked incredible controversy between NASA, scientists, and the public. The black speckled dots in the image are data errors
https://en.wikipedia.org/wiki/Cydonia_(region_of_Mars)

The **"Face on Mars"** was first discovered by **Dr. Tobis Owen** and **Gerald Soffen** at NASA in the same year the **Viking 1 and 2 space probes** started orbiting and photographing Mars. However, the discovery by Owen and Soffen of the "Face" seem to have been quietly filed away in NASA's photographic archives. First interpretations of the image would seem to support the "trick of light and shadow" official position of NASA and only further photographing of the area would confirm or refute NASA's position on the matter. Fortunately, there were other photos taken of the area which clearly showed a second image, in photo 70A13 **"the Face"** was seen

16

clearly again staring back up into space and was acquired by Viking, 35 orbits later at a different sun-angle from the original 35A72 image.

This latter discovery was made independently by **Vincent DiPietro** and **Gregory Molenaar**, two computer engineers at NASA's **Goddard Space Flight Center**. It was DiPietro and Molenaar who discovered the two "*misfiled*" images, **Viking frames 35A72 and 70A13**, while searching through NASA archives. It would appear as if NASA had deliberately hidden these photographs from any public scrutiny, except that DiPietro and Molenaar were NASA employees and they weren't about to play ball with NASA.

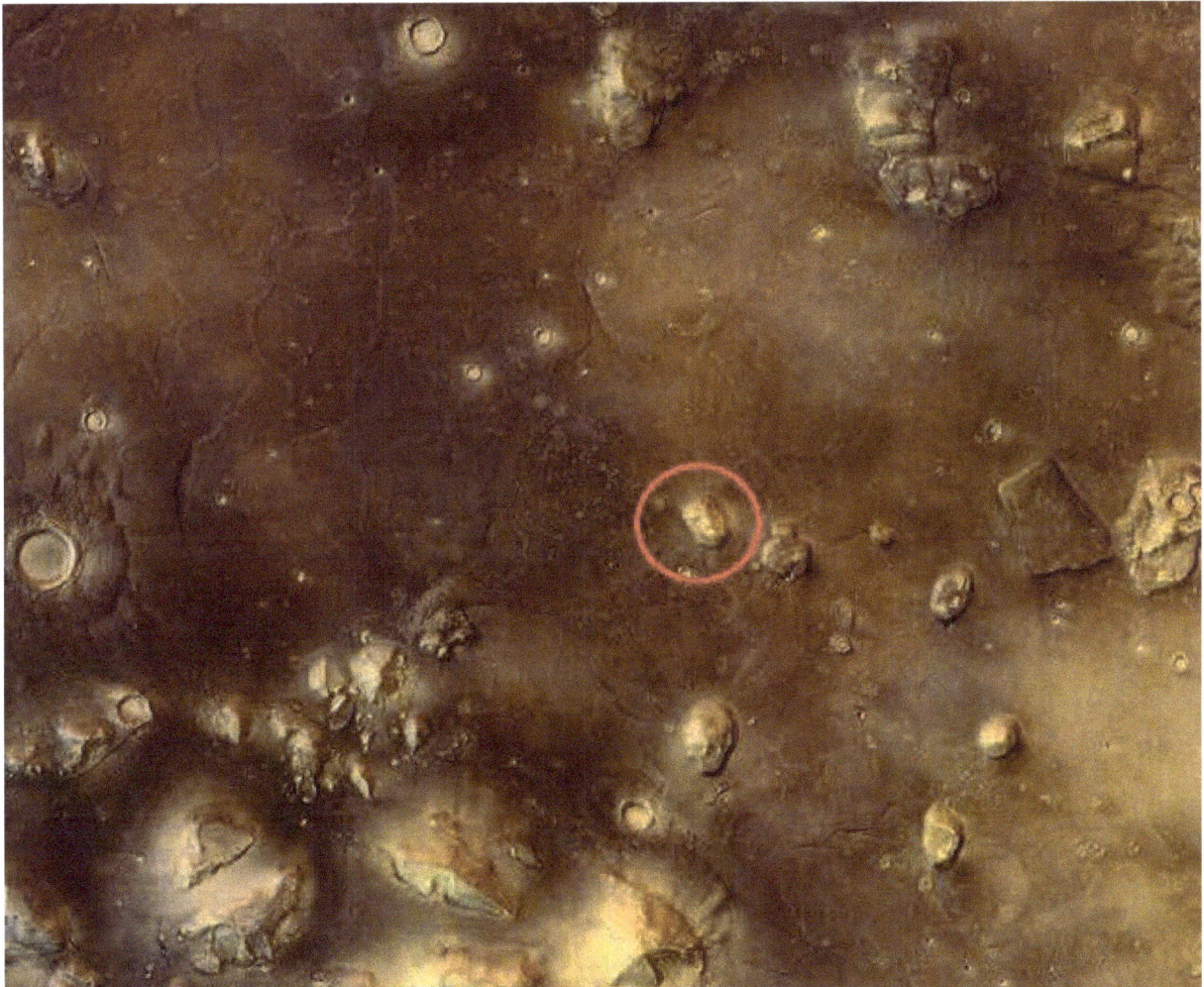

High-resolution view of the Cydonia Region obtained with the HRSC instrument on board the ESA Mars Express on 21 Sept. 2006. The Face on Mars lies just right of center
http://sci.esa.int/mars-express/40012-cydonia-region/

There were aspects of artificiality in this unusual hominid looking mesa that appeared to be more than just a natural geological formation due to either wind erosion or a past diluvial current sweeping through the area.

In fact, their discovery spurred them and other scientists to investigate the whole **Cydonia region** for other Martian anomalies and their efforts were soon rewarded. They found in their photographic search, an image of a very large five-sided pyramid on the plains of Cydonia about 10 ten miles south of the "Face" which seem to have buttressed or fortified corners. It was massive and in partial ruin or collapse on one side. It has become known as the **"D&M Pyramid"** after its discoverers, DiPietro and Molenaar. The D&M Pyramid sides have lengths from 1.934 km to 2.735 km and the height is approximately 1.000 km. Such architectural marvels seem to indicate an ancient civilization employing highly advanced technology in its monolithic constructions.

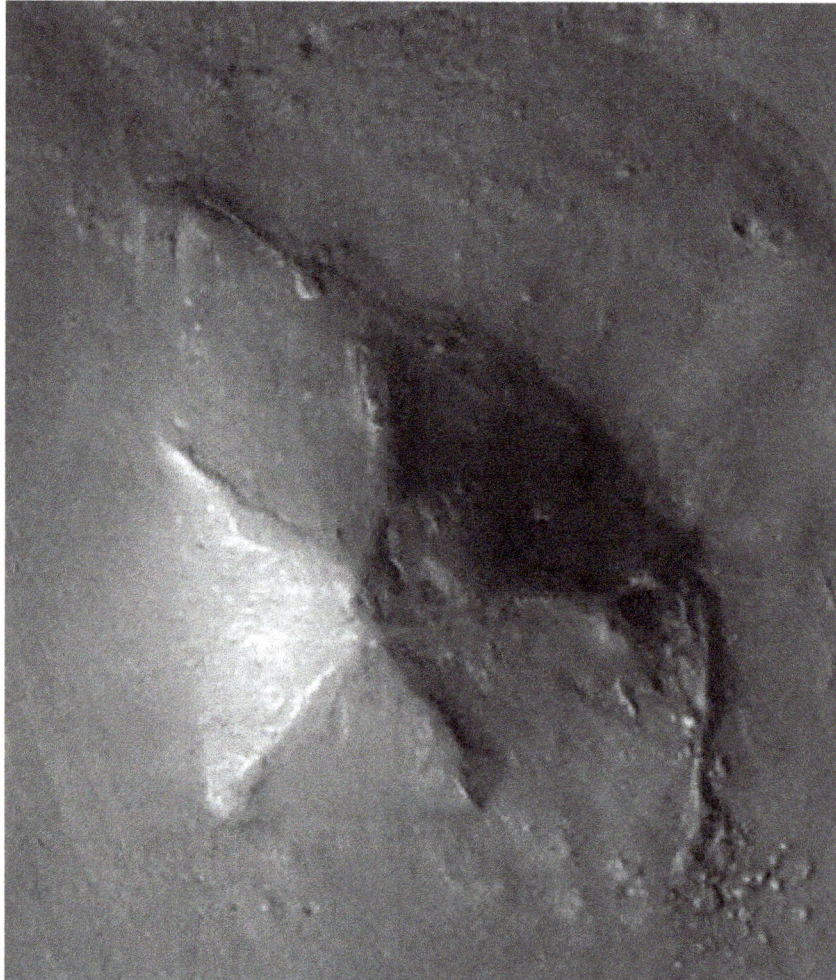

A five-sided pyramid on the Plains of Cydonia, often referred to as the "D & M pyramid" named after DiPietro and Molenaar
https://en.wikipedia.org/wiki/Cydonia_(region_of_Mars)

All these structures were built on a massive scale that was beyond anything constructed on Earth. This level of architectural sophistication on a scale of monolithic proportions found in the Cydonia region and as would be discovered elsewhere on Mars, had been seen before namely, on our closest planetary body… the Moon!

Could the same ancient civilization that once existed on Mars have built the massive domes and tower structures that were discovered on our Moon during the time of the Apollo Moon missions? Were they the same Extraterrestrials who moved about the Solar System establishing colonies and outposts? Were they from this Solar System or from some more distant star system within our galaxy?

The answers may lie on the surface of Mars waiting for us to uncover them and to read their ancient message. Perhaps, we will discover that it is we who are the Martians, we who are the Extraterrestrials!!

With the exhumation of photo images containing the "Face" from the NASA archives, along with the release of the pyramid discovery by **DiPietro** and **Molenaar** to the news media, these startling finds on Mars suddenly took on a life of their own. It became the hot topic among NASA engineers and scientists, as well as in astronomical circles, and of course among UFO researchers, investigators, and the general public. The Face is a mile-long, 1500-ft high humanoid "face" discovered in a northern Martian desert called **"Cydonia".** In its immediate vicinity have been identified other "anthropomorphic objects": most notably, several "pyramids", the largest was the five-sided **D&M Pyramid**. NASA however, was ignoring the evidence in true NASA- bias fashion holding steadfastly to the official "trick of light and shadow" mesa concept.

The public was not buying the official scientific babble-speak from NASA. They felt that the Viking images of the "Face" were just that, a face without any far stretch of the imagination. People were demanding a re-imaging of the "Face" and the Cydonia area of Mars to obtain higher resolution photographs that would settle the matter of the artificiality found in this Martian region. NASA however, wasn't budging from its position that this was nothing more than a very natural, common geological formation, a mesa.

It had been nearly 17 years later, since the **Viking I landing**, when NASA launched the **Mars Observer** on 25th. September 1992, the first of an "Observer series" of missions for planetary exploration, that the public had another chance to view the enigma of the "Face". The principal objective of the mission was to gather information on the geology and climate of Mars. Mars Observer was scheduled to reach its destination in August 1993 and begin its "mapping phase" by November 1993. If it was possible, NASA said it would try to re-image the Cydonia region but, it was not on their priority goals for this mission, nor would course corrections be sent to the spacecraft to do so. The Mars Observer space probe would either be over the area of its own initial orbit insertion or not; no allowances would be made by NASA.

During this phase, the Mars Observer camera would relay low, medium and high-resolution photographs of the entire planet daily for the purpose of securing geological and *albedo* (reflectivity) information on targeted areas of interest
 The Mars Observer was reported "lost" by NASA on 21st. August 1993, shortly after instructions for orbital insertion had been sent to the spacecraft. There was almost an immediate outcry by the public calling *foul* against NASA! It seemed too coincidental that NASA just happened to mysteriously lose its communication with the spacecraft, particularly when there was a high expectation from the public to re-image the Cydonia area again. The public's

suspicions toward NASA again had arisen further. Conspiracies started to fly with regard to NASA's ultimate reasons for its Mars exploration missions. It was even speculated that NASA had successfully maneuvered the Mars Observer into orbit around Mars and then, simply turned off the communication relay to the spacecraft, only to be turned back on at a later date and time to covertly re-image the Cydonia area and other anomalous areas on Mars.

To date, no communication has been reestablished with the spacecraft. It is not known whether the Mars Observer has gone into orbit around Mars or has passed the planet and entered a solar orbit.

Considerable furore arose in connection with NASA's policy for data release. NASA announced that in the case of the Mars Observer mission, unlike previous missions, there would be "No immediate transmission of photography to the public". Data may be withheld from the public for as long as six months solely at the discretion of the "Principal Investigators" holding private contracts with NASA. http://www.bibliotecapleyades.net/marte/esp_marte_41.htm

By June 1993, there were signs that NASA was caving into the ever-growing public pressure to modify this policy and it appeared that they were considering easing the restriction. All hopes, however, were dashed when NASA announced that only "selected" images were to be made available for viewing only at Pasadena, Washington or Houston with no guarantee that there would be images of the Cydonia anomalies or region and certainly with no release to the general public via NASA Select-TV.

NASA was apparently trying to create the impression of a more liberal policy on data release without actually making any significant change. The summary conclusion is that for the Mars Observer mission, NASA introduced a severe restriction on data release, providing a new potential for censorship, under the cover of a technicality. http://www.bibliotecapleyades.net/marte/esp_marte_41.htm

Independent researchers have concluded that the data from Viking I and 2 supported the possibility that some features at Cydonia may be the ruins of intelligently designed structures.

Some of the researchers like **Stanley V. McDaniel** and **Tom Van Flandern** (independent of each other) have gone further in their hypotheses regarding the Martian objects, and others have not.

McDaniel proposed the **AOC (Artificial Origin of Cydonia Hypothesis** claiming that the probability of there being artificial features is strong enough to make new high-resolution photographs a top priority for any future mission to the planet - including the **Mars Observer** mission, should communications with that spacecraft be restored.

The long-term purpose is to illustrate the political, ethical and scientific tension that arises when the potential of a discovery that could cause a major shift in our understanding of ourselves and our history comes up against the biases of individual scientists and the interests of a government bureaucracy. http://www.bibliotecapleyades.net/marte/esp_marte_41.htm

Tom Van Flandern, Ph.D. a former Chief Astronomer of the US Naval observatory approached the question of artificiality of the "Face" and the **Cydonia** area was one of healthy scientific skepticism and curiosity tempered with the Scientific Method.

Van Flandern assumed the **a priori principle** (based on hypothesis or theory) when dealing with this anomalous object which states that:

a priori: *The odds of something arising by chance are significant if calculated before any evidence of its existence is known*.

a posteriori: T*he calculated odds of something already found arising by chance apply only to the next instance.*

The Face when predicted in advance will factor in such aspects as size, shape, location and orientation: Eyebrow, Iris, Nostrils and Lips to which the Face has all these features in the correct proportions and location, except for the iris shape (a circle), no similar features exist in the background that would allow us to pick and choose.

Tom Van Flandern's conclusion regarding the Face is that the natural origin hypothesis is disproved at odds of 1000 billion, billion to one. (In other words *astronomical!)* Therefore, the artificiality of Cydonia is established beyond reasonable doubt!
https://www.youtube.com/watch?v=JPJRpBWuGbY and
https://www.youtube.com/watch?v=0DTfPW9hHb8

When it was first imaged, and into the 21st century, the "Face" is near universally accepted to be an optical illusion, an example of pareidolia, and theories that it was an artificial artifact were considered to be pseudo-science. After analysis of the higher resolution Mars Global Surveyor data, NASA stated that "a detailed analysis of multiple images of this feature reveals a natural looking Martian hill whose illusory face-like appearance depends on the viewing angle and angle of illumination". http://en.wikipedia.org/wiki/Tom_Van_Flandern

Yes, NASA in all its arrogance refused to recant its initial position, still holding on tenaciously to their official conclusion even, when it had been independently proven by outside scientific sources that the Face was, in fact, artificial in origin.

This conclusion pretty much re-affirmed what most of the general public already knew that the Face on Mars is artificial, an ancient monument constructed in the distant past by a Martian intelligence when the planet was more habitable. The public was now looking elsewhere for honest truthful answers, no longer were they just accepting any reports from NASA as the last word on the subject.

Van Flandern believes that certain geological features seen on Mars, especially the Face at Cydonia, are not of natural origin, but were produced by intelligent extra-terrestrial life; probably the inhabitants of a major planet once located where the asteroid belt presently exists, and which Van Flandern believed had exploded 3.2 million years ago. He gave lectures on the subject

(Mysterious Mars, a lecture in 2002 which can be found on YouTube) and at the conclusion of the lectures he described his overall conception:

> *"We've shown conclusively that at least some of the artifacts on the surface of Mars were artificially produced, and the evidence indicates they were produced approximately 3.2 million years ago, which is when Planet V exploded. Mars was a moon of Planet V, and we speculate that the Builders created the artificial structures as theme parks and advertisements to catch the attention of space tourists from Planet V (much as we may do on our own Moon some day, when lunar tourism becomes prevalent), or perhaps they are museums of some kind. Remember that the Face at Cydonia was located on the original equator of Mars. The Builder's civilization ended 3.2 million years ago. The evidence suggests that the explosion was anticipated, so the Builders may have departed their world, and it produced a massive flood because Planet V was a water world. It is a coincidence that the face on Mars is hominid, like ours, and the earliest fossil record on Earth of hominids is the "Lucy" fossil from 3.2 million years ago. There have been some claims of earlier hominid fossils, but Lucy is the earliest that is definite. So I leave you with the thought that there may be a grain of truth in The War of the Worlds, with the twist that WE are the Martians".* http://en.wikipedia.org/wiki/Tom_Van_Flandern

McDaniel points that many reputable scientists in several fields, including physics, astronomy, and geology, have expressed their confidence in the overall integrity of this report and have called for further investigations of these landforms by NASA. However, during the seventeen years since the controversial landforms were discovered, NASA has maintained steadfastly there is "no credible evidence" that any of the landforms may be artificial.

NASA's "evaluation" of the Cydonia area, says McDaniel is based on first impressions from unenhanced photographs, faulty reasoning, with a failure to apply any scientific methods of analysis. These initial impressions stem from flawed reports lack of any attempt at verification of the enhancements and measurements made by others and NASA's main focus has been on inappropriate methodology which ignores the importance of context.

Finally, **NASA** has based its evaluation almost exclusively on the alleged existence of disconfirming photographs which it has never identified and has recently admitted it is unable to identify.

In McDaniel's mind, there remains no scientific basis for NASA's position regarding the landforms.

As we have already come to know, in 1960, a report titled "Proposed studies on the implications of peaceful space activities for human affairs" was delivered to the chairman of NASA's Committee on Long-Range Studies. The report was prepared *under contract to NASA by the* **Brookings Institute**, Washington, DC
.

The report outlines the need to investigate the possible social consequences of an extraterrestrial discovery and to consider whether a discovery should be "kept from the public" in order to avoid political change and a possible "devastating" effect on scientists themselves - due to the discovery that many of their own cherished theories could be at risk.

The concept of withholding information on a possible extraterrestrial discovery conflicts with an understood NASA policy to the effect that information on a verified discovery of extraterrestrial intelligence should be shared promptly with all humanity. A report on the cultural aspects of the **search for extraterrestrial intelligence (SETI)** is presently being prepared for publication by the *NASA Ames Research Center*. In this report, the position that NASA would not withhold such data from the public is said to be strongly supported.

NASA's actual behavior in the specific case of the Martian objects, however, does not appear to be consistent with this policy NASA has regularly distributed documents containing false or misleading statements about its evaluation of the Face to members of Congress and to the public.

The absence of legitimate scientific evaluation of the landforms by **NASA**, its ignoring of the relevant research, it's apparently exaggerated warnings that such photographs would be extremely difficult to obtain, the possible sequestering of the data under the aegis of "private contract", and the ambiguous language used by NASA officials to generate a sense of complacency around the issue all support the suspicion of a motivation contrary to the stated policy. http://www.bibliotecapleyades.net/marte/esp_marte_41.htm

Instead of carrying out legitimate scientific inquiry, McDaniel says that NASA has regularly sent false and misleading statements regarding the landforms to members of congress and their constituents. NASA has condoned efforts to unfairly ridicule and discredit independent researchers, like **Hoagland, Carlotto, McDaniel, Van Flandern** and others and has insisted that there is a *"scientific consensus"* that the landforms are natural- despite the fact that the only real scientific study of the landforms indicates a clear possibility that they are artificial.

Of the various landforms investigated by the independent teams and individuals, the one that began the research, referred to as the "Face" because of its resemblance to the humanoid face, has undergone one of the most exhaustive series of tests for the evaluation of digital images originating from an interplanetary probe available to scientists today.
http://www.bibliotecapleyades.net/marte/esp_marte_41.htm

Into this tempest of NASA cover-up and controversy have entered other reputable scientists, some of them employees working for NASA who listened to the public outrage toward NASA and decided that a serious investigation of the Face and the Cydonia was warranted even if it wasn't sanction by NASA officialdom.

In 1985 **Dr. C. Wes Churchman**, a Nobel Laureate,, and professor Emeritus at the University of California at Berkeley founded the first **Independent Mars Investigation (IMI)** to more fully examine Viking photographs which showed an enigmatic **"Face on Mars"** and other *"unusual surface features"*. Dr. Churchman appointed a special commission under "Peace and Conflicts" to study the Viking images. Among original team members were; **Dr. Brian O'Leary**, astronaut, professor at Columbia and former President of the **President's National Commission on Space** during the Reagan era, **Mr. Vince DiPietro**, an image specialist at NASA's Goddard Space

Flight Center, **Dr. David C. Webb** - The **Presidents Commission on Space Exploration** during the Reagan era, **Dr. Mark Carlotto** - Image Specialist - TASC/NASA, **Dr. Randy Pozo's, Gregory Molenaar, Mr. Thomas Rauchenberg - Richard C. Hoagland**, former science advisor to Walter Cronkite and **Dr. John Brandenberg** of Mars Research, Washington D.C. as well as **Dr. Stanley McDaniels**. Included in this prestigious group, Dr. Churchman selected **Harry A. Jordan BFA,** a professional educator to be a member of the IMI team in order to construct accurate clay models and survey drawings of the entire Cydonia region.
http://www.tmgnow.com/repository/mars/mars_jordan.html

This dynamic group exposed many flaws in the scientific data that NASA had officially released to the US government and into the public domain. The constant debate between NASA officials and the **IMI scientists** and the public finally reached a boiling point. Backed by some of the members of the IMI team, many people decided to rally in protest, out in front of the White House in Washington, D.C. demanding that NASA either come clean on what it had found or re-image the "Face".

NASA in November 1996 launched their new space probe **Mars Global Surveyor,** which was headed up and under the exclusive control by chief scientist and CEO **Michael C. Malin** of **Malin Space Science Systems (MSSS)**, also a private contractor.

This same private contractor had been given sole authority to determine, not only what images would be released and when, but even what objects would be photographed by the high-resolution Mars Observer Camera. That contractor, Dr. Michael Malin, is an outspoken opponent of the hypothesis of possible artificiality. Dr. Malin's arguments against the hypothesis of possible artificiality have been uniformly fallacious.

Thus, the interests of the American public in relation to Mars Observer Camera data were effectively turned over to the evidently biased decisions of a private individual.

Finally, NASA submitted to the pressures of these scientists and the public requests to re-image the Cydonia area with emphasis on the Face but, sneaky old NASA had a couple of aces up their sleeve, one was Michael Malin was manipulation of the raw data. They planned to trump, any further public response or outrage to what their Surveyor spacecraft would re-photograph in the Cydonia area. They would use an old, tried and tested tactic of photo manipulation, distortion, filtering and sanitizing to alter the Surveyor's raw photo images. No way was NASA going to let the general public get what they wanted.

To an unsuspecting public and government, it would appear, however, that NASA had complied with the public's request in an atmosphere of honesty and disclosure. In reality, NASA would still stand by its original statement; that the Face on Mars was a natural geological formation, a mesa and nothing more. The re-imaging of the notorious Face on Mars would show a badly degraded image that has been described as the *"cat box" (kitty litter)*. The new raw images by Surveyor, in "high resolution" showed a mesa of rock outcropping that would seem to indicate that any previous photos taken by Viking 1 and 2 were merely an optical illusion, a **pareidolia** of a natural geological formation resembling a face.

Many of the IMI team scientists rejected NASA's newly imaged photos of the **Face** and spoke out publicly in news conferences and in their books on the cover up of Mars anomalies by **NASA**. **Dr. Carlotto, Dr. McDaniel**, and **Van Flandern** rejected the new Surveyor data proving that with proper shape and shading, 3D and fractal modeling and digital imaging correction, the colossal structure is still a Face, no matter how it is photographically imaged or in what direction the sun was shining on the **Cydonia** area. The Martian Enigmas, A Closer Look by Mark J. Carlotto; 1991; published by North Atlantic Books; ISBN 1-55643-092-2

Perhaps, the most outspoken of the IMI team has been **Richard C. Hoagland.** His curiosity was piqued by the **Face on Mars** and the whole **Cydonia region** that he wanted to find out what else was in this area of Mars. What was clear to Hoagland from the second re-imaging of the Face by the **Mars Global Surveyor (MGS),** even though there was higher resolution in all raw images of the area, when it came to the Face, the photograph released to the public revealed that NASA has deliberately altered the original raw data of the photograph. Half of the pixel data had been removed through high and low filtering, the contrast altered and brightened with an oblique distortion to the overall photo. NASA in false triumph seems to be pounding it collective chests, saying: *"we told you so and here's the proof!"*

**Richard C. Hoagland on CNN TV discussing the NASA photo
of the Face on Mars which NASA says is a trick of light and shadow**
Google image

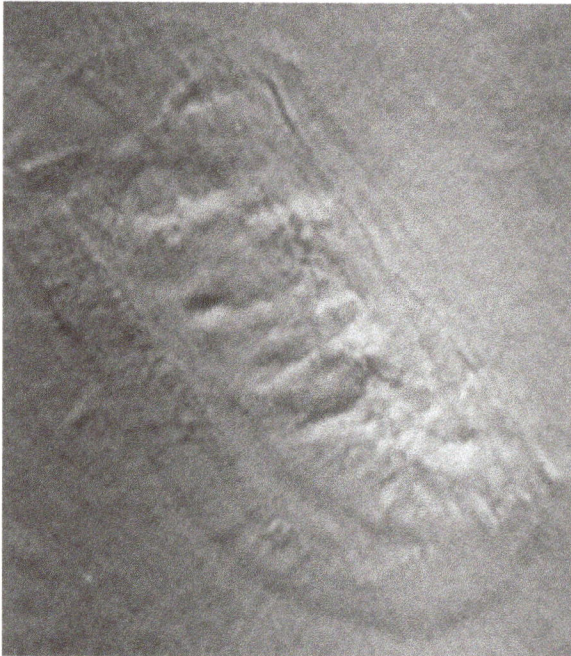

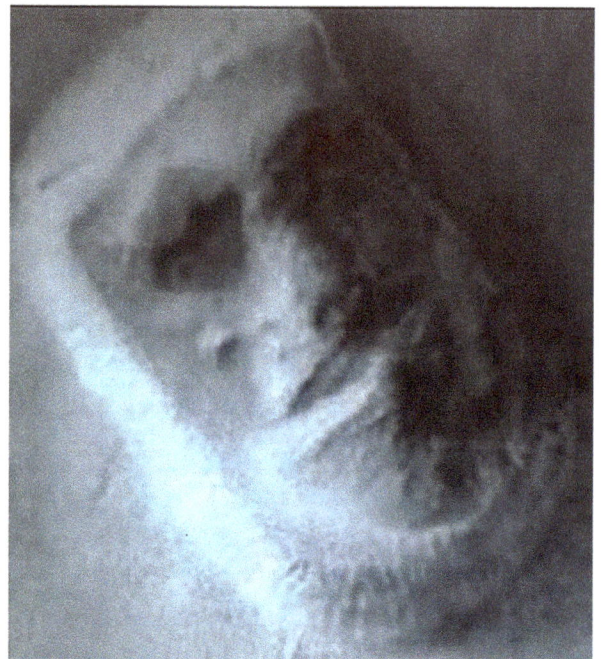

NASA's "high resolution", extremely filtered image of the mesa – the *"Cat Box"* redition of the Face (left). Tom Van Flandern's digitally corrected NASA image (right) with proper sunlight angle, still shows a humanoid face - Proof Positive of Artificiality!

http://mikebara.blogspot.ca/2013/10/there-is-no-such-thing-as-pareidolia.html and https://www.youtube.com/watch?v=0DTfPW9hHb8

The ace up NASA's sleeve, however, did not trump the public response; in fact, the Independent Mars Investigation team had called NASA's bluff with a reworking of the raw image data of the face and proved conclusively that it was indeed a face, even with half the pixels removed by NASA. The division between NASA and the public had grown to become a chasm!

Curiously, what was widely held to be a depiction of a ***humanoid face*** as originally imaged by the Viking orbiter back in 1976 because part of the face was in shadow now, clearly revealed that the right-hand (eastern) side, thought to be a collapsed portion of the monument was in fact ***"feline"*** in symmetry, similar to a ***lion!*** The western or left-hand side still retained the proto-human visage in which some people perceived the human side as being a pharaoh with a headdress! (See image below).

The problem with "collapse" as a natural mechanism for these features is that it just isn't very likely. For a long time now, the only active erosive mechanism on the planet Mars has been **"aeolian erosion" (wind).** And wind, especially when it carries large quantities of sand and grit, tends to blast away *outer* layers over eons, producing the pitted, cavity-strewn surface you see on the **Face's City** side (below left). The indicated areas on the right, on the other hand, seem to have collapsed *from the inside…* Without any fluvial or tectonic activity (for a *very* long time....) to induce such collapse, it is somewhat difficult to explain.
http://www.enterprisemission.com/catbox.htm

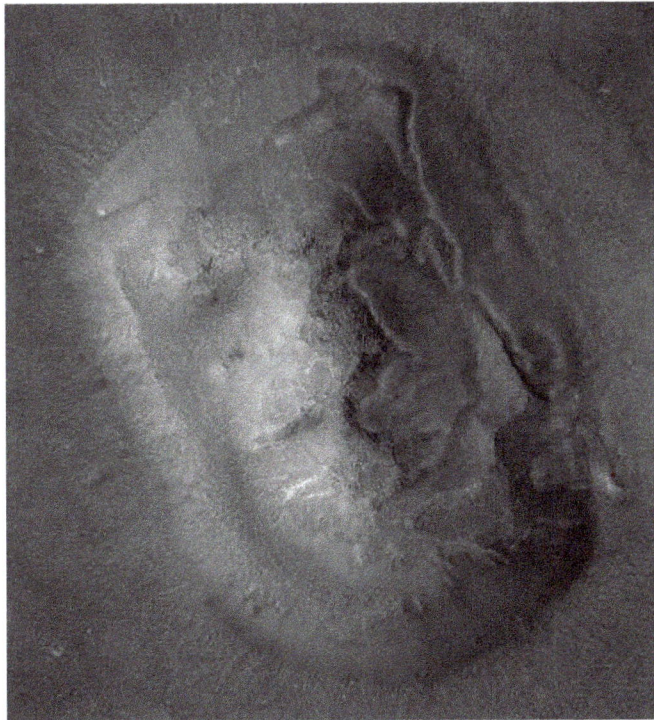

**The clearest picture yet of the Face on Mars without NASA's usual photo tampering
Photo was released May 24, 2001, by Dr. Michael Malin of MSSS**

https://en.wikipedia.org/wiki/Cydonia_(region_of_Mars)

c points out that many cultures around the world have tradition and iconic symbols that show hybrid humans and beasts, "split-faced beings" as if they have undergone some kind of grotesque genetic experiment. Yet, these icons appear to have religious symbolic significance, a reverence to something higher from a hallowed past that needed to be remembered, even worshipped, perhaps, as a way to commemorate some ancient event.

When the left side is mirrored image, one can easily see the human aspect of the monument and when the right side is mirrored imaged, the lion aspect is clearly evident. (See photo below).

Ancient cultures like the Egyptians were resplendent with all kinds of examples, the most obvious and the most famous is the **Great Sphinx** in Giza, Egypt. Mayans also had masks and statuettes as well as temple glyphs that depict split-faced gods having human and jaguar features. There are also, the many hybrid creatures in ancient East Indian mythology and the native Indians of West Coast Canada and the US had ceremonial masks and totems of split-face beings. Even the Bible has writings from Ezekiel that speak of the split-faced beings.
http://www.enterprisemission.com/catbox.htm

27

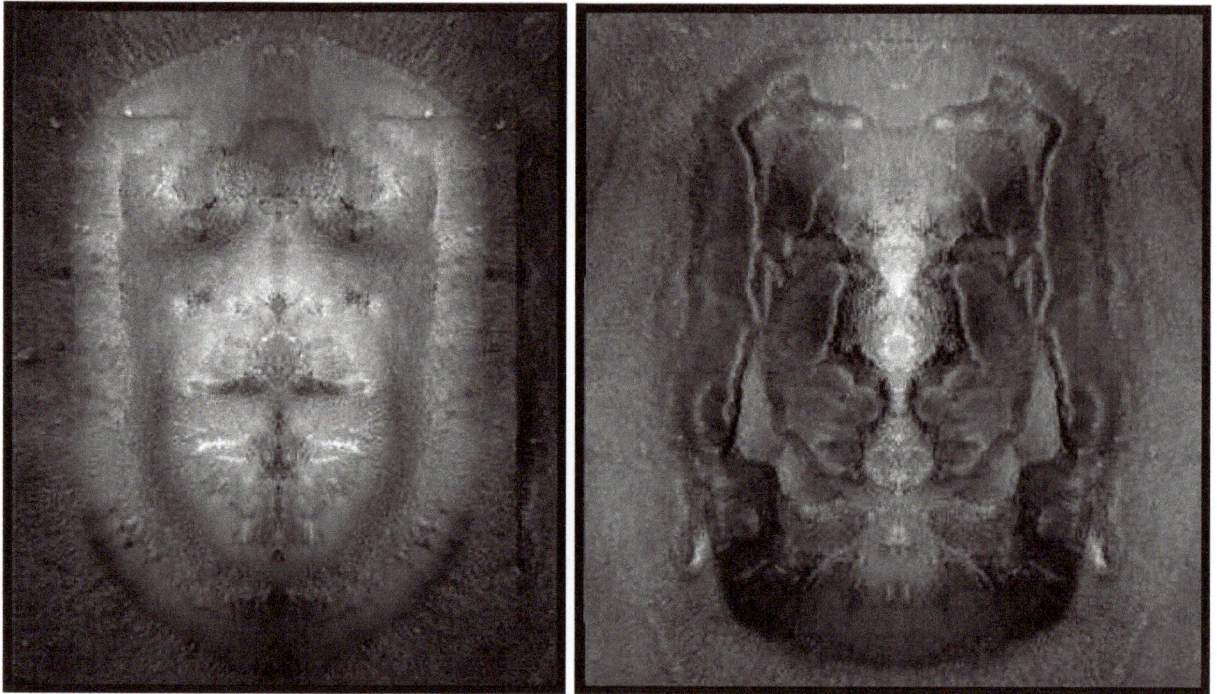

The left side mirrored image showing the human pharaoh and the right side mirrored image of the lion from the Face on Mars monument
http://www.enterprisemission.com/mola.htm

If we can show that this "alien artifact" has a fundamental "terrestrial connection" (as Hoagland has rigorously argued for years) -- both in form and fact --to the practices and rites of *ancient cultures here on Earth*, then we can go a long way to explaining how a **"Lion/Pharaoh Monument"** ended up on a nearby planet. Remember, our model -- shared by other researchers like **Michael Cremo** and **Graham Hancock** -- is that *all* of the ancient "advanced" cultures on Earth ultimately sprang (in the form of refugees) from the same *pre-diluvial*, truly *advanced* "root civilization." **The Golden Age of Science and Technology** that the Maya called **"the Fourth World,"** the Egyptians called **"Zep-Tepi" (the First Time),** and the Greeks called … **"Atlantis."**

Author's Rant: I have heard traditional stories by some of the West coast native tribes in British Columbia who refer to themselves a "Moon People" and their stories speak of a time when they came from the Moon on board "Thunderbirds" that roared when they flew. These say that the White culture has misunderstood the concept of the Thunderbird as not an actual bird or a mythological creature but, as a spacecraft originating from the Moon. They came to Earth thousands of years ago to avoid the "wars in the Solar System"!!!

So, there is a major human tradition -- across not one, but *several* human cultures -- that reinforces the notion that the apparent asymmetry of the Cydonia Face is, in fact, *intentional*. But we think even more important is the specific nature of that union -- the Man/Lion hybrid -- for it uniquely speaks to a very sacred, very ancient human religious tradition …
http://www.enterprisemission.com/catbox.htm

28

**The Face on Mars represents a "*Martian* Sphinx" -- the *first* Horus,
the equivalent to the Great Egyptian Sphinx on Earth**

How all of this terrestrial esoterica relates to a possible "monument" discovered by a ritually-bound space agency on Mars, is ultimately to be found in the true *meaning* of the Face on Mars. The now unmistakable "**Pharaoh/Lion connection**" at Cydonia -- and identical "dual imagery long present here on Earth -- was obviously *intended to express some deep, fundamental Message for the human species.* (Bold italics added for emphasis by author).

Even NASA's astonishing acknowledgement that this Cydonia enigma bore a strong resemblance to "an Egyptian Pharaoh", a sort of back-handed compliment to Hoagland's insight into the ancient traditions and mythology behind the similarity of the two Sphinxes that inhabit two different worlds. **http://www.enterprisemission.com/catbox.htm**

The Face on Mars along with the D&M Pyramid are the best-known anomalies yet discover in the Cydonia region and Richard Hoagland wanted to know if these oddities were singularly phenomenal or if there were others in the region. Hoagland's search using high-resolution satellite images turned up more artificial structures on the **Plains of Cydonia** and in other regions of Mars as well. As it turned out, there were other pyramids and dome-like structures in the Cydonia region, even a **"City of pyramids"** and an **Arcology** of hyperstructures (combining both *"architecture"* and *"ecology"*) reminiscent of the **ancient Egyptian cities** like Giza, Abydos, Heliopolis, Sakkara, and Thebes

Arcologies are an architectural concept termed by architect **Paolo Soleri,** to describe megacities for high-density population that are self-sufficient that contain a variety of residential, commercial, and agricultural facilities with minimum environmental impact. http://en.wikipedia.org/wiki/Arcology

South of the crater is a mound which has an apparent anti-clockwise spiral ascending it with a pyramidal structure on its peak. This spiral mound is a mile in diameter and is approximately 500 feet high. This is the "**Tholus**" at Cydonia is an isolated, semi-circular, mound-like landform, not a volcanic vent that lies in the cratered plain. The mound ("tholus" means "mound") is of relatively low relief in relation to the knobby terrain to the west. It appears to contain a thin depression or groove that circles the landform at its base and spirals up toward the landform's peak in a clockwise fashion cutting through the mound. http://www.greatdreams.com/geology.htm

There is the "**Cliff" or "Wall"** (is actually an elongated mesa overlaying an ejecta blanket of a 3 km impact crater. The Cliff contains a thin, almost linear central ridge running the length of it) to the northeast of the Face and the existence of the Wall or Cliff must post-date the laying down of the ejecta blanket, hence, proof of artificiality. The "**Tetrahedral Pyramid**" Crater beside the Cliff that has a small three-sided pyramid (a tetrahedron) on its rim; all these anomalies are signs of intelligent construction.

Located to the south-east of the city is the large five-sided D&M Pyramid. **Erol Torun**, a geomorphologist at the American Defense Mapping Agency and one of many scientists who have worked on the Cydonia subject has stressed that an object with five straight sides cannot be formed, or at any rate, cannot be maintained by the action of wind and weather. For the force that is sharpening one face will at the same time be causing any existing opposite straight sides or edges to erode. http://www.aulis.com/mars.htm

There is an apparent connection between the complex where the Face is to be found on Cydonia, Mars – located at 41°N latitude and the Avebury complex, just north of Stonehenge, England.

In a brilliant piece of deductive reasoning with the use of geometric proportionate overlaying **David Percy**, a British planetary anomaly researcher with a website called AULIS Online from which some of this material originates, discovered in 1991, a remarkable similarity between the **Avebury Circle**, with its earthen rampart and ditch, is a depiction of the **Cydonia Crater** on Mars, suggesting that **Silbury Hill** along with other features in the area might be an analog model or 'mirror' representation of the key features of Cydonia. Using a 14:1 reduction ratio, the satellite photo image of the Cydonia region was a perfect fit over the Avebury /Silbury area map.

Historically, the **Avebury Circle** is an ancient Briton earthwork consisting of an outer earth rampart, an inner from whence the earth material was excavated and the inner rings formed from circles of standing stones.

30

The 5000 years old ancient Briton earthworks of Avebury Circle (left) and the Silbury Hill (right) are the exact counterparts to those anomalies found in Cydonia , Mars
http://www.aulis.com/mars.htm

Nine million cubic feet of material was moved to construct Silbury Hill by hundreds of people over a period of many years to build the hill, a mammoth task not unlike similar accomplishments of the mound building Indians of America. Interestingly, ancient Britons used to dance up the hill to a maypole on the top – in a spiral. Moreover, it is possible Silbury Hill was constructed as a spiral. Silbury Hill is not a burial mound, and the entire purpose and significance of the Silbury/Avebury complex remains, at present, a complete mystery.
http://www.aulis.com/mars.htm

On the Ordnance map of Avebury rampart, there is a small mark exactly where the tetrahedral pyramid occurs on the crater rim in Cydonia. This mark is located above a gap in the Avebury rampart, just as the Mars pyramid is above a gap in the Cydonia rim. It is the only mound indicated on the Avebury rampart and is clearly an analogue of the tetrahedron located in the corresponding spot on Mars.

The Avebury pyramid on the rim of the rampart has been considerably eroded over a period of nearly 5,000 years, and must once have been a still more impressive feature. The tumulus is indeed tetrahedral in structure but not regular, with two long sides and one short, so that it projects almost lozenge-shaped in one direction. These very same features also appear to be duplicated by the Cydonia rim pyramid. http://www.aulis.com/mars.htm

The chances of two similar geographical features having been constructed in exact proportions to each other and positioned in near equidistance relationship to each other for possible ceremonial usage on two different but, neighbouring planets is astronomical beyond belief unless it was deliberately designed or influenced by one and the same intelligence.

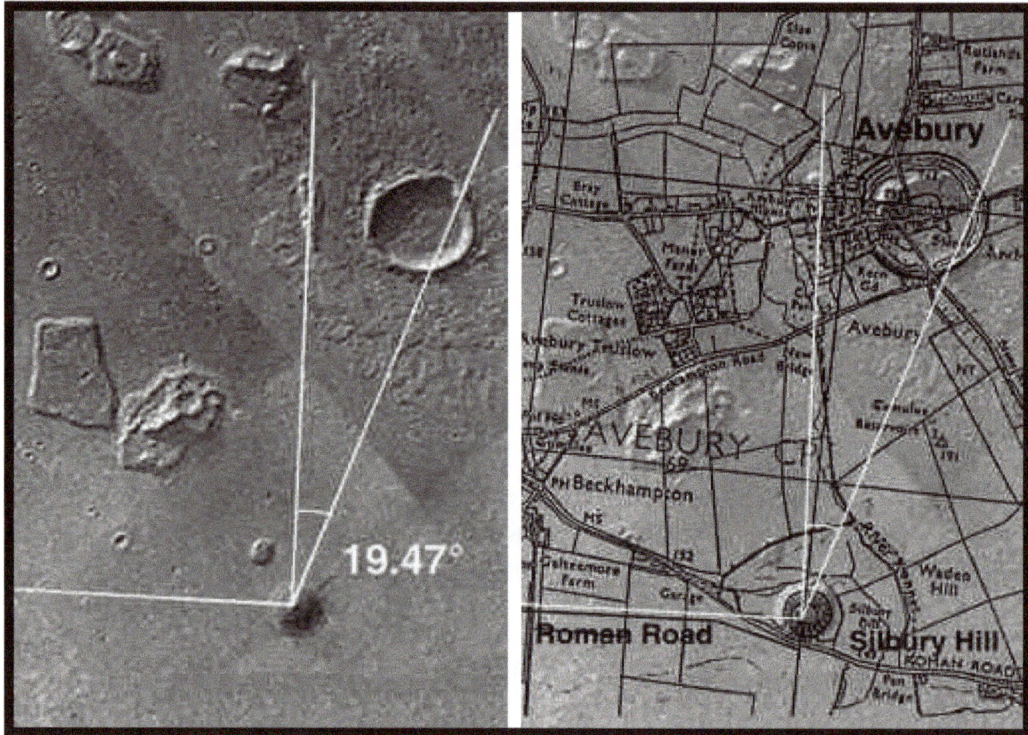

**Spiral Mound Tholus and the Cydonia crater with Tetrahedral Pyramid
on Mars (left) and Silbury Hill**
http://www.aulis.com/mars.htm

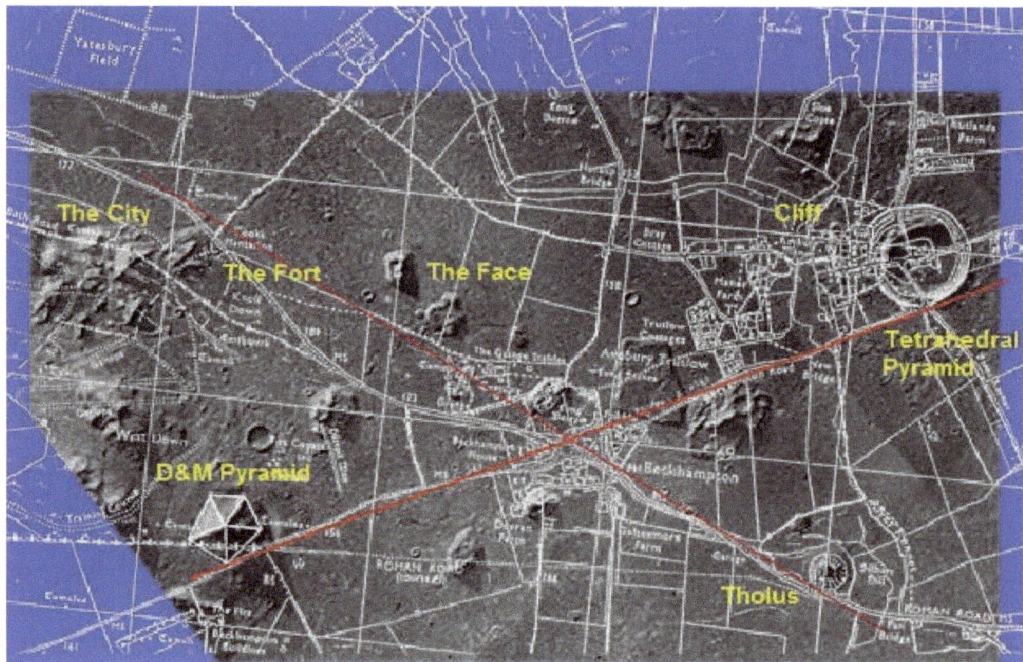

**The four roads on the Avebury complex form tangents to the structures
on Cydonia. The map overlay of Avebury/Silbury on the Cydonia
region is an exact proportional match**
http://www.aulis.com/mars.htm

"The Fort" is another collapsed or incomplete pyramidal structure west of the Face is also a major arcology structure as is most probably the large and small pyramidal structures in **The "City"** and the D&M Pyramid. Most of these structures were built on a massive scale that was beyond anything that has ever been constructed on Earth. The Pyramids in Egypt or in Central America or even in China pale in size next to these Martian monuments.

According to Richard Hoagland, all these megalithic structures were constructed by an intelligence in a layout of geographical and astronomical precision with the D&M Pyramid as the cornerstone monument connecting to or intersecting with other anomalies like the Face, the Tholus and the Pyramid Crater in the Cydonia region forming the ancient sacred geometry of Cydonia and ultimately leaving behind a message for mankind or to whoever should find it.

Hoagland says that the message is encoded in the **internal geometry of the Martian Monuments** where certain trigonometrical alignments to these monuments and structures precipitate angles of redundancy. Repeating universal constants such as **"e"** and **"π"** and where **"e/π = .865"** (where **"e"** is the limit of $(1 + 1/n)^n$ and is equal to **2.71828182846** and where **"π"** is about equal to **3.14159265359**), indicate that a **hyperdimensional physics** is reflected in the repeating tetrahedral geometry used in the construction of the monuments of Mars. This internal hyperdimensional/tetrahedral geometry sometimes referred to as the Cydonia mathematics is, says Hoagland, an integral part of every planetary and stellar body throughout the universe. http://en.wikipedia.org/wiki/E_%28mathematical_constant%29

In 1992 Hoagland gave a U. N. presentation in which the location of the **Martian "Pyramids"** and **"Face"** of Cydonia on Mars were connected for the first time with the Egyptian location of the Pyramids and Sphinx on Earth. The *arctangent* **(.865)** of the **"D&M Pyramid"** is the geodetic Martian latitude of **(40.87 N.)** on which the Pyramid lies, expressive of the Arctangent equivalent of "e/pi": 40.87 degrees = ArcTan 0.865 which is the same (within one part in a 1000) as the *cosine* of the **Sphinx's** current latitude on Earth in Egypt. **(Cydonia: Tan Lat. = 40.87 N = e/pi = .865)**

These identical, *"shared"* latitude relationships are, in turn, reflected in the internal geometry of key pyramidal and *"sphinx-like"* structures located at the Cydonia and Giza sites. They are also reflected in the repeating tetrahedral geometry identified between structures located at each respective site. **(Great Pyramid Internal e/pi geometry: 51.85 Degrees 51.85/60 = .865)** http://www.enterprisemission.com/ken2b.html

The angles, ratios, and trig. functions below are derived from the angles derived and measured from the D&M Pyramid (as seen above) to all other monuments and structures found in the **Cydonia Mensae** (see below). The repetition of the angle 19.5 degrees is found in all structures in the Cydonia Complex as to become redundant yet, the message of tetrahedral geometry deliberately intended to be redundant as it was meant to be understood in a larger context not merely in the construction process of the Cydonian monuments but, in all interstellar bodies in the universe!

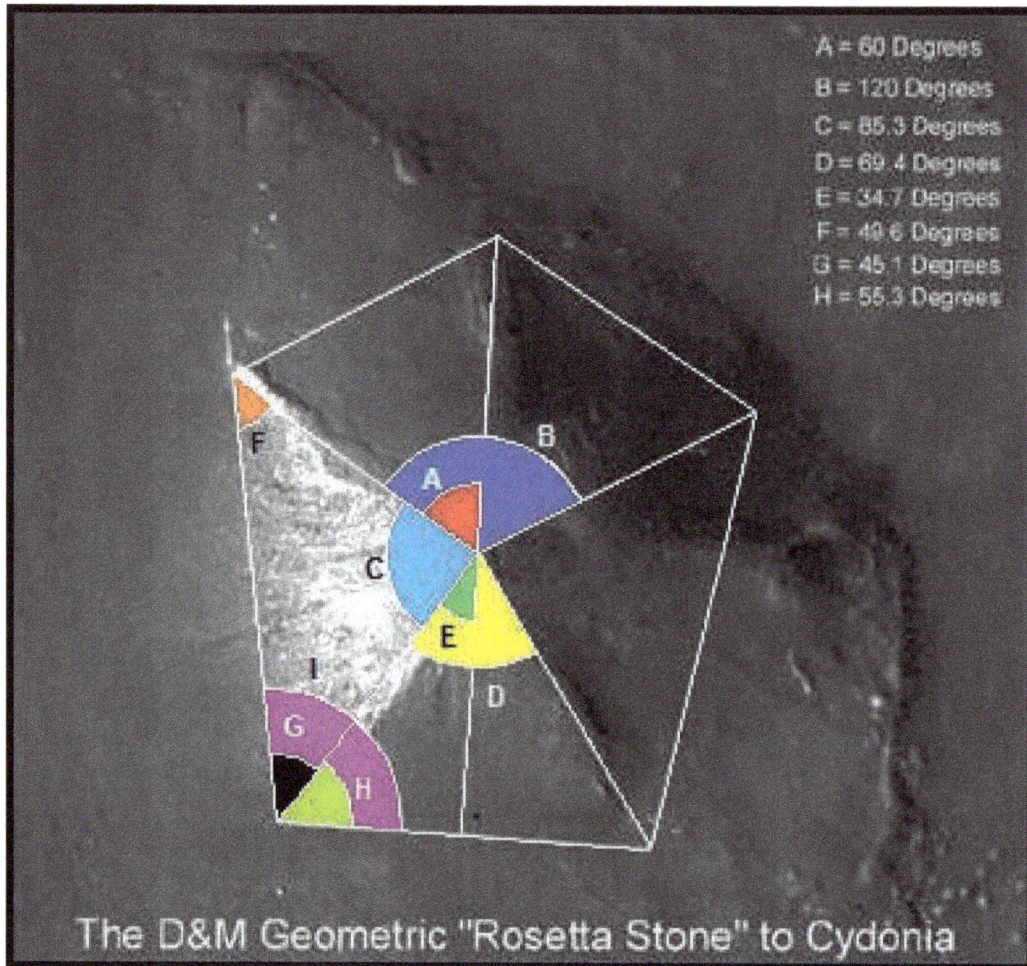

The D&M Geometric "Rosetta Stone" to Cydonia

A = 60 Degrees
B = 120 Degrees
C = 85.3 Degrees
D = 69.4 Degrees
E = 34.7 Degrees
F = 49.6 Degrees
G = 45.1 Degrees
H = 55.3 Degrees

The D&M Pyramid is the cornerstone or "Rosetta Stone" for determining the internal tetrahedral geometry between structures throughout the Cydonia Region

http://www.enterprisemission.com/sheep.htm

Angles		Angle Ratios	Trig. Functions
Degrees	Radians		
A = 60.0	= $\pi/3$	C/A = $\sqrt{2}$	TAN A = $\sqrt{2}$
B = 120.0	= $2\pi/3$	B/D = $\sqrt{3}$	TAN B = $\sqrt{3}$
C = 85.3		C/F = $\sqrt{3}$	SIN A = e/pi
D = 69.4	= $e/\sqrt{5}$	A/D = e/pi	SIN B = e/pi
E = 34.7		C/D = $e/\sqrt{5}$	TAN F = π/e
F = 49.6	= e/pi	A/F = $e/\sqrt{5}$	COS E = $\sqrt{5}/e$
G = 45.1		H/G = $e/\sqrt{5}$	SIN G = $\sqrt{5}/\pi$
H = 55.3		C/B = $\sqrt{5}/\pi$	SIN C = 1
I = 100.4		D/F = $\pi/\sqrt{5}$	TAN G = 1
			TAN I = .2e
			TAN 40.87 N = e/pi

34

Since the latitude of the entire **Cydonia Complex** seems to have been carefully chosen to reflect the Arctangent of this circumscribed tetrahedral "message, it occurred to the **Hoagland** and **Bara** that "something important might lie at the LATITUDE represented by the vertices of a circumscribed tetrahedron -- placed 'inside a planet'." This would represent the most elegant expression of the Arctangent trigonometric function emphasized repeatedly within the Complex of the specific Martian latitude: 40.87 N.

In working out the several possible implications of such geometry, it was discovered by **Errol Torun** (a team scientist member with Hoagland) that if a circumscribed **tetrahedron** is placed inside a globe representing a gridded planetary surface, with one vertex located either on the geographical "North" or "South" polar axis, the resulting latitude TANGENT to the other three vertices will lie at **19.5degrees N. or S. 120 degrees of longitude apart.**
http://www.enterprisemission.com/ken2b.html

Even more remarkable is the fact that the "*Message of Cydonia*" is encoded not only in the internal structure of the **D&M Pyramid** but can be found in the trigonometric angle of 19.5 degrees that connects the D&M Pyramid to the **Pyramid Crater**, **The Tholus** and **The Cliff** but, is also redundantly repeated and connected with the Face and the City structures. The **Cydonia Complex** when considered in toto is a model in miniature of a larger planetary message encoded within Mars itself, similar in aspect to that popular movie, "Indiana Jones and the lost Ark" where the city of Thebes in the Egyptian map room was a miniature version of the larger city of Thebes emulating the architectural characteristics of the larger city outside. The Cydonia Complex in like manner is, therefore, because of its latitude of Lat.40.87 N is one point in the internal tetrahedral geometry of Mars that can be found at latitude 19.5 degrees N in the geological formation of **Olympus Mons**!

As Hoagland quickly realized the internal tetrahedral geometry of Cydonia was a miniature expression of a much larger mathematical expression tetrahedral model embedded in all planets and stars in the universe. It can be found on Earth in the volcanic area of Hawaii, at the poles of Saturn, the geographical position of the **Great Red Spot of Jupiter,** the **Great Dark Spots on Neptune and Uranus**, and even, in the **sunspots of the Sun**.

It should be understood, as Hoagland and Bara point out, in order to avoid any confusion that there is no "tetrahedron buried inside each planet!" The "**tetrahedral geometry**" embedded at "Cydonia" is a model for a higher-order mathematical topology: i.e., a vorticular "two-torus" energy flow and internal fluid dynamics, equivalent to tetrahedral mathematics. That an internal "vorticular pattern" is mathematically well-known **(Porteous, I. R., 1981)** as per Hoagland is a "geometric abbreviation" underlying some physical manifestation of tetrahedral mathematics left at Cydonia for us to find, is its raison d'être. http://www.enterprisemission.com/ken2b.html

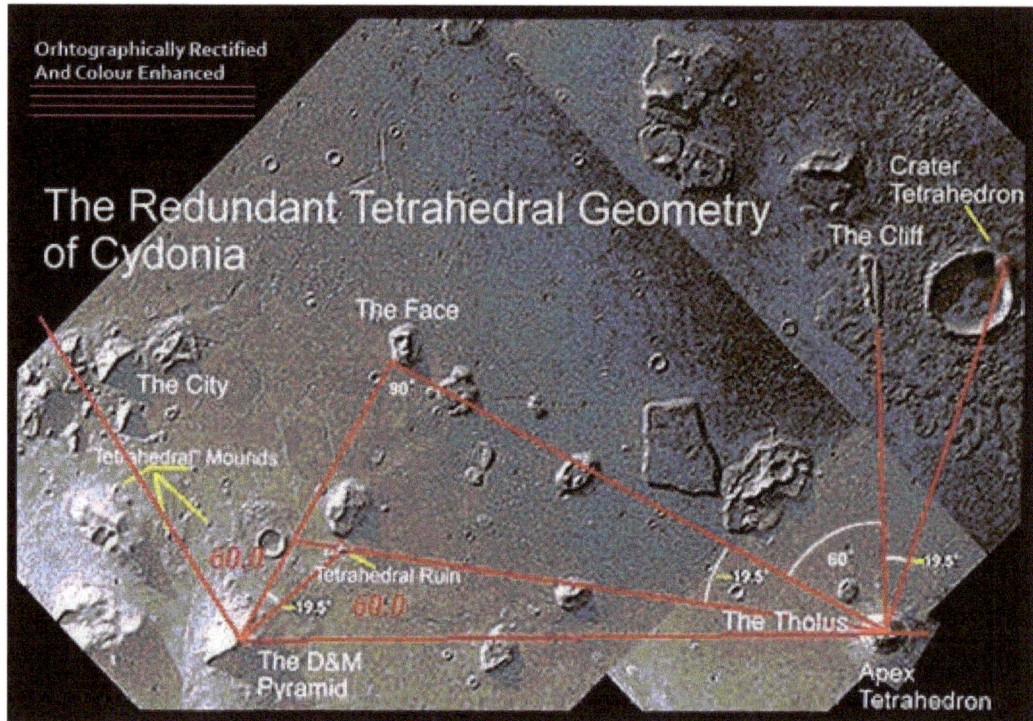

This map of the Monuments of Cydonia displays only some of tetrahedral geometry which does not include the "The City" or other artificial structures in Cydonia
http://www.enterprisemission.com/Curiosity-White-Crow.htm

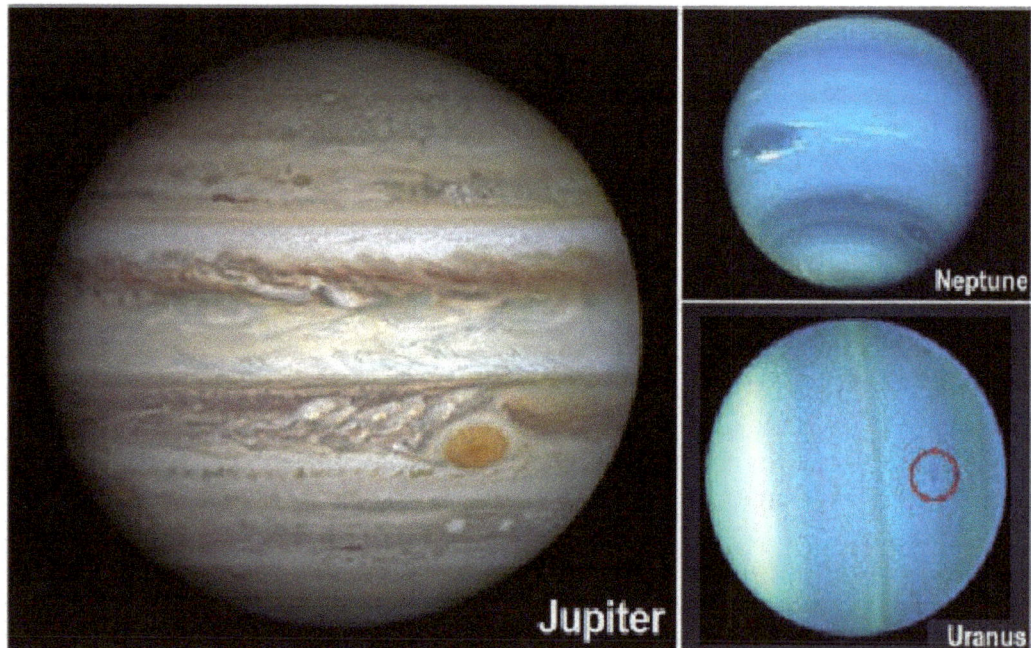

The Great Red Spot of Jupiter and the Great Dark Spots on Neptune and Uranus all located at 19.5 degrees latitude
https://www.nasa.gov/press/2014/may/nasas-hubble-shows-jupiters-great-red-spot-is-smaller-than-ever-measured/ and
http://bilinkis.com/tag/columnista-invitado/ and

Planetary Latitudes of Emergent Energy Phenomena

Object	Feature	Latitude	Comment
Earth	Hawaiian Caldera	19.6 N.	Largest shield volcano
Moon	Tsiolkovskii	19.6 S.	Unique Farside "mare-like" lava extrusion
Venus	Alta Regio	19.5 N.	Current volcanic region
	Beta Regio	25.0 S	Current volcanic region
Mars	Olympus Mons	19.3 N.	Largest shield volcano
Jupiter	Great Red Spot	22.0 S.	Vast atmospheric "vorticular upwelling"
Io	Loki (2)	19.0 N.	(Voyager 1 & 2 volcanic plumes) Presumably driven by intense Jovian tidal forces and/or Jovian magnetic field
	Maui (6)	19.0 N.	
	Pele (1)	19.0 S.	
	Volund (4)	22.0 N.	
Saturn	North Equatorial Belt	20.0 N.	Region of "storms" observed from Earth
	South Equatorial Belt	20.0 S.	Same as above
Uranus			(Voyager 2 IR Observations)
	Northern IR 1-2 K "dip"	20.N.	Presumably, deep "upwellings", creating high-altitude clouds
	Southern IR 1-2 K "dip"	20.0 S.	Same as above
Neptune			(Voyager 2 Imagery)
	"Neptune Great Red Spot"	20.0 S.	Presumably same as Jovian counterpart

Source: NASA and U.S. Geological Survey http://www.enterprisemission.com/message.htm

One more remarkable area on Mars incorporating these occult –like numbers of 19.5° and 33° is the **Ares Vallis** region of Mars situated at 19.5° N Lat. by 33° W Long about 1000 miles away from the **Cydonia Monuments** (i.e. the Face on Mars). **Pathfinder** landed in this area on July 4, 1997, amid what appeared to be a rocky terrain with a couple of low profile mountains nearby. To the anomalist community, this landscape nearly went unnoticed until it was re-examined and discovered that the mountains in the background of the photo image a (spliced mosaic composite) snapped by Pathfinder had distinct pyramid shapes. They also noticed there was also a large object "sitting" just in front of these pyramid structures.

Its reclining pronation appearance resembles a feline structure. Enlarging the image revealed the equivalent of one of Egypt's famous landmarks of antiquity, a **Martian Sphinx**! Beside it can be seen a large temple structure with a massive gated entrance (see image below. The pyramid behind the sphinx appears to be built as a stepped ziggurat or built upon a flat mesa; the flat hill next to it is a second pyramid but, its construction is incomplete without its apex.

Original photo image taken by Pathfinder in the Ares Vallis region on July 4, 1997. Right of centre are two pyramids and a dark sphinx-like structure can be seen in front of it
https://mars.nasa.gov/MPF/parker/highres-stereo.html

A split image view of the above area with right side colour corrected. Note there are many square blocks of stone and foundation areas in the foreground right up to the Sphinx
https://mars.nasa.gov/MPF/parker/highres-stereo.html

**Close-up and colour enhanced view of Martian Sphinx with pyramid in background.
Note that pyramid appears to be either a ziggurat or built upon a mesa**
https://mars.nasa.gov/MPF/parker/highres-stereo.html

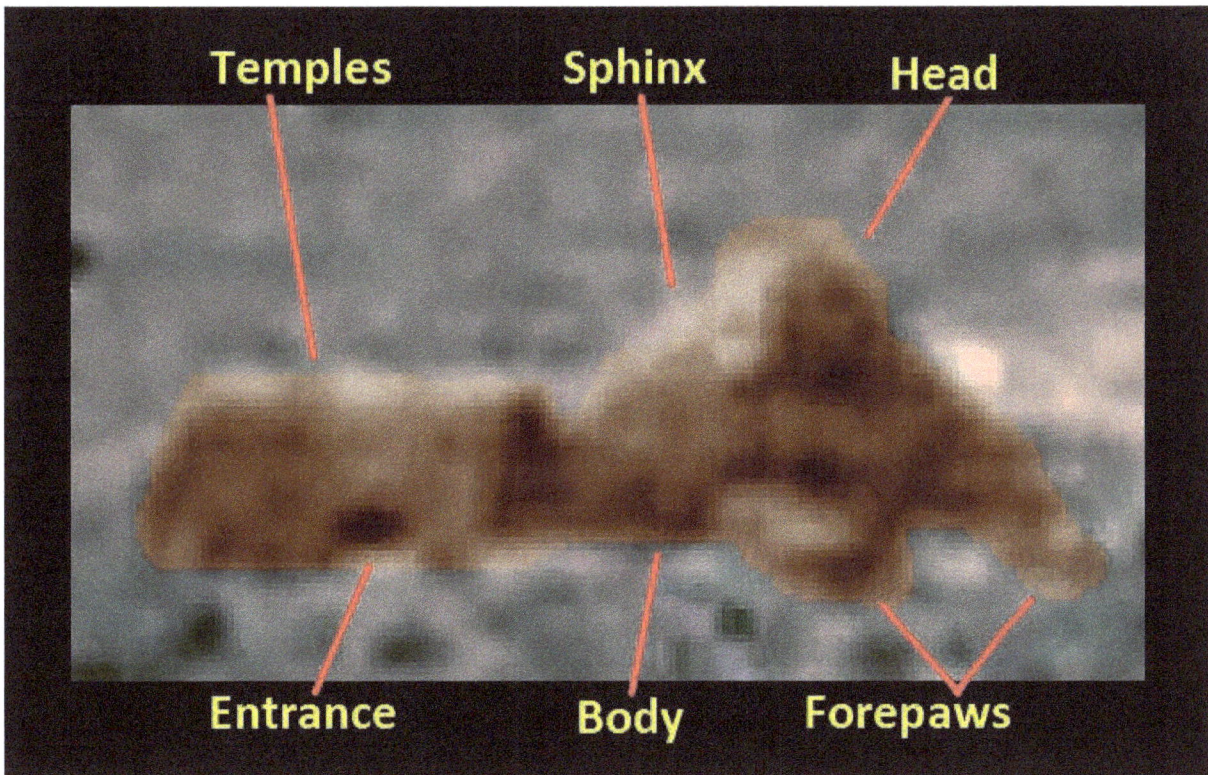

Enlarged image (limit of photo definition) of Martian Sphinx
https://mars.nasa.gov/MPF/parker/highres-stereo.html

**The Great Sphinx of Giza, compare the head, the forepaws, and body with
the Great Martian Sphinx above and note that each is located nearby to pyramids**
http://www.touregypt.net/featurestories/sphinx2.htm

Right away, when the Martian Sphinx is compared to the Great Sphinx at Giza, it becomes obvious that they share many of the same construction aspects. They share the same reclining repose to guard temples, tombs, and pyramids. Both have their forepaws extending forward as if to pounce upon any intruders or unfaithful and both Sphinxes share a human head, either a male or female head with a lion's body. Each Sphinx's head is adorned by the traditional pharaonic banded *memes* **headdress** emulating the lion's mane.

In the years since this area was initially photographed by the Pathfinder's panoramic camera, resolution and their analysis has greatly improve to the Super Resolution algorithm that should this area be re-photographed again, the detail from such images would remove any qualitative doubts as to whether this Sphinx is indeed and artificial structure and an association with the Greater Sphinx in Giza. **http://www.enterprisemission.com/Path-sphinx.html**

The Cydonian monuments on Mars are ***the conclusive proof of artificiality*** that NASA has been searching for in its space explorations. It exemplifies a highly developed Martian civilization employing advanced creative architectural and mathematical design concepts into megalithic structures able to withstand the eons and ravages of time for the sole purpose of transmitting a message to any passing intelligence or to a future civilization that may arise within the Solar System. Its very existence is undeniable evidence of an Extraterrestrial presence in our Solar System which when coupled by the discoveries made by the Apollo Moon Missions of artificial structures found on the Moon, from a time so ancient that no civilization on Earth had even arisen to take its rightful place as an intelligent species, that any opposing arguments or future debates are moot and invalid, relegated to the waste heap of history's foolish precepts and misconceived notions!

40

NASA, however, remains adamant in its position that no signs of life have been found and that the human-like face on Mars that they say even looks like an Egyptian Pharaoh is really just another mesa. NASA is in full denial, either ignoring the obvious evidence before their eyes or vainly attempting to promote false and misleading statements and pseudo-scientific explanations that Mars is a dead, dry and barren world bereft of any signs of water or life.
http://science1.nasa.gov/science-news/science-at-nasa/2001/ast24may_1/

Regardless of NASA's position on the subject of artificiality and possible life on Mars, there is no slow down in the image analysis by private citizens and scientists of the photographs taken by the orbiters and rovers around and on Mars.

Forced by public pressure over many years and by government order that all photos taken in the last few years had to be released into the public domain, NASA in May, 2001through JPL released over 65,000 never before seen photographs taken by the Mars Global Surveyor revealing many more unusual Martian anomalies. A pantheon of other monuments and pyramids had been discovered with the latest release of NASA images including a large three sided pyramid and a second face that has been carved out on the side of a cliff. (See images below).

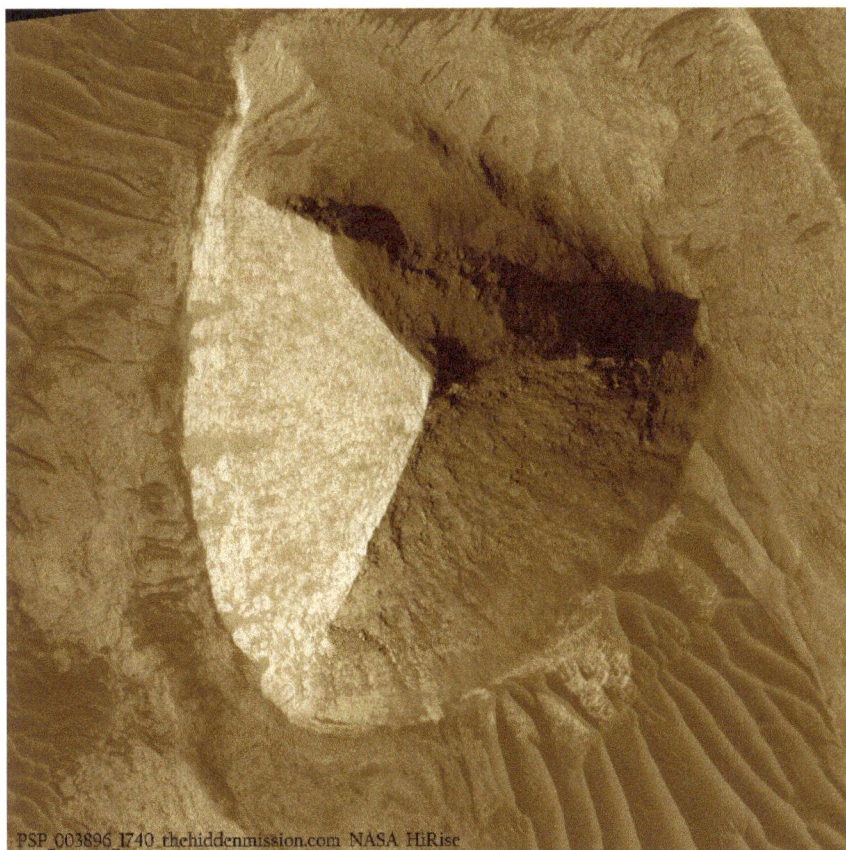

West Candor Chasm Tetrahedron Pyramid taken by HiRise PSP 002841 1740
http://www.thehiddenmission.com/CandorTetrahedron.html

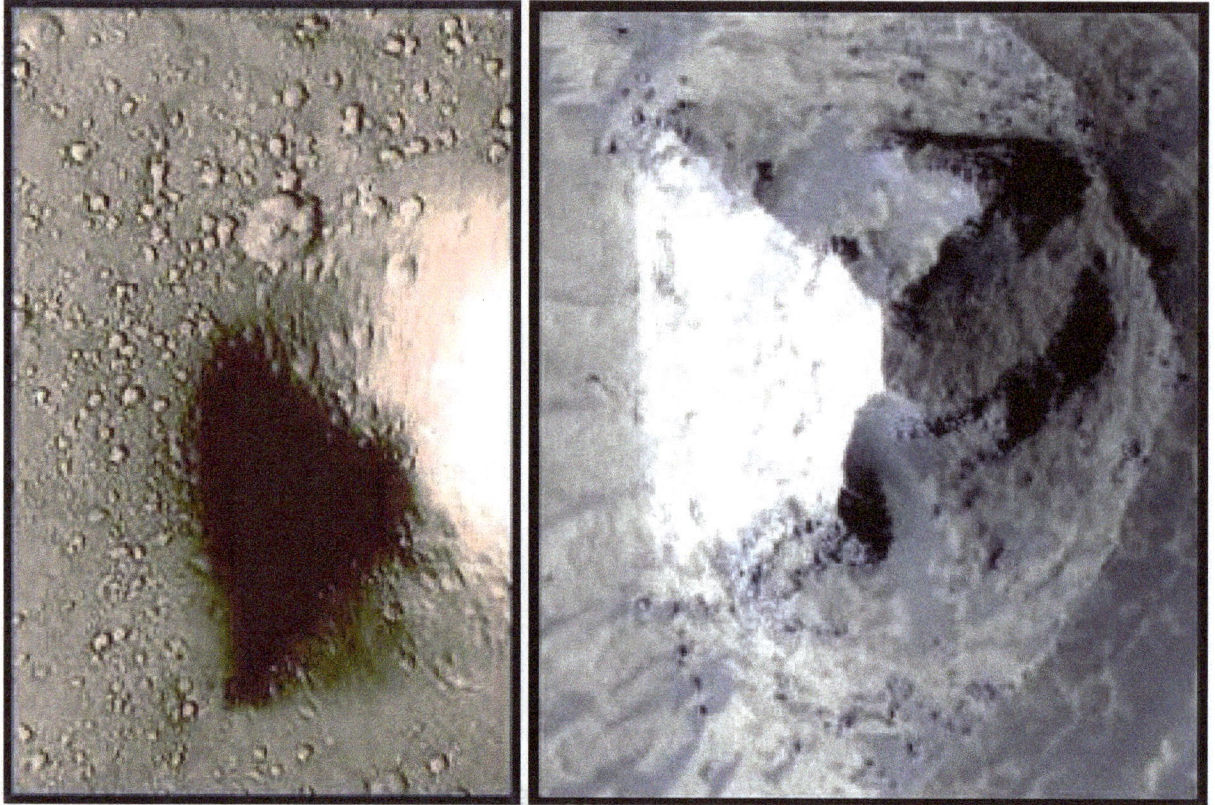

Cerberus Plains Pyramid - MOC m0701415 (left) and
Noctis Labyrinthis Pyramid - HiRise PSP_006679_1680 (right)
http://www.marsanomalies.com/pyramids and http://www.marsanomalies.com/pyramids

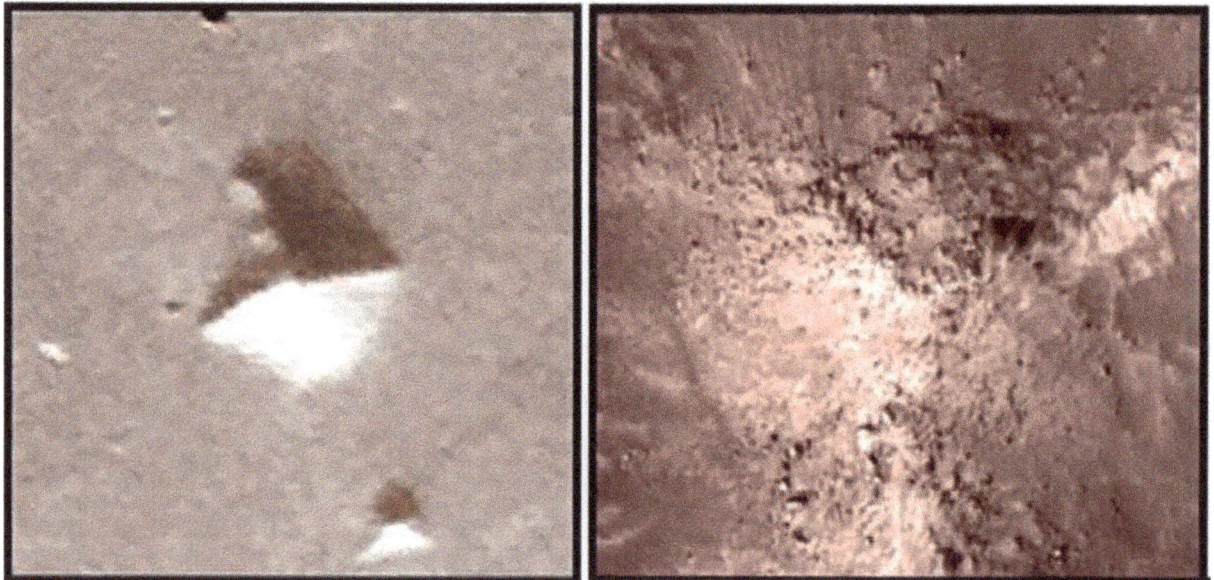

Elysium Pyramid - Viking and Cydonia Laney Pyramid Mound - HiRise PSP_005924_2210
http://www.marsanomalies.com/pyramids

SEMBY9474OD Mars Express recorded a 62-mile-wide swath of the Reull Vallis east of the Hellas basin from 170 miles overhead. The valley may have been carved by flowing water. Note three-sided pyramid lower right

http://sci.esa.int/mars-express/34508-reull-vallis/

Giant Geoglyphs of Mars

At a press conference in New York on May 8th of that same year, one of the images presented was this extraordinary new face, which is a mile and a quarter wide and is located near **Syrtis Major/Libya Montes** (about 3,000 miles from the famous Sphinx-like face photographed by NASA in the Cydonia region of Mars). The new face depicts a weather worn yet, very discernible visages of a strikingly noble appearing male adorned with what looks like a crown.

The announcement was made by **Tom Van Flandern**, a former Chief Astronomer for the United States Naval Observatory, who is president of **Meta Research** based in Washington, D.C., **Brian O'Leary**, a former Apollo scientist-astronaut trained for America's first manned mission to Mars, and **Michael C. Luckman**, director of the **New York Center for UFO Research**. http://www.bcvideo.com/bmars.html

Flandern created quite a storm of controversy at the press conference when he suggested that Mars may be the equivalent of a **planetary Disneyland** in which many monuments were created to depict animals, pyramids, half human – half animal glyphs that could be viewed from a high altitude or from space. Mars could be considered a sort of interstellar tourist sightseeing planet for Extraterrestrials flying through our star system. This theory of interstellar sightseeing could be labelled the **ET Tourist Hypothesis.**

Thomas Van Flandern in 2007
http://en.wikipedia.org/wiki/tom_Van_Flandern

The original image taken by MOC was barely discernible and ever since its discovery, in the summer of 2000 by fellow Mars researcher **Greg Orme**, this complex facial monument has received a lot of attention among the anomaly hunting community. Unfortunately, it would take a little over ten years of image requests and patiently waiting before NASA would release a new image of this intriguing formation. Well, finally in the summer of 2010 The University of

Arizona released a HiRISE image of the formation that confirms all the facial features of this amazing geoglyph. http://herotwins.hypermart.net/Crowned/CrownedFace.htm

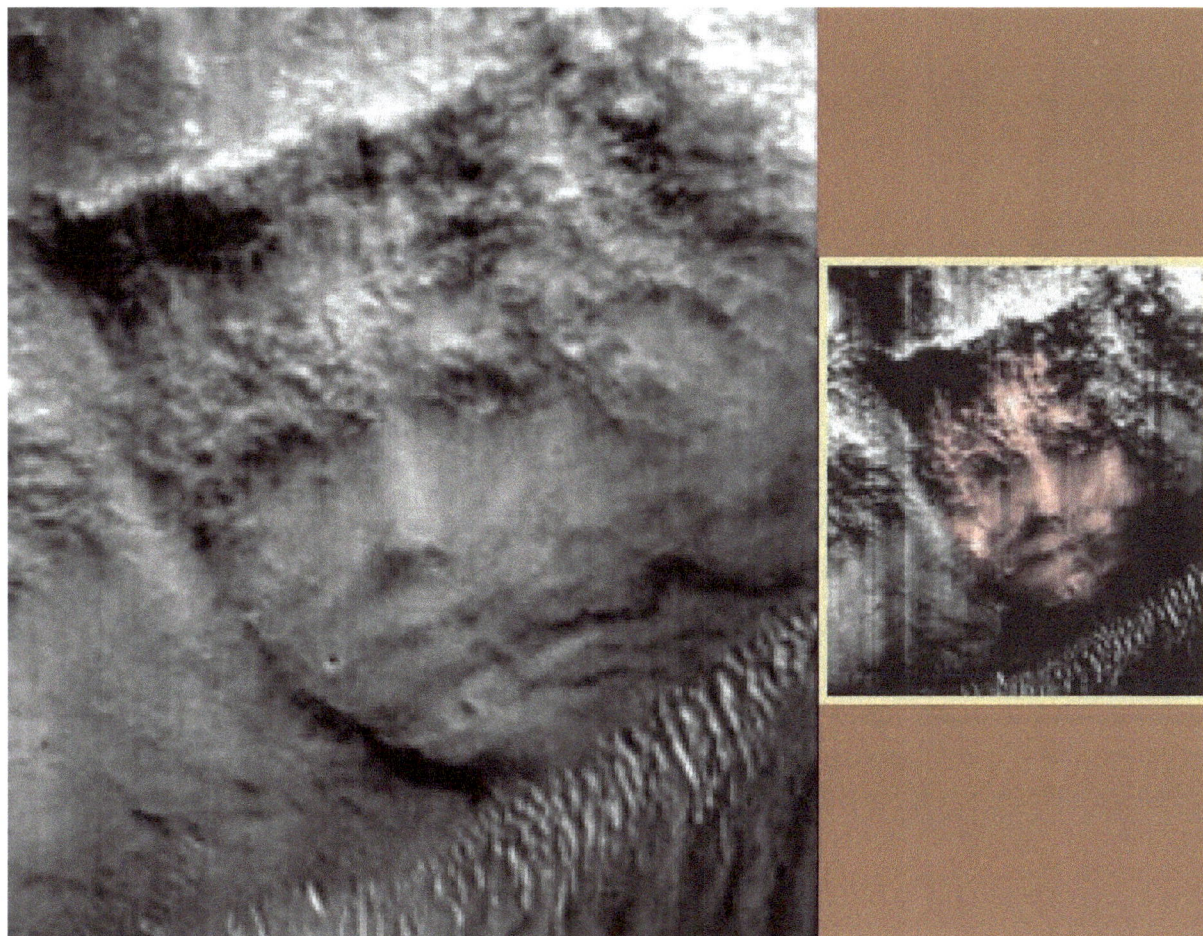

A new megalithic face, the Crowned Face (MOC image M02-03051) or the "King Face" (contrast and colour enhanced right) discovered in 2000 near Syrtis Major by MOC
http://www.bcvideo.com/bmars.html

In examining the facial features of the Crowned Face, its structural dimensions appeared quite accurate although there is no defined border framing the right side of the face. The face is embedded within the surrounding ridgeline of a sloping cliff that flows down into a sandy valley of dunes. The crown appears to be formed within the natural ridgeline that extends beyond the face. Within the face are two dark eye features. Although the left eye is aligned in the proper orientation to the nasal bridge line, the right eye is offset. This gives the impression that both eyes are to be seen as left eyes. Following the nasal bridge down the face, there are suggestions of nostrils and below the nose is a soft parted mouth that completes the face.

When the left side of the **Crowned Face** is mirrored a somewhat *feminine* visage appears to take form (see tri-photos below). The overall face takes the shape of a fault line that frames its internal features. Notice the winged headdress and the textured, lattices pattern that forms across

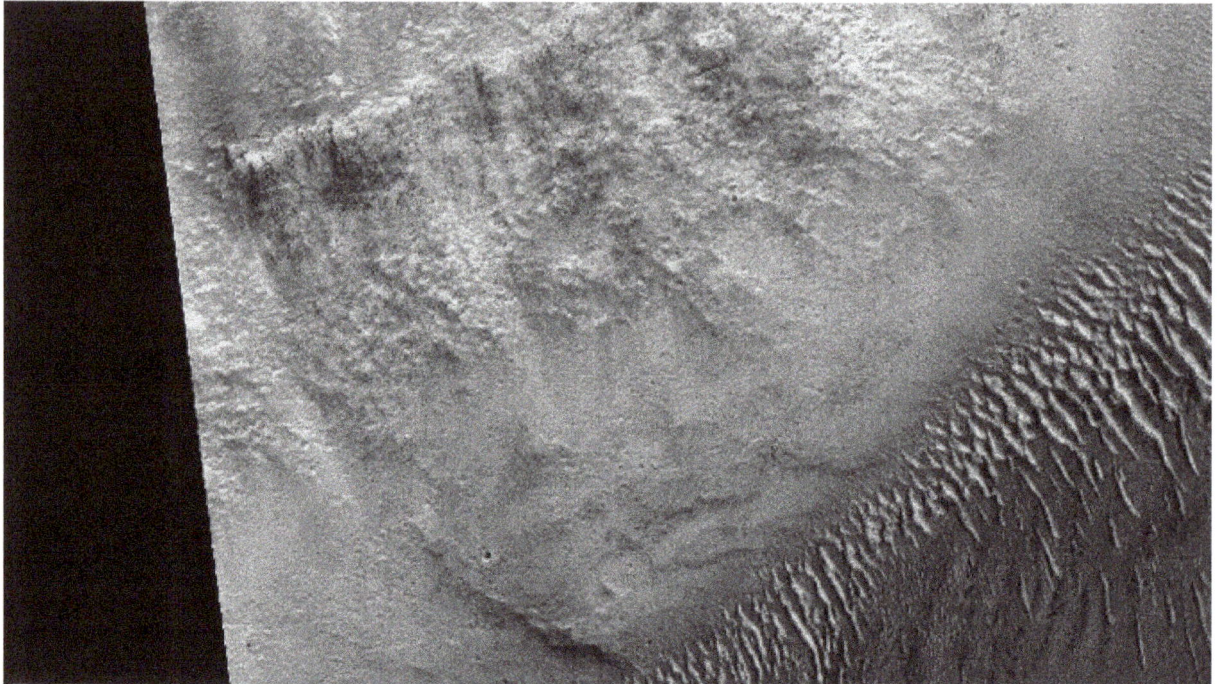

This image of a second face taken by Mars Obiter Camera in the Libya Montes area is a deliberate altered image by NASA to hide what was another sign of intelligence
http://thecydoniainstitute.proboards.com/thread/43/crown-face

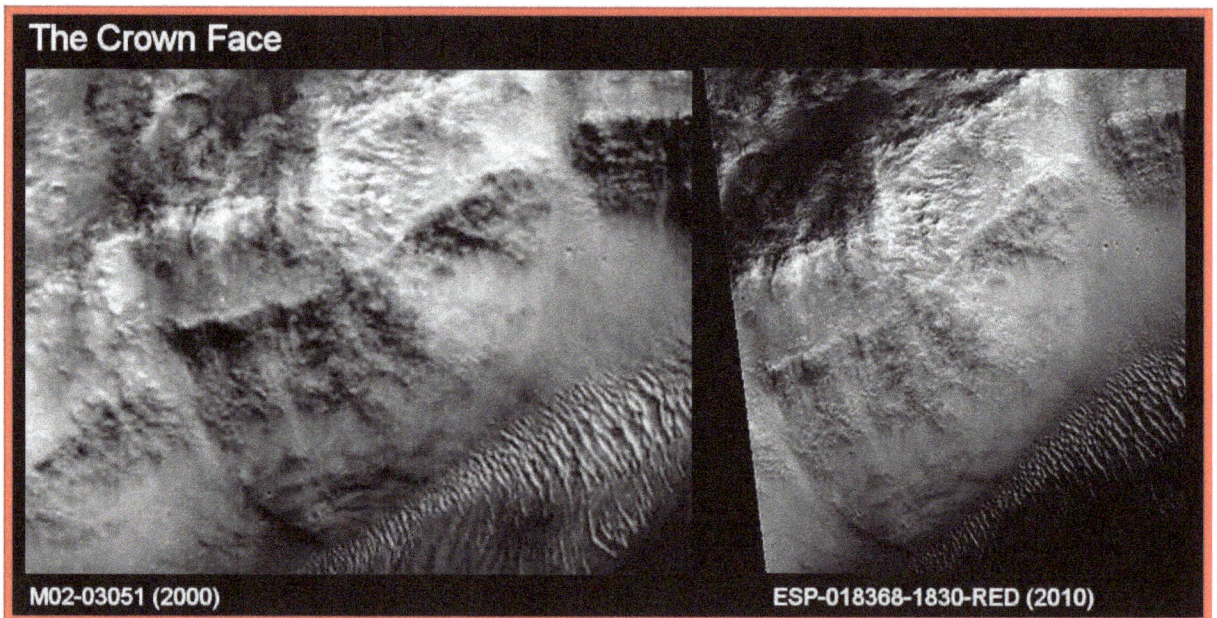

The Crown Face

M02-03051 (2000)

ESP-018368-1830-RED (2010)

A second re-imaging of the King Face glyph in 2010 confirmed the facial features of the original (above) almost unrecognizable image that was deliberately altered by NASA
http://thecydoniainstitute.proboards.com/thread/43/crown-face

the forehead creating a decorative crest. At the center of the crest is a V-shaped emblem formed between two elaborate flaming eyebrows. Note the wing formation of the central emblem forms

a **Phoenix Bird** with flaming wings. Below the flaming eyebrows the face has shadowy deep-set eyes, a nose, lips and a chin ornament. Below the deep-set eyes are two pockmarks or holes with connecting shafts that extend below the checks forming tear bands.
http://herotwins.hypermart.net/Crowned/CrownedFace.htm

The proportions are distinctly human in appearance which makes the viewer wonder if the human visage and body form, a universal body type found throughout our Solar System or the universe or whether this obvious resemblance is merely shared between the Earth and Mars. Is the intelligence of one civilization exactly the same on both planets because humans originated from one planet and migrate to the other planet in very ancient time, long forgotten or is there some other unknown reason?

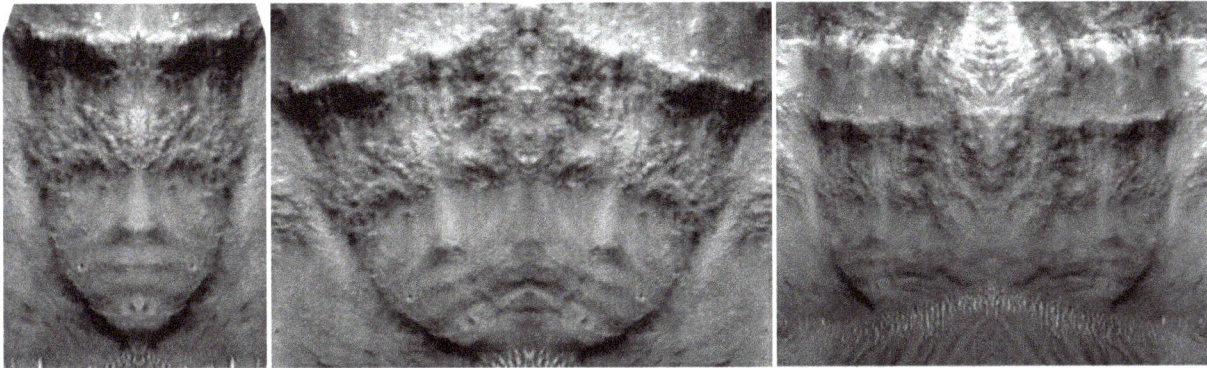

These photos were mirrored along natural demarcation lines which produced the crowned human visage (left), the demi-jaguar or were-jaguar (middle) and the moth or butterfly (right) with smaller feline or wolf demon above
http://thecydoniainstitute.proboards.com/thread/43/crown-face

When the right side of the second face is mirrored, a **Were-Jaguar face** is revealed with the Crowned Face mask framing it on either side (see above middle photo). Notice the elaborate crown formation with a small inset mask, the feline-shaped eyes, the pug nose and snarling aspect of the muzzle.

When the third face on the far right side of the composite mask is mirrored, along the third demarcation line, a moth shaped Mardi-Gras style mask is revealed with a totemic demon (feline or wolf) mask inserted in the center of the headdress (see left photo above). Note the large compartmentalized wing shaped grid forms two wing-shaped eyes. Below the cross section of the wings is a small nose and puckered lips that form the lower segment of the body. There is also a set of feathered antenna extending from the wing, framing the demon mask.

Moths and butterflies they are seen as symbols of metamorphic transformation from caterpillars in many cultures, especially throughout Mesoamerica. When one considers that a half faced human and a were-jaguar mask flanks a Moth mask, the same message of human and feline transformation that is embedded within the Face at Cydonia becomes quite plausible. And just as the Face at Cydonia has a direct relationship with a pair of masks found at the Ceros Mexico, when the **Moth Totem on Mars** is compared to a tri-faced Aztec mask known as the **Three**

47

Faces of Life (see below), a repetitive theme of transformation can be established within their common segmented composite design. While the Moth Totem mask deals with the transformation of human to feline, the Aztec mask deals with the transformation of youth to death. http://herotwins.hypermart.net/Crowned/CrownedFace.htm

Another example has been referred to as the **Martian Nefertiti,** a landform inscribed into the Martian landscape much like the **Nazca Lines** of Peru discovered again by Tom Van Flandern.

Aztec Mask: The Three Faces of Life. Note the progression of the faces from the youthful face in the center to the split elderly face over that, to the split death mask on the outside.
https://www.guggenheim.org/arts-curriculum/topic/mexico-tenochtitlan

48

Nefertiti (c. 1370 BC – c. 1330 BC) was the Great Royal Wife (chief consort) of the Egyptian **Pharaoh Akhenaten**. Nefertiti and her husband were known for a religious revolution, in which they started to worship one god only. This was **Aten** or the sun disc. Nefertiti had many titles; for example, at Karnak are inscriptions that read Heiress, Great of Favours, Possessed of Charm, Exuding Happiness, Mistress of Sweetness, beloved one, soothing the king's heart in his house, soft-spoken in all, Mistress of Upper and Lower Egypt, Great King's Wife, whom he loves, Lady of the Two Lands, Nefertiti.

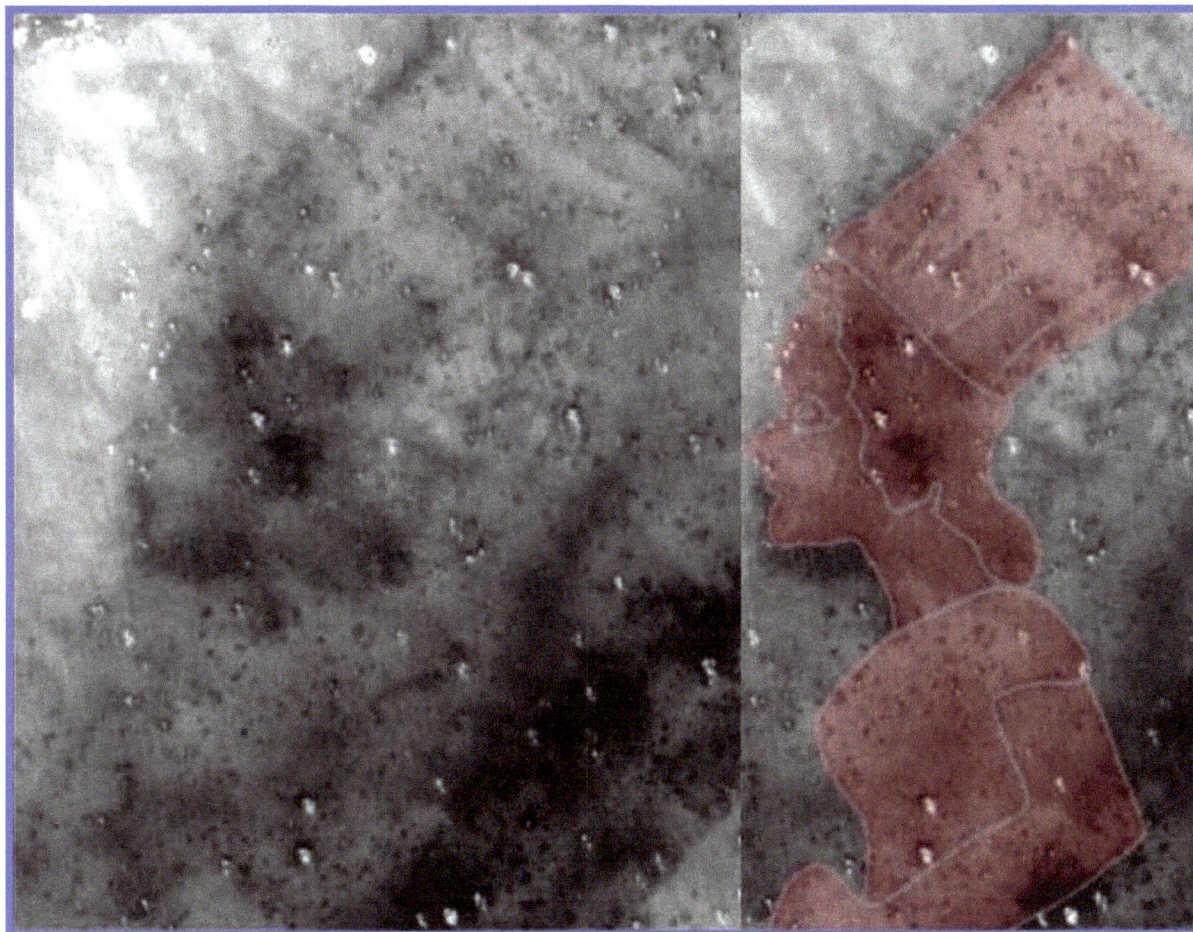

Another Mars "Face" discovered by Tom Van Flandern appears with an Egyptian style headdress similar to Nefertiti of Egypt (Image at right is colour enhanced)
http://www.coasttocoastam.com/pages/nefertiti-face-on-mars

What is evident from the above photo and those below should be obvious by to the reader which is the human-likeness of the Martian Nefertiti and other humanoid monuments on Mars to human beings of Earth, as well as the similar wardrobe or costumes worn by both cultures in ancient times. Such almost identical similarities, begs us to ask the question: what is the reason for this similarity and are we connected or related in some ancient way?

Nefertiti of Mars (left and middle) and Nefertiti of Egypt, Earth (right)

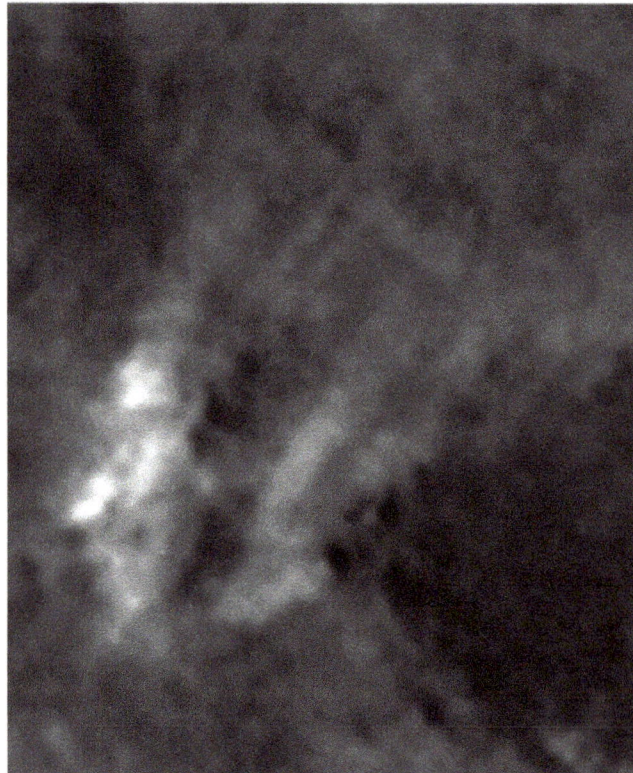

A companion formation to the profile of the Nefertiti Face is located on the opposite side of the planet in the Nili Fossae region, located in the Northeastern hemisphere of Mars

Faded topographical lines on the Martian terrain and hilltops reminiscent to the lines in Nazca, Peru or the Outbacks of Australia or even the white chalk lines on hills found in the UK are also indications of intelligence designed to communicate a message to the any passerby.

John Levasseur, discoverer of the **"Nefertiti"** formation (see above) and member of the **Society for Planetary SETI Research**, has developed a practical method of determining the probability that a given Martian surface feature is artificial. The *a priori* argument presented by Levasseur's offers the most stringent criteria yet for objectively assessing the problematic "profiles" that litter the Martian surface.

Levasseur found a large scaled rendering of a cat in the **MGS Image #M0202619** seen upon the Martian surface that was 6 km or so in length. Based on the impressionistic hypothesis surface features consistent with feline features outside the 3 km wide frame could be predicted. The impression of a big cat was strong having a well-defined head with eye, ear, lips and a nose with nostril, there is a shoulder with extended left front leg and torso. The two right legs appear to cross as if in a trot. There is a long back and a belly and it appears to be domesticated as it is wearing a collar. http://www.mactonnies.com/imperative42.html

Levasseur's suspicion that there were probably more hills, mounds, and raised areas continuing out of frame that would complete what appears as large-scale art can be tested with new satellite imagery. If there was a feline left claw, either in a closed or open position, a hind leg and haunches, and most importantly, a long tail of some kind either extended or curled existed in the new unseen image with the terrain having the same type of mounds and hills then, it would prove conclusively his cat art hypothesis and provide further supporting evidence to the artificiality of the **Cydonia Face** and the **Nefertiti Face**.

If the out of frame surface features do not complete the pattern of the cat, then the cat art hypothesis will be in question. Of course, it probably won't be conclusive because impacts and/or erosion could have disfigured the pattern, or it may be found that the cat has a bobbed tail.

The chance of acquiring an image that would satisfactorily provide a test for this hypothesis in the near future is remote.

Levasseur posits that there is the contention that we should not be seeing artistic renderings of what we see on Earth today, that the existence of artistic patterns depicting human faces and animals of recent evolution on Earth are ludicrous and embarrassing claims. But these arguments rest on flimsy origins-related assumptions, like that any artificial Martian objects would be built only in the very distant past by an indigenous civilization, one that should exhibit (ironically) earth-like infrastructure.

Skeptics have for decades made similar non-arguments regarding the humanoid **Cydonia Face**. But there are many other possibilities that cannot be ruled out at this stage. For example, some scientists seriously consider the interesting idea of ETI periodically passing through the solar system (something predicted statistically by Sagan over thirty years ago) and leaving its mark on the planet. In this case, the renderings need not be of them but maybe of us and other Earth images. They could have been made at almost any time, and need not require infrastructure such as roads, which would not have been necessary if the builders are flying and not living there. As far as I'm concerned, this idea is as plausible a scenario as any at this point because we are not in any position at this stage to say what or what cannot be found on that planet or how it got there. Humanoid faces and "modern" cats aren't necessarily ruled out. There are many origins ideas. Arguing that these different scenarios are implausible doesn't make the evidence go away. http://www.mactonnies.com/imperative42.html

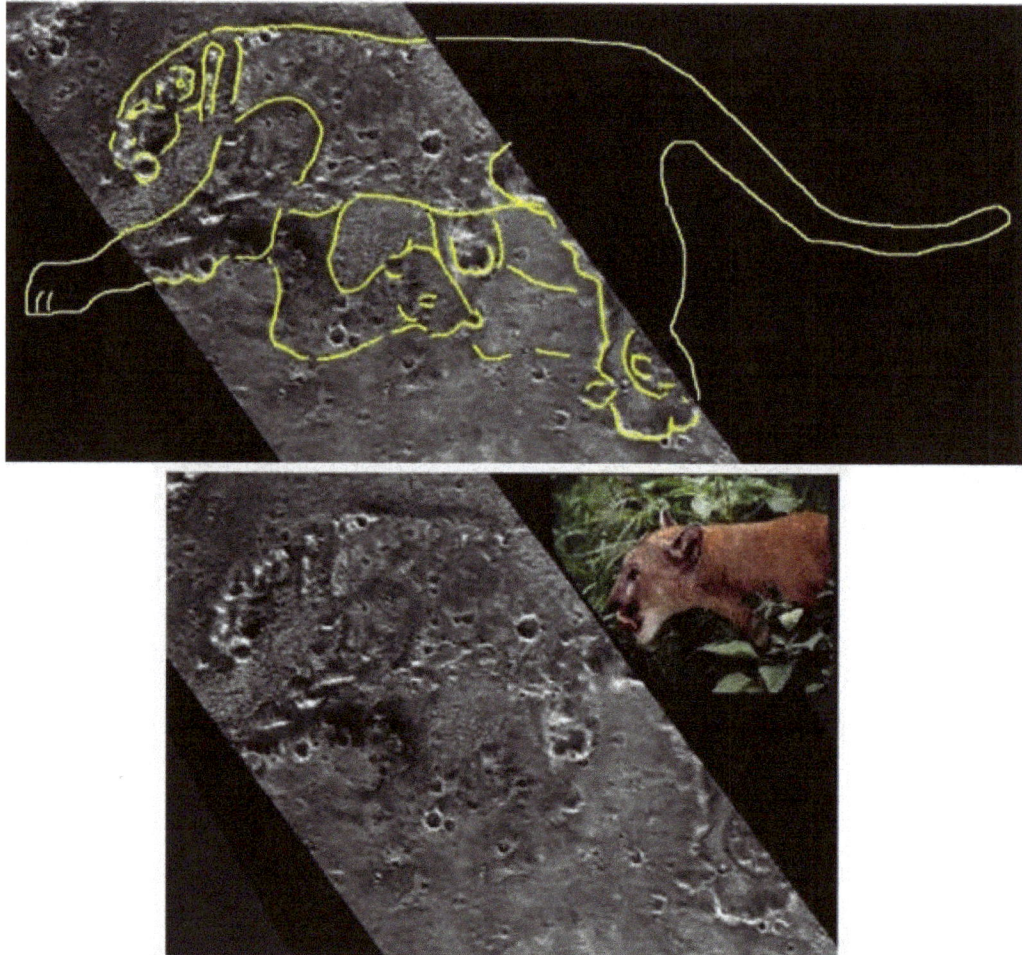

MGS Image #M0202619 Puma Predictions. Out of frame features can be predicted on the impressionistic conclusion that the object is a large-scale rendering of a big cat
http://palermoproject.com/lowell2004/legacy4.htm

Fortunately, new images were taken of the area in question however, the resolution was disappointingly low but, they did appear to show a feature consistent with the "tail" predicted in **Levasseur**'s original article. Levasseur's *a priori* out-of-frame method thus adds credibility to the idea that the "puma" may be more than random geomorphology. The discovery of the "tail"-like feature -- while certainly debatable -- challenges our criteria for potentially artificial Martian surface formations.

52

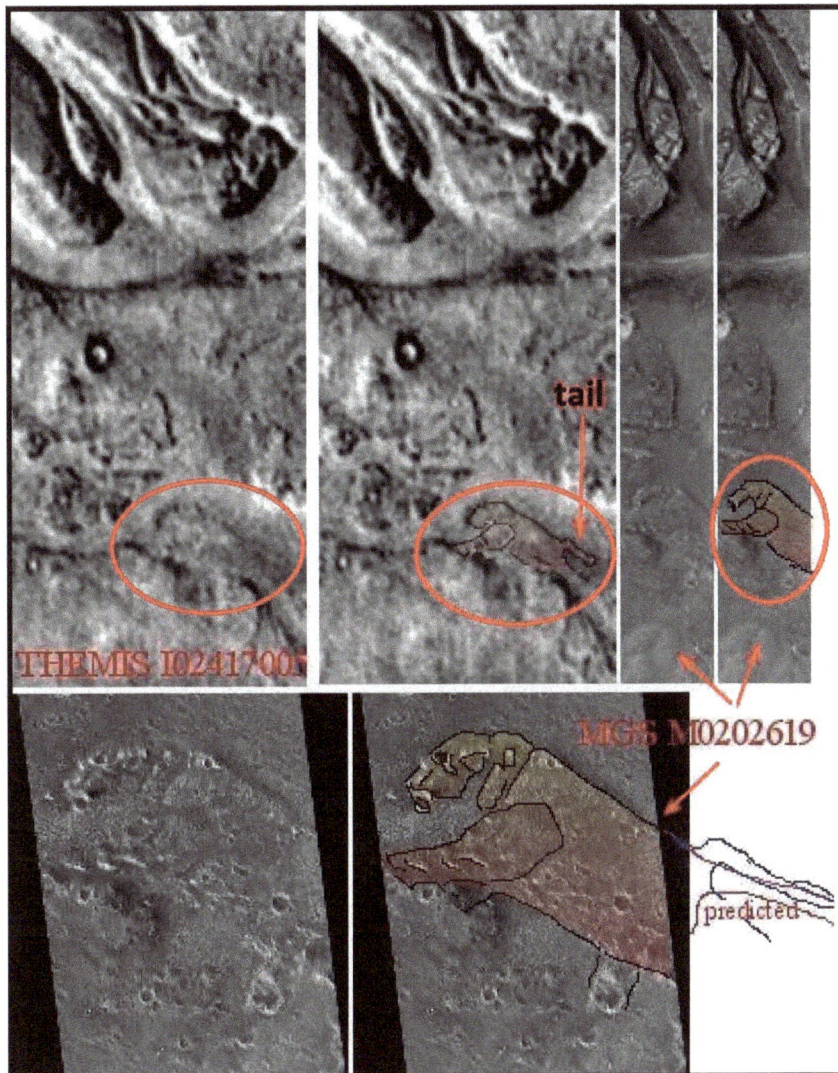

**The puma's tail is verified and adds one more supporting proof
of artificiality, ego Martian intelligence once existed on Mars**

Most planetary SETI efforts involve analysis of upward-facing formations such as **The Face** and **D&M Pyramid**, asymmetric profiles tend to be brushed aside as fanciful **Rorschach "ink blots".** Anomaly researchers with a "wannabe" attitude to become someone who made the first discovery of a new artifact on the Moon or on Mars and thus, become a publicly respected peer among the Ufologists in the UFO community incessantly contend with fragile claims of Nazca-like Martian "birds and seahorses, faces, statues and crashed alien space vehicles". **Lausch** does not endorse the "puma" as proof that a Martian civilization once used the planet's surface as a geological canvas but he is intrigued with **Levasseur's** research techniques. "Proof" of such isn't likely to manifest so conveniently. And it should be noted that some anomaly researchers inclined toward the possibility of megalithic structures on Mars voiced dissatisfaction with Levasseur's original, tailless "puma" -- specifically, pointing out anatomical flaws such as an apparent "club foot."

I regard the out-of-frame method as an intellectual exercise that planetary SETI must deal with in its pursuit of hard evidence of intelligent extraterrestrial design. The out-of-frame technique's ability to isolate moments of genuine anomaly is largely untested because partially perceived likenesses simply aren't liable to come under scrutiny; familiar-looking shapes tend to catch the eye because they're generally *intact*, leaving out-of-frame prediction useless. To my knowledge, the "puma" is a lone exception. Regardless, Levasseur's unique approach is liable to ignite the philosophical foundations of the planetary SETI inquiry.
http://www.mactonnies.com/imperative42.html

As has been previously stated, anomalies seem to pop off the Martian surface as more anomaly researchers investigate the release of NASA photographs that have been accumulating in their archives and it may be a few more years before new announcements from *"old discoveries"* are released to the news media and the public. So, it is that we must content ourselves with old NASA photo images to prove our case for the former existence of an ancient civilization on Mars.

A pharaonic-like statue discovered by Mars rover Opportunity on the side of a crater compared to Egyptian statues of the pharaohs, the resemblance is amazingly similar
http://www.abovetopsecret.com/forum/thread356078/pg1

The pharaonic statue photo images above re-enforce the hypothesis that an Egyptian-like culture (although, far more advanced than the ancient Egyptian culture on Earth) once existed on Mars, especially when we compare these images with the **Nefertiti** pictographs discovered by **John Levasseur** and the **"Face"** on Mars monument. There are even images on the Martian terrain that depict the sacred image of the "Eye of Ra" or "Horus" that find its equal on Earth in many ancient Egyptian cities.

The image below was also found in 199 at the Cydonia Mensae area along with the Face and the D&M Pyramid which shows a large Dolphin intaglio complete with a crater as a blowhole! Are these pictograms above and in the images below pareidolia- an illusion or aberration of the mind, then why do we marine life and other land animals familiar to us being repeated over and over again in the shifting sands of Mars? The recurrence of the cetacean theme across the terrain of Mars bears further investigation, as it can't be just happenstance that nature decides to reproduce in geomorphologic formations animals that just happen to exist on Earth and maybe at one time on Mars.

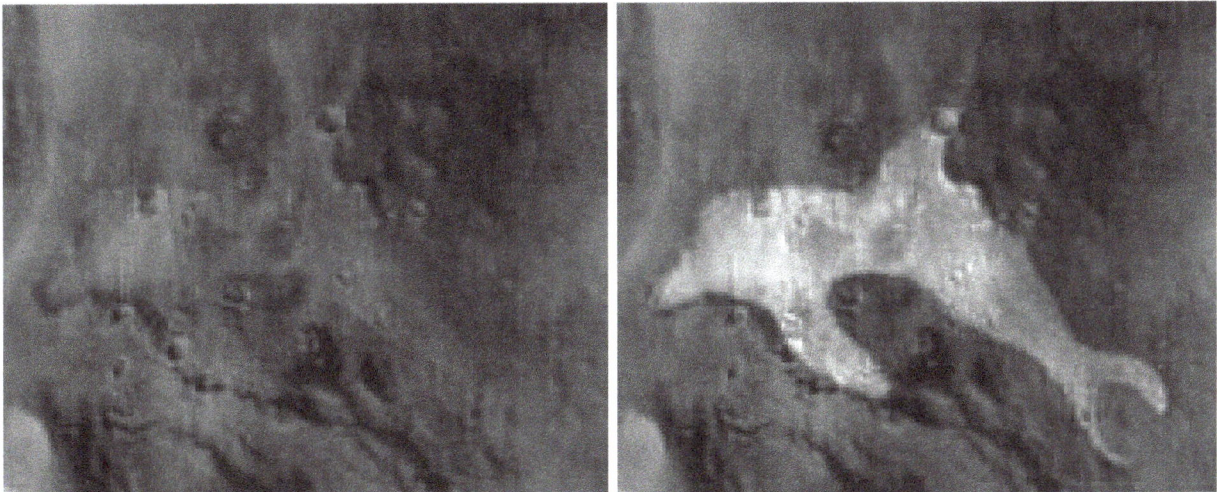

**The Dolphin of Cydonia Mensae pictogram left and whitened at right
was discovered in 1998 near the Face of Mars**
http://palermoproject.com/lowell2004/legacy3.htm

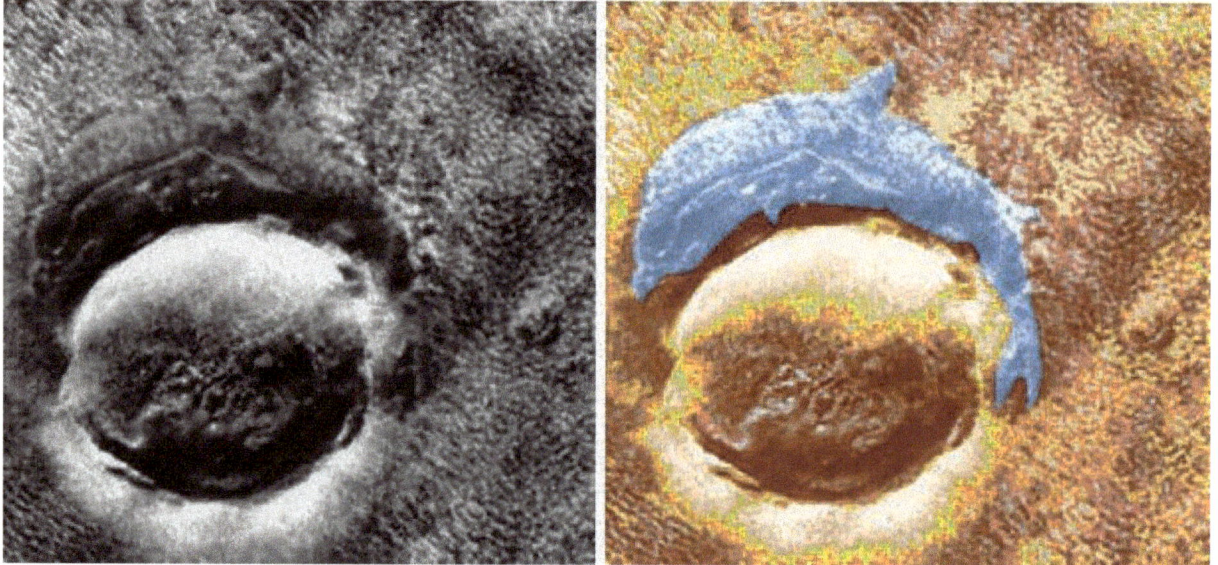

**Dolphin pictogram from MOC Image M1501765 left is original image
and right is colour enhanced to bring out detail**

**Compare these dolphins cousins of Earth to the dolphin
and whale pictograms of Mars above and below**

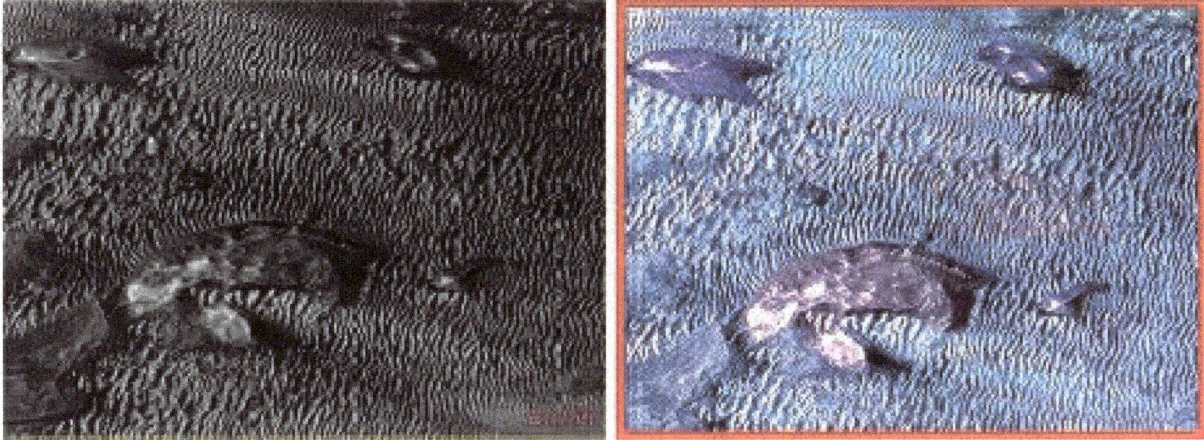

MOC Image E12-0072 Mega-Dolphins and Mega-Whales "Swimming" in Martian Sand

http://palermoproject.com/lowell2004/legacy3.htm

The above photo image of what appears to be whales swimming in a sea of Martian sand is too eerie for windblown sand dunes to align themselves perfectly into recognizable features like sea creatures that can be photographed from space. This type of pareidolia that appears so common across the surface of Mars suggests that the planet spirit of Mars is having some fun with the imagination of humans or alien civilizations that just happen to be touring the solar neighbourhood. If the reader is inclined to believe that planets have spirits or souls then, this would seem to be a logical answer to the mimicry of animals and humans seen on Mars.

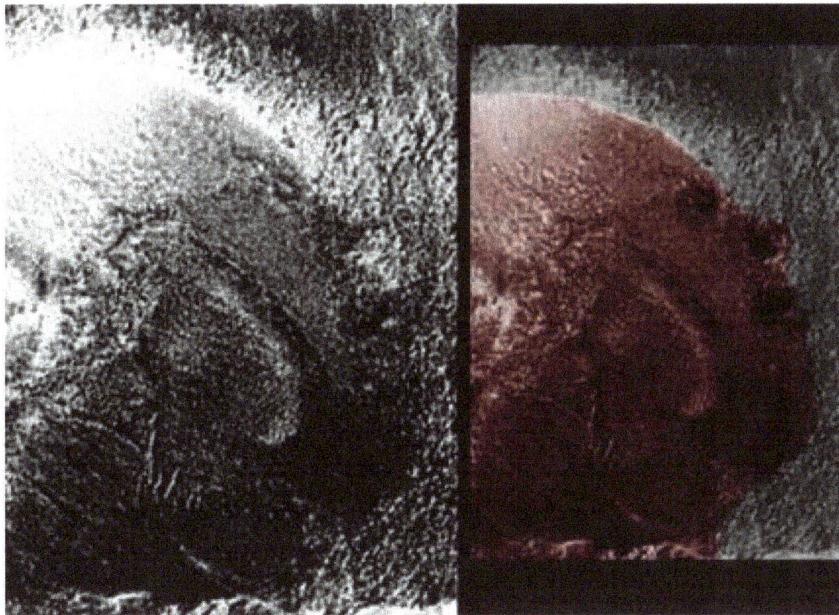

Another Profile Face in the Cydonia region

http://raphaelonline.com/profileface.htm

However, the more likely explanation is that these topographical geomorphs are the handiwork of intelligent design which means an Extraterrestrial civilization built them long ago with a deliberate intention to communicate concepts, ideas, and messages to whoever will visit the planet. One obvious message being communicated is a *love of life in all its bio-diversification with a conservational reverence and management of the environment that once existed on Mars.* Perhaps, Mars is an interstellar Disneyland-like planet after all!

Yet another 'Profile face' with forehead, eye, nose, mouth, chin, neck, and ear can also be found in the Cydonia area. Its appearance is similar to the ancient Olmec culture of Central America, particularly when it and the famous **Face of Mars** are compared to the Olmec colossal head monument in Mexico. (See below).

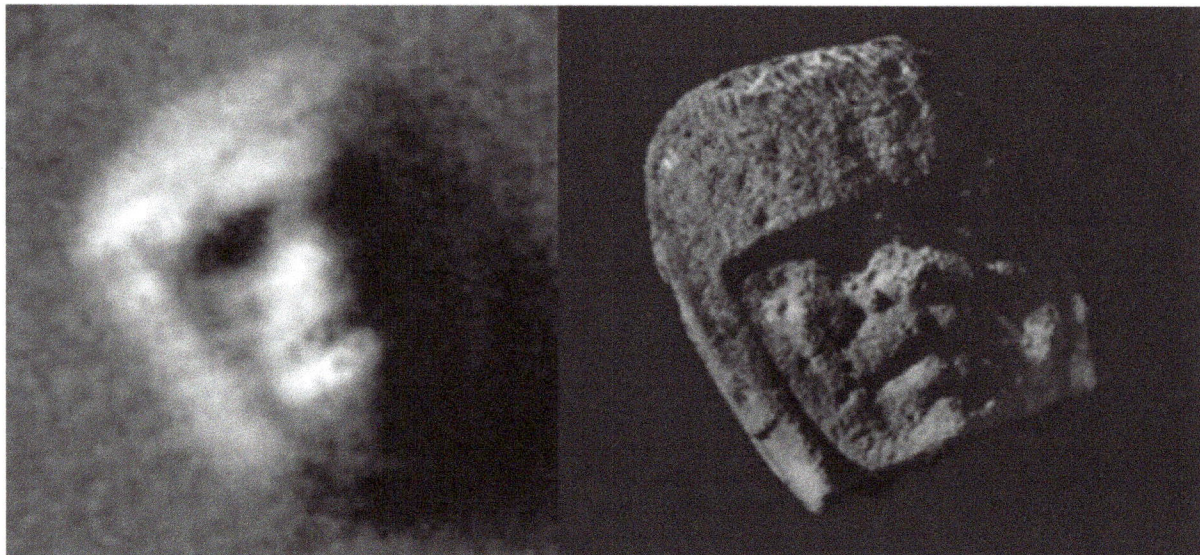

1976 Face on Mars compared to an Olmec colossal head
http://thecydoniainstitute.com/The-Face-on-Mars-wears-an-Olmec-Headress.php

Once again, these photo of geoglyphs on the Martian terrain begs our attention and consideration that these are not natural landforms but constructs of artificial design that indicate an intelligence once existed on Mar. It is one more proof of an Extraterrestrial presence in the Solar System besides our selves.

"We Have Worm Signs the Likes of which Not Even God Has Ever Seen"

With the release of the tens of thousands of NASA images in 2001, there were photos showing long, strange and winding tubes partially buried in the Martian ground and also exposed to the surface terrain. At first appearances, these snake or worm-like tubes had an organic quality to them, in fact, many people thought that these were the tell-tale signs of gigantic earthworms or should that be Mars worms, much like what was seen in the movie **"*Dune*"** burrowing into the earth and then exploding up through the surface. Some thought that these worm-like anomalies were just another unusual landform which Mars seem to have an abundance of across its landscape. However, upon closer inspection of the raw photographs, the images revealed the unmistakable characteristic of an artificial manmade or Martian-made construction!

NASA says that these are sand dunes in narrow valley floors and that the perception that they are tubes is an optical illusion due to sunlight angle and shadows, much like craters are perceived to be domed or convex depending if the photo is upside down or upright or if viewed from the right or the left. In other words, like the Face on Mars, it too is a trick of light and shadow!

According to **David C. Pieri, Ph.D.** of Earth and Space Sciences Division of the **Jet Propulsion Laboratory**, after only examination of one photograph posted by Richard Hoagland on his *Enterprise Mission* website of the glass tube anomaly concluded that the **"Martian Glass Tunnel Anomalies"** are nothing more than simple **"Martian Dune Trains"**.

Initially, it appears that he may be right, particularly when one examine the many photographs that show sand dunes and ripples across much of the Martian terrain, mars is, after all, a desert planet, even though some photographs do appear to show worm-like structures that seem to have trails slithering across the Martian terrain. They are in reality dune trains and not worm signs of some gigantic organism out of a science fiction movie. But, these are many exceptions and these are easily spotted even by the novice researcher particularly, if you have some basic high school education in science like physics or biology.

Richard Hoagland in his search for past intelligence on Mars discovered these Martian Glass Tubes or Tunnel anomalies while sifting through the MSSS raw image data back June 2000 and one of these glass tubes stood out from the rest. The photo (see below) showed a megalithic structure that had evenly spaced ribbing of uniform size measuring hundreds of feet in width running approximately over a mile in length on the Martian surface and between open chasms and rifts on the floor of the **Acidalia Plains**. It had other connecting tubes that intersected and seemingly disappeared into the ground but, one major portion of the tube stood because it had a bubble-like protuberance with a glint of light coming off it or from within it.

Very few natural geological formations like sand dunes reflect sunlight or light of any kind; this had to be yet another artificial structure on the surface of Mars. A three dimensional imaging of the anomaly rendered from a *shape from shadowing* process showed a glass structure that flowed up and down; in certain places, it was actually anchored to the walls of the chasms and not always resting on the floor bottom. Hoagland believed that it was conceivable that these gigantic glass tunnels were some type of car or rail transportation system which ran all over the surface and underground connecting the many cities on Mars.

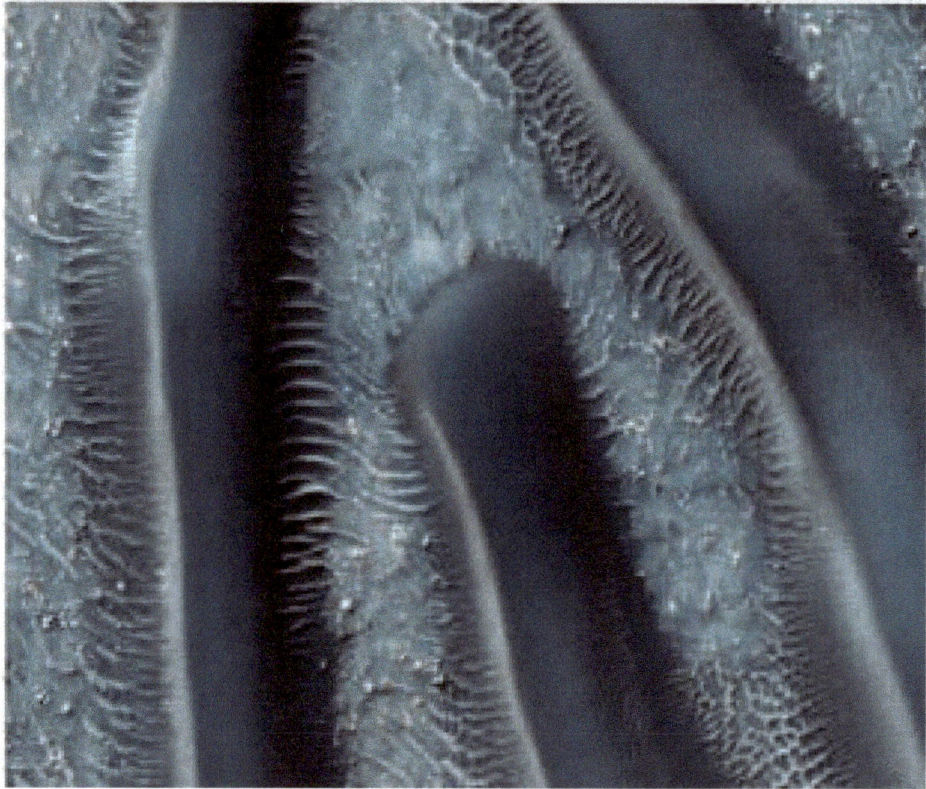

**Two very distinct and different Martian geo-forms, the Glass Tunnel Anomaly (top) is
artificial in appearance while the Dune Train (bottom) is a geological formation**
https://pianetax.wordpress.com/canali-di-marte/ and https://www.pinterest.com/davidmmerchant/astro-4-mars-dunes/?lp=true

Tom Van Flandern points out that there some obvious aspects that refute the **Sand Dune Train Hypothesis** of Dave Pirei of JPL:

a) *The glassy tubes have distinct outlines.* Nothing about the "dunes" hypothesis requires that the extremities of the dunes be connected by an outline, yet the "flat view" interpretation of the glassy tubes is that they are outlines paralleling or connecting the extremities of the dunes.

(b) *The glassy tubes appear to be translucent.* In many places, one sees portions of faint white bands between the bright ones, as if seeing through translucent tubing. In isolated places, one sees complex structure faintly between white bands.

(c) *A glassy tube appears to produce a specular reflection of sunlight.* In one case, a roundish spot of saturated white light appears on a glassy tube (see below). It is positioned on the side toward the Sun and is positioned such that *a specular reflection of the Sun is a possible explanation.* As a singular, very bright spot near the end of a section of glassy tubing, no other obvious explanation suggests itself, and any invented for the purpose would be ad hoc. Because natural terrain scatters sunlight, if this spot is reflected sunlight that would be consistent with tubules of a glassy or plastic-like quality.

(d) *The glassy tubes cast shadows.* Where shadows can be seen, they are consistent with the glassy tube interpretation, but not always with the "dunes" interpretation. In some cases, such as the glassy tubes on the "Cliff" at Cydonia, the shadows are clearly cast by tube-like features. The shadows are beside the tubes on the side opposite the Sun, parallel the tubes, and narrow when the tubes narrow. No "dunes" or other alternate interpretation to the tubes is available for such cases. http://metaresearch.org/home/viewpoint/archive/010313GlassyTubes/Meta-in-News010313.asp

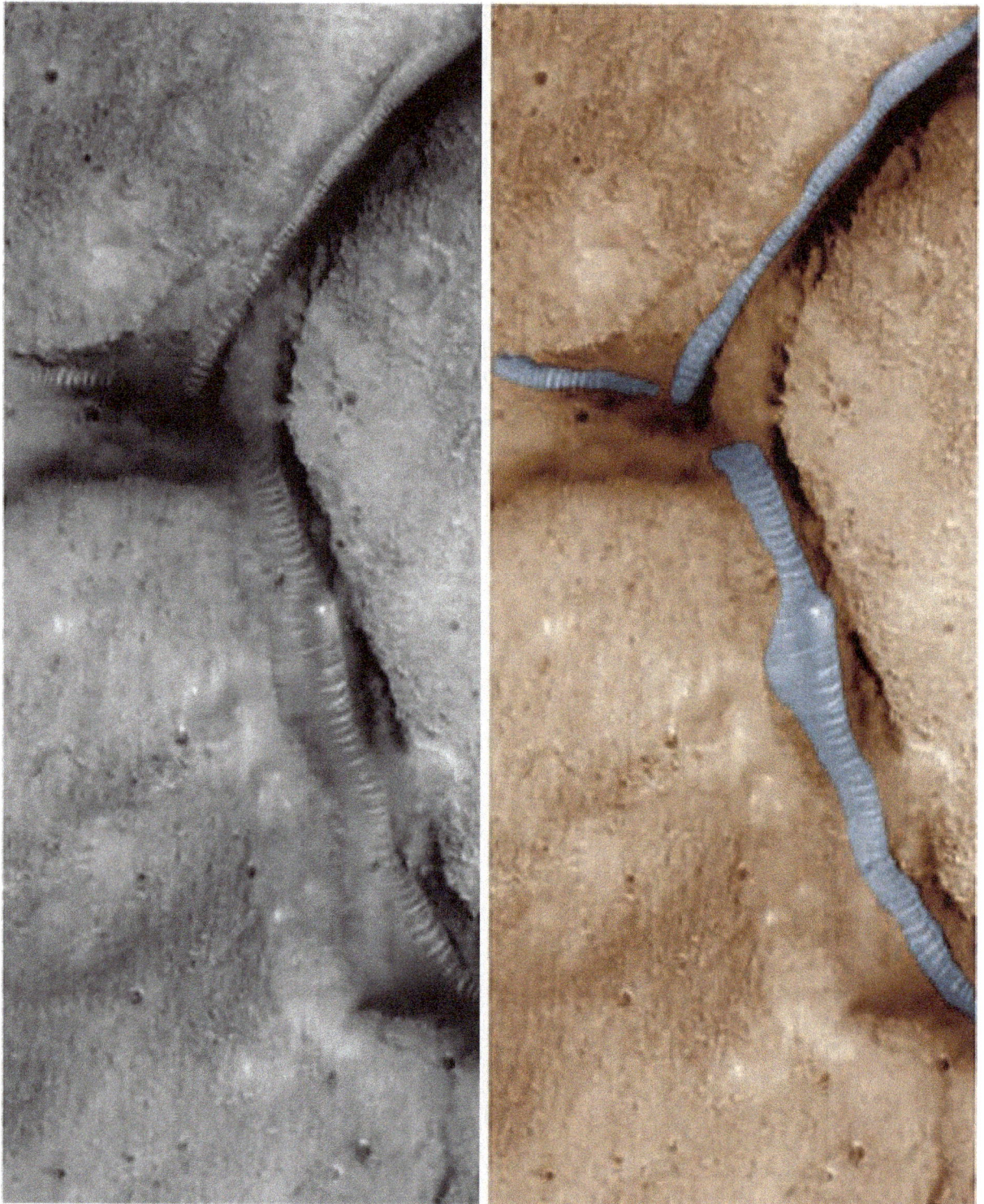

**MSSS MOC image m0400291a. Are these Martian glass tubes used as a
vehicle or rail transportation system or used for some other purpose
such as a massive water irrigation/transport system?**

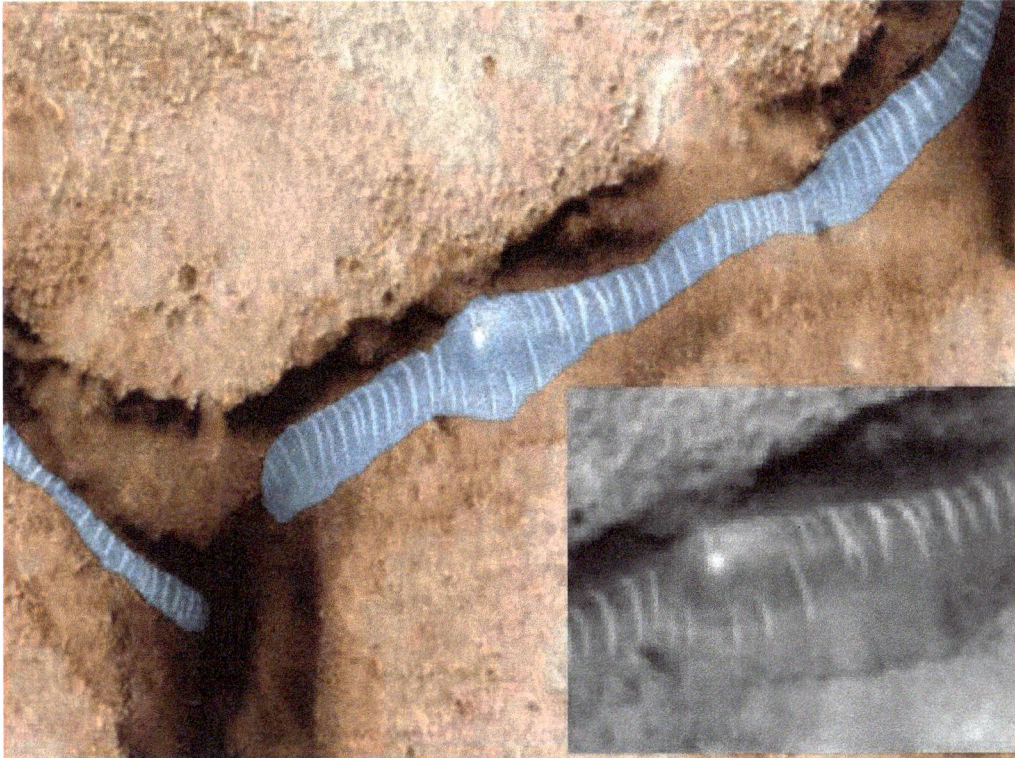

Close up of the dome area of Glass Tunnel, note also "sunlight" reflecting off its surface with something that seems to be inside the tunnel near the dome area

http://www.thelivingmoon.com/43ancients/02files/Mars_Images_11.html and
https://www.pinterest.com/pin/186758715775848073/

Below are more photographs of these mysterious **"Martian Glass Tunnel Systems"**, their appearance shows translucency and the reflection or glint of sunlight off their surfaces indicating artificiality. Hoagland believes that the purpose of these tubes or tunnels may be as a multi-level transportation system or as a water delivery and irrigation system to move water to various desolate areas of Mars.

Until astronauts actually go Mars and start to explore its surface with particular attention to these topographical features, we can only speculate as to the true function of these artificial anomalies.

The two photos below show the complexity and extent in which these glass tunnels course through and over the planet. Many of these tunnels may have been exposed on the surface of Mars but or eons of time have become covered and buried by massive wind storms. Many tunnels may have collapsed from sand storms or degradation from metal fatigue and stress or just from the ravages of time.

Many tubes run close to or near other anomalous objects like dome or convex craters. Some of the craters have geodesic design and structure and could have been used as small villages or industrials sites. These areas need to be re-imaged, particularly with infrared cameras, ground penetrating radar, as well as high definition cameras to bring out greater detail of these structures and the ground around them.

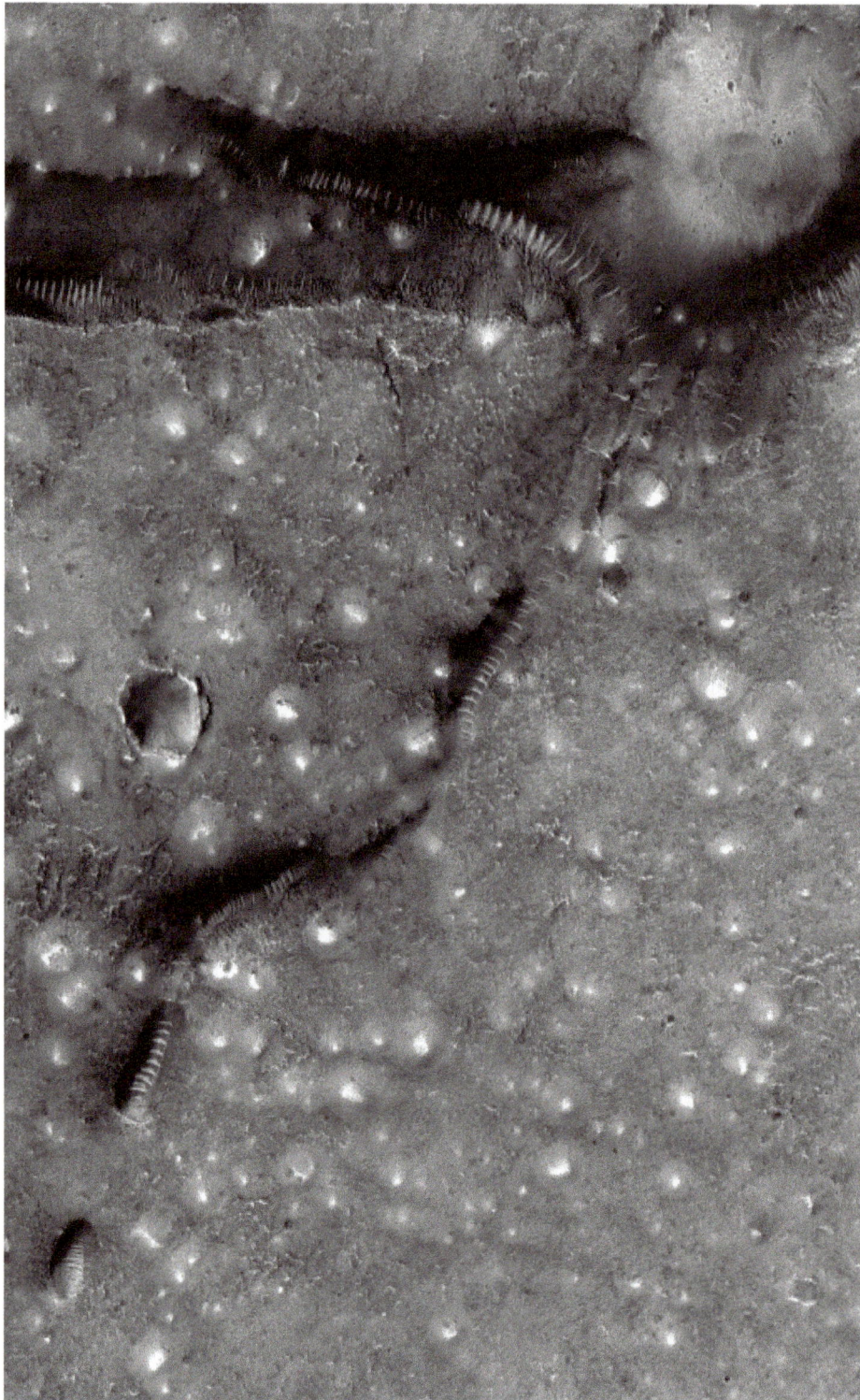

MOC image E020647. A glass tubular structure twists its way over and under the Martian surface. Partial tunnel system collapse and debris (top) can be seen on chasm floor

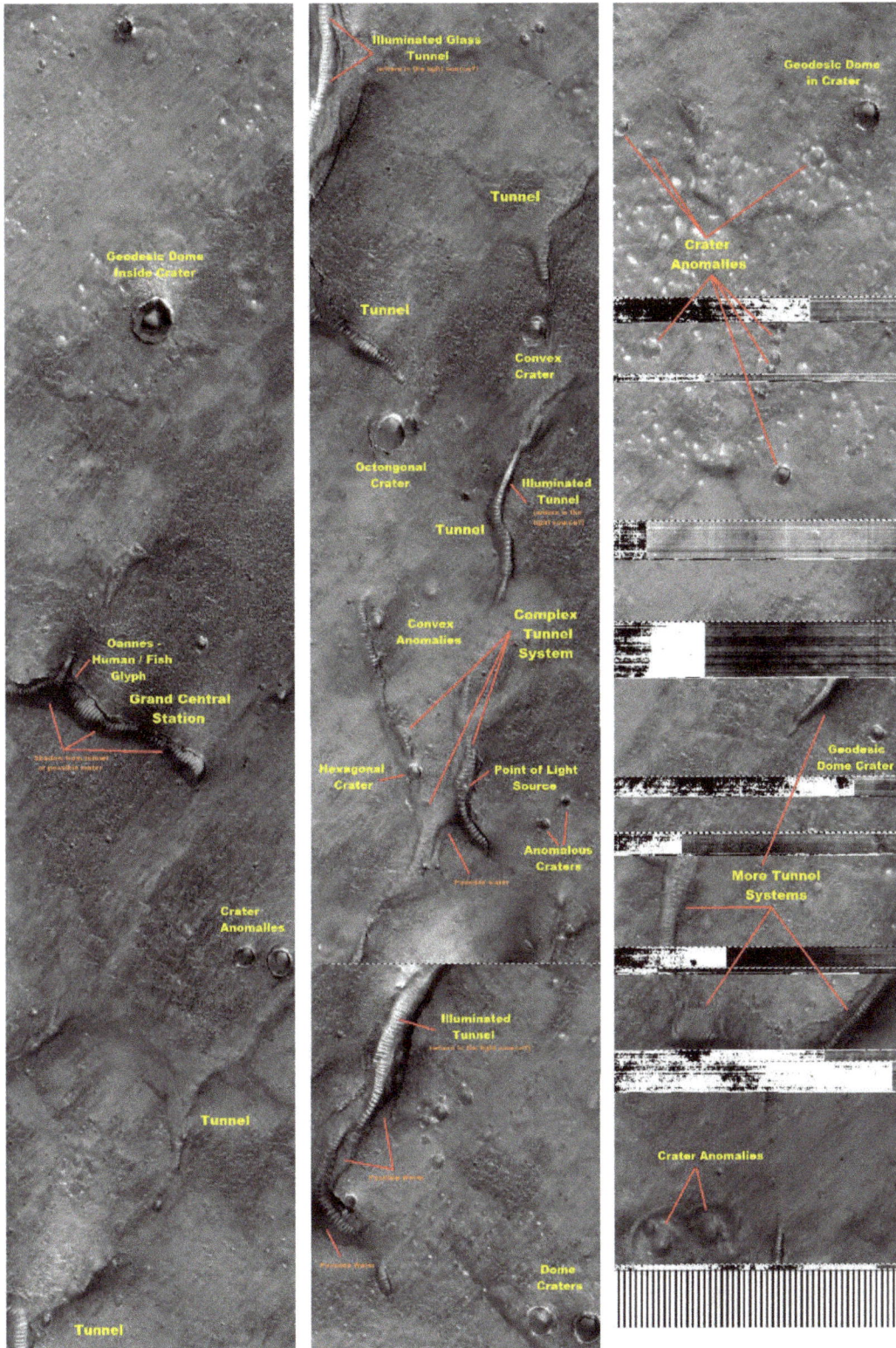

In this segmented MOC (Mars Orbiter Camera) image strip m1501228 there are many lighted glass tunnels and craters with domes, proof of artificiality

http://palermoproject.com/lowell2004/grandcentral.htm

In the MGS MOC image m1501228b (see below, **Eric C. Lausch** of Auburn, a layman Martian anomalies researcher did some great research work on one of the Martian Glass Tunnel Systems called the **"Grand Central Station"**. The image originally comes from *Johnny Danger's Dangerous Mars Site* website; Mr. Danger's **"Grand Central Station of Mars"** image can be seen immediately below. http://www.viewzone.com/marsobject.html

After careful examination of all pertinent data such as *Distance from Target Center, Emission Angle, Gain, Image Description, Incidence Angle, Latitude, Longitude, North azimuth, Offset, Orientation, Phase Angle and Spacecraft Altitude,* Lausch proves conclusively that the anomaly is an artificial construct of intelligent design.

The scale of the structure (top of Forehead) is approximately 331m high with (width of humanoid head) at 183m wide (1085 feet by 600 feet). The overall distance from the "water" to the top of the head is approximately 1920 feet. The entire anomaly (humanoid and fish sculpture) is approximately 585m high by wide 393m or (1920ft by 1289ft).

That would make this edifice around three-eighths of a mile high! A construct larger than either **Mt. Rushmore** or **Crazyhorse Monuments** in South Dakota or even **Stone Mountain** in Georgia.

The roof of the "tunnel" entrance to the right of the opening is approximately (733feet) across at the leftmost arrow location. If this is indeed the "roof" of a structure the area enclosed is enormous. The ribs of the tube at the two locations shown in this crop are (557 ft) and (469 ft) respectively (left to right) at the arrow locations.

This Martian edifice (see below) towers over one-half mile above the northern lowlands of Mars and like the Cydonia Face, this effigy appears to be another example of the "split faced" gods similar to ones depicted in the Mayan ceremonial masks. The effigy contains elements of both human and fish and it has been suggested by Eric C. Lausch that this is a representation of the ancient **Sumerian god "Oannes",** who was half fish and half man and came from beneath the ocean to teach the Sumerians about science and mathematics. Predating the Egyptian cultures by two millennia, the Sumerians were the earliest recorded culture to utilize the written word (**Cuneiform**), design and build architectural wonders (**ziggurats**) and implement a system of laws and government. http://www.viewzone.com/marsobject.html

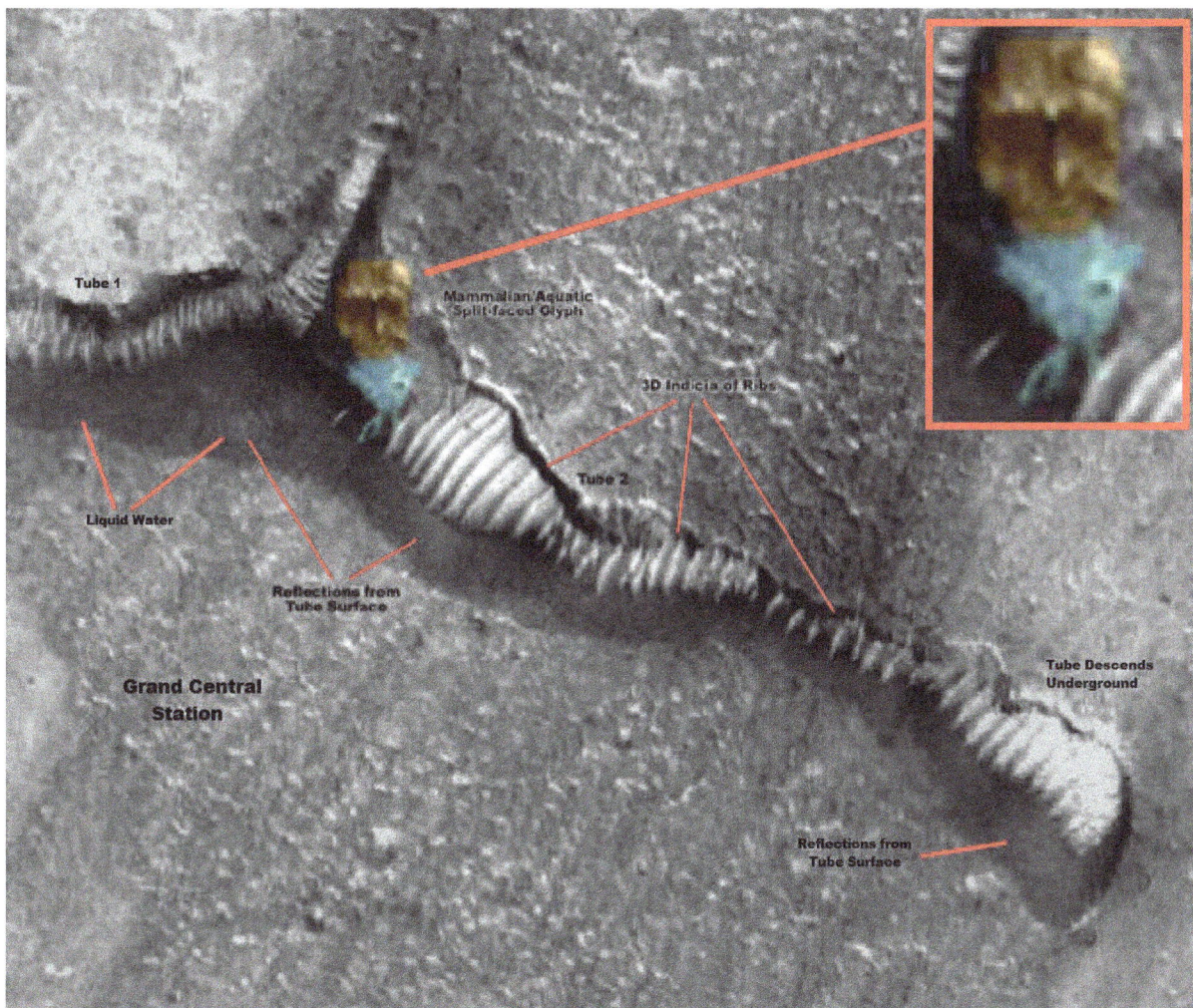

**Mars Global Surveyor MOC Image m1501228b – "Oannes effigy at Acidalia Planitia "
shows an artificial glass tube structure also called the "Grand central Station" with a
split-faced glyph (colour enhanced) architectural design of a human head/fish
which harkens back in time to the ancient Sumerian god, Oannes**
http://www.viewzone.com/marsobject.html

In many of the museums of Europe and the Middle East, effigies and ancient tablets depicting
Oannes and the Sumerian culture still, survive today. Using a **"Von Daniken mindset"**, one can
draw parallels of similarity to modern scuba gear with that of the costume of the Sumerian god,
Oannes. The fish's head and gaping mouth (shown in profile of the Sumerian plate below) does
look similar to a diving mask when pushed to the top of one's head. The humps and scales on
Oannes back could be some type of breathing apparatus. The description of Oannes having "a
man's feet within his fish's tail" provokes images of divers removing their flippers.
http://www.viewzone.com/marsobject.html

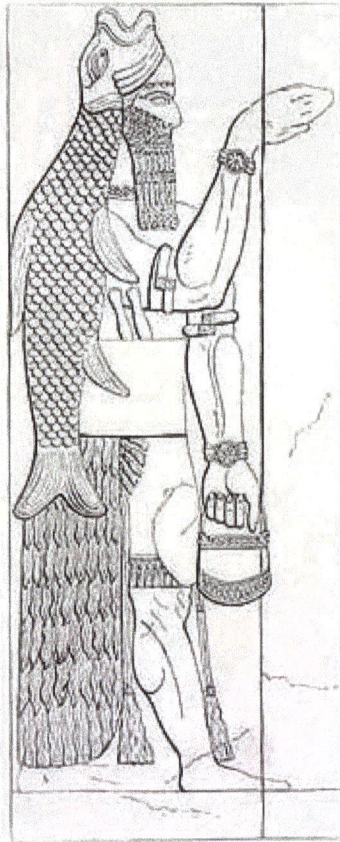

> "At Babylon there was (in these times) a great resort of people of various nations, who inhabited Chaldaea, and lived in a lawless manner like the beasts of the field. In the first year there appeared, from that part of the Erythraean sea which borders upon Babylonia, an animal destitute of reason [sic],* by name Oannes, whose whole body (according to the account of Apollodorus) was that of a fish, that under the fish's head he had another head, with feet also below, similar to those of a man, subjoined to the fish's tail. His voice too, and language, was articulated and human, and a representation of him is preserved even to this day.
>
> "This Being was accustomed to pass the day among men; but took no food at that season; and he gave them an insight into letters and sciences, and arts of every kind. He taught them to construct cities, to found temples, to compile laws, and explained to them the principles of geometrical knowledge. He made them distinguish the seeds of the earth, and shewed them how to collect the fruits; in short, he instructed them in everything which could tend to soften manners and humanize their lives. From that time, nothing material has been added by way of improvement to his instructions. And when the sun had set, this Being Oannes, retired again into the sea, and passed the night in the deep; for he was amphibious. After this there appeared other animals like Oannes."
>
> - Berossus, from Ancient Fragments (Isaac Preston Cory)

The story of the Sumerian god, Oannes
http://palermoproject.com/lowell2004/legacy4.htm

If this type of anomaly was a single topographical construct of artificiality found in just one area of Mars, it could easily be dismissed as a unique landform that may have been created by past diluvial events and/or current planet-wide high wind erosion. But, these structures are found all over the planet which run in and out of the ground and up against cliff walls or hanging over precipices, where natural geological formations of this construct never or rarely occur in this manner.

One remaining possibility in support of natural geological formations comes from a set of photos sent to Lausch by a friend from Vancouver. The photos show walls of ice carved by powerful wind action high in a mountainous region somewhere in British Columbia. The ice forms are similar to the glass tubes or tunnel structures found on Mars which are also reflective to sunlight at certain angles during the day. What this indicates is that ice can become covered up by snow or in the case of the planet Mars, by sand from the wind action of massive sandstorms. Therefore the water can be as NASA has been stating for decades trapped beneath the surface of Mars. Could these glass-like tubes and tunnels really be frozen water (ice) trapped beneath a few kilometers of soil, sand, and rock or are they as some scientists are saying **Glass Tunnels**?

Massive high wind-sculpted ice formation that shows ribbing and sunlight reflection off a translucent blue colored curvilinear structure photographed in British Columbia

http://www.viewzone.com/marsobject.html

This geological formation found on Earth, however, may give a clue as to the reason of why the artificial glass tubes or tunnel structures on Mars may have been constructed.

In a paper by **Laurie A. Leshin** titled: ***Insights into Martian water reservoirs from analyses of Martian meteorite QUE94201,*** he states that Mars may actually be capable of storing more water in its crust by as much as 5.2 times more than terrestrial (Earth)!! This means that stone, rock, and soil in the Martian crust is capable of absorbing and storing water than previously thought. Water, therefore, is not only likely to be present but, that it is active on the planet in terms of ebb and flow, evaporation and condensation and in some form of conservation!

"The Martian atmospheric D/H value of 5.2 times terrestrial is significantly higher than any found on Earth and has been ascribed to preferential loss of H (Hydrogen) relative to D (Deuterium) from the atmosphere through Jeans (thermal) escape over time. Here, based on ion microprobe analyses of apatite grains from Martian meteorite QUE94201, it is shown that the pre-Jeans escape Martian water reservoir has a D/H value ~twice that of terrestrial water, rather than the "terrestrial" value that has been assumed in prior work. The data support a two-stage history for Martian volatiles in which early hydrodynamic escape enriched Martian water to ~2x terrestrial D/H values. Subsequent Jeans escape to produce the current atmospheric values has thus been responsible for less D-enrichment than previously thought. A Martian crust containing 2–3 times more water than previously proposed is implied by the results". **Insights into Martian water reservoirs from analyses of Martian meteorite QUE94201 by Laurie A. Leshin; December 1, 1999; published in Geophysical Research Letters Volume 27, Issue 14, pages 2017–2020, 15 July 2000**

What this means to NASA is that the certainty of finding water on Mars has been established and a news release indicating a discovery or confirmation can proceed without further delay. The announcement, however, will open up another can of worms for NASA that will cause them more headaches with public outcry and debates that being, if there is water on Mars in the soil substrata or possibly on the surface then, the possibility of life of some kind must also exist. This brings us back to the purpose of why those glass tunnels and tubes were constructed in the first place.

Lausch and his *"Dangerzone Team"* of investigators suggest that with some logical argument, mathematics, physics and planetary geography that the megalithic glass tunnel structures were built to transport huge volumes of water from s sources like, beneath the Martian crust, the polar ice caps and other water reservoir areas of Mars, some perhaps underground, to the drier desolate areas of Mars as a kind of mega-irrigation system! The proof he says can be found where many of the glass tubes abruptly end at the craters, either on the surface as visible channels or in underground openings, where there are crater-reservoirs filled with water!

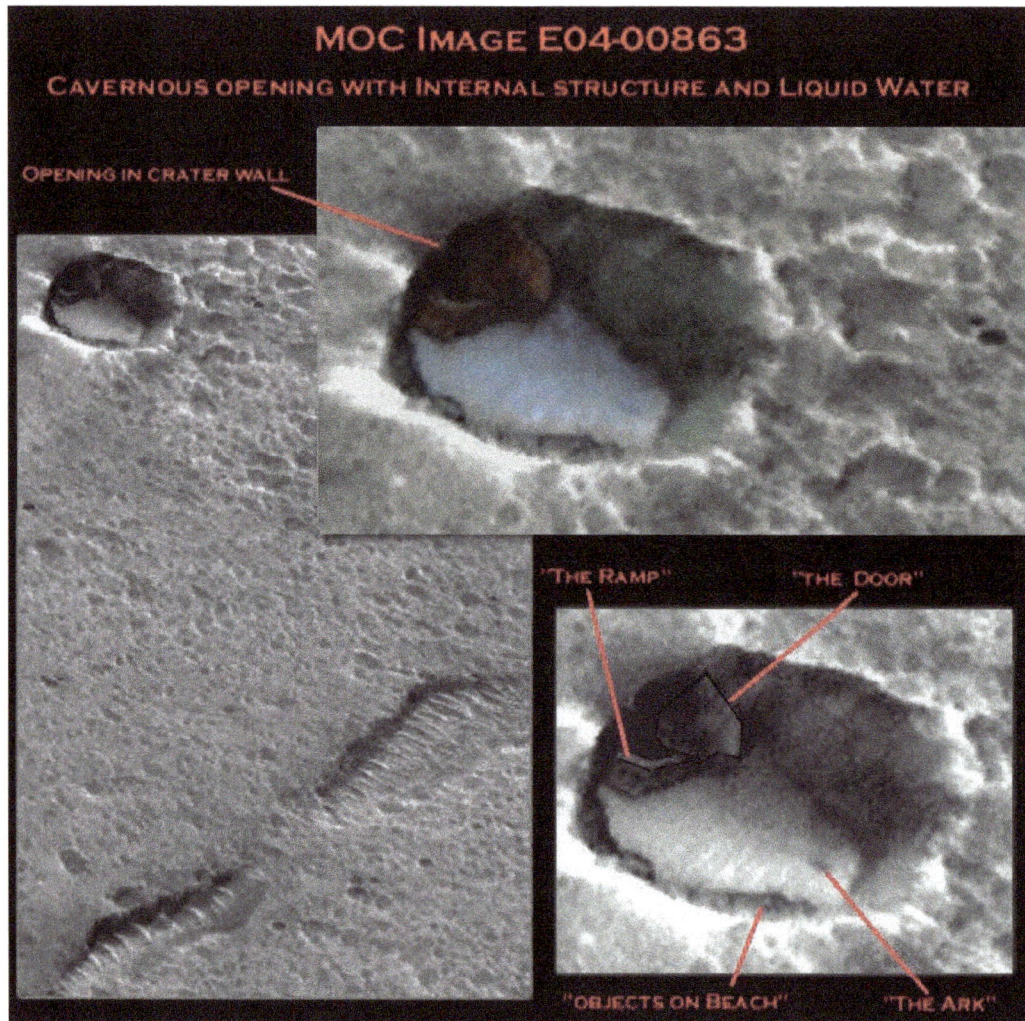

MOC IMAGE E04-00863
CAVERNOUS OPENING WITH INTERNAL STRUCTURE AND LIQUID WATER

OPENING IN CRATER WALL

"THE RAMP"   "THE DOOR"

"OBJECTS ON BEACH"   "THE ARK"

**The crater at top left (also colour enhanced with inscribed highlights)
indicate that this crater is used as a water reservoir filled seasonally
or as needed by the glass tunnel water irrigation system**
http://www.webring.org/l/rd?ring=mars;id=30;url=http%3A%2F%2Fpalermoproject%2Ecom%2Flowell2004%2Fsite%2Ehtm

If this is the true function of the **Martian Glass Tunnels**, as a massive planet-wide irrigation system then, it goes a long way to explaining the seepage of water emerging from cliff tops, faces, alluvial fans indicative of river deltas, alluvial flows and fluvial deposits, *"dried river beds",* sand rippling on crater floors and the occasional water-filled craters.

such crater reservoirs would require constant monitoring and control for possible evaporation loss into the atmosphere, therefore, some of the geodesic or convex domes seen in some craters (as indicated in the above photo images) would potentially serve to act as protective covers against the constant evaporation process as well as the seasonal and cyclical planet-wide sand storms that occur from time to time.

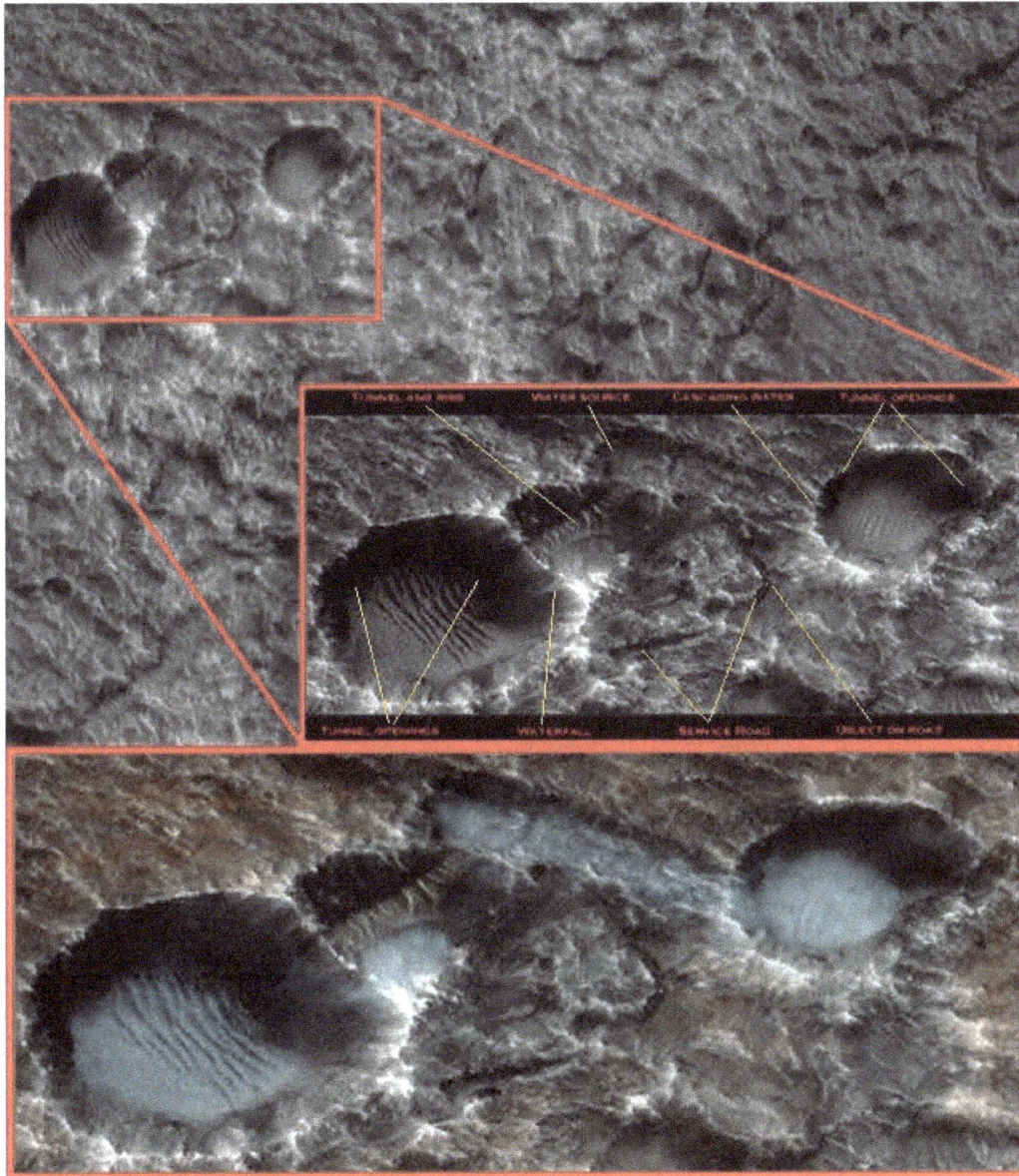

MOC image E0400863 with close-up inserts and colour enhancement indicating water filled double crater reservoirs with waterfall and service road with possible vehicle on road

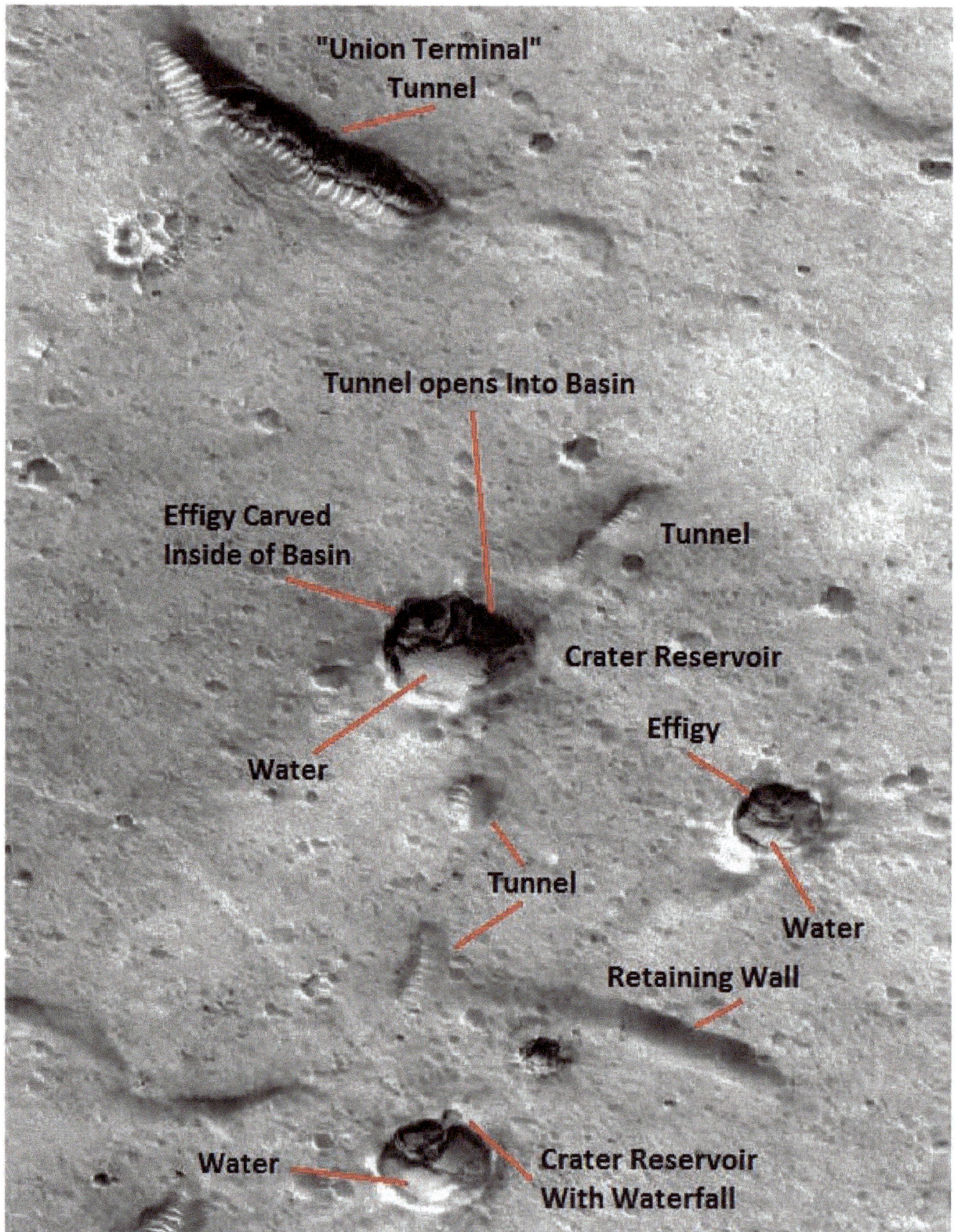

MOC image E0400863 showing artificial tunnel structures and crater reservoirs

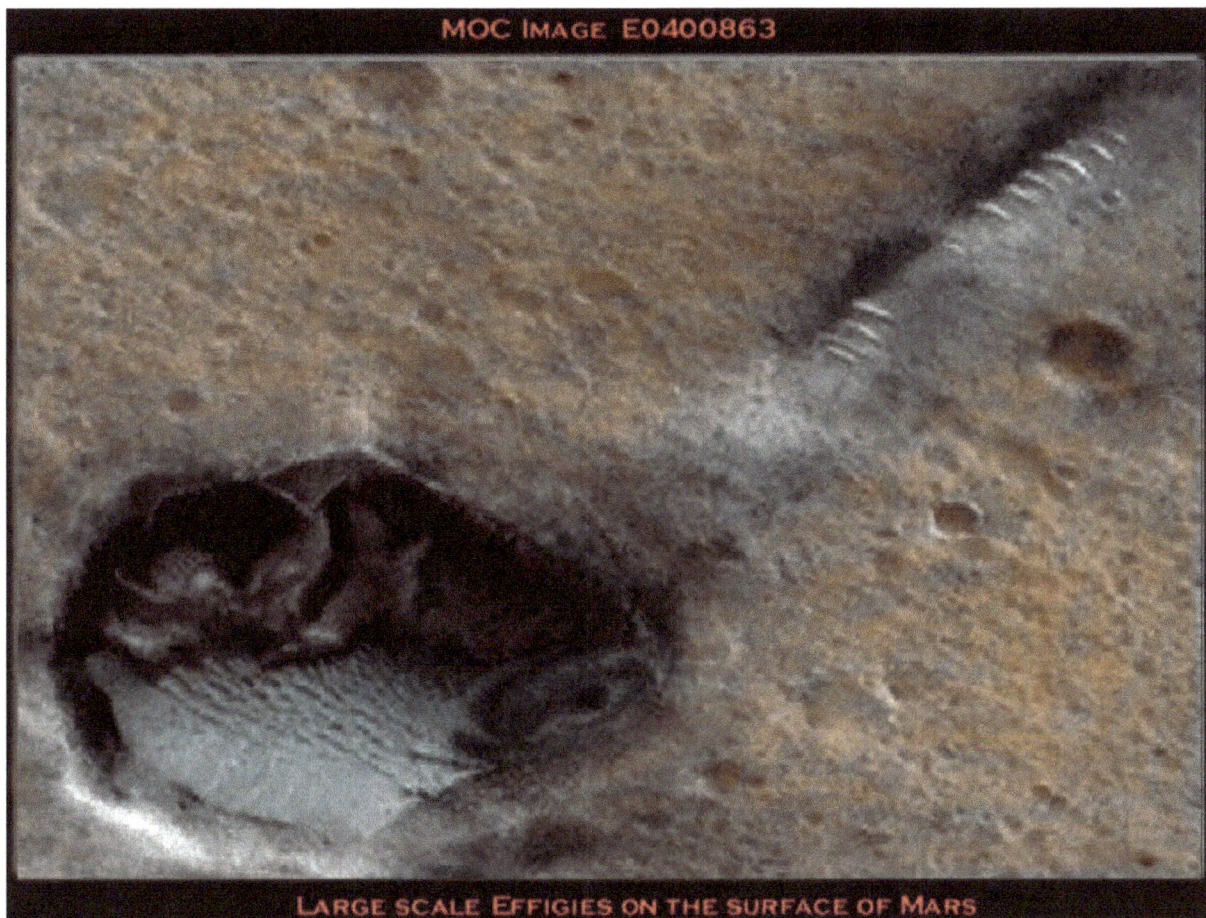

MOC IMAGE E0400863

LARGE SCALE EFFIGIES ON THE SURFACE OF MARS

A close up, colour enhanced image of the above crater reservoir showing water and effigy

http://www.webring.org/l/rd?ring=mars;id=30;url=http%3A%2F%2Fpalermoproject%2Ecom%2Flowell2004%2Fsite%2Ehtm

The reader should recall the "shiny spot" on the dome of the tube that originally caught the eye of researchers in MSSS MOC image m0400291a. There has been conjecture on the nature of the shiny spot debated on the various Mars anomalies websites except by the very people who should be discussing it, the good people of JPL/MSSS who response is: "no comment"!

Contrary to Hoagland initial hypothesis that the glass tubes are a possible transportation system which cannot be completely ruled out at this time, the most obvious explanation remaining for the shiny spot on the glass dome is that it is being created by a light source from within the tunnel. An illuminated light source that was imaged (MOC image m0400291a) in the year 2000 and still appears to be illuminated in 2003 as seen in the re-imaged photo E21-01421, exactly in the same spot and probably still currently shining, must be a light shining from within the tunnel and not due to a reflection of sunlight which changes from season to season and even year to year.

The implications of this more tenable hypothesis are profound on many levels. A light source implies a state of high technology; it implies an operational power generation source still in use on Mars today. A light source implies the intervention of an intelligent presence with a determination of purpose. It also implies a deliberate message meant to be seen in much the same

73

way that city lights on Earth can be seen from space at night: *there is life down here on this planet!*

Once again, we must ask, why does NASA continue to deny such compelling evidence of extraterrestrial life? It has been often insinuated that NASA's refusal is a refusal to investigate or comment on such anomalies is simply to avoid being caught in denial. Public emails to MSSS on such issues often go unanswered. The American society must be asking itself is this our hard earned tax dollars at work?

Like so many areas on Mars that are being added to a steady growing list, this area also requires a thorough investigation with a full complement of cameras and instruments on the next generation of ESA and NASA Mars spacecraft with all data acquired being disseminated as widely as possible for independent analysis.

This massive planet-wide irrigation system should now be considered as the **Martian Irrigation Hypothesis (MIH)** supportive in retrospect with the findings of **Percival Lowell**. Lowell was a visionary scientist, a researcher, and astronomer of the latter part of the 19th century, who predicted *"it probable that upon the surface of Mars we see the effects of local intelligence."* a concept that he postulated in his book "Mars", published in 1895.
http://palermoproject.com/lowell2004/legacy.htm

Lowell believed that **Schiaparelli's "canali",** a term used by the Italian astronomer two decades earlier to describe apparent "geometric grooves" he observed on Mars during favorable viewing conditions were indeed just that, conduits for water circulation on an arid world.

Lowell would spend much of his time during the Mars 1894 opposition, scrutinizing the Martian surface through his telescope, meticulously sketching a series of plates detailing the complex hub and spoke systems of interconnecting "canali" which he believed was intelligently designed and built for the purpose of circulating water to all corners of Mars. It would not be until the later part of the 20th century and the beginning of the 21st century that with the use of contemporary imaging and spectrometer data would prove Lowell's assumptions correct, that Mars has vast regional aquifers containing enough water to cover the entire planet to an approximate depth of 50 to 500 meters as well as surface ice and liquid water existing at some locations of low elevation. **Insights into Martian water reservoirs from analyses of Martian meteorite QUE94201 by Laurie A. Leshin; December 1, 1999; published in Geophysical Research Letters Volume 27, Issue 14, pages 2017–2020, 15 July 2000** and
http://palermoproject.com/lowell2004/legacy2.htm

Enough water to justify construction of a planetary water system to procure this vital resource for the needs of its builders. There are indications that Mars is still hydrothermally active, if so, given Lowell's penchant for speculation, Percival probably would have hypothesized a thriving ecosystem filled with alien beings under the surface of the red planet traveling about on their daily business hurtling down the glass tunnels of Mars, weary travelers on an alien interstate. This phenomenon that astronomers observed on Mars can hardly be attributed to pareidolia as it does not resemble any type of physiognomy, but is more readily likened to a map of the hub and spoke system commonly used to route commercial air carriers or possibly to some type of large transportation system. http://palermoproject.com/lowell2004/legacy6.htm

Percival Lowell at his telescope in 1914, observing Venus in the daytime with the 24-inch (61 cm) Alvan Clark & Sons refracting telescope at Flagstaff, Arizona.
https://en.wikipedia.org/wiki/Percival_Lowell

Lowell's protégé' **Antonaudi** reportedly captured images of the **Martian Canali** (circa 1910) by photographic means. It is interesting to note that Antonaudi's map of Mars, replete with canali was still in use by the US Air Force as its official map of Mars until the late 1950's. From the 1850's to the 1920's, one can see from the various Mars renderings as recorded by numerous researchers via telescopic observations over that time period beginning with Schiaparelli to Antonaudi, a steady continuous alteration of the Martian topography. After many years of sketching the observed Martian features, Antonaudi pronounced them an illusion after they were no longer discernible. Perhaps, Antonaudi was not aware of the cyclic planet-wide dust storms that tend to obliterate much of the planet's surface detail for months on end, even with the use of

a telescope. Antonaudi lived long enough to see the Mars of his youth come to resemble the Mars of **Carl Sagan**, a sterile lifeless orb, pockmarked and desolate.
http://palermoproject.com/lowell2004/legacy2.htm

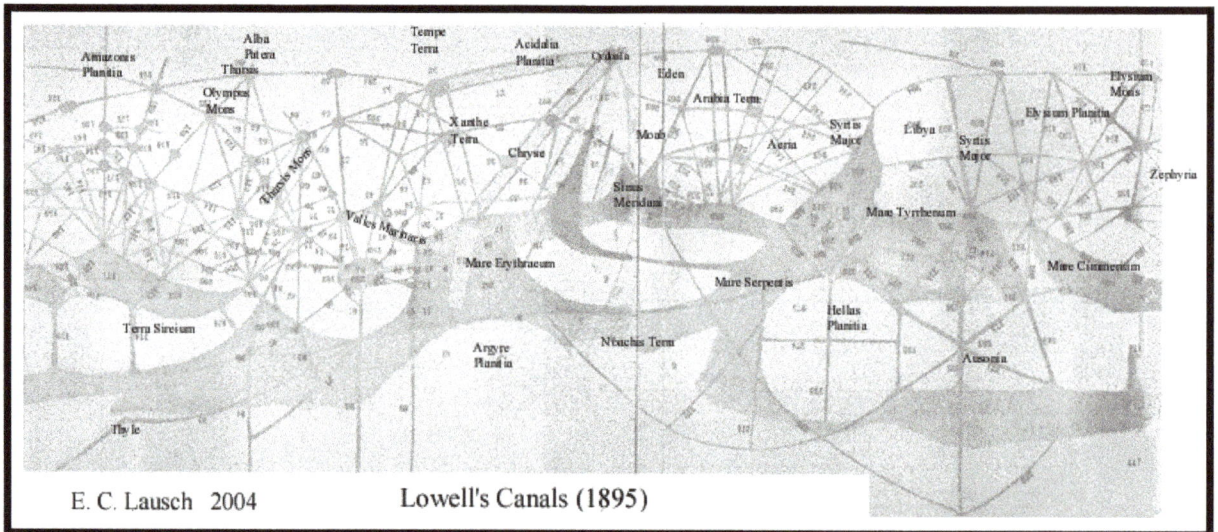

Lausch's revised map of Lowell's Martian canals
http://palermoproject.com/lowell2004/legacy2.htm

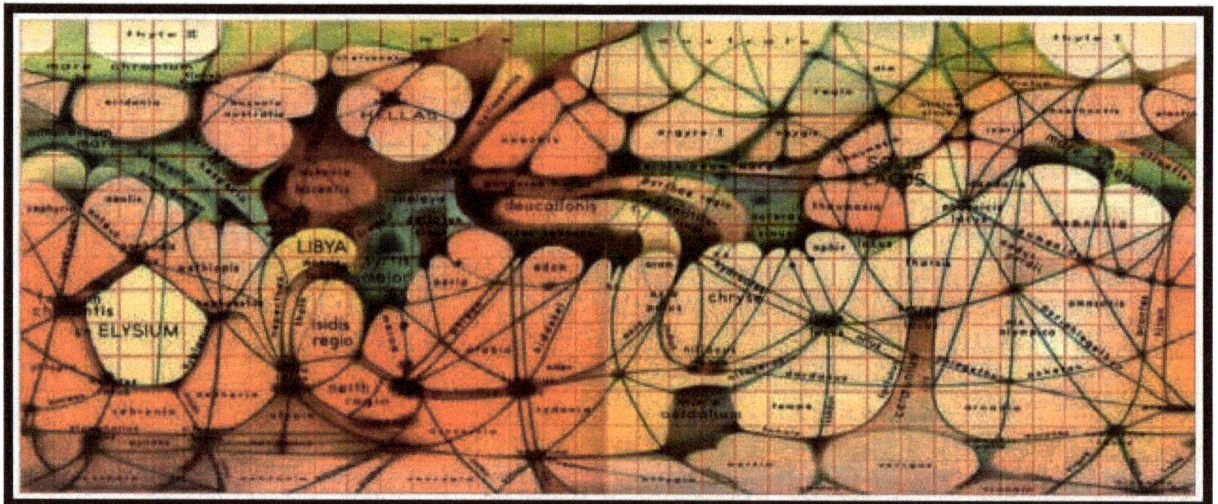

**Antonaudi's colour map of Mars topography that was used up and until
the 1950s by the US Air Force and became outdated by recent
Hubble telescope images and orbiting Mars satellites**
https://www.flickr.com/photos/ares2000/27892653

In the photo below that compares the Antonaudi Mars globe with the recent Mars image taken by the Hubble telescope, there can still be seen basic similarity in the symmetry of lines, shades of light circular areas, dark shadow regions, including the north polar cap region. The difference is in sharpness of detail factoring in a "Dune-like" planet that is battered by constant windstorms that cover and bury surface detail.

With the fine analytical work of the *Dangerzone Team,* the **Martian Irrigation Hypothesis** gains support from the discovery of glass tunnels that may be a planet-wide irrigation system and/or a mass planet-wide transit system. This, in turn, validates the concepts and theories originally posited by Percival Lowell that Mars is not a dead planet but a planet struggling to maintain a viable life-sustaining ecosystem against immeasurable adverse conditions.

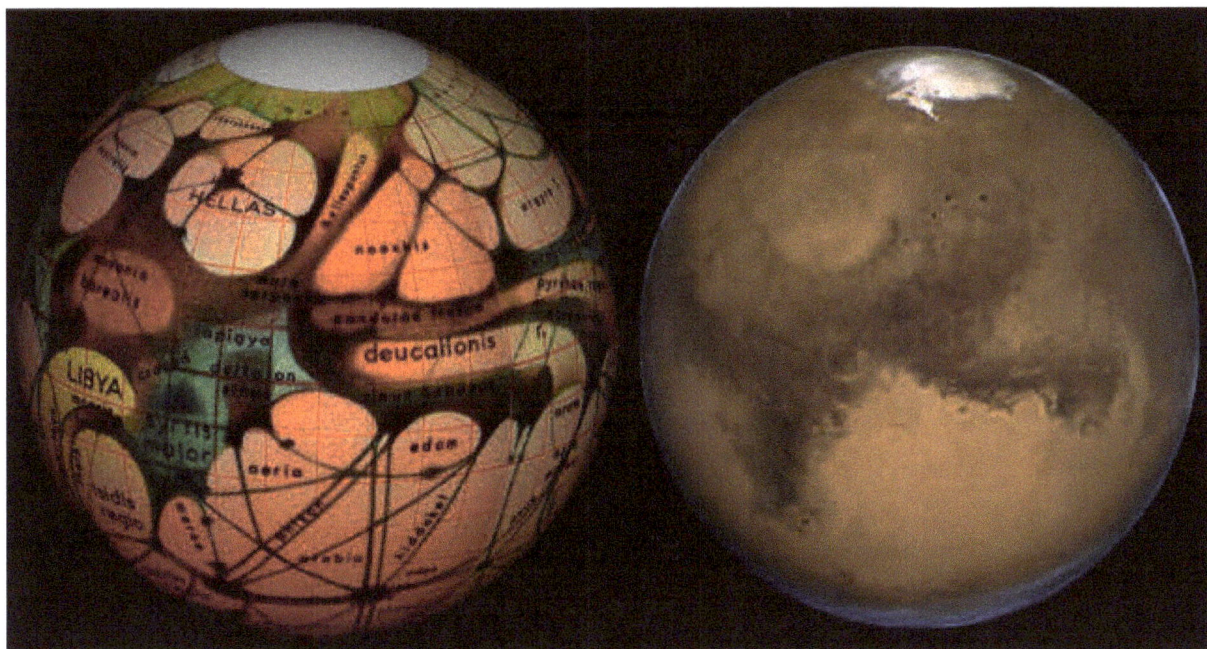

Antonaudi's Mars as compared to Hubble images
(Image credit: Tom Ruen, Eugene Antoniadi, Lowell Hess, Roy A. Gallant, HST, NASA)
http://antwrp.gsfc.nasa.gov/apod/ap031112.html

With the strong probability of a massive planet-wide irrigation and transportation system on Mars admittedly in some serious need of maintenance and repair, on more insurmountable barrier has been swept aside in establishing proof for the presence of an Extraterrestrial civilization on Mars.

The Monolithic Monuments of Mars

The infamous "They" say that "the devil in is in the details" or should that be "that God is in the details"? Either way, we will examine further proof, evidence of such megalithic, monumental proportions that are so awe-inspiring by the massive size that very few things on Earth are even remotely comparable in their engineering design and construction, all of which establishes beyond any doubt that a former advanced Martian civilization once existed on Mars millions of years ago, and may still exist today!

What are the chances that a freak windstorm can carve an image easily recognizable from the vantage point of space that resembles the visage of a majestic creature familiar to all people on

Earth? All ancient civilizations, current aboriginal cultures as well as new-age groups have traditions and beliefs that state that every planet has a spirit; a living energy form that is more than the physicality of its being which is conscious, intelligent, sentient and interactive with other intelligences upon it surface. Earth's spirit is known as **Gaia**! What is the spirit of Mars?

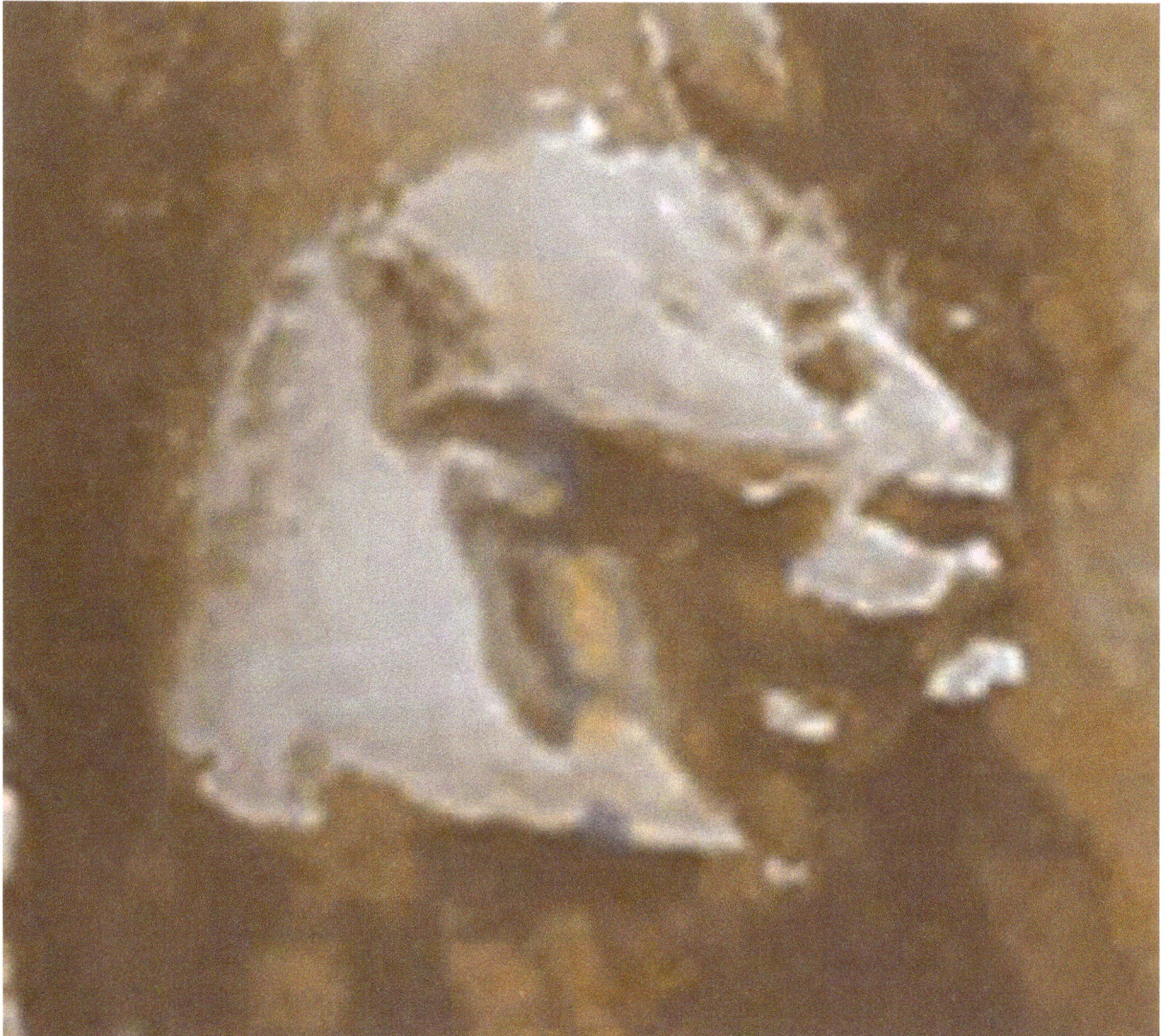

Is this an example of pareidolia or a one-time freak of nature mimicking an animal with Lionesque likeness near Mars' North Polar Region or a deliberate topographical design engineered by Martian intelligence?
http://www.msss.com/mars_images/moc/may_2000/n_pole/

Could image above resembling a lion's head seen in profile be the handiwork of an intelligence inhabiting Mars or the result of natural forces from a planetary spirit? The latter answer is the most likely explanation from a non-metaphysical position as science does not recognize matters that are spiritually related. Natural forces or geological forces from a science perspective is the obvious answer for several reasons. There is nothing intrinsically unique about the "lion's head" when viewed in geological context, it is a single component in the vast canvas of swirls that is Mars' northern polar cap, therefore, at least one portion has a terrestrial likeness.

Compounding the argument against intelligent manufacture is the fact that the "lion's head" is seen in profile. When reviewing a profile image for signs of artificiality, suggestive properties such as bisymmetry are unavailable for analysis. This doesn't necessarily imply that all profiles are natural formations, but it certainly makes proper scientific assessment much more difficult, if not impossible.

It should be pointed out, however, in this argument against artificiality that nature can play tricks of perception every day upon people, like seeing the image of **Jesus** in a piece of toast or the **Virgin Mary** in a tree knot. To the dutiful religious worshippers of **Christianity,** these natural occurring icons are miraculous symbols reassuring the faithful of their belief. In science, we can accept such cases of pareidolia as nothing more than limits of one imagination based upon the individual's upbringing, conditioning, education, and environment. But, on Mars where the symbology of so-call natural geological formations repetitively reproduces itself in recognizable geomorphs, such as in the feline or human or aquatic forms then, we must question the odds or chances of such formations occurring naturally through a geological process or from deliberate intelligent design.

The *a priori* prediction for intelligence would require that repetitive engineered designs of animal bio-forms or humans be easily recognizable from a great distant and still be recognized as being engineered when in close proximity to the artifact as per **Tom Van Flandern**.

As discussed elsewhere, the Cydonia face was intended to represent two separate visages, as supported by **Richard Hoagland's** *a priori* prediction that the two halves were designed to encode hominid/feline imagery.

Eric Lausch contends that some commentators suggest that the Face was meant to represent something more, and cite mirrored, "upside-down" graphics such as the one below as evidence that the **Face** includes a (very) cleverly inserted portrait of a **"Gray" alien**.

"I don't think the "alien" likeness passes scientific muster. The impression requires too much imagination on behalf of the viewer and too much speculative "restoration" in order to be seen in the first place. (In the numerous "Gray" reconstructions, the Face's hominid side is mirrored and the large groove corresponding to a "mouth" is bifurcated to produce questionable "eyes" such as the ones seen on the cover of Whitley Strieber's 1987 book "Communion.")"

Lausch further contends that we are culturally conditioned to expect the visage of a big-headed, big-eyed **"Gray"** when confronted with photos alluding to extraterrestrial intelligence.

Author's Rant: The reason for this conditioning toward expecting to see aliens as diminutive Grays in the news media, in news articles, on posters, in children's cartoons and particularly in TV shows and documentaries, Hollywood movies, in UFO books and magazines and at UFO conventions is to indoctrinate the public into essentially believing that the only ET that might be visiting the Earth is this type of alien. This is the alien being

behind the abductions and as such must be feared and with this constant barrage of the little Gray ET image into the public consciousness, it is hoped that a negative and fearful mindset will be engendered subconsciously. With a negative mindset instilled upon the public, if and when the day comes that a false flag alien invasion scenario unfolds in our skies around the world then, the public will respond in a predictable manner by demanding that the military step in to deal with the invasion and in doing so we will give up more of our rights, privacy and financial resources to be protected from the little "Gray buggers"! We will give the Military Industrial Complex whatever they need or ask for in return for their protection. In the "dirty 30s" this type of protection came from the Mafia at the end of a gun barrel!!!

"That we managed to find a "Gray" lurking in the curves of the enigmatic Face on Mars comes as no particular surprise, especially after realizing the demanding steps required to bring out the supposed "alien" image. On the other hand, if we admit that the Face was perhaps designed to encode hominid and feline forms, as argued by Hoagland, then who is to say with absolute certainty that the "Gray" likeness is purely fanciful?"

Lausch rates the appearance of a **"Gray"** as dubious at best; its predictive scientific merit is zero.

If the features in Cydonia are artificial, we will likely find additional structures elsewhere on the Martian surface. In the meantime, the search continues unabated.

In the photo images below taken from the official NASA/JPL/MSSS MGS MOC number M11-00099 image strip (http://www.msss.com/moc_gallery/m07_m12/images/M11/M1100099.html), we discover another very anomalous area on Mars that abounds with structures of artificiality. There are many domes or oval convex structures of various sizes dotting the landscape with one interesting feature standing from among the rest. It shows what appear to be two nozzle-like structures with long mechanical looking pipes and one is in the process of spraying some liquid material or possibly water onto the surface.

Joseph Skipper of *"Mars Anomaly Research"* website believes that what we are seeing here is the possible construction of a new oval dome structure for housing Martian community or for storage of water or material. Skipper believes that the liquid being sprayed from the one active nozzle may be a type of concrete foam **(Gunite)** which is widely used in various curvilinear construction projects on Earth or the material used on Mars could be a type of resin, either of which would hardened very quickly, usually within a 24 hour period and reach maximum strength and solidity within a week.

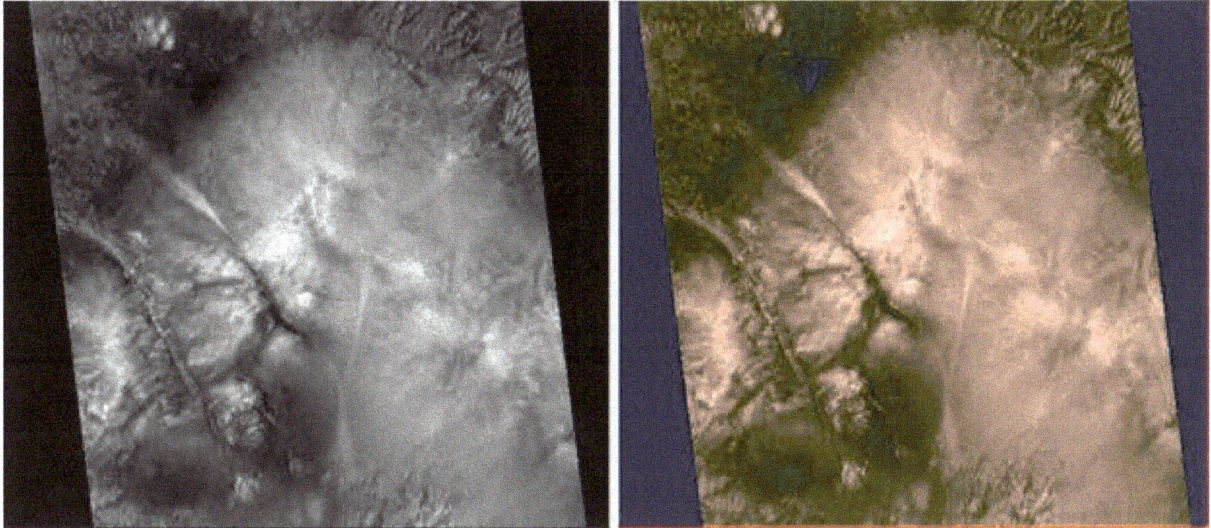

NASAJPL/MSSS MGS MOC number M11-00099 (B&W and colour enhanced) show huge gun nozzle blasting out a liquid spray material, perhaps water or concrete foam (Gunite)

https://www.bibliotecapleyades.net/marte/marte_structuresanomalies03.htm

What is interesting in the photo images is that the outer appearance of the dome nearby to the nozzle is very smooth and somewhat translucent allowing for light to either enter into the structure or for internal light to be reflected outward. If the sprayed liquid material is a type of concrete foam or resin as **Skipper** suggests then, its outward appearance could be rough or mottled looking, providing an overall camouflaged appearance, if necessary to blend into the surrounding terrain. Such structures would be almost armour-hardened like wartime bunkers to a certain degree if their purpose was built for protection from a possible enemy or merely as protection from the harsh environment of Mars. It would be of great interest for the **Mars Orbiting Camera** to fly over this area and re-image the construction project to see if the dome building had been completed and it final appearance. Skipper thinks that several different interstellar ET species may be co-habiting Mars.

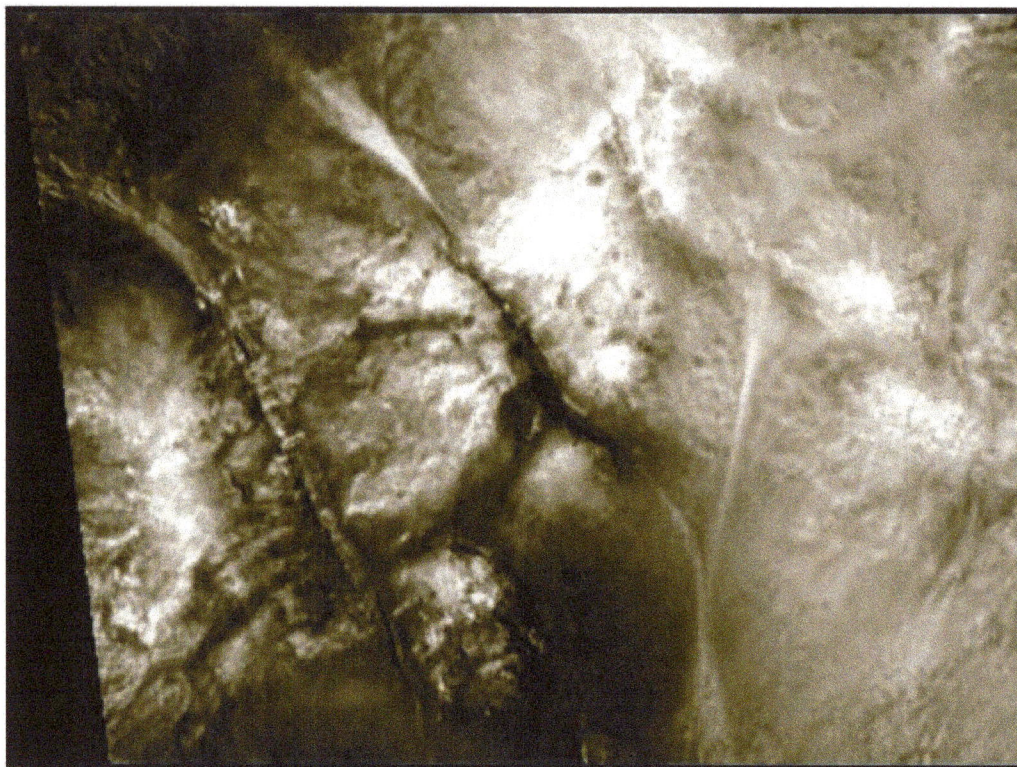

Enlarged colour enhanced photo of above image. Liquid spray can be seen pouring out of nozzle and a possible waterfall or stream can also be seen off to the right

https://www.bibliotecapleyades.net/marte/marte_structuresanomalies03.htm

Speculation as to the usage of the dome structures in the region is any body's guess. They could be as communal villages for diminutive ET beings as speculated by Skipper, as some of the domes seem to cover over rock and low-lying terrain but, they may also have excavated interiors as well which means that the ETs may not necessarily be small in stature, so speculation abounds as to the function of these constructs.

Water can also be seen to flow along, down and over the Martian terrain nearby; in and around these oval dome structures with their associated rib or ridge forms which according to skipper is also indicative of the presence of an indigenous ET civilization and provides camouflaging with the natural surroundings. http://www.marsanomalyresearch.com/evidence-reports/2001/029/huge_nozzle.htm

This photo is from the same black and white MOC number M11-00099 image strip that clearly show other similar domes of various sizes in the region, like pebbles on a beach

One more curiosity in this area can be found at the top of the image strip which caught my attention. A pareidolia, perhaps, but on Mars, things are not always what you initially perceived them to be. I was about to dismiss this oddity as just another topographical feature of mounds, sand dunes, and crater, but on a second closer inspection, something about it jumped out in familiarity.

It appears to be another split-glyph depicting a male lion and its offspring kitten! When mirrored toward the left a large male lion can be seen in a somewhat menacing visage and when mirrored toward the right a cute and gentler feline kitten appears to be sleeping. It could also be argued that the mirror image on the right is a wolf with its eyes also closed. In fact, closer study reveals many feline faces and animals within the mirrored lion face including where the nose is situated a small cat can be discerned. You be the judge! (See photos below).

The Feline and Canine images found on Mars in many monuments and pictoglyphs seem to be a universal theme that is constantly displayed over and over again. On Earth, we view these two species as being antagonistic toward each other. *Perhaps, there is more to this than just mere monuments of animals familiar to us; perhaps, there are intelligent ET beings that resemble these creatures as has been reported in some UFO sightings and ETI encounter cases!*

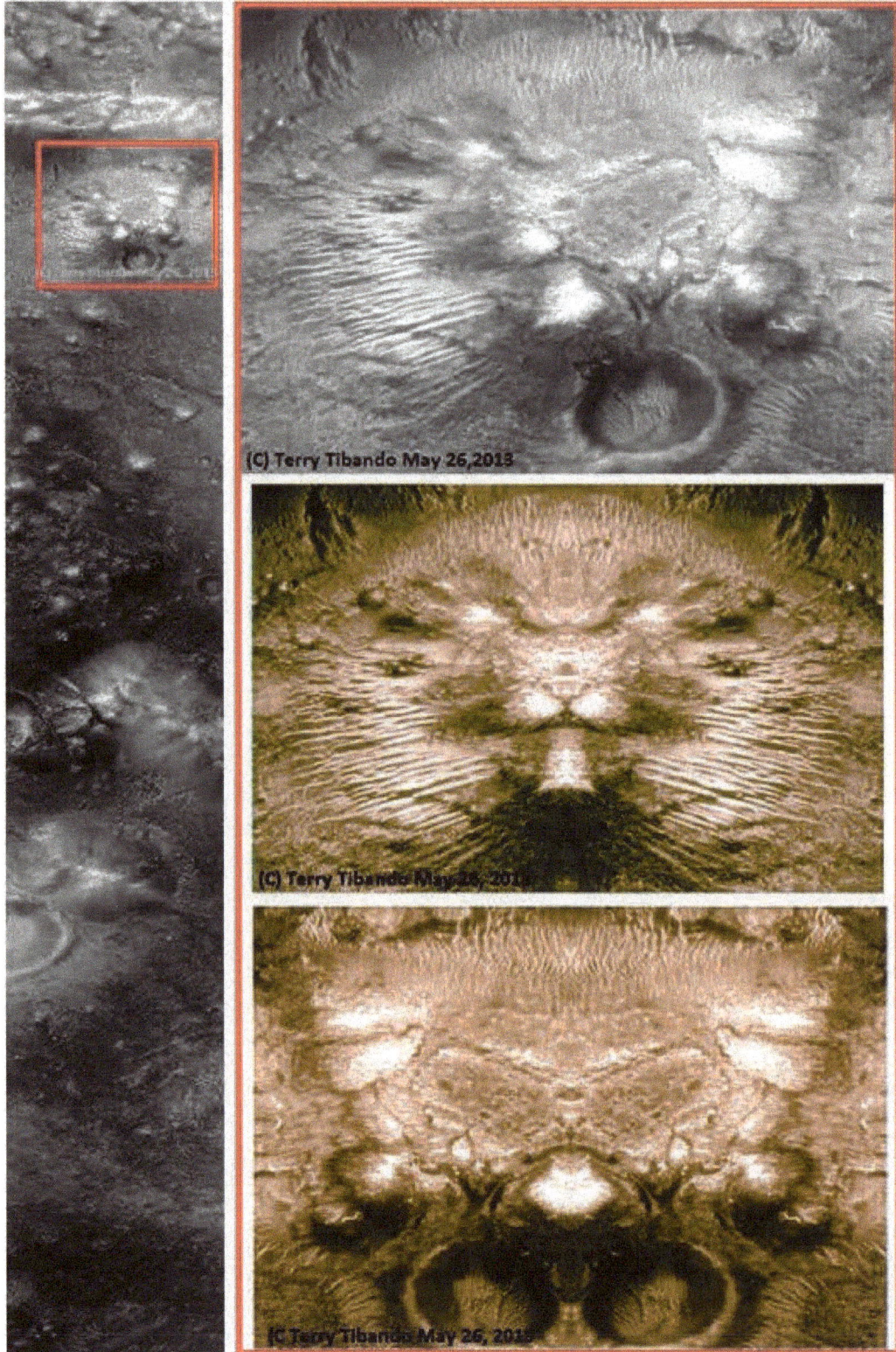

"The Lion Sleeps Tonight ... with One Eye Open". The MOC number M11-00099
(partial strip) depicting a half lion/feline and wolf/canine (or is it a kitten?) split-glyph

Cynocephali (the Dog-Head Men)

The next image (below) is considered by NASA as a sand dune on the slope of a cliff perhaps covering a rock outcropping which gives this pareidolia it unusual alien or canine features. On first inspection, the anomaly one perceives some type of being with a flat head, a protruding brow ridge over what appears to be the eye areas of the face. The nose is large with nostrils but, it overall appearance is distorted as if tampered with and the face is extended in a muzzle-like fashion similar to a canine or dog. The open mouth is box-like or rectangular and the lips conform to the shape of the mouth with a chin that slightly slopes to a strongly pronounced jaw line. The ear on the side of the head is oddly defined as a short or "snipped" earflap but, it could also be argued that the ear is covered with some sort of "electronic type earphones"! The whole head rests upon a very strong and wide neck and muscle structure that tapers down to narrow shoulders that are barely visible.

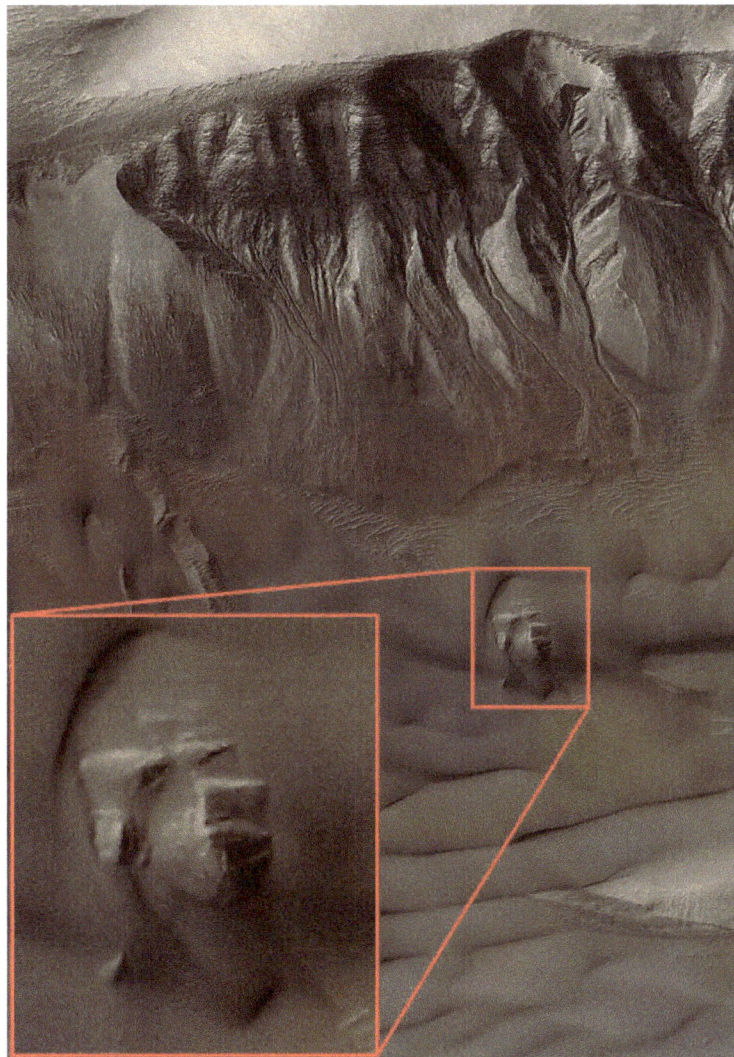

NASA/JPL/MSSS E03- 02550 The sand sculpted Canine ETI being of Mars found in the Newton Crater photographed by MOC
http://www.msss.com/mars_images/moc/e7_e12_captioned_rel/

There is a 3D quality to the overall appearance as oppose to the 2D profile images of **The Puma** or **Nefertiti"** sites. There appears to be something else buried or cover up just above the left side of the head by sand flow from the cliff. What it may be is anyone's guess, it could be just rock.

Anomaly researchers have likened this sculpture as a version of a ***Martian Picasso of a dog being*** which this author is inclined to believe as well. In fact, it is the author's contention that we are looking at a sculpture of a **Canine Extraterrestrial being** which many researchers may think is absurd except that canines on Earth as personal pets are highly intelligent and their outward appearance varies greatly. There have been reports of Feline-like ET beings seen near UFOs on Earth, so it's not a far stretch of the imagination for there to exist **Canine ET beings** as well.

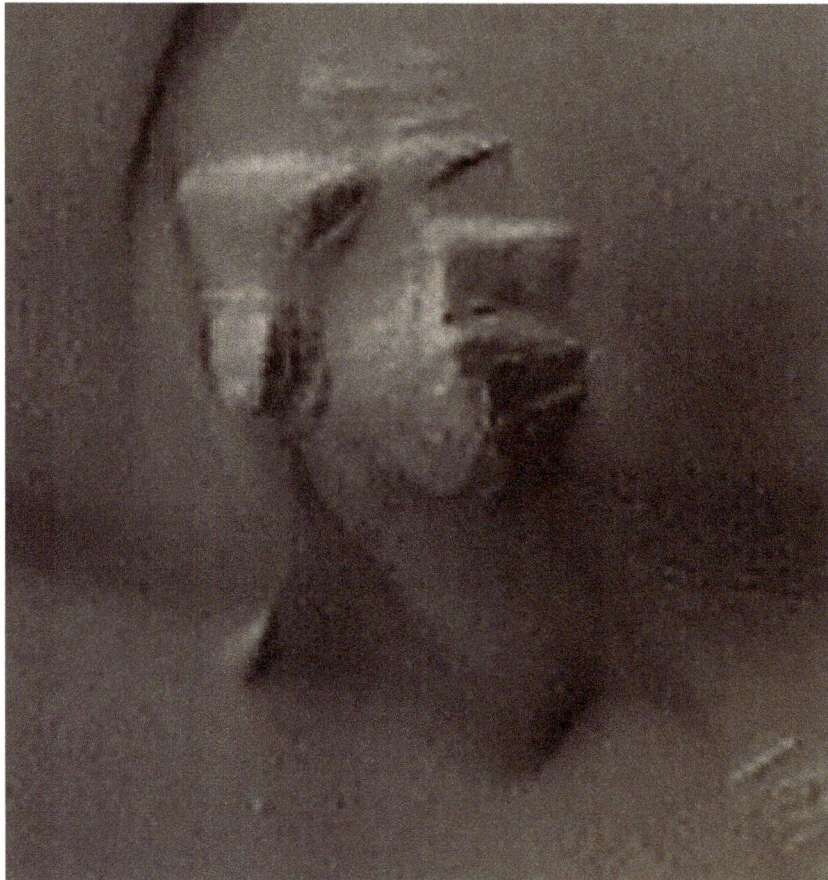

An enlargement of the Martian Dog-Face sand sculpture
http://www.msss.com/mars_images/moc/e7_e12_captioned_rel/

There was a few years ago, an unusual art sculpture exhibit on public display throughout Europe with many unusual, oversized displays that all had in common, a life-like hybridization quality to them. Many of the bizarre sculptured pieces possessed both human and animal aspects with the endearing attributes of gentle motherly nurturing of smaller associated displays.

The art piece that intrigued this author was the human/canine hybrid being, a female dog creature suckling her human-like pups! (See photo below and compare it to the Martian dog creature).

86

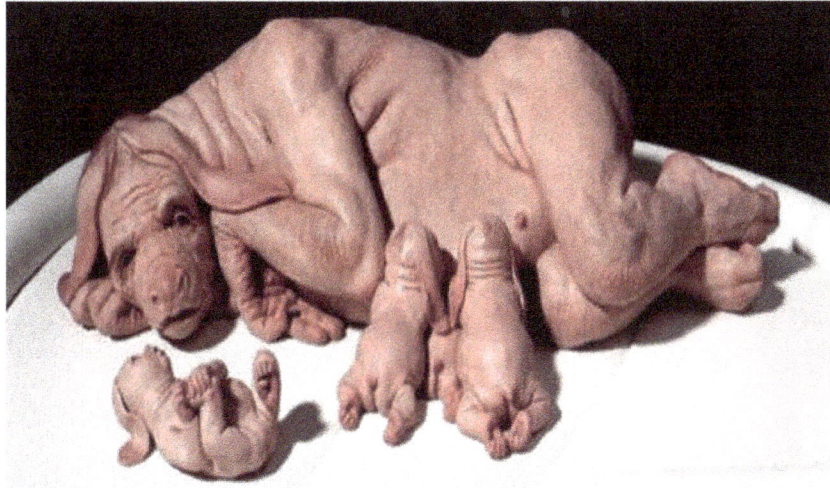

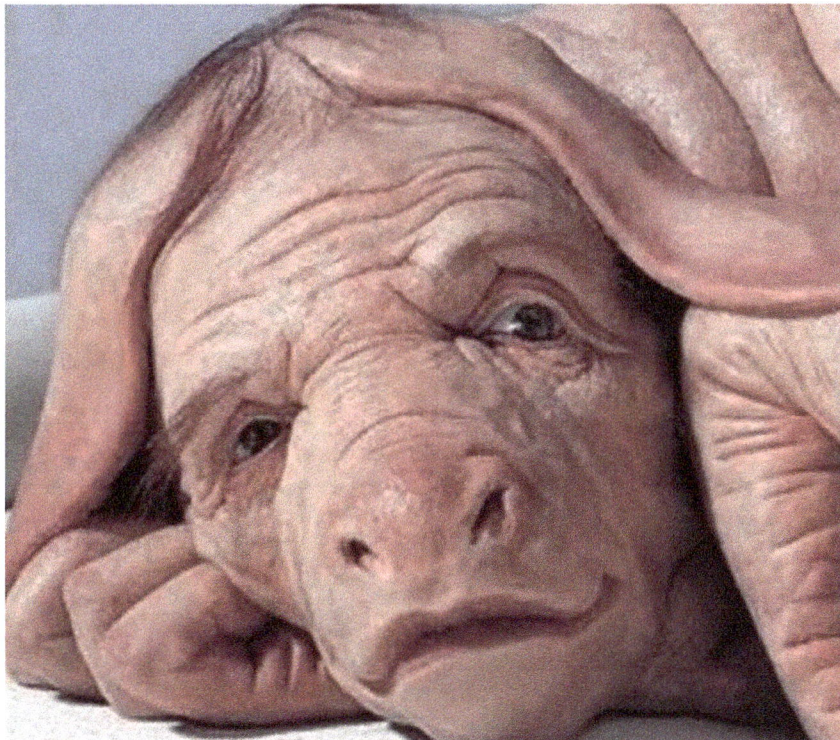

The human-dog hybrid and her human pups with close up of the face.
Sculpture was created by Patricia Piccinini

The human-dog hybrid sculpture with her human pups created by **Patricia Piccinini** are similar to the Martian sand dune sculpture in certain aspects lending a quality of subtle credibility to this type of being's viability. This author has had a lucid dream about such creatures who offered the author a virtual trip into space (it is a dream after all)! A coloured sketch is shown below of the author's lucid dream ET beings. Keep in mind that lucid dreaming is as subjective as remote viewing or foreseeing the future until such psychic aspects are proven by an unfolding real time event.

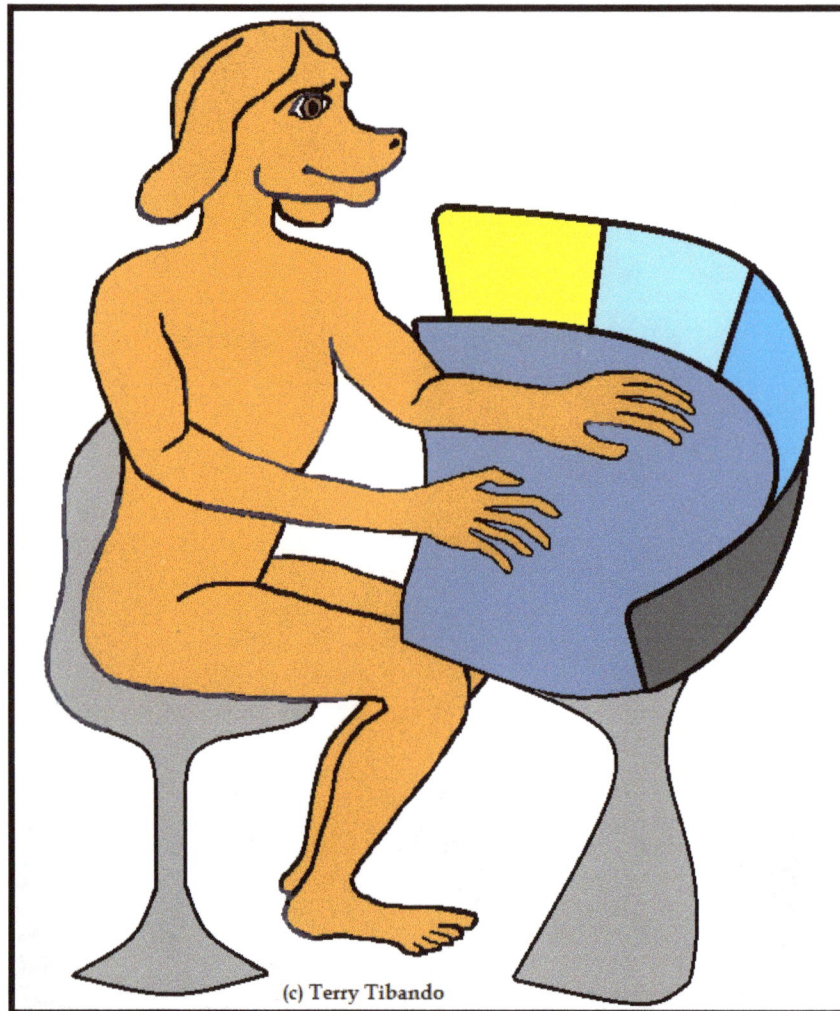

**A drawing from a lucid dream of an ETI dog being at a control panel
who offered the author a virtual ride on his spacecraft**
(c) Terry Tibando

Sculptures, dreams, and interpretative perceptions are one thing but they are not proof that such creatures exist, however, there are many accounts throughout history on this planet that do indicate that there is precedence for this type of being!

Most people have heard of werewolves, but few know of the ancient race of dog-headed men, better known as the **Cynocephali**. A **Cynocephalus** was essentially a man with the head of a dog. They could understand language but had no ability to speak. Though they are sometimes depicted in artwork as being civilized, they were by all reports savage beasts who lived to hunt and to kill.

While this may sound like just another mythical creature, there are very good reasons to believe that the Cynocephali may have actually existed. Above all of them, these dog-headed men were described in reports by famous explorers such as **Christopher Columbus** and **Marco Polo**!
http://www.marsanomalyresearch.com/evidence-reports/2001/029/huge_nozzle.htm

The earliest accounts of Cynocephali can be found in Egyptian history with **Horus** the jackal-headed demi-god and the dog-headed people who taught ancient Egyptians about mummification and helped dead people transition over to the afterlife.

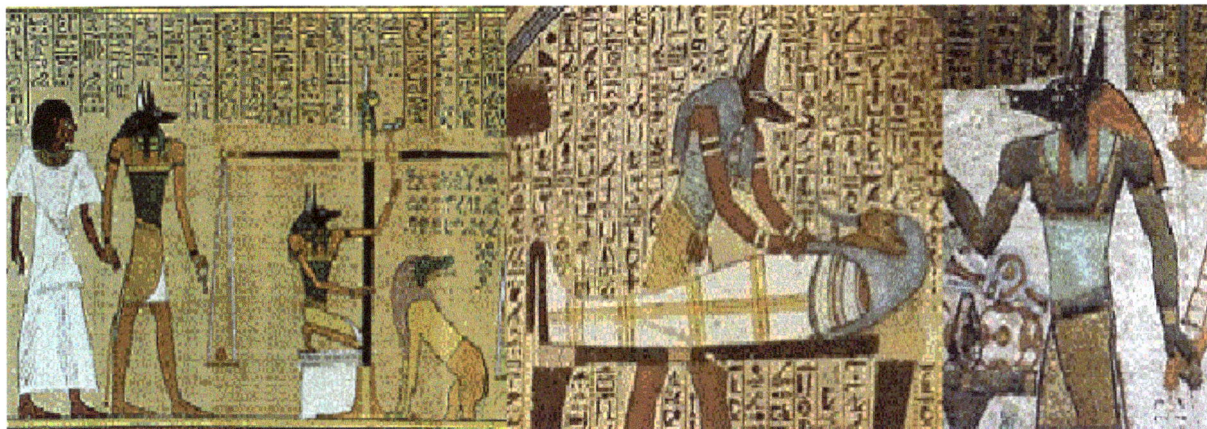

**Dog-headed people taught ancient Egyptians about mummification
and helped dead people transition over to the afterlife.**
(Google Images)

Around 400 B.C. the Greek physician **Ctesias** wrote the following passages (translated to English from Greek), describing the tribes of Cynocephalus:

"They speak no language, but bark like dogs, and in this manner make themselves understood by each other. Their teeth are larger than those of dogs, their nails like those of these animals, but longer and rounder. They have tails above their hips, like dogs, but longer and more hairy. They inhabit the mountains as far as the river Indus. Their complexion is swarthy. They are extremely just, like the rest of the Indians with whom they associate. They understand the Indian language but are unable to converse, only barking or making signs with their hands and fingers by way of reply... They live on raw meat. They number about 120,000." http://www.gods-and-monsters.com/cynocephalus.html

Depending on the location, the legend of dog-faced men either lead simple lives or one similar to humans, they were either fierce or a just race of people

The **Cynocephali** do not live in houses but, in caves on inaccessible mountains and sleep only upon leaves or grass. They are skilled hunters but are also farmers rearing sheep goats, asses for milk. Though they do not practice any particular trade they do trade simple fruits and spices with the king of India. *"They exchange the rest for bread, flour, and cotton stuffs with the Indians, from whom they also buy swords for hunting wild beasts, bows, and arrows, being very skillful in drawing the bow and hurling the spear."*

The clothes of men and women wear very fine tanned skins and the richest wear linen clothes, but they are few in number. Wealth is considered by the number of sheep in one's possessions, and so in regard to their other possessions. *"They are just, and live longer than any other men, sometimes 200 years."*

The dog-headed men were a fierce warrior tribe, but they also traded with the few humans they trusted. They were reported to live primarily in India and Northern Africa but were seen in many places in between. http://www.gods-and-monsters.com/cynocephalus.html

**The dog-headed men were considered fierce warriors,
stronger and swifter than the common human**
http://www.mirror.co.uk/usvsth3m/st-christopher-head-dog-5172620 and http://www.gods-and-monsters.com/cynocephalus.htm

Perhaps the most famous **Cynocephalus** is Christianity's **Saint Christopher**, who was described in several texts as having the body of a man but the head of a dog. Not only that but originally the future saint was said to have been a wild and fierce warrior who was captured in battle in Cyrenaica. Not only was this creature a very large man with a dog's head, but came from a warrior tribe of dog-headed men who looked similar to him. According to Christian mythology, he eventually met **Jesus Christ** and learned the error of his former ways. He repented and became baptized and eventually received sainthood and the gift of a human appearance. Multiple historical images show Saint Christopher as having the head of a dog. http://www.gods-and-monsters.com/cynocephalus.html

Saint Christopher was formerly a dog-headed man before being blessed with a human appearance from his meeting with Jesus Christ
https://turkcetarih.com/turk-mitolojisinde-barak/ and https://www.pinterest.com/manga_wolf/untitled-project/

The great explorer **Marco Polo** mentions **Cynocephali** indirectly while describing his travels to the island of :

"Angamanain is a very large Island. The people are without a king and are Idolaters, and no better than wild beasts. And I assure you all the men of this Island of Angamanain have heads like dogs, and teeth and eyes likewise; in fact, in the face they are all just like big mastiff dogs! They have a quantity of spices; but they are a most cruel generation, and eat everybody that they can catch, if not of their own race."

No one knows for sure what happened to this small, but powerful race. It is believed that as the empires around them expanded they were killed off. They were most certainly a warring tribe and would have preferred death in battle to succumbing to another culture's ways. Either way, they have disappeared from human view. Perhaps there are still some of them living in caves awaiting a day where they may return to power.

One can't help but wonder if this ancient race of dog-headed humanoids are related to the various types of semi-wolf, semi-human creatures such as the **werewolf.** When considering the history of werewolves, this little-known creature may just prove to be a missing link in the mystery of their existence. http://www.gods-and-monsters.com/cynocephalus.html

Martian Cities and Ruins

The real proof of any intelligent species is not whether they build just monuments, pyramids and carve some interesting pictoglyphs into the landscape of a planet. Any intelligent civilization can do that on any planet that they happen to fly by on their way to some other star system, as a calling card to say in a cosmic graffiti fashion, *we were here*!

If that is all that remains on Mars, we would never know whether Mars had indigenous inhabitants or if some other star culture merely left their calling call to say hello. Real proof requires that an intelligent race would have built a society or a planet-wide civilization of towns, villages, and cities indicating a level of progress and development.

From all the satellite images and the photographs taken by the Mars rovers, it has become overwhelmingly evident that the Red Planet did indeed have a planet-wide civilization that endured up until the last few million years ago.

Richard Hoagland believed that whoever built the Face on Mars and the pyramid complex in Cydonia must have had a labour force that lived nearby to these megalithic construction projects. Since these immense monuments are found all over landscape of Mars, it stands to reason that cities or small towns to house inhabitants and provide infrastructure and commerce must also have been built and developed around these construction projects, much in the same manner that ancient Egyptian, Chinese, Indian, Sumerian, Mesoamerican and European cultures developed on Earth.

There would of course, also have been distinctive differences between Earth and Martian cultures and in the manner in which city building would have taken place, just as there were between cultures on Earth. However, there would also be expected to be many similarities in common with any planetary intelligent thus, the frequent familiarity in structural design of pyramids, geoglyphs and even in city planning and layout. The first and most obvious way to recognize if there is a city or town or even a building on the surface of Mars is the detailed inspection of all raw image data taken by orbiting satellites about the planet and then comparing the geometry patterns found with known geometry of Earth type cities, towns, and buildings.

This is what precisely Richard Hoagland and other scientists started to do in 1997 after the Cydonia region and many other areas of Mars were reimaged and NASA began releasing photographic images into the public domain. The **Mars Global Surveyor/Mars Orbiting Camera (MOC)** and the **Mars Odyssey/Thermal Emission Imaging System (THERMIS)** revealed a panoply of remarkable images that were exquisitely most telling. Among the numerous deliberate altered and tampered images of the "Face" and other areas on Mars, anomaly researchers were getting their first real look at Mars in high definition which for that time was truly astounding. https://www.youtube.com/watch?v=urdtDcua5ik

Beside ancient dry rivers, lakes, and old seas, beside nearby hills and mountains, out in the dry desolate desert areas and in the north and south polar regions of Mars, there are signs in high resolution that reveal the structural geometry of an organized layout of roads and buildings... cities indicating the former presence of an established and vibrant ET civilization.

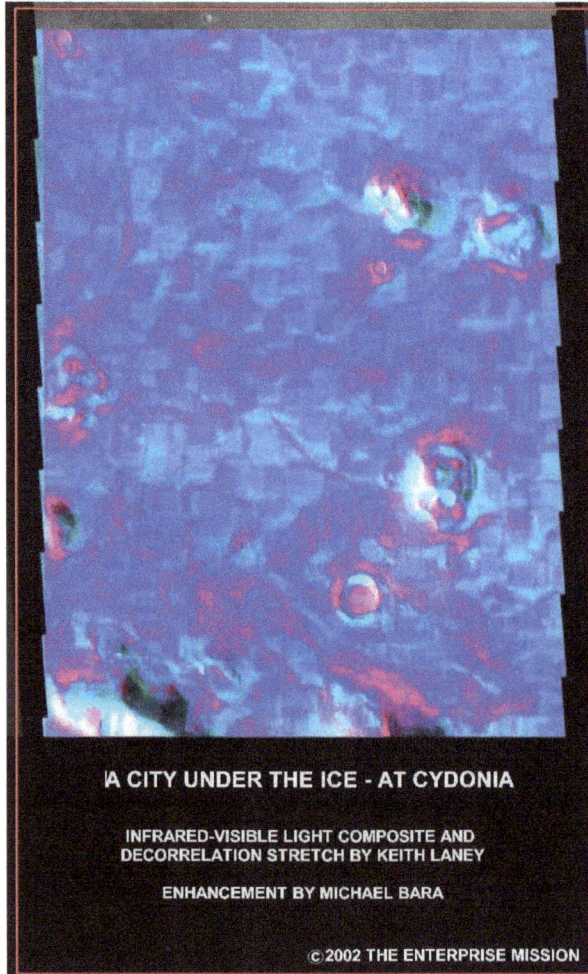

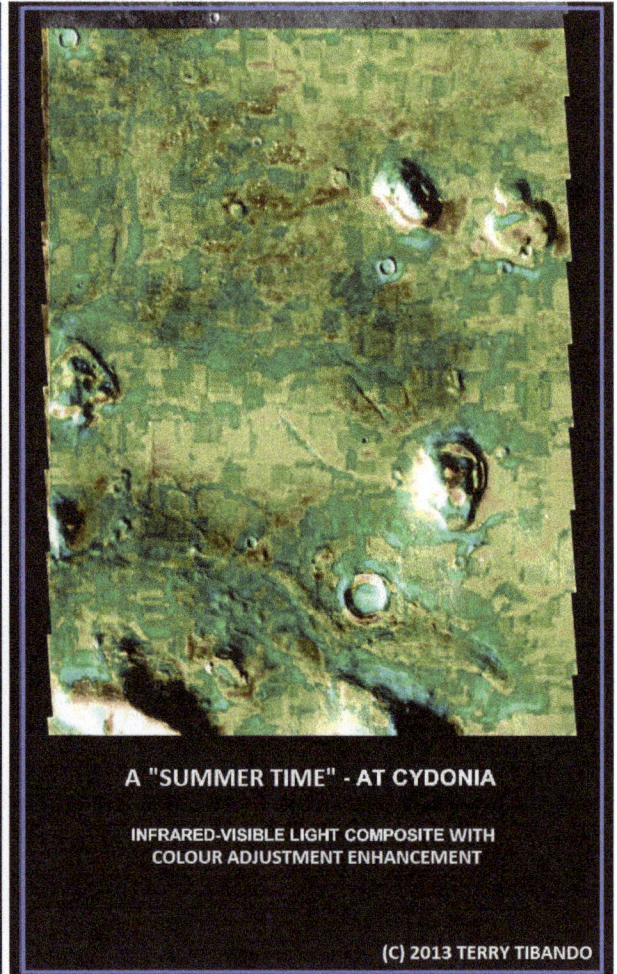

A CITY UNDER THE ICE - AT CYDONIA

INFRARED-VISIBLE LIGHT COMPOSITE AND DECORRELATION STRETCH BY KEITH LANEY

ENHANCEMENT BY MICHAEL BARA

(c)2002 THE ENTERPRISE MISSION

A "SUMMER TIME" - AT CYDONIA

INFRARED-VISIBLE LIGHT COMPOSITE WITH COLOUR ADJUSTMENT ENHANCEMENT

(C) 2013 TERRY TIBANDO

Both images in infrared light show detail of a city structure hidden under the desert region. Left image gives thermal values of hidden structures and right image conveys a possible summertime quality to Cydonia, similar to areas in southern California (below)
https://www.bibliotecapleyades.net/marte/esp_marte_39.htm **and (c) Terry Tibando**

Imperial Valley farm areas in Southern California
https://ucrtoday.ucr.edu/37984/view_from_above_in_the_usa

In MOC m14-02185 image strip, one of these cities can be seen surrounded with some amazing features sculpted into the natural formations of the Martian Terrain. Hoagland has named this Martian city "Argyre City" because it is situated on the south-facing slope on the massif in West Argyre rim region. This city appears to have many small and large rectilinear features that look like blockhouses with connecting roads and streets, all situated by an ancient dried up lake that has old wharfs and shipping docks protruding out from the city to the lake area. There is a feature within the city that looks like a racetrack and a large covered oval dome that could pass as the "Seattle SkyDome" near the lakeshore! (See photo images below).

If the **Pyramid City of Cydonia** is the typical architectural city model that we can expect from other city areas on the Red planet then, the city of Argyre does not disappoint. In true Martian architectural fashion this city is surrounded with massive landforms. One landform as seen photographed from space has been sculpted to resemble a humanoid face in profile called the "Guardian" that faces south toward the city harbour. South of the city is another massive geomorphic feature which taking advantage of particular terrain topography like hills and mesas has been sculpted to look like a bird. It has been called the "Parrot" because of its obvious appearance. (See below).

As seen in other massive pictoglyphs and monuments there is a great reverence for animal, reptile, bird and aquatic life that is repeated in the **Argyre City** and the surrounding terrain. Large satellite images reveal other bird species on the massif Argyre rim region; the reader is invited to explore the MSSS satellite images of this site for himself.

Photo image of MOC m14-02185 image strip shows a Martian city called Argyre surrounded by topographical geomorphs of a parrot and face
http://parrotopia.org/The-Anomaly-Hunters-Roundtable-Study.php

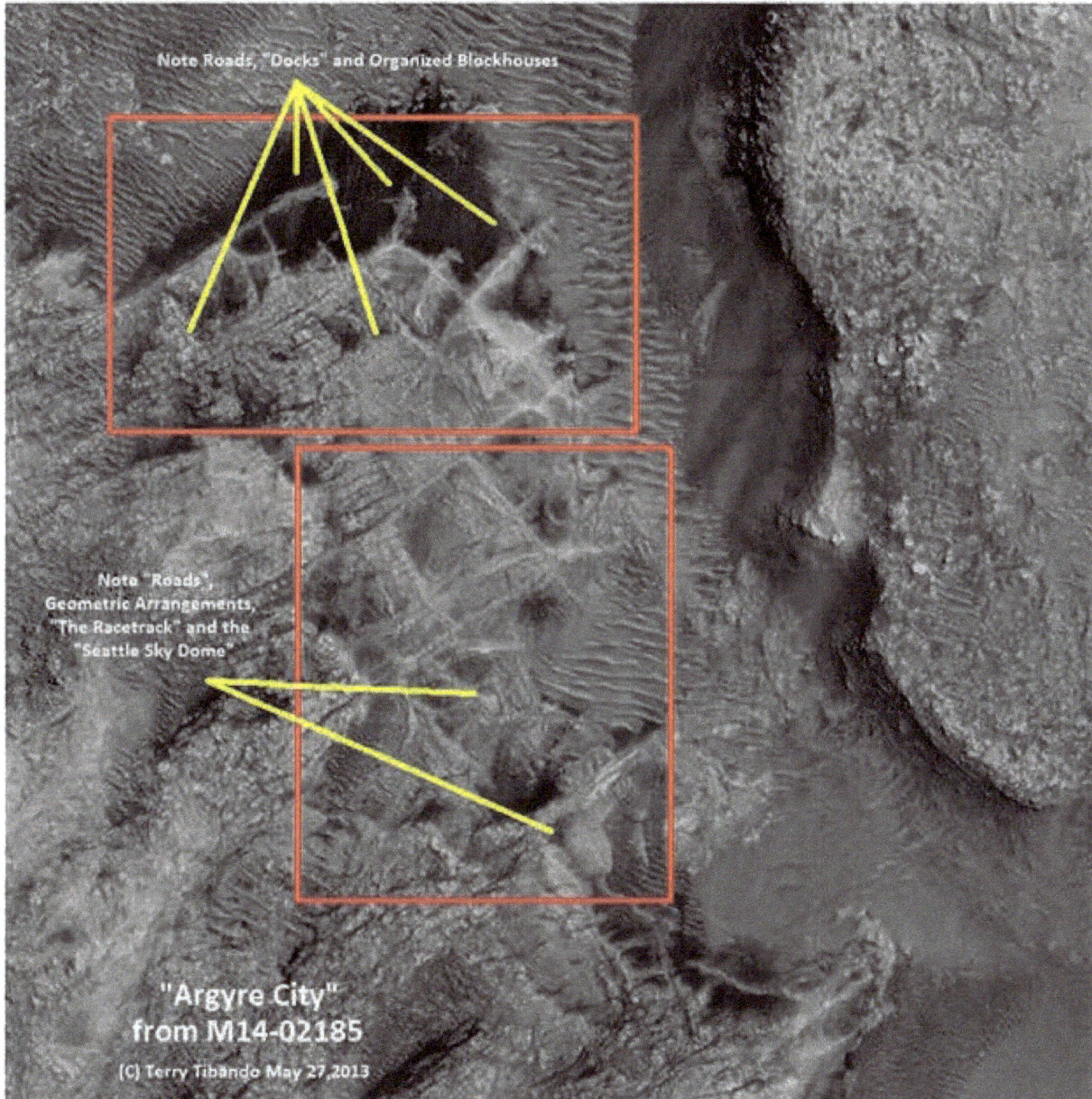

Close up of the Argyre City with roads, blockhouses, and harbour docks
all typical features that can be found any lakeshore city on Earth
(c) Terry Tibando

Below is a drawing of a hammered copper plaque of an avian form that was produced by the **Hopewell Indians** of Ohio about 400 BC (**Thomas, 1994**). It is presented here as a comparative image for the avian feature found at the edge of the Argyre Basin on Mars. In reviewing this Hopewell plaque, avian specialist and author **Orosz** in **2006** identified the form as representing an indigenous parrot. She acknowledges the overall profiled posture of the Hopewell *parrot with its **extended wing motif** is reminiscent of the design expressed within the avian feature on Mars.* She also notes the shape of the parrot's head and beak shares a common form with the avian feature on Mars, while the shape of the clawed foot, the round belly, and the stylized tail feathers are also analogous.

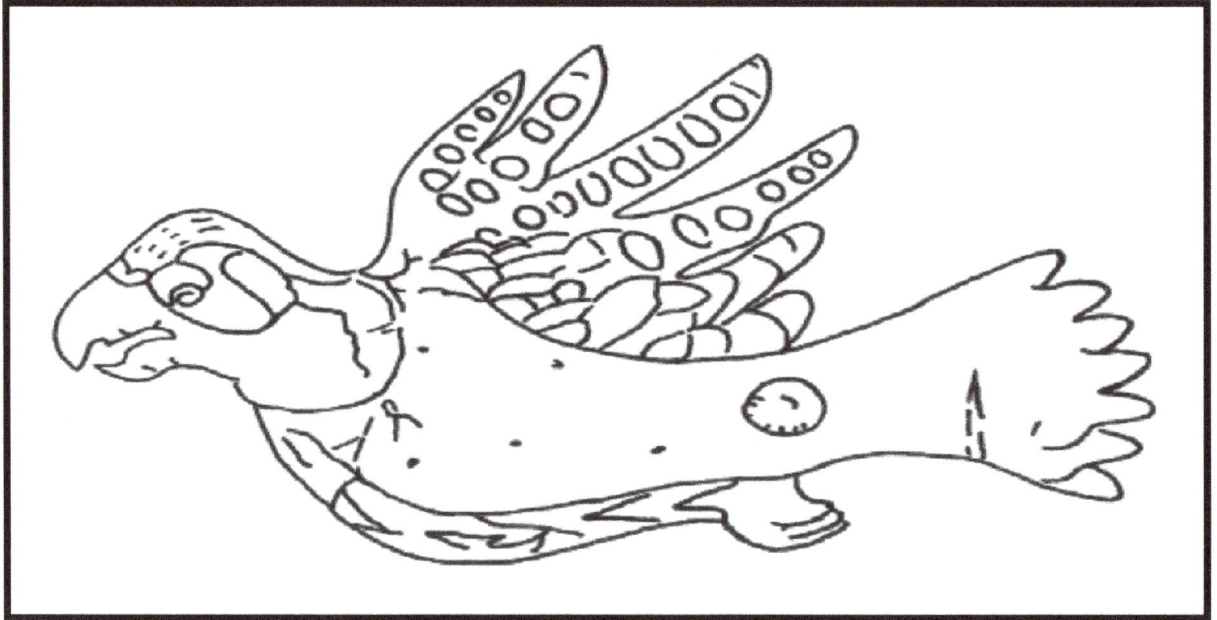

The Hopewell Indian hammered copper Parrot plaque as drawn by George J. Haas with it, extended wing motif is reminiscent of the design expressed within the avian feature on Mars as seen in the above image
http://parrotopia.org/The-Anomaly-Hunters-Roundtable-Study.php

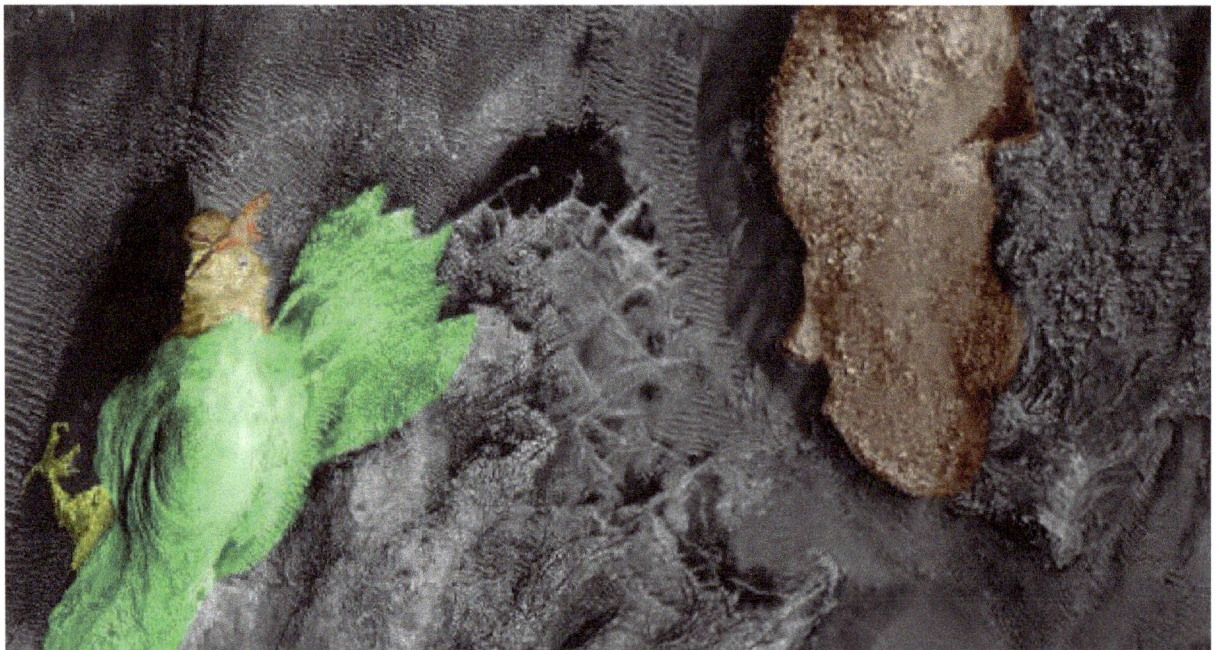

The Parrot glyph embraces the city of Argyre and the Guardian keeps a watchful eye on the city from across the harbour
http://parrotopia.org/The-Anomaly-Hunters-Roundtable-Study.php

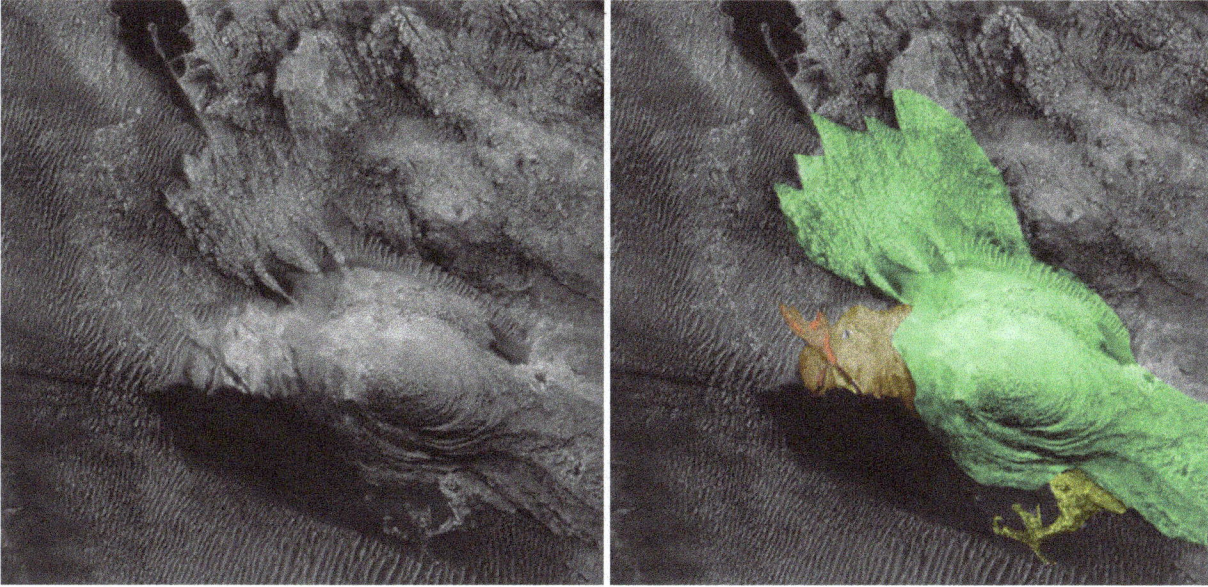

The Parrot in the above B&W image is fairly obvious but it is colour enhanced on the right to show detail. Compare this to the Hopewell Indian copper plaque below

The majority of comparative examples of manipulated terrestrial geology come to us in the form of earthworks that were created by ancient cultures throughout North and South America. These huge mounds and earthworks were shaped like animals and geometric symbols, while others were formed like ceremonial platforms and step pyramids. It is estimated that the amount of earthworks found throughout North America number in the hundreds of thousands. However, over time almost all of these monuments have been either destroyed by natural erosion or by the rapid expansion of rural and urban development. Because there are a limited number of examples of animal and figurative earthworks in the available database, only two meet the criteria of this study with comparable detail and content.

The first is a 5,000-year-old, eagle-shaped geoglyph located in the town of Eatonton, Georgia. At the site, an eight-foot-high bed of white quartz stones forms a silhouette of an eagle hovering within a circular mound. The apex of the mound forms the eagle's abdomen, which creates a similar elevation as seen in the mound-shaped abdomen of the avian formation on Mars. The body of the eagle effigy measures more than 100 feet from head to tail and has a wingspan of more than 120 feet (**White**, 2002). The overall shape of the eagle effigy is symmetrical in design, featuring a set of outstretched wings, tail feathers, and a head that faces eastward. As seen in the illustration, its contours project only the simplest form of a bird without providing additional details.

A second example of an avian earthwork is etched on a hillside in the Peruvian Andes, not far from the famous **Nazca lines** (**Longhena & Alva**, 1999). The Peruvian pictograph is formed by a set of conjoined lines that create the impression of a standing bird. Although the awkward shape of the Peruvian pictograph is not proportioned or anatomically correct, the overwhelming consensus is that it indeed represents the generic form of a small bird.

Accepting the consensus that this simple mound and hillside rendering are accepted as intentional works of art by the limits of aerial observations, it would be reasonable to suggest that the formal organization expressed within the avian feature on Mars conflicts with the randomness of mere chance. *There are no terrestrial geoglyphs that induce such a visual impression that approaches the refined modeling of relief sculptures as seen in the avian formation at Argyre Basin*. "Avian Formation on a South-Facing Slope along the Northwest Rim of the Argyre Basin" by Michael A. Dale, George J. Haas, James S. Miller, William R. Saunders, A. J. Cole, Joseph M. Friedlander, and Susan Orosz; 6/10/2011; Journal of Scientific Exploration, Vol. 25, No. 3

In the photo below, **the Face (Guardian)** and **the Parrot** have been coloured highlighted to show detail and the harbour city Argyre is situated between the geoglyphs. It can be seen that the city has many rectilinear structures in its appearance but, what is often overlooked with this photo image is that there is a massive city complex on and right behind the **Guardian Face** as well that is full of rectilinear features that resemble buildings, roads, and streets. This **Argyre Basin** is home to twin or neighbouring cities separated by and ancient lake or sea.

Joseph Skipper also points out that the city has many other glyphs of avian and reptile creatures when the image is enlarged and examined. Skipper also thinks that this city may also be home to an ancient mining program or archeological dig site which is located near the large oval dome structure. http://www.marsanomalyresearch.com/evidence-reports/2009/167/parrotopia.htm

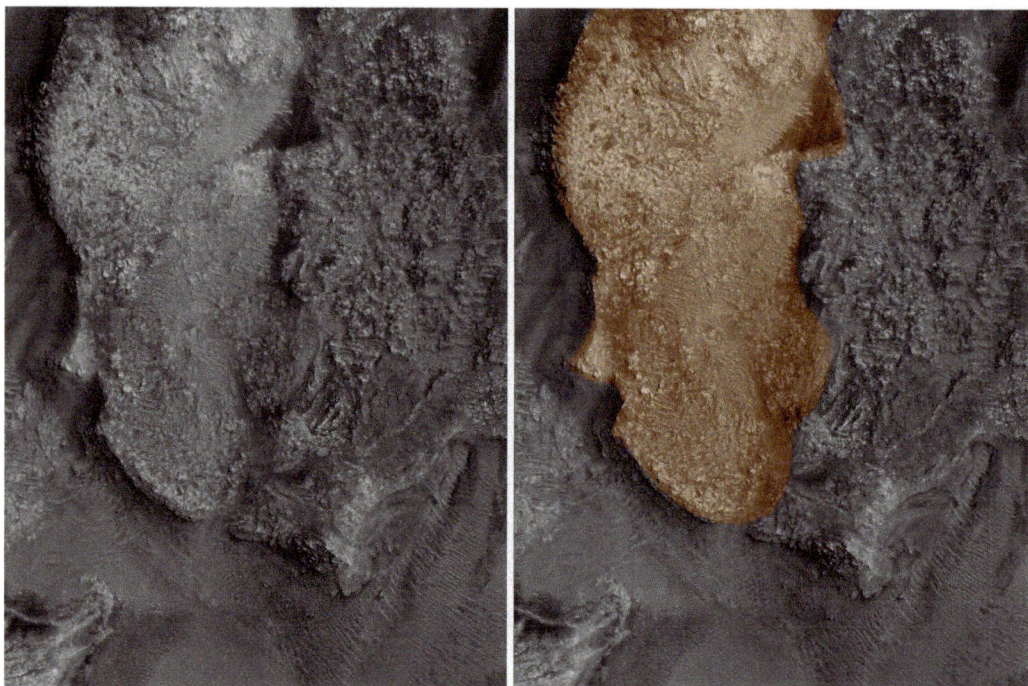

**The Guardian city complex with its numerous rectilinear features
keeps a valiant eye on its sister city, Argyre City**
http://parrotopia.org/The-Anomaly-Hunters-Roundtable-Study.php

The next area on Mars to draw public attention is a desert area called the **"Inca City"** of Mars which was first photographed by Mariner 9 in 1972 and then by Malin Space Science Systems with the **MGS MOC MOC2-319** on August 8, 2002. It was called the Inca City by NASA Mariner 9 scientists for the set of intersecting, rectilinear ridges that are located among the layered materials of the South Polar region of Mars and which bear a strong resemblance to Incan cities in Central and South America.

The Inca City on Mars, note that this area is approximately 86 km in diameter. Compare this area to the photo of Khorezmian Fortress Koy-Krylgan-Kala before and after the 1956 excavation (below).

http://www.msss.com/mars_images/moc/8_2002_releases/incacity/

The Inca City is situated on the circumference of a large circular area that measures 86 Km across. There are no Earth cities whether in ancient times or in the present that are as large as this Martian city and as can be seen in the photo image much of the city structure is buried under sand dunes. Their origin has never been understood; most NASA scientists thought they might be sand dunes, either modern dunes or, more likely, dunes that were buried, hardened, then exhumed. Others considered them to be dikes formed by injection of molten rock (magma) or soft sediment into subsurface cracks that subsequently hardened and then were exposed at the surface by wind erosion.

It is possible that this pattern reflects an origin related to an ancient, eroded meteor impact crater that was filled-in, buried, and then partially exhumed. In this case, the ridges might be the remains of filled-in fractures in the bedrock into which the crater formed or filled-in cracks within the material that filled the crater. Or both explanations could be wrong. While the new MOC image shows that "Inca City" has a larger context as part of a circular form, it does not reveal the exact origin of these striking and unusual Martian landforms.

A close-up view of the Inca City showing detail of the rectilinear structure
http://www.msss.com/mars_images/moc/8_2002_releases/incacity/

However, another possibility is the comparison with other ancient Earth settlements particularly those found in Iran and Iraq has raised the question that the massive structure of the Martian Incan city may be artificial and not a natural formation that was formed over millennia. Like everything on Mars, that is illustrated in this book, the true nature of all these structures will require investigation by mankind setting foot into these areas and performing archeological site excavation.

The **Khorezmian Fortress Koy-Krylgan-Kala** in Iran is a great example of what looks like a small crater impact in the desert that has over time been covered with blowing sand nearly hiding all traces of its existence. When archaeologists investigated the site they discovered that it was, in fact, an ancient fortress circular in appearance with concentric rings containing many cells or compartments with entry/exit roads facing in the four compass directions.

The Khorezmian Fortress Koy-Krylgan-Kala in Iran appeared as an impact crater and after excavation in 1956, its true artificial nature was revealed
http://www.abovetopsecret.com/forum/thread483164/pg1

This Anasazi (a Navajo Indian word for "the ancient ones") Pueblo Bonita (great house) is an example of cellular compartments found also at the Inca City on Mars
https://www.pinterest.com/pin/552253973026373046/

When this stunning **"Martian ruin"** (below in enhanced colour) is compared to similar structures here on Earth as in this 1936 aerial photograph of a long-abandoned "eighteen hundred-year-old **Sasanian Palace**" in Iran (below), the eerie geometric similarity is instantly apparent.

103

CRINST. AE NEAR SARVISTAN, IRAN. THE SASANIAN PALACE, LOOKING APPROXIMATELY NORTHEAST FROM AN ALTITUDE OF 427 METERS ON MARCH 30, 1936

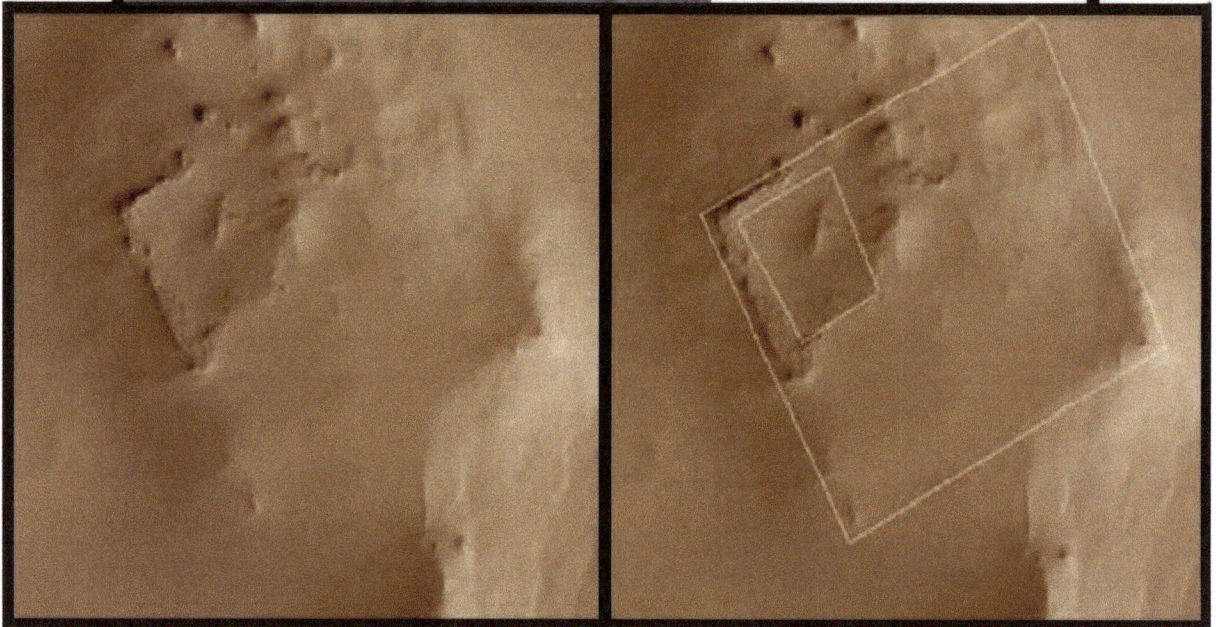

A buried ruin on Mars from MOC E1000462 image strip (colour enhanced)
compared to the ancient Sarvistan Palace (above) in Sarvistan, Iran
http://www.enterprisemission.com/LostCitiesofBarsoom.htm and http://mars-civilisation.e-monsite.com/pages/page-12.html

There are many Martian ruins that seem to have their counterparts on Earth which can be found in the Middle East, chiefly in Iran and Iraq as can be seen in the following photos below. Any archeologist must wonder at the apparent similarities of ancient city structures found Earth and Mars as more than just mere coincidence found on two neighbouring planets. How is it possible that the ancient Sumerian culture that suddenly sprang up in these countries 6000 to 10,000 years ago, seemingly out of nowhere, suddenly developed advanced civilized skill sets in writing, mathematics, astronomy, agriculture, city building and many other social developments which prior to that time, the people were only a prehistoric nomadic and cave dwelling culture. What could have taken place in this desert region to ignite such a blazing development in civilization?

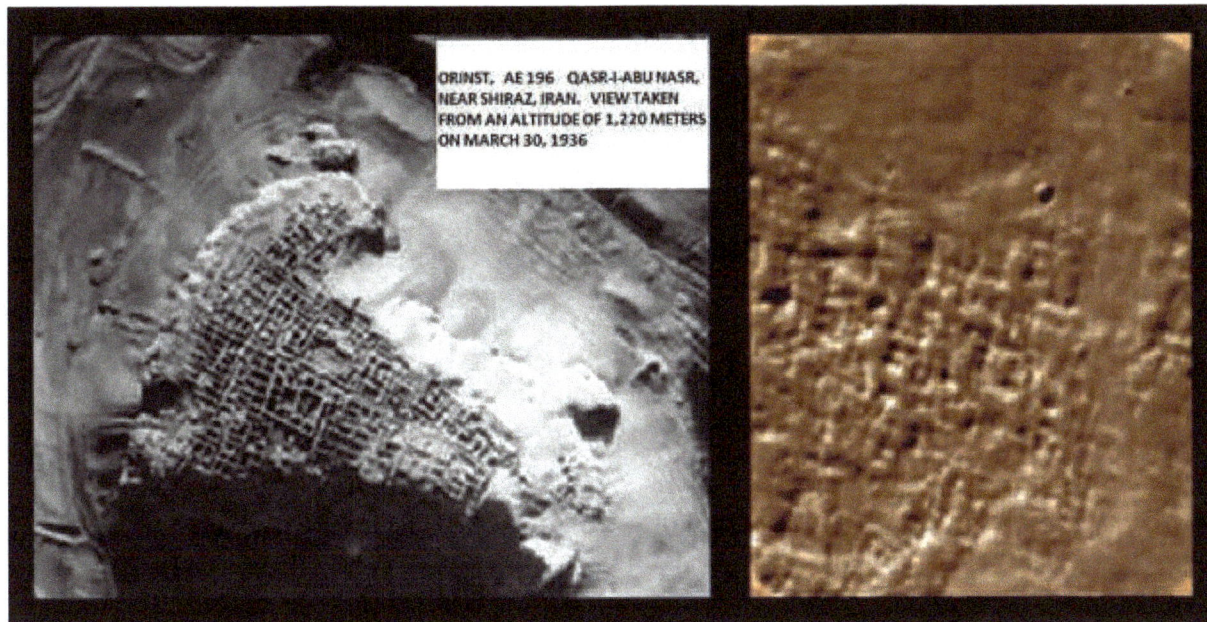

ORINST. AE 196 QASR-I-ABU NASR, NEAR SHIRAZ, IRAN. VIEW TAKEN FROM AN ALTITUDE OF 1,220 METERS ON MARCH 30, 1936

The ancient ruins of Qasr –I –Abu Nasr, Iran on an isolated mesa bears remarkable similarity to geometric constructions found on Mars
http://www.enterprisemission.com/LostCitiesofBarsoom.htm

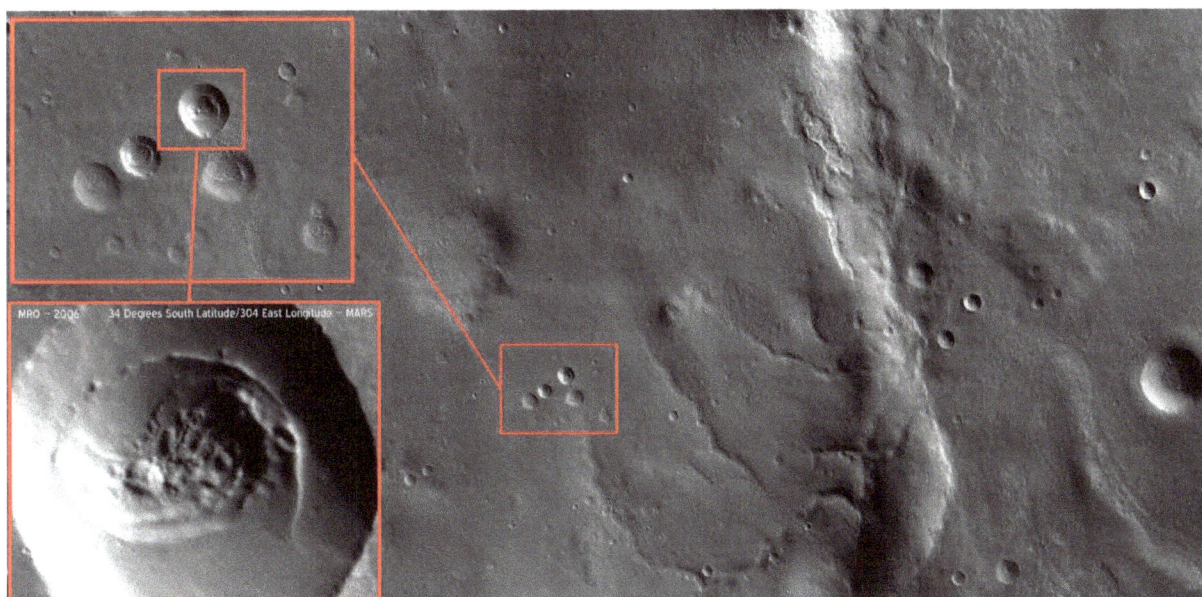

MRO – 2006   34 Degrees South Latitude/304 East Longitude - MARS

PIA08014: Detail of First Mars Image from Mars Reconnaissance Orbiter (MRO) with HiRISE Camera reveals in stunning detail many cities on Mars
http://www.enterprisemission.com/LostCitiesofBarsoom.htm

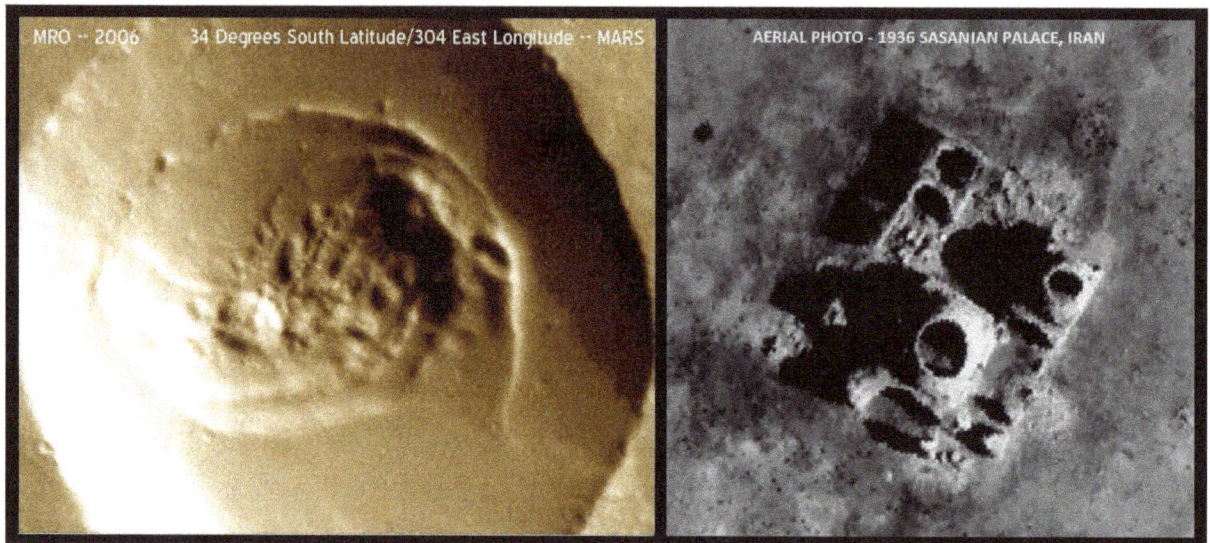

MRO -- 2006   34 Degrees South Latitude/304 East Longitude -- MARS

AERIAL PHOTO - 1936 SASANIAN PALACE, IRAN

**A close up view of the crater (above) containing a small village
complex similar to the ancient Sasanian Palace in Iran**

http://www.enterprisemission.com/LostCitiesofBarsoom.htm

Some researchers who follow the **Ancient Alien Hypothesis** speculate that the **Sumerian** and
the **Mesoamerican** cultures, as well as other ancient cultures, may have all been influenced by
an alien intervention possibly from Mars or beyond this Solar System. The more we delve into
the raw photo images from NASA and examine the ancient anomalies, artifacts, the massive
monuments, and city-like structures, the more it becomes evident that the Earth and Mars have
shared an ancient connection in civilization and cultures.

https://www.youtube.com/watch?v=RxbCtBjkW98

CHAPTER 82

THE EVIDENCE IS CONCLUSIVE - MARTIANS DO EXIST!!!

At this juncture in our search for life on Mars, *a priori hypothesis needs to be established that the intelligence on Mars and Earth are one and the same species with the possibility of one or more additional intelligent civilizations having shared co-habitation on one or both planets, either simultaneously or at different times.*

The proof for this hypothesis to become self-evident requires the discovery of actual buildings or building materials and smaller constructed artifacts manufactured by the planet's inhabitants that perchance may be found lying around or in such buildings. The proof would have to come from the raw photographic images that have been taken by the Mars-orbiting satellites and the Mars rovers that are currently ravelling across the terrain. Such photographic evidence will most likely be of recognizable artifacts similar in appearance to what we would expect to find in modern day Earth societies as well as what would be unearthed in archeological digs of ancient cultural sites. In other words, the artifacts would appear to have been built by humans or humanoids.

Buildings that were built several millennia ago, even millions of years ago in order to still be standing or at least somewhat still visible and recognizable require conditions that favour their preservation and not their deterioration. Corrosive elements from Martian weather would need to be minimal or non-existent, terraforming from earthquakes and crustal displacement would have to be infrequent and low Richter level which means Mars would have to have no functioning internal magnetic field to generate in order to effect magma turbulence and crustal plate upheaval and finally with some view of optimism, the occurrence of bombardment by meteorites and asteroids upon the **Red Planet** would also have to be random and extremely rare. Fortunately, Mars weather conditions are cold and dry where winds seem to be the main corrosive element on Mars. Mars seems to have lost its magnetic field or it is very weak at best having been so for tens of millions of years. The reason for its loss may have been due to the meteoric bombardment that seems to have taken place throughout our Solar System in its early formative development and perhaps, later as a periodic solar shower of material residue still settling out from the birth of our solar system. Whatever, the reason, Mars did undergo destruction on a massive scale that was planet-wide in its scope. Mars, however, did seem to foster an advanced intelligent species which did build and thrive for a period of time and there are remnants of that former Martian civilization, NASA's raw photo data prove that its existence is real.

In the photo below, **Mars rover Spirit** snapped a black and white image that shows a rectangular stone block in the foreground and a square faced block in the middle of the picture. Other stone blocks can be seen scattered about in the picture and beyond that can be seen two or more rectangular foundation sites indicating the former existence of some buildings that once stood in that area. https://mars.nasa.gov/mer/gallery/panoramas/spirit/2005.html

Stone blocks in the foreground and building foundations in the distance can be seen in this photo taken by the Mars rover Spirit "Postcard Above Tennessee Valley"

https://mars.nasa.gov/mer/gallery/panoramas/spirit/2005.html

A closer inspection of some of these square faced stones and blocks would help to determine to if they have been hand-formed or machined or whether they are just natural rock formation. iIdeally a discovery of a building, even in partial ruins, but with some of its walls and rooms intact would provide NASA and investigators of Mars anomalies with measurements from which height and proportions of the building's occupant could be deduced.

The **Mars rover Curiosity** which is a nuclear-powered vehicle about the size of a family car has according to NASA performed amazingly well meeting all of NASA's expectations so far as it travels and explores its way across the **Gale Crater**. Anomaly researchers examining raw photo images of the Martian surface have also been excited by what they are finding even if NASA continues to remain silent without explanation to the many oddities that Curiosity has been photographing.

Below is one of these square block-like artifacts that appears to be a box of some kind in appearance and one would expect that NASA would have moved the rover over for a close photographic look but, instead NASA has decided it not worth further investigation or a high-resolution photo image to obtain more detail from it. Such apathy by NASA makes one wonder if NASA is truly looking for life or the various elements that contribute to sustaining life something else.

Curiosity finds a box-like structure but fails to investigate it further for its unusualness. Perhaps, NASA should rename their Mars rover "Apathy" instead of "Curiosity" to better reflect NASA 's attitude toward its exploration of Mars
(Google Image)

Recent photos released by NASA has sparked Richard Hoagland to add a new diatribe expose with photos to his Enterprise Mission website titled "***NASA Goes Apartment' Hunting"!*** His informative website certainly adds balance to NASA's official *"Never A Straight Answer"* position on Mars and its current space exploration agenda. His recent point of view on Curiosity's mission caught this author's attention to examine the main photo image on this subject as posted on his website.

In the original photo image below taken from Curiosity's panorama camera on Sol 120 in Gale Crater (from which this is a portion of that panorama image), NASA says this is an area (dubbed "**Shaler**") of shale rock which is evidence of stream flow. Hoagland's perception of the same photo was that it looked more like a collapsed apartment.

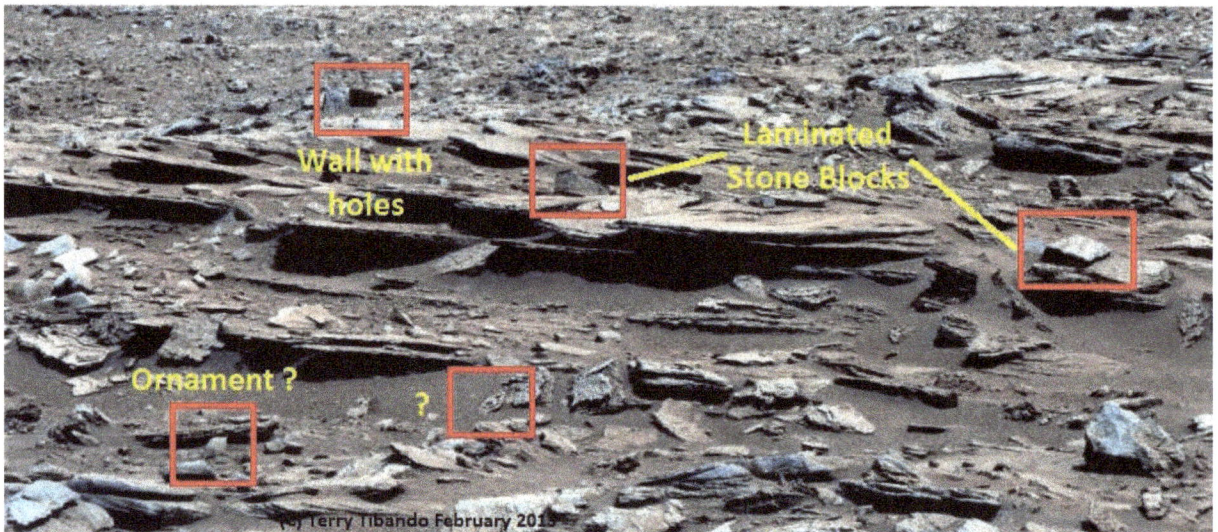

A small section of the "Shaler" area (colour corrected) from a much larger panorama image shows many anomalies that are not natural geological formations

https://mars.jpl.nasa.gov/msl/multimedia/images/?ImageID=4937

A small wall with windows/Doll House and a laminated masonry block (one of many)

https://mars.jpl.nasa.gov/msl/multimedia/images/?ImageID=4937

A possible gasket or animal mask and an ornament with pointed ends with leg supports

https://mars.jpl.nasa.gov/msl/multimedia/images/?ImageID=4937

110

NASA's "Shaler" outcrop area also referred to as the "Martian Apartment" by Richard Hoagland, photographed by Curiosity on Sol 120. Note anomalous debris and rubble containing artificial geometric symmetry

https://mars.jpl.nasa.gov/msl/multimedia/deepzoom/

The first reaction by most investigators to this photograph would be to agree with NASA's official position, that this area was a natural rock formation of shale rock, evidence of a former stream flow and that Hoagland had simply got it wrong this time. However, upon closer inspection of the photograph, there are inconsistencies with NASA's shale explanation as many rock formations in the panorama did not seem to belong to this particular snapshot of the Martian terrain. There were too symmetrical features displaying both curvilinear and rectilinear aspects of artificiality contained within the rocks.

What Hoagland was seeing in the photo image were walls of concrete containing rusted rebar support rods, curved roofing tiles, stone foundation blocks and an assortment of objects owned by former residents. (See highlighted photo below compared to the debris from the Bangladesh collapsed garment factory from the 2013 tragedy).

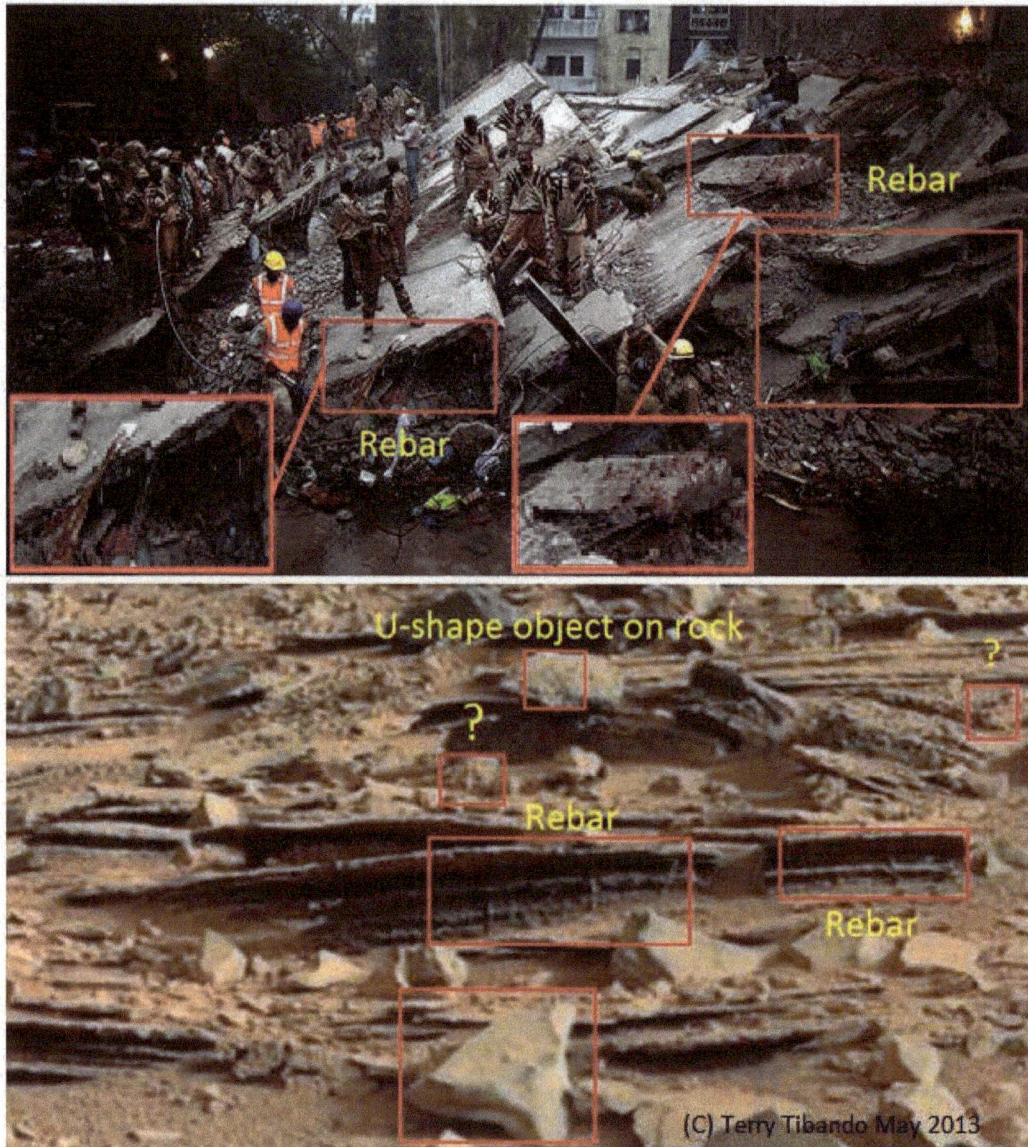

Compare the rubble debris in the collapsed garment factory in Bangladesh, 2013 to the "Shaler" area in Mars Gale Crater. Note the rebar in the concrete walls, the similar geometry of both structures and scattered personal items of former inhabitants
(c) Terry Tibando and https://mars.jpl.nasa.gov/msl/multimedia/deepzoom/

The reader's curiosity would naturally be piqued at this point from Hoagland's apartment interpretation to want to search for these smaller artifacts which were not obvious without enlargement of the photograph. One can see the concrete walls of a building and the rusted rebar embedded in the walls that had led Hoagland to deduce that this was a large collapsed residence that had at one time housed many people. However, similarities to an apartment building require residents or at least their personal belongings be present in the rubble of the collapsed structure to be proof that this was a place of habitation. (See above Bangladesh photo).

At this point, a ***guiding rule of thumb*** is necessary for any amateur anomaly researcher in their examination of NASA's hundreds of thousands of photo images. Unless, you are a professional in aerial photographic and satellite image interpretation who uses the following elements of photographic interpretation on a daily basis: Location, Size, Shape, Shadow, Tone/Color, Texture, Pattern, Height and Depth, and Site/Situation/Association, and can also spot a needle in a haystack in any raw photo image, ***always look for the obvious***.
http://en.wikipedia.org/wiki/Aerial_photographic_and_satellite_image_interpretation

Look for easily recognizable, objects that pop out of the photo, that anyone could see at first glance and recognize without doubting their eyes or understanding. Don't make your "discoveries" difficult by labelling things in photos that no one else can see that are merely wishful imaginings of your own thinking and certainly don't alter or add things to a photo that simply not there in reality.

In others words, don't be adding **CGI effects** in order to claim some limelight of fame as a "wannnabe discover" or to achieve some infamous notoriety! Nobody is going to slap you on the back and congratulate you for your work. You're only doing everyone, who seriously wants to investigate the UFO/ETI phenomenon a disservice and setting UFO/ETI research backward with hoaxed investigations and false discoveries unless, that was your original intent to start with, in which case consider yourself an agent of disinformation!

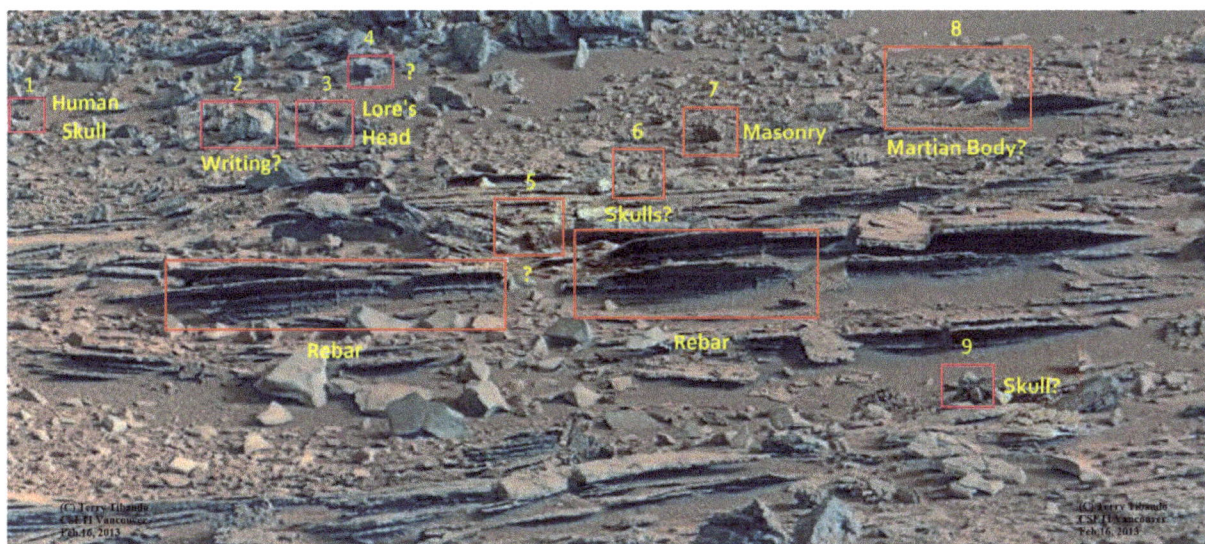

(Colour corrected) Debris field containing all kinds of artificial and organic anomalies that Richard Hoagland calls an "apartment". (See enlarged photos below for details)
(c) Terry Tibando and https://mars.jpl.nasa.gov/msl/multimedia/deepzoom/

Author's rant: In my own search, I was looking for objects that did not require rendering or stretching one's imagination to recognize the familiar, unlike the pratice that so many anomaly researchers would have the average person do in order to see what they thought was hidden in the rock formations. I was looking for the obvious, those things that were artificial in appearance with no ambiguity of their true nature that would not require any interpretation.

113

From the above photo of the Shaler area, many artifacts became apparent when the images were enlarged, as can be seen below in the following images that display prominent artificiality.

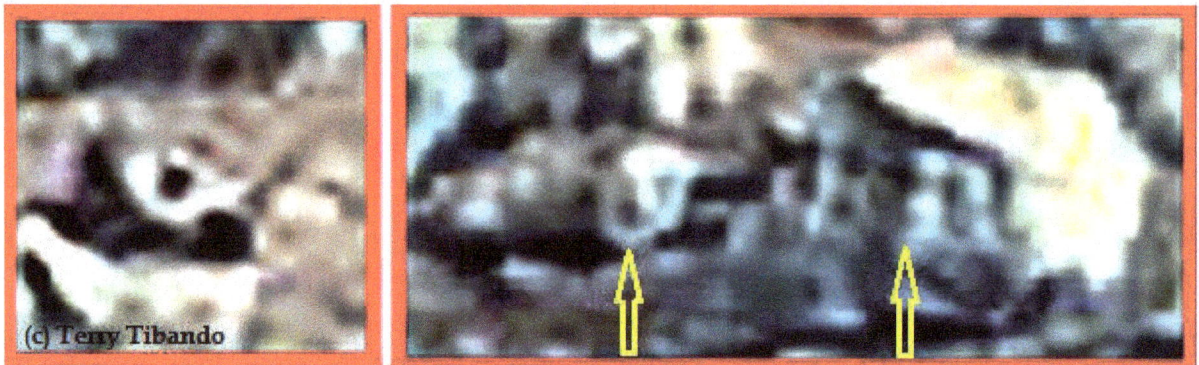

(1 & 2) Left, object appears to be a humanoid skull, possibly robotic as its surface is reflective indicating a metallic finish. Right, two rocks, one cylindrical in shape with a bulge in the centre and the other with inscribed lettering on its surface
(c) Terry Tibando and https://mars.jpl.nasa.gov/msl/multimedia/deepzoom/

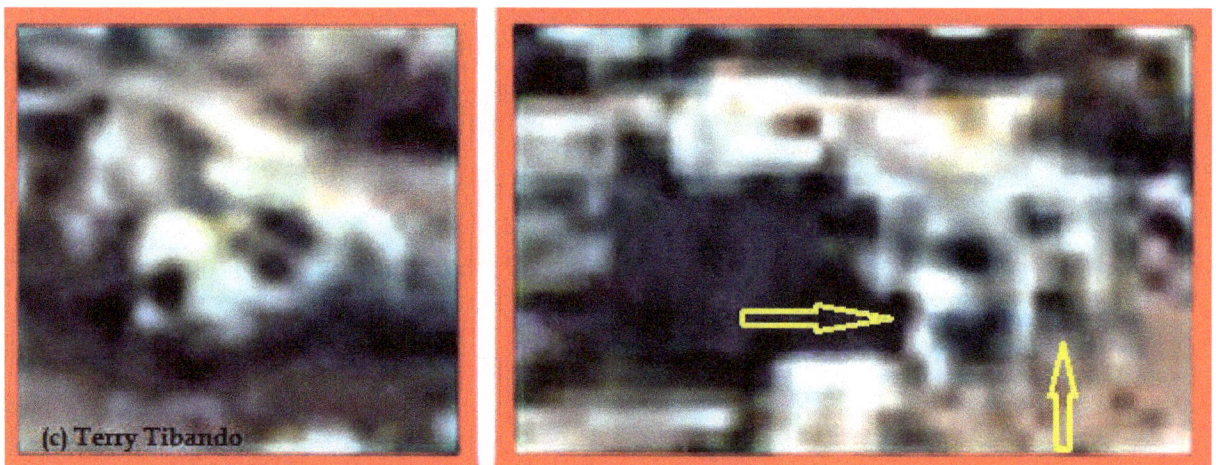

(3 & 4) Left, robotic-looking head or mask dubbed "Lore's Head" (brother to Data in Star Trek NG). At right, a piece of machinery with four symmetrical machined holes Compare this Martian head to "Data's Head" photographed on the Moon
(c) Terry Tibando and https://mars.jpl.nasa.gov/msl/multimedia/deepzoom/

It would seem that whenever any of the Mars rovers come across a debris field of pulverized and scattered rock rubble, in all likelihood, that pile of rock debris may actually contain the remnants of animal life, pieces of machinery as well as signs of past intelligent life that once inhabited the planet. Such areas are an archeological gold mine of antiquity that crosses the borders of many disciplines of science and as such deserve the greatest scrutiny possible by NASA and other investigating scientists. Literally, no stone should remain unturned in these debris fields. To do so, would be critical in our understanding of the nuances of Martian life from its simplest forms to its most complex and intellectual life forms.

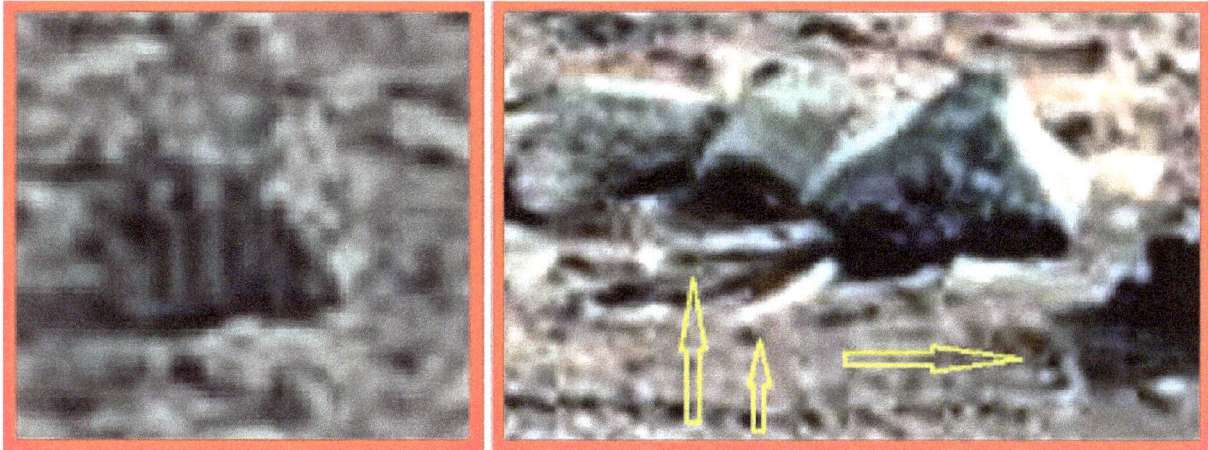

(7 & 8) Left, another piece of anomalous rock which may be a piece of machinery judging from the evenly spaced symmetrical hollows in the structure. Right, could this be the legs and an arm of a humanoid skeleton crushed by a large rock?

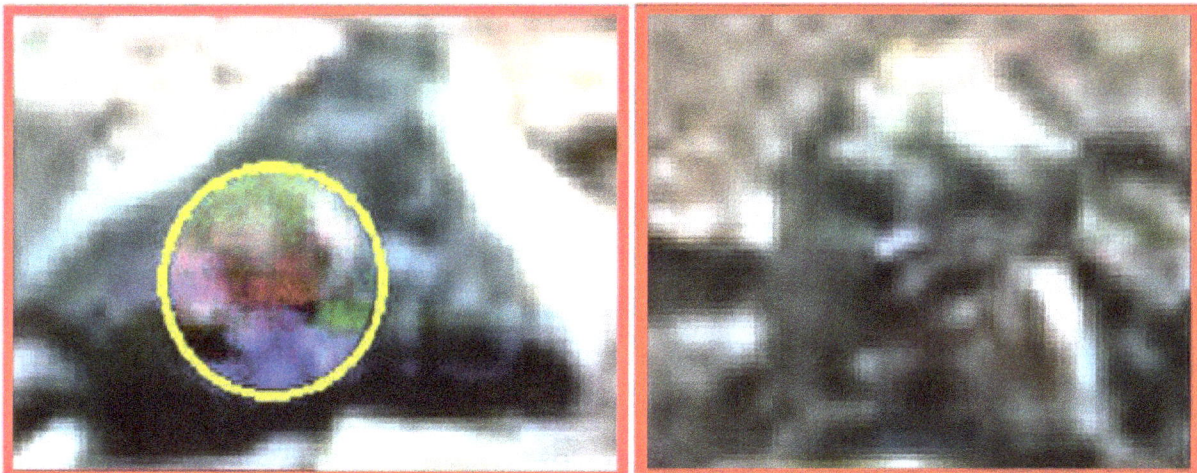

(8 & 9) Left, close-up of the rock (auto-ehanced). Note the "Red Heart" painted on its surface. Right, a human skull can be seen "peeking out" from behind a rock

Close-up view of the rebar support rods in the apartment wall

It now becoming obvious why NASA chose the Gale Crater as a landing site for their latest Mars rover Curiosity, they must have known from their orbiting satellites that this area was home to numerous artifacts indicating that a long extinct civilization once inhabited this crater. The proof was in the photos as can be seen in the close-up images above and below. A vast Martian civilization must have thrived across much of the landscape of Mars and judging from the numerous NASA images, something catastrophic must have occurred millions of years ago that virtually destroyed the eco-system of the planet and killed off most life forms with the exception of a few of the most hardy of creatures.

The close up of photo #3 which shows a head or mask is without exception, human in appearance and one cannot help but compare this with the robotic head dubbed by Richard Hoagland as 'Data's Head". The resemblance to "Star Trek Next Generation" robotic character Data and also its resemblance to the "Star Wars" robot C3PO were profound and evident that a high technology from a former ET presence, once existed on the Moon as discovered during the Apollo 17 mission crew and now, its counterpart had been discovered on Mars by Earth's own robotic vehicle, the NASA Mars rover Curiosity.

The remarkable resemblance of Data's Head from the Shorty Crater on the Moon and Lore's Head found in the Gale Crater on Mars seem to be virtually from the same manufacturer! The intelligence that built the robotic head found on Mars, by its existence and its similarity to the robotic head found on the Moon implies that they were created by one and the same intelligence who were ergo, space travellers from an ancient time!!

Comparison of the facial features of both robotic heads are very uncanny. The cheek structure is identical, the mouths are similar and the eyes are both circular in shape. The minor differences in heads could be accounted for by custom manufacturing. A second head/skull which is highlighted in the above red border photo and again below also appears to be metallic in construction due to the high reflectivity coming off its shiny surface, it too suggests that it may be robotic or a cybernetic organism based upon a human resemblance.

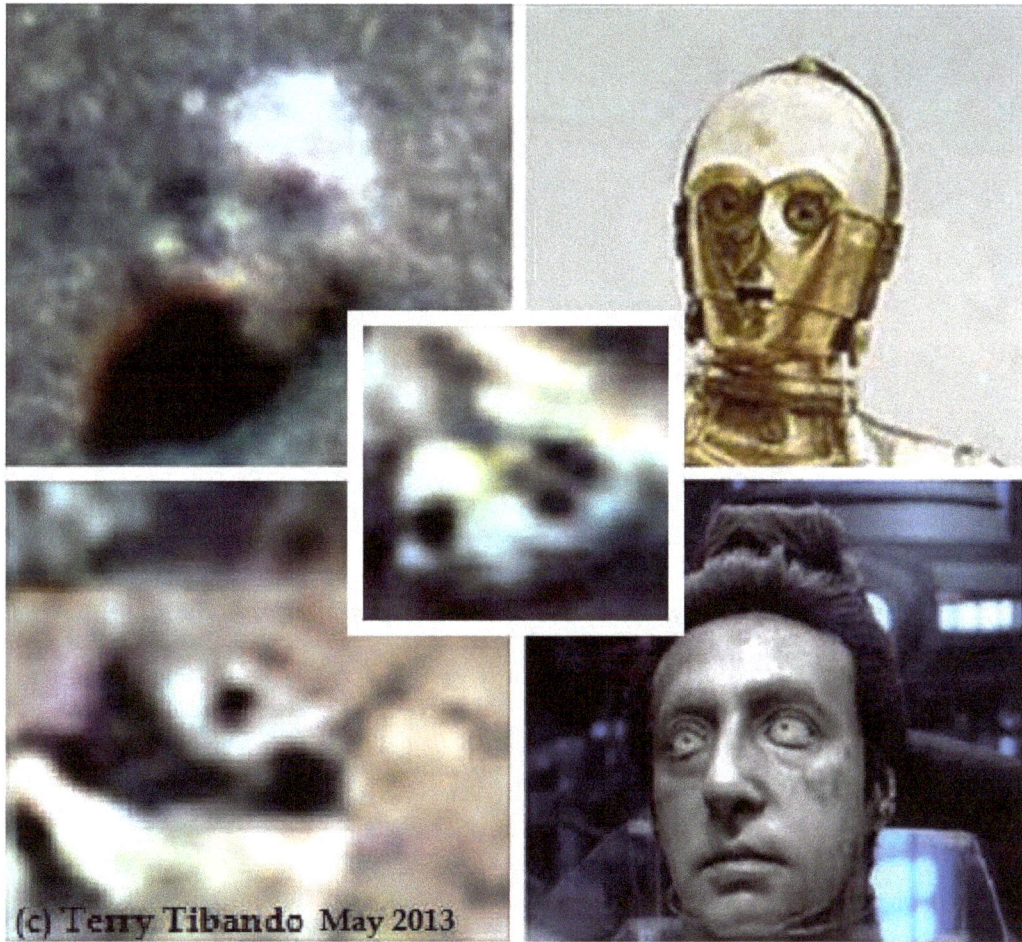

The Robotic Heads of science and science fiction. Top left, the robotic head dubbed "Data's Head found on the Moon during the Apollo 17 mission with the head of C3PO from Star Wars at right. Below left, the robotic head dubbed "Lore's Head (by the author) found on Mars with the head of Data (STNGat right. Centre, the head or face mask found close to Lore's head

(c) Terry Tibando (Google Images)

(C) Terry Tibando May 2013

**This second head was also found in the Shaler Martian Apartment area
Note the similarities to the other heads above**
(c) Terry Tibando

The implications of such discoveries on two planetary bodies is staggering, to say the least; it is one more major proof of a previous Extraterrestrial presence in our Solar System.

But we are not done yet! The evidence so far indicates a humanoid intelligence creating technology in its own image as well as in visages familiar to Earth humans. The parallels are overwhelming that Martians may very well resemble human beings! There is yet, still more proof which will finally nail the proverbial NASA coffin closed on the subject of whether there is life on other planets!

NASA comes through yet again, in mute silence and no doubt with a silent outrage and frustration as the public discloses what it has failed to do with more raw data and photographic images. One can only imagine the free fall in spin doctoring any future public press releases of information or disclosure.

It would seem the public is doing the disclosure of information of the new discoveries that Curiosity and the other rovers are finding and photographing. In some ways, if the public release information over the internet and on YouTube, there is a strong possibility that such information release will contain errors and with many errors, NASA can still come back and re-assert itself and its position of official authority on matters of space exploration and on any discoveries found on Mars.

We have seen many things that in this author mind are artificial beyond any doubt or contestation that must be taken seriously by the scientific community for further investigation by the roaming

118

Mars rovers and by any other country wanting to send their own land roving autonomous vehicles to Mars.

We have poven with photographic evidence taken from NASA's very own photographic files from JPL and MSSS that an intelligent civilization once existed on Mars creating megalithic structures miles in size, mammoth size pictographs that can be easily seen by space satellites, there have been satellite photos of desert and polar Martian cities that resemble many of the ancient cities found on Earth in Egypt, in the Middle East, and in Mesoamerica. Many of these Martian cities are connected to ancient civilizations found on Earth but,even more amazingly, these ancient cities on both planets, also share an interstellar connection to certain constellations and star systems that are easily seen in the night skies, whether on Mars or Earth!

Constellation maps (when proportioned) of the Pleiades, overlay perfectly with the **Pyramid City of Cydonia** and certain pyramid in Cydonia and the Giza Pyramids align not only with each other but, also align perfectly with the the three stars of the "belt in the constellation of Orion!

It is a concept that moviegoers will recall from the movie "2001, a Space Odessey" where the *monolith* is used as a marker and silent educator which leads mankind from the Earth to the Moon, where another monolith is unearthed that points to Jupiter and beyond. These ancient planetary pyramidal and megalithic structures on Earth and Mars were constructed to the point of redundancy. These were massive mnemonic devices constructed to jog humanity's memory to recall its very ancient past and as celestial markers guiding mankind forward from one planet toward another and onward to star systems within major constellations.

Everything we have examined so far has been postulated to prove that there is an Extraterrestrial presence in the our solar System. There is only thing remaining is should be axiomatic to any reasoning individual who has followed the evidence contained in this book. The final proof of a Martian intelligence is the presence of their physical being, whether dead or alive.

It would seem that there is hardcore physical evidence to support both states of reality!

A Martian Gives NASA, the Finger!

The **Mars rover Curiosity** in its continuing travels across the ancient seabed terrain, now known as **Gale Crater** routinely stops and photographs its surrounding and samples some of the Martian soil in its search for water, microbial life, and rare and common earth elements. This time, an astute and keen-eyed anomalist researcher spotted something that may indicate a humanoid presence on Mars. Below the right front wheel of Curiosity among the small pebbly rocks was something that looked out of place and somewhat organic in appearance.

Closer inspection revealed what looks like a human digit, to be more precise, a human or humanoid finger! (See images below). **Mars Rover Curiosity** Mast Cam manages to capture a couple of pictures of the "finger" with a "nail".

Using Curiosity rover wheels as a base of measurement which are 29.52 Inches wide, a human finger is about 3.5 inches long therefore, the fossil finger appears to be *roughly 5-6 Inches in length* which means if this was from either a living Martian or statue of a Martian, it would stand somewhere between *10 to 12 feet in height!*

Curiosity's Mast Cam photographs what looks like a humanoid finger in its exploration of the Gale Crater terrain. Is this a fossilized finger from a human being or Martian entity or is it from a statue?
http://www.ufo-blogger.com/2012/08/mars-nasa-curiosity-photographed-fossil.html
and https://www.youtube.com/watch?v=iZ3jcbKY2q4

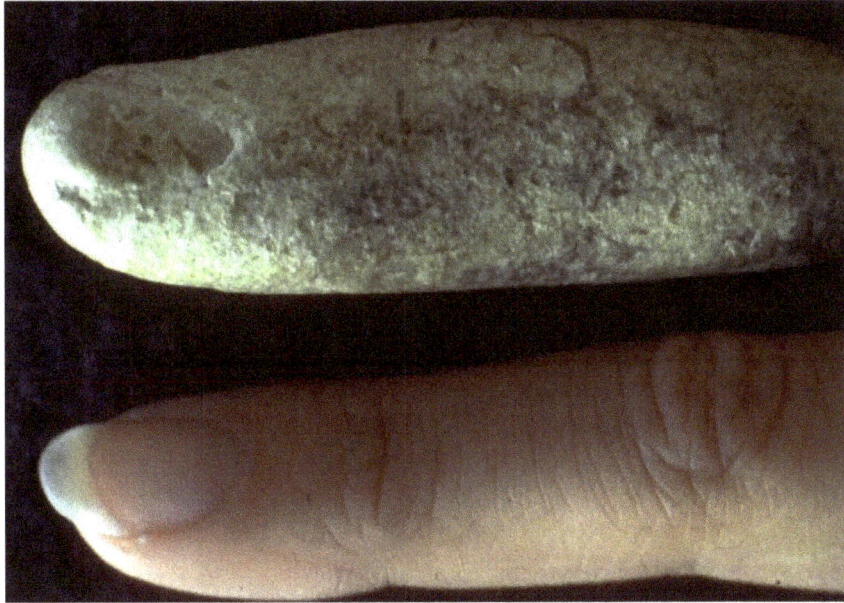

A human finger next to fossilized human finger. Now compare this to the above photo images of the Martian finger

-

Bearing in mind that our proof comes from raw scientific data and photographic imagery of the Martian surface taken by orbiting satellites and Mars rovers from space agencies like **NASA**, the **European Space (ESA)** and the **Russian Space Agency (RSA).** Any book or website on the subject of life on Mars is based chiefly upon photographs that the investigator was able to obtain directly from these space agencies or from other researchers. It is this author's opinion that the best photographic evidence on the many subsets covered on this topic has not been disclosed to the public as yet. Such evidence unquestionably will contain high-resolution photo images revealing in exquisite detail, artifacts both organic and inorganic, that would lay to rest any doubts that life existed on Mars not only in ancient times but may in fact, still be in existence today.

Humanoid Skulls Litter the Martian Terrain.

The evidence for an intelligent being, a Martian having lived on the **Red Planet** comes from the skeletal remains found by the various rovers that have travelled across the Martian terrain photographing everything in sight. Without actually seeing a Martian corpse, it is a sure bet that they appear to be human or humanoid in appearance based on the many monuments, glyphs, and statues that reflect their visage as photographed by the Mars rovers and orbiters.

The next set of photographs display images of what appears to be humanoid skulls which in comparison with typical terrestrial human skulls appear to be much larger. This may confirm the theory that Martian ET beings are as tall as 10 to 12 feet in height.

121

The **Spirit Rover** and its brother rover, **Opportunity** were the first twin rovers to have simultaneous missions traveling across the Martian terrain imaging many features that NASA thought interesting enough to get a closer look at. About 15 months into its journey Spirit took a photograph that appeared to be a human skull. Staring up out of the Martian soil was a half-buried head in a debris field that should not be there, yet there it was caught by the onboard Spirit camera on **Sol Day 482**. The black and white photo image below tells the tale or maybe it doesn't because NASA official will tell us that this is merely a trick of light and shadow, mere pareidolia. Nothing to see here folks! Move along now!!

As it turns out, this wasn't the only image of a human skull that Spirit would photograph. On **Sol 513** near **Gustev Crater**, another humanoid skull was found. Although, the skull has the characteristic dome feature of the head with round eye sockets and there is a hint of a nose that protrudes partially above the mouth area. But, what makes this skull a little unusual are the two cylindrical protuberances, one on either side of the lower jaw area. This gives the overall appearance to the skull as being, perhaps a statue or a perfectly formed helmet with breathing apparatus attached to its sides.

Photo taken by NASA's MARS exploration rover SPIRIT Sol 482
Image courtesy NASA/JPL-Cornell

Spirit Rover captures an out of place image of a human skull on Sol 482
https://mars.nasa.gov/mer/gallery/all/2/p/482/2P169153230EFFAAB2P2417L7M1.HTML

Joseph Skipper of **Mars Anomaly Research** thinks that the photo image has been tampered with to remove incriminating evidence of a possible "flight suit" buried or covered up with a diffused fuzzy quality using a software application by JPL. If you look just in front of the skull in the enlarged photograph, the earth appears darker and fluffy indicating tampering to cover up more of the "body or flight suit" which would certainly be proof of artificiality or proof of the organic nature of a human body. Either case would be positive proof of a former Martian life upon the planet. (See below). http://www.marsanomalyresearch.com/evidence-reports/2006/102/mars-humanoid-skull.htm

This tampering and smudging or smearing of objects in many of NASA raw photographic data makes one wonder just how many photographs may reveal artifacts of artificiality or of an organic nature. Unfortunately, we will never know as the original, unadulterated photo images have not nor is it likely, will ever be released into the public domain. Censorship is in full routine practice at NASA's JPL photo labs and this allows them to sanitize any revealing pictures before the public has access to them thus, fulfilling its obligation by government order to release all images taken by all rovers and orbiters. The problem for the public investigator and researcher is which photos have not been digitally tampered with.

Black and white photo appear as rocks and debris; auto colour corrected image right and below shows the skull with dark brown earth in front of skull indicating image tampering
https://mars.nasa.gov/mer/gallery/all/2/p/513/2P171912249EFFAAL4P2425R1M1.HTML
and http://ancientmistery.weebly.com/mars--1.html

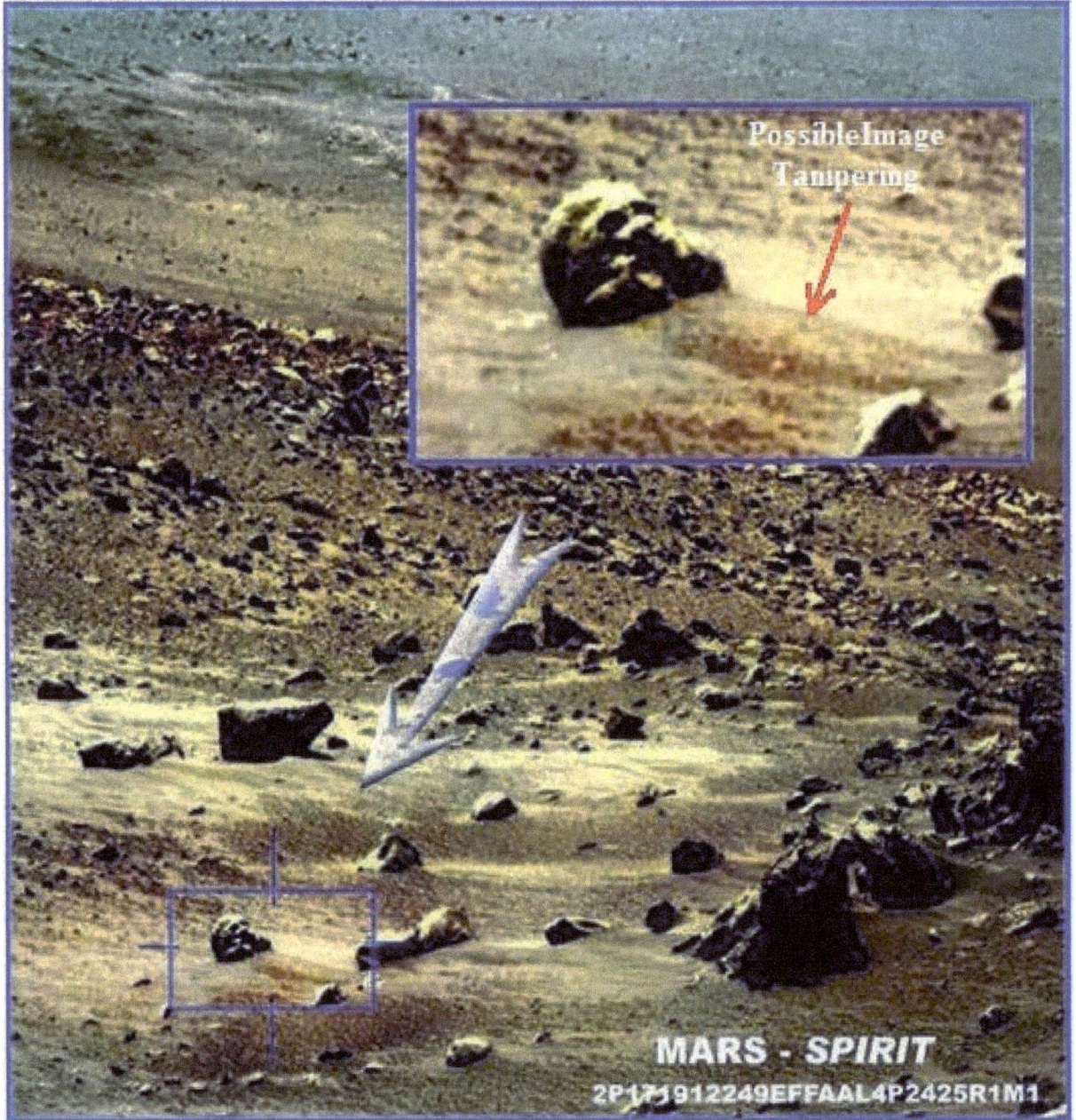

An enlargement of the "skull/helmet" showing NASA photo tampering in front of the skull

http://ancientmistery.weebly.com/mars--1.html

August 23, 2012, Mars rover Curiosity looks across the Gale Crater terrain toward Mount Sharp and captures another human skull with its Mast Camera.
https://mars.nasa.gov/multimedia/images/?ImageID=4565&s=2 (c) Terry Tibando

There are many images of human-like skulls being discovered by the Mars rover, some images are questionable as to their authenticity; others are unusual as they appear to be humanoid but, not human. Perhaps, there was more than one ET race co-existing on Mars with a human-like race in ancient times. What is undeniable is that this is proof positive that Mars did support an intelligent life form in its past.

The photo image below comes from the NASA - JPL website but was first found on a Chinese website and has created a firestorm of controversy since it went public. It shows a clothed humanoid appearing to be walking downhill; other researchers suggest it is a statue in a posed position. Whatever it may be, it represents either artificiality of intelligent design or an actual Martian although, skeptics in typical debunking fashion would state that it's just a strange rock formation. Speculation abounds! https://www.youtube.com/watch?v=RacHZpYXhU8

A panorama image taken by Spirit on Sol 1366 – 1369 of "Home Plate" plateau inside Gusev Crater also know as West Valley. Is this a Martian ET in a hooded robe or a small statue created by a sentient being of Mars?
https://photojournal.jpl.nasa.gov/jpeg/PIA10214.jpg

Author's Rant: The best evidence has been saved for the last! To date, these photos have not been posted on any website, blog site, or on YouTube or anywhere else at the time of this writing so, I believe that this is the first time it has been revealed to the public with the exception of what may be a Martian spacecraft or land speed cruiser. If this author has

correctly interpreted the raw photographic data released on NASA's JPL website then this is the "smoking gun" or the "atom bomb" that proves we are not alone in the universe or in this solar System!!! Nevertheless, this author is open to other researchers' input and further independent re-evaluation of the photo images.

Needless to say, the naysayers, professional debunkers, and skeptics will come out of the woodwork to discredit this evidence and any official, scientific rhetorical explanation by NASA, no doubt will be used to disprove this author's claims and conclusions as pareidolia or something quite natural, non-organic or non-artificial in nature. Given, however, NASA's credibility these days and the growing public mistrust for any honest disclosure by NASA with its less than truthful explanations of recent discoveries, I am confident that my personal discoveries will hold up under close scrutiny.

With so much counterposing evidence by UFO and anomalist researchers supported by many independent scientists, all refuting NASA's official position, the overwhelming evidence is conclusive, this is the *"final nail in the coffin"* of NASA!!! (See images below).

A Red-Letter Day for NASA – Discovery of a Martian Cadaver, Alien Technology, and Martian Creatures!

The images below taken by the Mars rover Curiosity found in **Gale Crater** on Sol 107 could unequivocally be said to represent a *"red letter day"* in NASA's space exploration on another planet with profound implications for this world in general. This discovery also has historical implications for all of humanity that are far reaching. It not only answers that haunting question that has followed us down through the ages: *"are we not alone in the universe?"* it confirms that *we are not and never have been alone in the universe!* It also confirms what astronomers from ancient times up to and including the last two centuries have always wondered and speculated that Mars is the twin to the Earth because they are so vastly similar to each other in so many ways. More importantly, the images presented indicate that Mars once had a thriving highly intelligent civilization that may have rivaled our own current terrestrial civilization, but it may have actually exceeded it by hundreds of thousands or even millions of years!!!

There is also every indication based upon NASA's raw photographic data that Mars still hosts many living plant and animal life forms, a hard concrete fact that contradicts the hypothesis which means that Mars is not a cold, dry, barren, dead planet devoid of life as was once thought but, that Mars is still inhabited and home to a race of highly intelligent, sentient beings that we call the Martians!!!

Martians or whatever, they call themselves are no longer exist just in the imaginative realm of science fiction writers but, they exist in the stark naked truth of reality!

Examine the photographs carefully; note the comparison images of Earth type beings and technology and draw your own conclusions. The images are real and undeniable, however, further investigation and analysis are still required by all fields of science to conclusively prove that Mars is a living planet inhabited by a highly intelligent civilization!

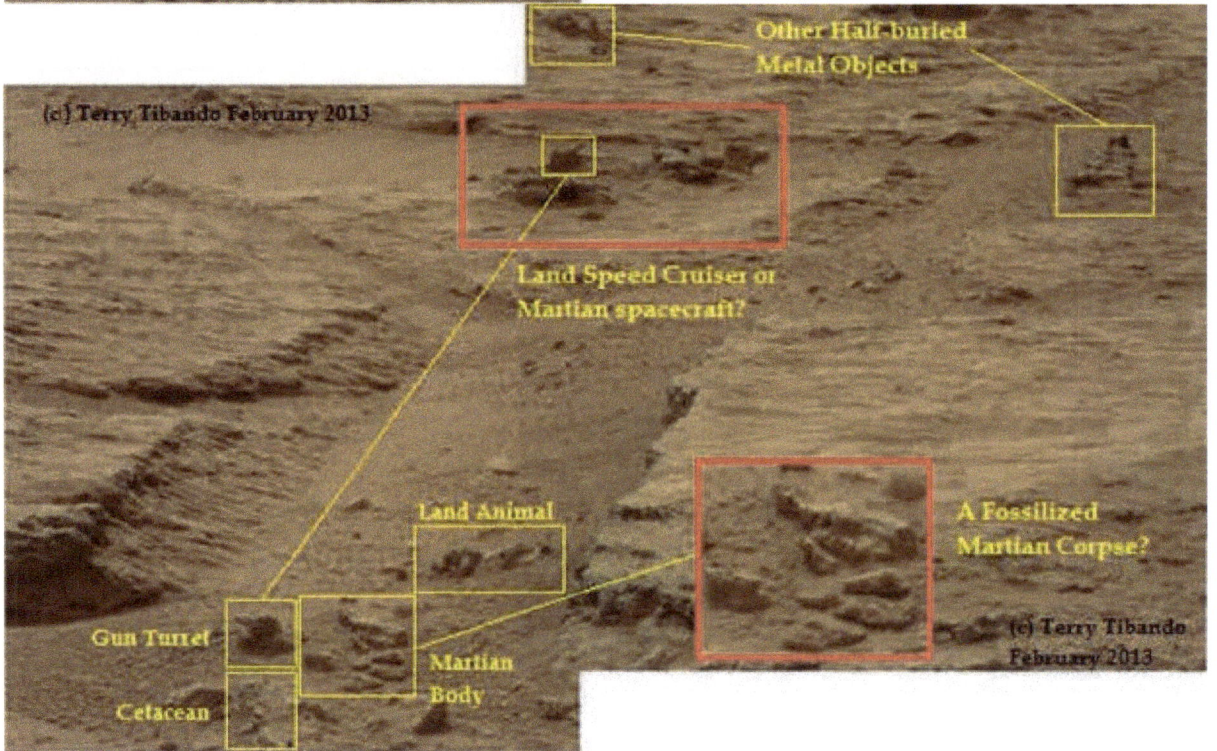

The Mars rover Curiosity on Sol 107 photographs more than just rocks.
Note the metallic objects in the background. The real discovery
in the foreground is a mummified humanoid cadaver!
(c) Terry Tibando

A Martian Land Speed Cruiser or a Spacecraft or Possibly a Mini-Sub?

In the above mosaic photo images, the first thing that captures our attention is some dark anomalies in the background that have various curves and straight lines that seem to be out of place with the natural surroundings of the dry lakebed of **Gale Crater**. It screams out at us to look closer for signs of artificiality. When the photograph is colour corrected and enlarged, we can see that the dark anomalies are twisted and corroded pieces of metal of considerable size constructed with an intentional functionality of design. The metal appears to have suffered from extreme heat over its surface causing the metal to bubble, melt and become distorted.

Many things jump to mind as the first impression that one gets from the image is that it is some sort of transportation vehicle. As to what type of transportation vehicle this could be, given its final resting place is located in a large crater that use to be an ancient lake or sea bed area, this could be a boat or a submersible craft, perhaps a one man or two man submarine.

There is some kind of turret at the front of the device and a type of rocket-like nozzle or engine in the rear of the device. Opposite to the turret in the forefront is a curious square hole which looks as if something is missing from it. (See images below). This is an important aspect to keep in mind as we will return to it shortly.

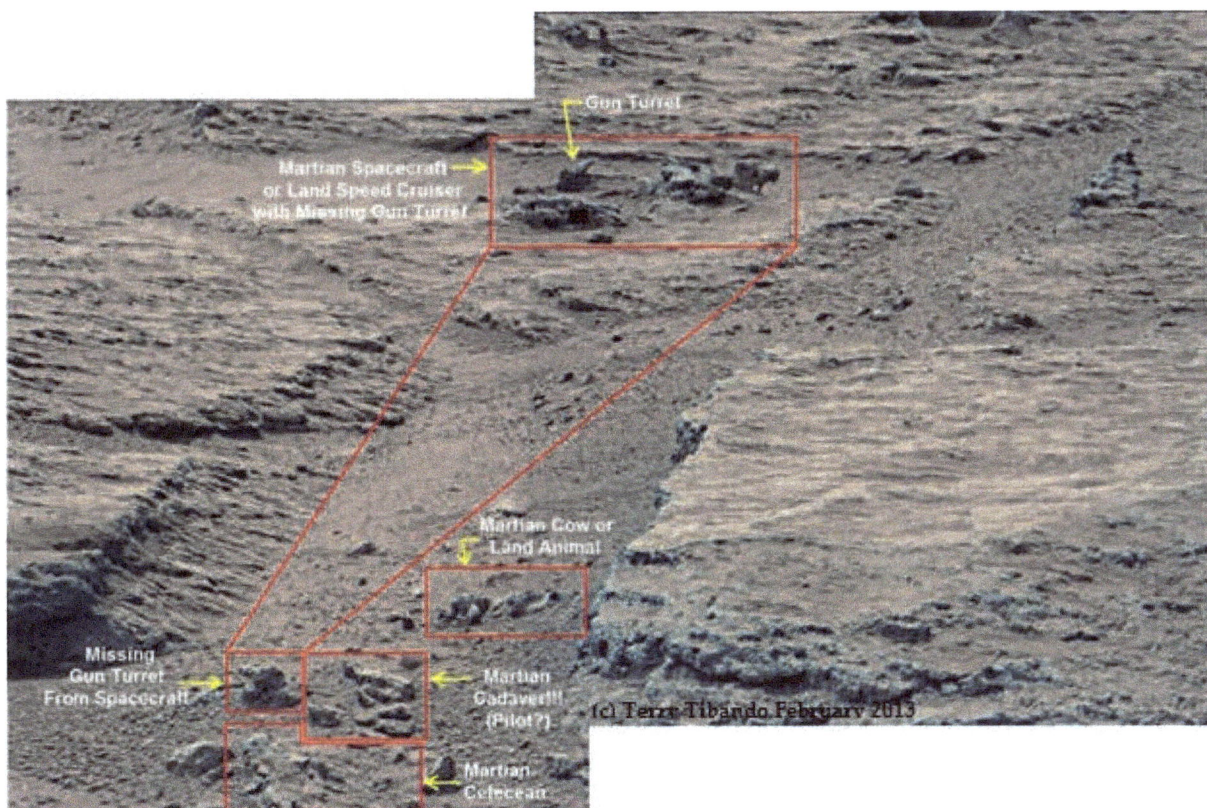

Auto colour corrected image of lakebed in Gale Crater. It shows a possible spacecraft and mummified human remains indicative of high intelligence and technology, all in one place!
(c) Terry Tibando

Note what looks like the engine or propulsion system in the mid-rear section and the nozzle-end appendage that resembles a jet or rocket nacelle. In the middle section, something appears to be missing or eroded away or has disintegrated upon impact. Toward the starboard bow or right front of the craft there is a "gun turret" protrusion that may be a weapon (cannon or laser?) of some kind and on the port bow or left front of the craft there is a number of holes and one large square orifice, an obvious geometric shape of intelligent design. Within this square hole, a smaller square or rectangular shape can be seen. This opening suggests that something fits in or over the aperture which is now missing. Below, we will discover that missing piece of the craft beside the Martian cadaver!

Is this rusted metal debris a spacecraft, a land vehicle or a submarine? Note the tail section appears to have a type of nozzle similar to a rocket nozzle and the gun turret object on the north side of the craft and the square opening on the side
(c) Terry Tibando

A colour corrected image provides greater detail indicating that this melted and twisted object was manufactured by an intelligence many millennia ago.
(c) Terry Tibando

The overall impression is that this chunk of twisted metal is some type of vehicle that resembles the land speed cruiser from the first **Star Wars** movie that carried Luke Skywalker and Obi Wan Kenobi across the desert toward Mos Eisley on the planet Tatooine. (See image comparison below).

This altered coloured image shows clearly the small light coloured square within the large square aperture where some part of the craft is missing
(c) Terry Tibando

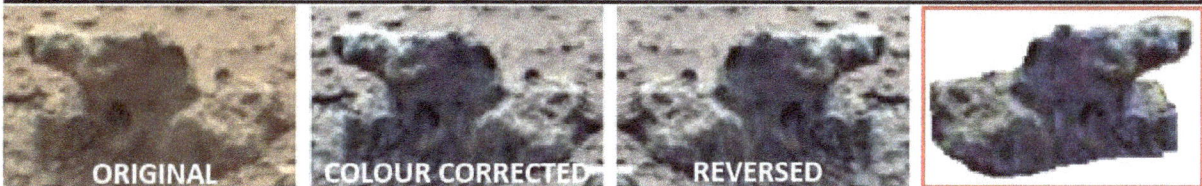

When the turret- like object located by the Martian corpse is colour corrected, horizontally rotated, then cut and pasted into the square aperture of the "spacecraft", the new fit appears to be a natural component of the craft adding to its symmetry!
(c) Terry Tibando

Compare Luke Skywalker's "land speed cruiser" from the movie Star Wars to the above images of the Martian submarine/land cruiser/spacecraft above.

https://brainsnorts.com/2010/07/13/a-small-star-wars-issue/

However, the vehicle could just as easily be an aircraft or a possible a one or two-man spacecraft, one individual as the pilot and the other as the craft's weapons specialist as there appears to be a "**gun turret**" at the starboard bow of the craft. The craft may possess the capability of all three modes of travel... submersible, land and air/space similar to the aerodynamics and functionality of ET spacecraft reported on Earth that can fly in space, hover above the ground and submerge into the ocean, this craft may also possess those same qualities. Bottom line is that this is a craft designed by some high intelligence for the purpose of transportation and travel.

Now, if we go back to the mosaic panorama photo image above we see that NASA's original orange (false colour) photo images makes things difficult to perceive their true reality. Without some sort of colour correction or alteration, the tendency is for the eyes to miss the foreground objects and simply conclude that they are just another bunch of rocks. The author missed these anomalies on the first inspection of the image due to NASA's constant insistence of making all raw photo images a false orange- reddish colour to hide anomalies that could have been easily spotted if they were displayed in their true colours. This deliberate ploy by NASA keeps many obvious anomalies hard to hide but by altering the overall colour one can easily *"hide things in plain sight"* which has become a routine practice by NASA. In this author's mind, there was something familiar about these "rocks" which the subconscious recognized on some basic level due the symmetry and pattern of lines.

This ploy of hiding things in plain sight may explain why no one else to date has reported this particular anomaly or posted it anywhere on the internet and why this author discovered it back in February 2013 and has sat on it for two and a half years, waiting to see if anyone would confirm the same discovery; to date (July 2015) they have not.

Again, when we adjust the colour and enlarge the photo image, the anomalous "rocks" take on their true characteristics that being that some have organic structure along with a surprising

132

metallic object nearby. The apparent "rocks" in the middle of this small debris field have definite recognizable structure, uniformity, and consistent symmetry indicating an organic quality. It is, in other words, a body of an intelligent humanoid being, that of a fossilized **Martian corpse**!!!

Surrounding the mummified Martian corpse, there is another metal object to the left of it that looks a lot like another "gun turret" sitting on the ground, could this be the *"missing something"* from the spacecraft in the above photo? ***Comparison of the two "gun turret" objects shows amazing similarity to each other***. By repositioning the gun turret found beside the fossilized Martian body next to the square hole area of the spacecraft, we see it seems to be a natural fit adding to the basic overall symmetry of the spacecraft. It could also indicate a second craft may be buried somewhere nearby or that it came from the same craft in the above image. There is reason to suspect that other devices or craft are buried under the sandy topsoil as there are metal structures that can be seen poking through the surface. (See the highlighted colour corrected panoramic photograph above).

A Martian Land Animal?

To the top right of the corpse (see photo below) is a very unusual shaped anomaly that seems to be organic in nature than it does as some kind of artificial manufactured device. There appears to be two appendages at what may be the head area at the left which could be horns or these appendages could be hindlegs, in which case the head is on the right side which seems to have a snout and large ear cavities. The author photographic analysis is that this is some sort of land creature and not just a pile of some unusual rocks. Comparison of this **Martian creature** to many similar terrestrial animals, we find the closest resemblance is to that of a **South American Tapir**!

(c) Terry Tibando February 2013

Could this be a large, fossilized land-dwelling animal (Tapir?) with head it's to the right and its hind legs to the left or maybe, it's the other way around or is it an aquatic creature?
(c) Terry Tibando

Compare this large Tapir from the South American jungles to the Martian creature in the above photograph. Is it possible to have two similar animals living on two neighbouring planets? What could be the reason for their similar evolution?

http://subject.com.ua/textbook/geography/7klas_1/22.html

A Martian Cetacean?

Below and slightly to the right is another creature that seems to have a marine quality appearance, unlike the first creature which has vertebrae and skeletal structure to it overall appearance. This creature looks flat with perhaps, a cartilaginous structure giving the impression that it is a marine animal. It has appendages that could be flippers or fins at one end, presumably near the head and the ridge area along the back could be a large dorsal fin, long since decayed away. The head area is unusual in that it appears to have a face or sorts but this could be due to decomposition of the animal. Could this be a **Martian cetacean**?

(c) Terry Tibando February 2013

Could this be a Martian cetacean or a pile of rock? This fossilized creature appears to have a dorsal fin, flippers and a possible blowhole on the top of its head, its basic shape looks similar to a pilot whale as seen below
(c) Terry Tibando

A dead pilot whale washed up on the shore on the east of the United States
Google Images

A Giant Mummified Martian Humanoid Cadaver!

This leaves us with the most important feature of this incredible find and that is a giant mummified **Martian Cadaver** in the centre of the photograph which from appearances seems to have suffered a lethal blow causing immediate death from dismemberment and thorax damage that split the rib cage apart exposing the thoracic organs to the elements. The lower section of the body from the abdominal region and waist area including the lower extremities, the legs and feet are missing or incomplete perhaps, due to animal predation.

What remains of the humanoid is head and neck, the right humerus, the left thorax breastplate of the chest cavity and partial abdominal region, the right side of the rib cage blown away but, near the right arm, bits, and pieces of leg bone mixed with body parts from the cetacean. The head is complete with eye sockets, nasal area, mouth and lower jaw supported by a long, strong neck structure. The impression is that this humanoid is probably 10 feet in height, even though the legs are missing to give a more accurate estimate! There is no doubt that other body parts are in this area but, they are probably buried by silt and sand from blowing winds. This area should be re-imaged again with high definition cameras either from another rover or on a future mission using a remote control aerial device or plane.

The odds of finding an advanced piece of technology like a spacecraft or land vehicle still partially intact is extremely high, but the odds of finding the pilot of the craft along with marine and land animals all together in one site, along with other devices of highly advanced technology nearby is absolutely astronomical!

A question that immediately arises is there an association of this Martian being with the **"spacecraft"** that rests a short distance behind it? Could this ET being be the pilot of this craft and if so what happen to him? What circumstances caused a Martian in a flying machine to end up in an area along with a large marine creature and a land animal, only to meet their demise by some terrific external force? What was the cause of his death and what happened to cause the massive devastation in this area and on the rest of Mars?

The answers to these questions may reside with the artifact that lies close beside the Martian cadaver that being the "turret" like object which looks remarkably identical as the "gun turret" attached to the "spacecraft"!

As stated before, when we cut and paste this turret device onto the spacecraft, we see that it seems to be a natural part of the craft. This means that the shock waves and shearing forces generated externally that were required to disintegrate the craft, blowing pieces of it in every direction including a lethal blow to the pilot, sending his bodily remains out and away from his spacecraft had to be of immense power. We can deduce (although, somewhat prematurely without any hands-on forensic investigation) that part of the force that devastated the planet came from the north by northeast direction based upon how the Mars rover **Opportunity** took the photographs. However, if the crater was a still an active lake or sea then, the blast and the resulting debris may have been carried by current and/or tidal forces distributing the spacecraft debris over a wide area on the sea floor. This would include the death of any land animals that were knocked off their feet and hurled into the water, ending up near other dead aquatic life

A fossilized Martian cadaver with body parts lie in a dry lakebed surrounded by other creatures and a turret-like device similar to the turret on the spacecraft above
(c) Terry Tibando

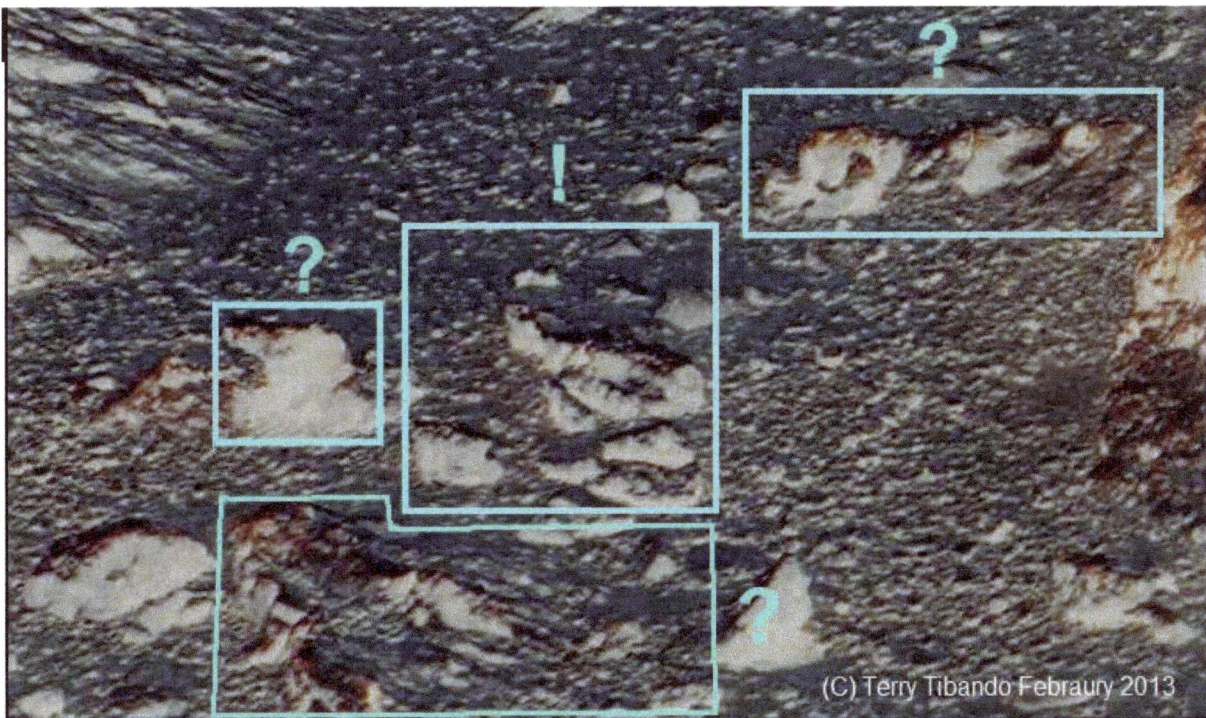

This negative colour image shows shape and shadow detail of the mummified corpse and the other creatures. (c) Terry Tibando

forms and the remains of a humanoid pilot and his spacecraft. Speculation, certainly! But what other reasoning logically fits this debris field location and other debris fields in **Gale Crater**?

The set of photo images both above and below have been enlarged, colour corrected and have also been colour inverted (negative colour) images of the Martian humanoid mummified skeleton to show body detail. Included are photos of human skeletons and a mummified skeleton presented for comparison purposes. In the photo of the prone mummified skeleton, one should note that there is still flesh intact upon the hands and legs indicating that this individual probably died in a very dry area where moisture and predator animals were not present when the person died. This would also appear to be the case with the deceased Martian with the exception that animals were present when this being died.

 It is also worthy to note that **Joseph McMoneagle** who was one of the remote viewers of the secret military SRI **Project Stargate** had been given a set of coordinates in which he remote viewed the "**Face**" on Mars as well as tall beings who still live on the planet. He drew a pencil drawing of what he remote viewed and this drawing is included beside a sculptured bust of a Martian being and it in turn is beside The raw **NASA** image of the mummified skeleton found by this author. The similarity of these images is more than coincidental but indicate a synchronicity between the power of the mind and actual tangible photographic evidence from the US agency, Nasa.

Frequently, NASA is not forthcoming on what it finds on the Martian surface from its many Martian rovers and orbiters, but it is obligated to place its raw photographic images out into the public domain. These images may be smeared, digitally altered by "rock placement" over an object of interest, or frequently, altering the Martian terrain and sky with a false red or salmon pink colour, in conjunction of hiding things of interest in plain sight, sometimes with blurring of the image. Such practices by NASA make anomaly investigates overlook something or create a situation of **pareidolia** of some rock formation which doesn't exist or is use as a cover-up of an anomaly of interest.

The reader or anomaly investigator should be very careful before jumping to conclusions as to what they think they have discovered and compare it to more familiar terrestrial things before announcing their discovery.

It was for this reason that comparison of a human cadaver was necessary to illustrate the similar anatomy with the Martian humanoid cadaver to show this was not a mere pile of rack or pareidolia.

The reader should remember that the Martian atmosphere is very dry and very cold and if the bacteria and microbial organisms of Mars are extremely low or non-existent then, these factors offer the perfect conditions for long-term preservation of organic materials, perhaps for tens of millions of years. With a very slow decomposition process taking place as compared to the quick decomposition of organic materials here on Earth, a human body or in our case of a Martian body would decompose very, very slowly. Without an insulating atmosphere and extreme fluctuations in high and low temperatures on Mars, bacteria would stop their processes and mummification in the dry climate would take over. Thus, the high incidence of the Mars rovers discovering, whether NASA admits to it or not, fossilized or mummified organic sea life, animals and the former presence of an ancient and highly intelligent Martian civilization.

Close up (auto-enhanced) of mummified Martian corpse and body parts
(c) Terry Tibando

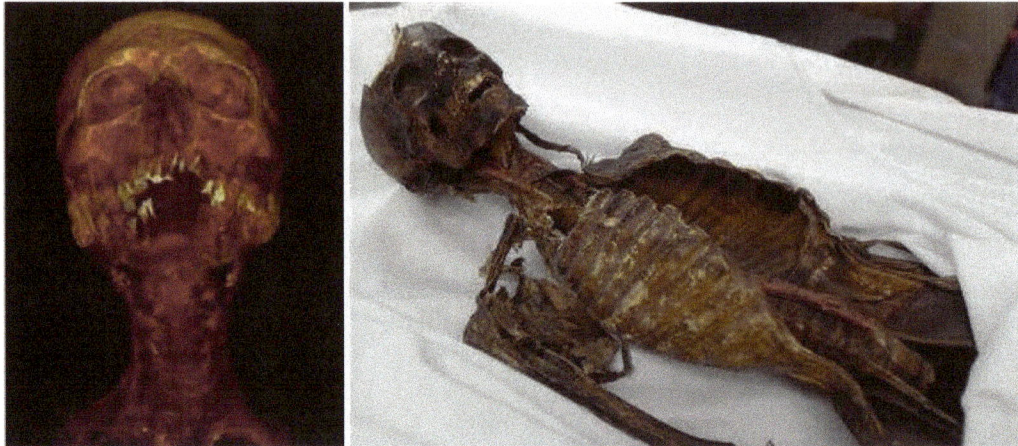

Human skeletons of head and torso. Compare this to the Martian skeleton above
(Google Images) and https://www.youtube.com/watch?v=rc4wvBqoQhE

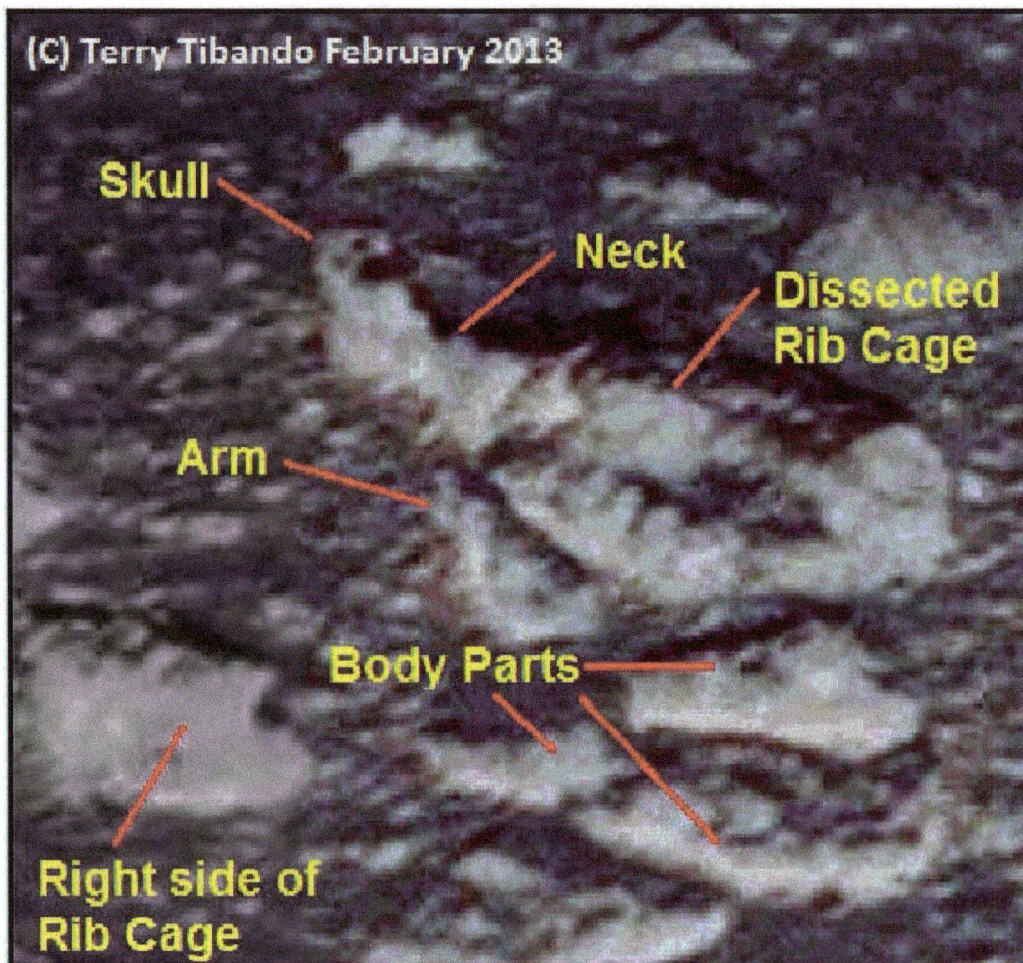

Inverted colour image detailing humanoid skeletal torso and mummified body pieces: an exposed rib cage, head with neck, a right arm beside the corpse with part of the rib cage close to the arm and other body parts
(c) Terry Tibando

**A mummified human skeleton, compare this with Martian mummified skeleton above.
Note the flesh is still intact on the hands, legs and some of the chest area**
(Google images)

A Second Tall (Giant Size) Martian Humanoid Skeleton is Discovered!

A few years after this author found the above tall humanoid Martian skeleton, a second discovery of a giant size Martian skeleton was made by the NASA Curiosity rover on Sol Day 1441in 2016 by the Mastcam near Pahrump Hill.

In the Gigapan high resolution image, there are small hills and broken rock and rubble strewn about. To the left of Pahrump Hill and to the right side of that hill is the location of the second skeleton. Upon zooming into this area can be seen a huge skeleton stretched out against the hillside.

The huge skeleton consists of a head, neck, large rib cage, spine, pelvis and long legs that can be seen which appears to have been bleached white by the sun.

The question to ask here is how did this tall Martian humaniod die? What were the circumstances that brought about the death of this intelligent being?

Based on the overall panoramic scene, where there is a vast amount of cracked rock and rubble surrounding small hills and buttes under a dry near airless climate, we must assume that either a natural catastrophe occurred or a colossal cataclysmic global war that took place among the Martian inhabitants or an interplanetary war with another extraterrestrial intelligence!

When we compare the two giant skeletons and their surroundings to the place in which they died, especially in the Curiosity photo taken on Sol Day 107, where there is a destroyed transportation device and other dead fossilized creatures, all in the same vacinity, the cataclysm was swift and decisive in its finality!

Hopefully, the implications for such remarkable similarities between the **Martian skeletons** and with the human skeleton would not be lost on anyone examining the above images. The search for ancient sentient life forms or the possibly of extant Martian intelligent life is a key element to any future Mars missions regardless of which space agency sends spacecraft, satellites or even manned missions to that planet.

141

**The discovery of a second tall Martian humanoid skeleton by
the curiosity rover on Sol Day 1441 at Pahrump Hill** uploaded by Keith Laney

With numerous humanoid skulls and body parts scattered around the Martian terrain that the NASA Mars rovers have photographed over the years since they first landed on Mars and now, two giant intelligent being have been discovered on different occasions, We must conclude that Mars once harboured intelligent humanoid life and so in the mind of this author there is not doubt that intelligent life existed and may still exist elsewhwere in outr Solar System and Mars is the proof of it existence.

When we started our investigation of the environment of Mars, we operated on a priori:
That any stable star system similar to our Solar System would have a number of orbiting planets and every planet would have creatures or life forms inhabiting it, whose number would be beyond human computation.

Such a controversial hypothesis as postulated in this section has still, yet to be proven or disclosed by science. Nevertheless, with this, a priori firmly in mind, our investigation of the Mars environment was to prove and establish that there was a variety of life forms that were or are still present on this planet. The proof of our hypothesis came with the discovery of ancient artificial structures indicative of construction and engineering by a highly sophisticated civilization, which we called the Martians. These highly intelligent sentient beings were a planet-travelling, star-faring species, whose appearance resembles us in many ways. In a long forgotten time, many millennia ago they frequented our Moon and built massive city settlements which are now in a steady decline of degradation. They also visited the Earth and may very well have altered or influenced the natural order of our evolution. They may have also created genetic enhancements or interbred with terrestrial **Homo sapiens** thus, ipso facto, this would make us the new Martians!

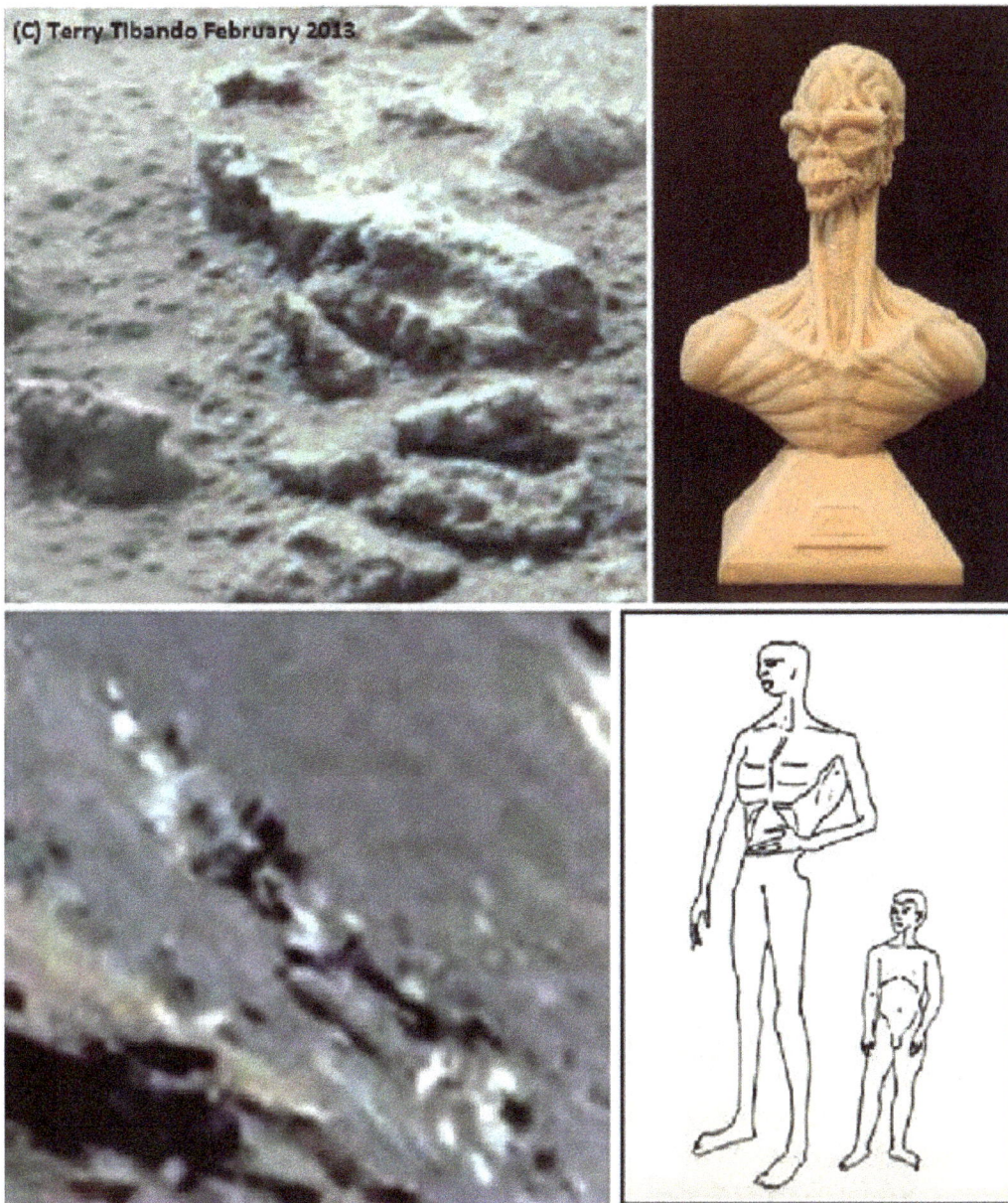

The two giant Martian humanoid skeletons taken by the Curiosity rover compared to an artist's bust of a Martian being and a sketch of a 10 foot-tall Martian beside a human by remote viewer Joseph McMoneagle from the book, Mind Trek (1993)
(c) Terry Tibando and Google Images and Joseph McMoneagle

It is no coincidence that the structures on the Moon and the ancient cities on Earth found around the planet, merely happen by freakish astronomical chance to resemble the structures found on Mars. They travelled throughout our Solar System establishing their presence on many planets and moons, where feasible to their survivability. They even travelled beyond to the stars; they were an interstellar civilization of explorers and **"colonizers"**.

This may, however, have been their downfall as a civilization!

When we look at the evidence found in the hundreds of thousands of NASA and ESA photographs of the Martian surface, one is struck by the uniformity of barrenness upon the planet with the exception of a few remote semi-desolate areas. Mars appears to have been thrown into an apocalyptic upheaval in a planet-wide catastrophe which may have been due to a natural but, unavoidable catastrophic astronomical event or due to a planetary war, either through self – annihilation in a global war or coming from another interstellar civilization, or perhaps from some massive science experiment gone horribly wrong. So devastating was the destruction from whatever was the source of the cataclysm, as to literally pulverize the the Martian terrain leaving barely any traces of its former civilization.

Author's Rant: My personable hope as an eternal optimist is that something astronomical occurred which was unavoidable and that it was not because of an interstellar war.

Whatever occurred on Mars, apparently left its devastation on other planets and their moons within our Solar System. One of these scenarios is an undeniable fact! Regardless, of which scenario occurred, there has been an ET civilization in the past on Mars and this too is now an undeniable fact. Could we be the new Martians as science fiction novelist, **Ray Bradbury** suggested in his classic novel, "Martian Chronicles"? Every piece of evidence uncovered on Mars seems to suggest that **"we are the Martians"**, if not by origin then, at least by association! We have covered a lot of territory in hunting for proof of life on Mars. We have examined scientific data and raw photographic images from **NASA, ESA and the RSA** to answer the question of the feasibility that Mars *was* and *is* a planet capable of sustaining life. The evidence we've provided runs in juxtaposition to everything the public has been told and lead to believe by NASA's official authoritarian position. It feels counter-intuitive to what we feel and inwardly know that Mars is not a lifeless, cold, dry, barren, desert, "dune-like" planet as posited by NASA.

Using NASA's own data and carefully scrutinizing the evidence,the conclusions reached were exactly the opposite from NASA's official findings. NASA's long-held position of being an agency of honest scientific information has being challenged. The moniker of **"Never A Straight Answer"** was rightfully earned. Everything we thought we knew and was being told was really nothing more than disinformation or obfuscation or just outright lies and denials. Taking a page out the CIA's or any other intelligence agency's playbook on maintaining national security, disclosure to the public news media of recent NASA discoveries were often spin-doctored or played down as just routine exploration with explanations of geological features encountered.

In reality, the evidence found and gathered from all the space agencies proves that Mars has a reasonably stable and functioning atmosphere that can sustain life, even in rudimentary form.

Methane in high quantities had been discovered in the atmosphere which was a major clue indicative of an active life form on the planet.

The evidence found, also indicates that there is water on and under the Martain surface with even more water trapped within the structure of Martian rock, more than there is on Earth. The age-old maxim: *"where there is water, there is life"*, still holds true, even on Mars! We have found water in places where NASA stated there wasn't but, only surmised may have existed at one time in the distant past.

The new maxim as we have repeatedly stated throughout this volume and in other volumes in this series that should be seriously considered especially in **exobiology**:
 "Know thou that every fix (stable) star hath its own planets and every planet its own creatures whose number no man can compute." ~ writings of **Baha'u'llah**

With two important conditions needed for the viability of life to thrive on Mars under our belt, we searched the data further and found microbial life in fossilized form and that even larger aquatic life right to large cetaceans, once existed in ancient seas, lakes, and rivers and there were also indications that rudimentary crinoids and "blueberries" may still flourish in certain regions of wet Martian terrain.

The spirit of adventure to find more complex life forms now became the driving force behind many amateur Mars anomaly researchers, as we came upon many images of sand dunes in the polar areas of Mars which NASA said had active carbon dioxide ice or water ice geysers and at first observation that is what they appeared to be except for one glaring detail. Where some areas certainly were active with CO2 geysers, other areas displayed organic structural symmetry resembling various types of trees one would expect to find on Earth. From **gymnospermous** (conifers - pine trees) to **deciduous** (broad leafed and shedding) to **tropical** (with bulbous trunks) in appearance and they were not all in small groupings but, in massive forested areas, some beside lakes!!

Thus far, our discoveries in which NASA remained silent on were now taken on their own momentum as we probed for possible sea life and animal life. If there were seas, oceans and lakes then, there must be sea life. A simple enough proposition. **Crinoids** were the first clue to possible sea life followed by sea shells and trilobites, commonly found on the beaches of Earth. When NASA found one of these crinoids, it decided to grind it out of existence with the drill from its Mars rover! No crinoid, no evidence of sea life!! But such nefarious actions cannot hide large creatures like fish skeletons and even larger fossilized cetaceans found in Gale Crater and other old dried Martian seabeds.

We then, went searching for evidence of animal life and almost immediately, we discovered a debris field that could only be best described as a bone yard or a graveyard of animal corpses and skeletons. Animal skulls and pieces of bone ranging from cats or felines, reptiles, birds to simian and hominids. Such areas would be any biologist's dream work site, yet NASA moves silently on. We then came across large reptilian skeletons reminiscent of ancient dinosaurs unearthed on our own planet. Such finds were staggering in their implications and we had proven that not only simple aquatic and land-dwelling life existed but, highly evolved complex life forms swam in ancient seas and strode across the Martian landscape.

If this was a baseball game, we would be batting and pitching a perfect game!

We can now take a measure of pride and prove what NASA could not officially disclose, that ancient life did exist and that some types of sea and plant life today, still survive. We also considered the possibility, could actual living creatures still inhabit the Red Planet? The answer came when the Mars rover **Opportunity** touched down in a sandy crater on the **Meridiani Planum** landscape and photograph a small white object that was dubbed the "**White Bunny**" because it looked like a rabbit with long ears.

NASA explained that it was a bit of wind-blown **Kapton tape** from the airbag of the lander. The problem with this explanation was that the "tape" had a habit of moving around and multiplying in number and then completely disappearing altogether. When it reappeared again, NASA decided to run it over with the wheels of Opportunity trying to remove its presence once and for all! If it wasn't a piece of airbag tape then, why was the "White Bunny" photographed hi-tailing it out of the way of the wheels of the rover?

This author quickly realized that this was a live Martian animal similar to an **Australian Bilby** found on Earth and because this Martian cousin to an Earth rodent had long "ears", it was re-named the *"Martis, Animalia, Chordate, Mammalian, Marsupialia, Peramelemorphia, Thylacomyidae, Macrotis, M. Leucrura, Tibando"* or the *"Tibando Long-Eared Martian Bilby"* also known as, the *"Long Ear Martian Bunny"),* after the author.

This was probably the first real evidence of a living animal discovered on Mars which soon became a serious search for other possible living creatures. This was quickly followed by another Martian desert inhabiting creature, the *"Martis, Animalia, chordate, Mammalia, Marsupialia, Australidelphia, Notoryctemorphia, Notoryctidae, N. Typhlops (N. Caurinus), Tibando"* or the **"Tibando White Marsupial Martian Mole"** aka. the **Martian White Marsupial Mole.** It wasn't long before small lizard creatures were found; even a questionable rodent and other unusual slithering creatures. This was now looking real bad for NASA's credibility that so many amateurs were making discoveries from their armchairs at an expense of pennies an hour in research time compared to NASA's billions of dollars spent on land roving vehicles with high-tech cameras that supposedly had not turned up anything unusual enough for NASA to announce to the public or press.

The big question on the minds and lips of all anomaly researchers and UFO/ETI investigators: Is there any signs of a former intelligent life on Mars?

So far, Mars seems to be "living up" to our challenge and is delivering on all fronts of our investigation, regardless of what it is that we were looking for. The signs of intelligent life on Mars turned out to be monumental, in the very literal sense of the term.

We've discovered many megalithic artificial structures built by a highly advanced civilization, some of which appear to be still operational, others are in a state of disrepair or abandonment.

They range from the great **Face on Mars** in the area called **Cydonia**, the **D&M Pyramid**, including the **Cydonia City of Pyramids**, the **Martian Sphinx** in Ares Vallis, the **Split-Face Pictoglyphs** dotting many areas of the planet, the planet-wide **Glass Transportation Tunnel Systems** and the **Water Irrigation Tubes**, the desert and polar cities of Mars and the many collapsed apartment buildings containing many personal objects and items from former residents.

Here was evidence that was literally in your face, undeniable proof that simply could not be ignored, nor hidden physically other than by digital means. Here was evidence that shouted artificiality, all created by the hands of a high intelligence, by an advanced civilization not too dissimilar to our own.

As if this overwhelming accumulated evidence was not enough to prove a former ET presence once existed on Mars, anomaly researchers through patient and diligent search, found photographic evidence of skulls, bones and body parts of what can only be described as human beings or humanoids!

It is the final ***body of proof*** in our search for intelligent life on Mars that suggests that Mars and Earth once shared a common ancestry a long, long time ago! ***The evidence points to "Us as being the Martians or maybe we are "Comic Cousins!"***

For NASA, this evidence could be the final nail in their coffin, unless, they are at a point in their agenda to come forward with a major announcement that they have discovered life on Mars, whether existing from ancient times or currently extant.

What is odd is NASA's position of giving minimal information of its progress on Mars by continually saying, they are still looking for water and/or microbial life, when there is abundant evidence to support all kinds of life forms including evidence of an ancient intelligent life that once existed on Mars.

Even more puzzling, throughout our armchair investigation, is the deafening mausoleum-like silent from NASA on all these discoveries that have been posted on the internet in places like YouTube. What is going on with NASA and it so –called exploration for life on Mars?
It is a reasonable assumption that if amateur planetary scientists and anomaly researchers can find all these amazing artifacts and evidence of life on their own then, it stands to reason that NASA, ESA and the RSA, given all their sophisticated cameras and onboard lab equipment of their orbiting satellites and rovers, must also de facto, have found these same astounding pieces of evidence. It's no secret that these space vehicles employ some of the best military cameras and hardware to obtain greater detail with higher resolution than the public is being let on to know. Current military surveillance and reconnaissance satellites orbiting the Earth are capable of reading newsprint from hundreds of miles out or even further and can see through clouds! NASA, therefore, would have found what we have discovered long before we were able to, with their sophisticated equipment.

The Search for Extraterrestrial Technology - NASA's Real Mission!

It is no accident or random choice by NASA to set down their Mars rovers in specific locations on Mars and it doesn't appear to be in search of alien life, as they seem to go out of their way to kill or obliterate any such evidence that they literally "run across" on the Martian terrain.

With each step forward in trying to understand the nature and reality of Mars, we are challenged once more, to ask a provocative and confrontational question, "If NASA and the other space agencies know that life exists on Mars, but refuse to disclose that evidence to the public then, what is it that they are searching for? Is it ancient or extant life, a thriving civilization or acquisition of alien technology?

The clue to that answer comes from the **Apollo Moon Missions**. Every Apollo mission has been couched in a quasi-religious mysticism supported by sacred geometry, numerology (eg.19.5 and 33) and chronometry whereby every space mission is timed to astronomical alignments or astrological events. Embellished within this mysticism of space exploration are its protocols, teachings, and traditions that are obligatorily upheld by its adherents and liberally embellished with appropriate flourishes of icons, symbols, and logos by the faithful.

Richard C. Hoagland and **Mike Bara** discuss at some length in their book, "Dark Mission" that many of NASA's employees and astronauts belong to the secret orders of **Freemasons**, the **Skull and Bones** and the **Illuminati**. **Buzz Aldrin** is probably one of Freemason's best-known followers, a 33rd level member who planted a Freemason flag upon the Moon right after **Neil Armstrong** planted the American flag. Aldrin and his fellow Freemasons follow a belief system set down in ancient times that worshipped the human god-like beings of **Osiris**, **Horus** and the whole pantheon of Egyptian gods which also have their co-existent counterparts in ancient **Sumerian** and no doubt are represented in other ancient Earth cultures. These god-like beings are believed to have imparted knowledge and wisdom to promising cultures and societies on Earth that originated with their arrival from the stars and from certain constellations like **Orion, Leo,** and the **Pleiades**. Dark Mission: The Secret History of NASA by Richard C. Hoagland and Mike Bara; 2007; a Feral House Book; Los Angeles, CA; ISBN: 978-1-932595-26-0

We've seen this type of occult fanaticism and mysticism before, during World War II being practiced in the belief systems and thinking in Nazi Germany with **Hitler** and his military generals!

NASA seems to go out of its way to ensure this occult practice is incorporated into every aspect of their space program, as if to honour and worship these ancient deities of Egypt and Sumeria perhaps, believing that they may still exist in a nearby constellation and will recognize their faithful followers upon their return to our Solar System. It's a big hopeful premise to believe you can buy your way into the good graces and patronage to the deities by displaying power, wealth and control over the masses around you. We all know what happen to Nazi Germany after WWII, is NASA doomed to follow in the same self-destructive footsteps?

Hitler sent hundreds of archeological teams searching the world over for ancient artifacts, manuscripts, religious iconography, relics of spiritual or metaphysical power and long forgotten, *ancient technology!*

It would appear that NASA is doing the same thing but, with an updated version of that agenda. This time around, the field parameters of search have expanded to stellar proportions, within our

148

Solar System, which include searching all satellite and planetary terrains like the Moon, Mars and the rest of the planets for one thing only**...** See the **"Why"** question below!!!

As absurd as this may sound, let's look at this provocative possibility. Does NASA have a hidden agenda?

Recall that NASA born out of the **National Advisory Committee for Aeronautics (NACA)** through the **National Aeronautics and Space Administration Act** which established the space agency subordinate its activities to the requirements of national security, with all its apparatus of secrecy.

Whether the public wants to acknowledge it or not, there has always been a continued military influence behind many of the missions throughout the **Space Race** and beyond. All astronauts had military backgrounds or were former military test pilots until the Space Shuttle program, and there is still a very large military component in the astronaut corps. Today, it no secret that many Shuttle launches carried classified cargos; outside of **Cape Canaveral** in Florida, the military since NASA's inception has had its own launch facilities and mission control center. And because NASA follows both the *protocols of the Brookings Institute Report and subordinates its activities to the requirements of National Security,* it is empowered by dutiful obligation to withhold any information, including photographs that might compromise *national security*. (Bold italics added for emphasis).

Could there be something on Mars that could threaten US national security? Since the space agency began in earnest to explore Mars, NASA has gone out of its way to diminish public interest in all things Martian and artificial, rather than instigate the public into urging for a manned Mars mission. This would seem counter-intuitive to NASA's goals of space exploration particularly when it relies so heavily on public interest for it financial support.

Some exaggeration as used by the military and the government to spark public interest in a program or agenda would go a long way unless, that agenda is to play it safe, particularly if it's covert and thus, Mars exploration could be claimed as routine and nothing extraordinary. Playing up the geological wonders and mysteries of Mars as a **"Dune-like"** planet would, it is believed, be sufficient to fuel public interest.

Yet, public support for the space program has floundered with indifference over the decades due in part to a lack of administrative proposals that truly inspire the imagination with a sense of adventure to pushes the boundaries of science and human endeavour, as it did during the President Kennedy years of the '60s. NASA's failure to disclose all its Apollo Moon Mission discoveries, its lies and obfuscation of the 1976 discoveries of microbial life made by the Viking 1 and 2 spacecraft along with the cover up of intelligent design in the creation of the Face on Mars monument compounded further with the strange actions of **Malin Space Science Systems** with their exclusive pictures from Mars Global Observer.

Most curious are unresponsive to requests to image the Face and other interesting features at Cydonia until pressured by Congress, they have released degraded pictures (such as the infamous **"cat box"** image of the Face). Despite promises to announce further imaging of the region, they have instead *secretly examined* the area without any announcement whatsoever and the photos images have yet to be released to the public.

The distrust toward NASA has enabled a public disclosure of sorts by amateur Mars researchers over the internet of other anomalies of artificial construction found in recent years, thereby circumventing NASA's inaction and their responsibility to inform the public of all discoveries. NASA's response is to further lie, deny and mislead or to remain silent.

In examining the many missions to Mars by spacefaring nations of Earth, one is struck by the strange pattern of numerous mysterious mission failures. Approximately two-thirds of the probes sent to Mars have either failed or disappeared at least this is what the world is told and why doubt this, as spacecraft failure occurs frequently with each new program.

The **Mars Observer**, the **Polar Lander**, and now Europe's **Beagle II** have been some of the probes to have vanished under strange circumstances and the excuses that pass as reasonable scientific explanation have been some of the most ludicrous indicating an act of expensive incompetence, than that of confident exacting professionalism." Mars Polar Lander, for example, was lost due to confusion between English and metric units. And when NIMA, a Defense Department agency tasked with examining maps, believed they had spotted not just the Polar Lander but its heat shield and parachute, NASA curtly informed them they were mistaken, blaming it all on bad data". http://archives.weirdload.com/nasa-shame.html

Then, in a sudden turnabout, whether in a guilt of consciousness or in a deliberate counter-retaliatory move toward fulfilling an alternative agenda, without fanfare they released tens of thousands of photo images of things that look like forests, pond scum, even archeological remains; all with continued silence and without explanation or even comment. Their only admission was the possibility that there was flowing water on Mars, when the evidence of seepage down the sides of craters and ravines became overwhelming, even to amateurs.

There has been so much cover-up of evidence of Martian anomalies and artificial structures on Mars that even, when the European Space Agency (ESA) was thought to be an honest agency of the European governments providing colored high-resolution photographs of the same areas originally imaged by NASA, the public thought that they finally had the truth to what lay upon the Martian landscape. Up to that point, the public was correct in their assessment of ESA, until in the last few years, someone got to them and influenced them to get their act in line with NASA and RSA. Then they too, began to release corrupted and manipulated data to the public showing in some cases photo digital alterations to Martian surfaces, adding artifacts and things that didn't even exist, an example was the bump on the forehead of the Face in Cydonia and the astrological signs appearing on the ground in the Cydonia Plain that weren't there before! It would seem that the public could no longer trust any space agency for honest and truthful disclosure of any Martian discoveries they came across or imaged.

One could only imagine or guess at who or what was the reason behind such covert actions by these powerful space agencies and what influences were being exercised over the other space agencies in order to control their actions and the release of information.

150

NASA no longer has a fleet of Space shuttles and relies solely on the RSA to take astronauts into space and bring them safely back to Earth. China continues to develop its manned space program under the watchful eye of the US military and NASA. One has to wonder if the NASA and the covert military industrial complex has decided to put into full operational mode their covert black space program, thereby, bypassing altogether the old outdated rocket-powered space program of NASA fame with the more highly advanced antigravity electro-gravitic propulsion powered saucers and the workhorse of their black space program, the huge space platforms, the size of aircraft carriers!

So the inevitable conclusion that we as amateur researchers and the public have reached was that *NASA does not want to admit that there is life on Mars!* But, there was more that is still hidden from our sight; what was the real agenda behind NASA/DOD's covert secretiveness?

A clue to the secret dimensions behind JPL came from an unexpected source, **Daniel Goldin**, former head of the Space Agency, publicly thanked **Admiral Bobby Ray Inman** of the **National Security Agency** for the "oversight" given JPL. Why would the spooks even be interested in space probes to distant planets, much less have to exercise some kind of supervision? What sort of national security considerations could possibly be involved? http://archives.weirdload.com/nasa-shame.html

The reason has to do with *"secret societies"* and *"technology"* which is where this topic left off! Secret societies which are fanatical, ritualistic, stepped in occultism and mysticism. NASA as stated before has its share of **Masons**, and no less than the co-founder of the Jet Propulsion Laboratory himself was a ceremonial magician who performed *"sex magick"* with **L. Ron Hubbard!** http://archives.weirdload.com/nasa-shame.html

In a world that is spiritually bankrupt and moribund, in desperate need of a divine physician to cure it of its ills, such practices, and behaviours by corporate and administrative officials high up in the NASA agency is in this day and age not totally unexpected! However, be that as it may, in this immediate context, we are still left with the question: **WHY?**

NASA's hidden agenda is not the search for Extraterrestrial life on Mars! It is in reality, only a secondary consideration and is only important if it leads to finding ...highly advanced alien technology!!! (Bold italics added for emphasis).

This means that if any of the orbiting space probes or Mars rovers come across a piece of technology, it will undoubtedly be imaged from all possible angles in high resolution and its latitude and longitude will be marked on a global map of Mars for future unmanned or manned retrieval.

We've already seen NASA's "high regard" for finding and obliterating any signs of ancient or current life on Mars, so their interest in finding evidence of life is only for photographic value, unless, they come across an intelligent sentient life form that confronts these autonomous vehicles! Such an incident should be a cause for re-evaluation of the whole Mars mission program and agenda. The last thing that we want is hostility between two intelligent species!!

NASA Spirit Rover drills, grinds, takes soil samples runs chemical analysis and determines the topographical and substrata composition of the Martian terrain. Every once in awhile in repeating the process over and over again, thousands of times, it may fortuitously come across an area that also happens to have some artificiality lying right beside the experiment field site, in fact within inches or centimeters of where it is grinding rock, as can be seen in the image below!

NASA –JPL Mars rover Spirit Sol 1220 pan cam photographs a disk object that has uniformity, round, flat and thickness similar to a coin

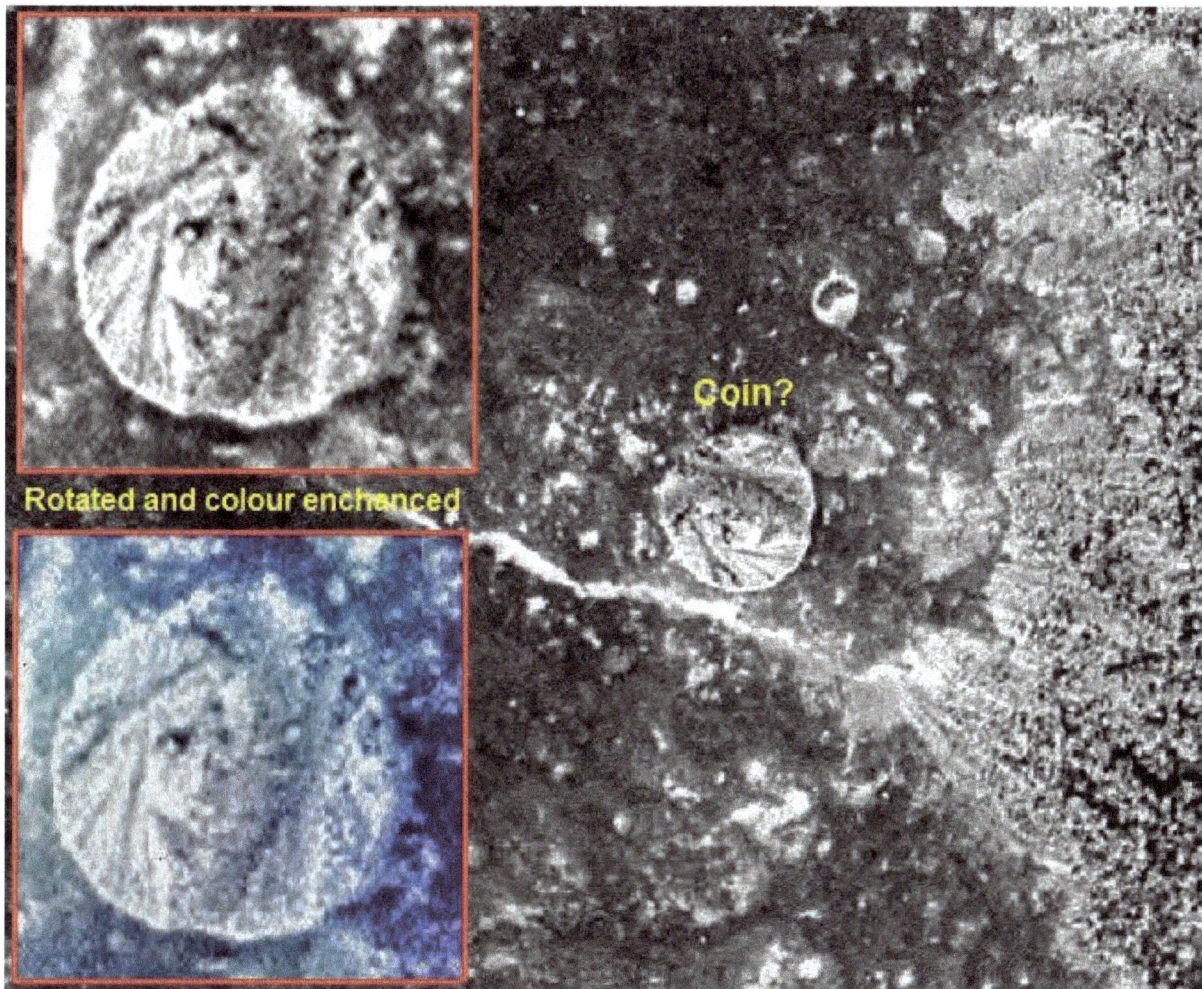

Spirit rover microscopic imager photographs a coin like object on Sol 1434
https://mars.jpl.nasa.gov/mer/gallery/all/spirit_m1434.html

If this is NASA's ultimate agenda, the ***search for alien technology*** then, what are the implications of such discoveries and whom does it benefit? Is it for occult-oriented secret societies and is this the primary reason for NASA's existence?

As any astute researcher of the UFO/ETI subject would have realized by now, the only logical conclusion that could be reached, given the years of NASA's often secretive agendas within what was supposedly routine space exploration programs and its less than honest transparency in disclosing what it knows to the public, is the same the conclusion reached by **Dr. Steven Greer** and publicly announced through the **Disclosure Project**.

NASA is just a front for the real space exploration program, a deep black budgeted program operated by the Military Industrial Complex. NASA is really nothing more than a public relations effort and has been since the early sixties when the US decided that it would go to the Moon.

153

An unusual metallic anomaly sticking out of the ancient seabed in the Gale Crater taken by the right Mast Cam on Curiosity on Sol 173 – (colour corrected image). To date, there are no explanations by NASA as to what it may be

https://mars.jpl.nasa.gov/msl/multimedia/raw/?rawid=0173mr0926020000e1_dxxx&s=173

The Curiosity Mars rover has also found some strange-looking little things on Mars – you've likely heard of the Mars '"*flower"*, the piece of benign plastic from the rover itself, and other bright flecks of granules in the Martian soil. Now, the rover has imaged a small metallic-looking protuberance on a rock, (see above). Visible in the image above, the protuberance appears to have a high albedo (light and shadow) and even projects a shadow on the rock below. The image was taken with the right Mast Cam on Curiosity on Sol 173 (January 30, 2013).

As to what it is, has everyone guessing including NASA, who eventually decided that it is just unusual rock outcropping that has been windblown over a period of millennia. ***Strange how the surrounding rock hasn't been effected by the same winds!*** This is an ancient seabed and most educated guesses say that something artificial and unusual is buried just beneath the surface! Is this one more example of ancient Martian technology that was created by Martian intelligence

154

and which unavoidably got destroyed from a cataclysm and ended up buried in the silt bed of an ancient sea?

NASA's Curiosity Right Mast Cam photographs mechanical object in Gale Crater on Sol 64. It appears to be either part of a plumbing system with an intake valve protruding out or a tank turret with a short gun barrel or something else (see images above)
https://mars.jpl.nasa.gov/msl-raw-images/msss/00064/mcam/0064MR0285006000E1_DXXX.jpg

Another truly anomalous object was photographed sitting on the Martian terrain out in the open its large size would be hard to ignore by any NASA scientist and would invite closer inspection by the Mars rover Curiosity. As indicated in the above photo image, this object looks like a rusted out tank turret or part of a larger plumbing or sewer system as there is a pipe protruding perpendicular to the main body of the turret-like object. There may be other related debris nearby but it is not evident in the photo. And as Curiosity makes its way across Gale Crater, there will no doubt be other unusual objects not natural to the landscape as the following photographs will attest.

Another close up photo image of the tank turret object and also colour enhanced in blue (right) to show greater image detail
https://mars.jpl.nasa.gov/msl-raw-images/msss/00064/mcam/0064MR0285006000E1_DXXX.jpg

Many of the mechanical objects and metallic artifacts are usually twisted and distorted beyond recognition, their appearance indicates that they were affected by some terrific external heat causing the objects to melt, bubble or blister leaving broken gaps or air pockets on the surface. Some of this technology looks like large cogwheels that were a part of some larger structure or device, maybe from a factory.

The purpose of these objects is unknown and can only be speculated as to their ultimate function, although one can extrapolate what these objects may have been by the type of terrain in which they were found, the debris field in which the objects were scattered about and where some of these anomalous artifacts ended up, sometimes beside body parts of human/humanoids and animals. There are other facts that would need to be taken into consideration much like the wreckage littering the debris field from an airplane disaster.

In some circumstances, it can be surmised that some objects are from collapsed buildings and homes and may be the internal plumbing structure of pipes, storm drains or metal support beams. There is even strong evidence that factories could account for the massive twisted metal particularly from I-beams, production machinery, generators and heating units or furnaces, etc. In other locations, there is clear evidence that some metal objects have a distinct vehicular function either as submersibles (submarines) because of fossilized aquatic life found nearby or as personal or family transportation vehicles (cars or land speed cruisers, etc.) because of land animals are found in the same area. Some devices actually appear to be some type of aircraft or spacecraft as parts of the device appear to carry aerial weapons –cannons or guns!

In a somewhat questionable NASA photo image (see images below and colour enhancement), can be seen the ruins of what appears to be a collapsed residential home or apartment building (note the organic like debris, the broken stone pillar, masonry and various personal resident items). Of obvious particular interest is the device that looks like a **"ray gun"** poking up through

156

the ruins as if it had just been placed there, (see enlarged image). This photo was almost completely dismissed as not being authentic as there was no NASA Mars rover image number associated with the image; it was merely posted on **YouTube** as a "leaked NASA image" video. However, closer examination of the dubious image revealed a small stone statue just behind the "ray gun" which seems to be more organic and natural to the environment than does the ray-gun which may be either a child's plastic toy or a real lethal weapon. Note there is a "trigger" on the "weapon". Is this part of NASA's disinformation ruse to release hoaxed images of Mars to see "who will bite" at the hoaxed images? Note there is no Mars Sol day or NASA image number to verify this photo as genuine, even the YouTube video of this "Ray Gun" is no longer accessible.

An 'alleged' NASA image supposedly taken by Curiosity showing what looks like a "Ray-Gun" and a statuette against a "very organic" Martian terrain.
(c) Terry Tibando

If the above photo image is a real ray gun then, you can bet that NASA and everyone else associated with NASA would sit up and pay attention to this discovery. This is exactly what NASA is hoping to find on Mars and would try to find some way to get their hands on it.

Another example of alien technology has been the discovery of robotic or cybernetic machines such as the robot heads dubbed **"Data"** found on the Moon and now the Martian equivalents dubbed **"Lore" 1 and 2** by this author (see images below) found in the **"Shaler"** area of **Gale Crater** by this author.

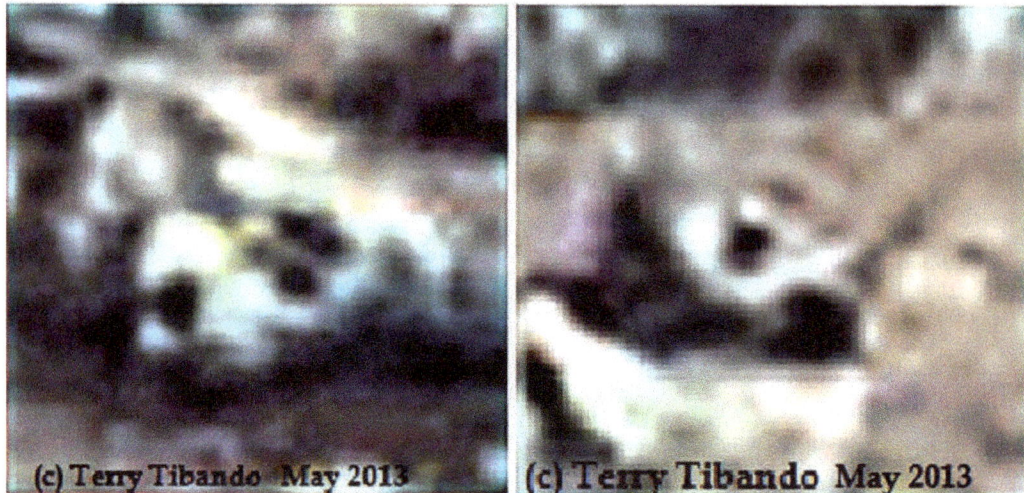

(c) Terry Tibando May 2013   (c) Terry Tibando May 2013

These two robot heads found in "Shaler" area of Gale Crater by the author, were dubbed "Lore" I and 2 in conjunction with Hoagland's robot head, "Data" found on the Moon, all named after the characters from the TV series, Star Trek Next Generation
(c) Terry Tibando

Close up of possible alien technology from above photo Note its geometric shape
https://mars.nasa.gov/msl/multimedia/rawimages/?order=sol+desc%2Cinstrument_sort+asc%2Csample_type_sort+asc%2C+date_taken+desc&per_page=50&page=0&mission=msl&begin_sol=597&end_sol=597

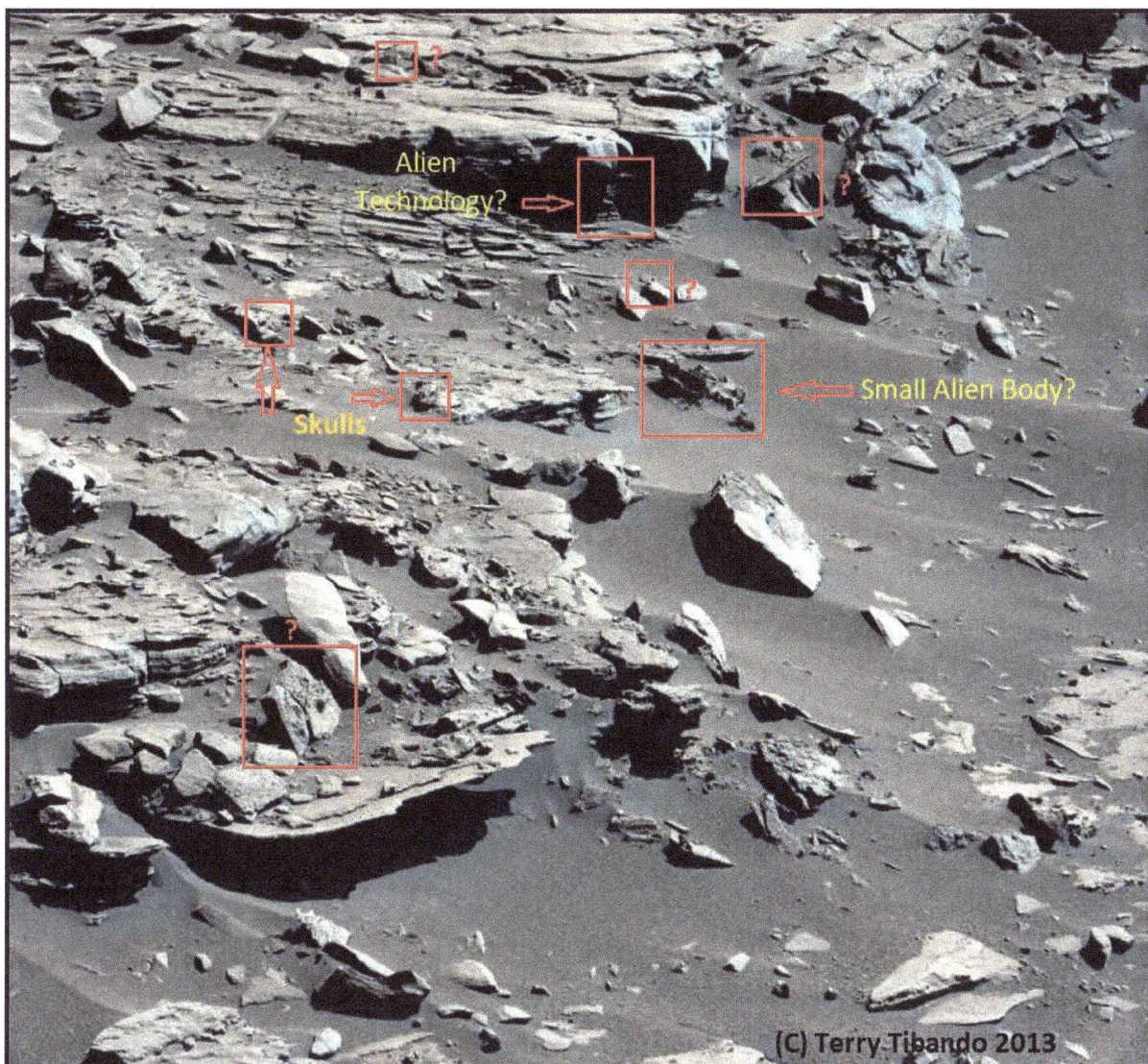

In this photo (Curiosity Sol 537) there what appears to be an alien technology a small alien body, and a few skulls See also, the above photo

The photo below has an unusual relief on a rock that was discovered by a Mars anomalist hunter that appears to be similar to a Celtic cross that was also found in Clonenagh, Ireland; it also has a rectangular relief beside it like the Irish Celtic cross. The colour enhancement brings out the shape for comparison with the Clonenagh cross.

A research would have to wonder how two objects on two different planets nearly a hundred million miles apart can look so similar and draw some conclusion of a possible common intelligence (human or Martian) created both or influenced the other intelligence on the other planet. If this is possible, then did humans visited Mars in the distant past or did Martian visited the Earth in its ancient past?

**Martian technology and artificiality abounds, a Celtic Cross or
a small cog in a larger cog wheel (colour enhanced)?**

CHAPTER 83

ARE MARTIANS WATCHING OUR MARS SATELLITES AND ROVERS?

ET Surveillance Orbs are Monitoring the Activity of NASA'S Mars Rovers

One of the repeated anomalous objects that keeps reappearing in many of the rover images are the strange little gold or silver orbs that are sometimes photographed by NASA's Mars rovers. They are often seen flying or hovering off in the near distance or just lying on the ground, sometimes beside seabed rock as seen in the photo image below. Other times, they are found in the **"Shaler"** area. An area in which **Richard C. Hoagland** has identified as a ***"collapsed apartment building"*** that is littered with residential apartment debris. Are they just "ornaments" or do they operate in some fashion for amusement or for a more practical purpose like a surveillance probe?

In many of the camera panorama and close-up shots taken by the various NASA rovers, these little orbs seem to be everywhere. They are certainly not rock geoids as many photo images show them to be floating or hovering above the ground sometimes in great height. If they were naturally formed rocks, we would have to re-evaluate our understanding of physical forces and properties associated with such small objects and their interaction with the gravitational forces of a planet.

They are also, not pixel degradation within the cameras or else they would show up in every picture frame taken by the rover's cameras and they do not. The fact is that they appear in different locations within those pictures that are fortunate enough to have captured them on camera.

They appear artificial and manufactured suggesting that they are made by some Martian intelligence designed with a particular purpose in mind. Since they appear close to the Mars rovers, it is a reasonable assumption that they are a type of surveillance device monitoring NASA's rover missions.

Such surveillance probes have been reported here on Earth, all around the planet by many UFO witnesses and have even been photographed as well. The fact that we are photo imaging similar or identical flying objects in such close proximity to witnesses or remote controlled cameras on two planets in our Solar System (one being our own home world) suggests that these devices belong to and are being operated by a highly advanced civilization, not of this Earth!

Whoever they are, has a real interest in monitoring human activity whether on this planet or on Mars or possibly any other planet that we are sending space probes to, in the name of space exploration. While we supposedly search for life in the Solar System and the rest of the universe, it has become evident that Extraterrestrial Intelligences have already found us not only on our home world but, have also found evidence of us exploring other planets, perhaps theirs, like Mars!

**Mars Rover Spirit photographs (Sol 1314) an old seabed terrain for "blueberries"
(possible ancient sea life) and for unusual rock. In the upper left corner
a metallic gold orb is seen; is this ancient Martian technology?**

http://alienanomalies.activeboard.com/forum.spark?aBID=47797&p=3&topicID=35916948

**Three small golden orbs found in Shaler area by this author.
What function do these little objects serve?**

(c) Terry Tibando

163

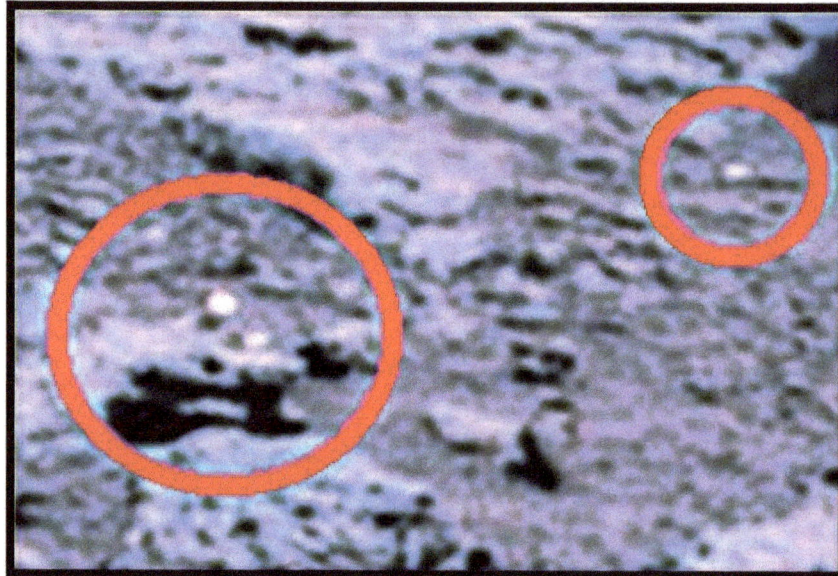

A close up of the highlighted area from above photo reveals possibly three orbs, the largest one appears to be floating and casting a perfect circular shadow on the ground!
(c) Terry Tibando

(C) Terry Tibando April 2013

Curiosity Mast Cam (MSSS-MALIN) image for Sol 107 in Gale Crater is a treasure trove of artificiality as seen in the above image. Note the large cogs and metal pieces and the half-buried head with a floating white sphere just above it.
(c) Terry Tibando

Curiosity had what appears to be a very busy day on **Sol 107** as another area that it travelled by was filled with anomalies like a huge half-buried stone head reminiscent of the ancient **Olmec Heads** of Mexico. Now this interpretation may appear subjective to others looking at the landscape but, what makes this really an unusual find is the strange little glowing white object that appears to be just floating above the stone head! (See images above and below).

Close-up of head and mechanical debris and large cog wheel in Gale Crater

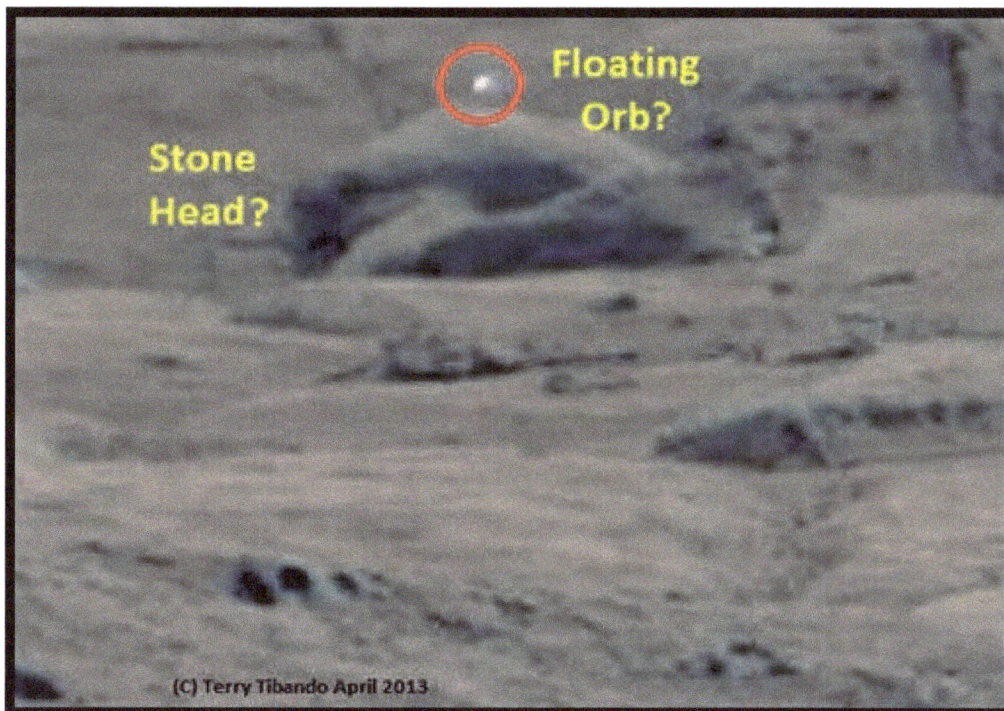

Colour corrected photo Curiosity Sol 107 of above images note the white orb object appears to float above a huge rock or statue that looks like a buried head
(c) Terry Tibando

Within this debris field depicted in the photo images above are large mechanical pieces of metal. Some are rectangular blocks with many holes in along the side of the object, others pieces are big wheels with cog-like teeth while other objects are twisted pieces of machinery that may have been a part of a Martian factory. The actual function of these pieces of metal is, of course, speculative; no one will know for sure until we land and set foot on Mars to see for ourselves.

In the case of the small gold and silver orbs that are frequently photographed hovering and flying so close to the rovers, they represent a miniaturization of anti-gravitational technology, something that any scientist or country would love to get their hands on as the benefits would be exponential. In these instances, this would be the type of technology that NASA would be searching for and the kind of things that would benefit a secret society or a covert black military force or some other rogue organization.

Another image captured by Mars rover Curiosity in 2014 shows a box-like structure with symmetrical indentations and close to it floats a curious little shining orb
(Google Images)

Any retrieval of these types of technologies would probably be scrutinized back on Earth in some secret, out of the way laboratory, away from prying eyes. Closer examination, analysis, further research and development and an evaluation of such alien technology could possibility result in a proposed integration into existing terrestrial technology or it may be completely reverse engineered to full operational development and production and kept secret for black project military use only.

In the red false coloured image, we clearly see another orb hovering above the Martian terrain monitoring Curiosity's journey across Gale Crater toward Mount Sharp. The bottom image is the auto colour corrected image, note the light blue sky
(Google Images)

This is a panoramic view from the 'Rocknest' area taken by the Mars Rover Curiosity. In section "B" of this mosaic to the far right, barely visible is another Martian Orb Monitor, the orb is one pixel in size

https://photojournal.jpl.nasa.gov/catalog/PIA16453

The above panorama photo image is a mosaic of images taken by the **Mast Camera (Mastcam)** on the NASA Mars rover Curiosity while the rover was working at a site called "**Rocknest**" in October and November 2012.

The center of the scene, looking eastward from Rocknest, includes the Point Lake area. After the component images for this scene were taken, Curiosity drove 83 feet (25.3 meters) on Nov. 18 from Rocknest to Point Lake. From Point Lake, the Mastcam is taking images for another detailed panoramic view of the area further east to help researchers identify candidate targets for the rover's first drilling into a rock.

The image has been white-balanced to show what the rocks and soils in it would look like if they

were on Earth. The raw-color version shows what the scene looks like on Mars to the camera.
http://www.nasa.gov/mission_pages/msl/multimedia/pia16453.html#.VO7PDuH7OVp

**Enlarged image of the Martian Orb that monitors and follows the Curiosity
rover around on its Mars mission to explore Gale Crater**
(c) Terry Tibando https://photojournal.jpl.nasa.gov/catalog/PIA16453

Acquisition of any alien technology that is at least hundreds or thousands of years ahead of
human science, if probably understood with some "outside the box thinking" and is reproducible
would give its possessor a new technology and a great advantage over one's enemies. Such
technology may decide an entirely new direction of an elite covert group or even a nation. It
could possibly even, determine the course of human evolution on this planet.

Martian UFOs Track the Mars Rovers and Space Orbite Missions

Perhaps, the pinnacle of all technological finds would have to be the discovery of an alien
spacecraft, hopefully, one that is still operational and without a pilot or navigational crew. As it
turns out, Mars seems to have it share of such artifacts in areas already photographed from

169

orbiting satellites and from the various Mars rovers. Some appear to be completely intact with little or no damage while others are severely damaged as if they had been engaged in a horrendous, all-out catastrophic planetary war or maybe, as the result of a natural global cataclysm!

Below is an example of one structure found by **Richard Hoagland's Enterprise Mission team**, that seems to have all the right details to indicate that its function is designed for operation in outer space.

This object appears almost dead center in the frame, indicating that it may have been the target of interest for this narrow-angle shot. It is resting atop a ridge which displays unusual geometric structure. The ship seems to be covered with a thick coating of dust but, it has an obvious perfect symmetrical oval shape, with a number of distinct features, including an angled front wedge, rows of parallel "windows" or notching, a faint hexagonal cell structure on the roof, and a "hatch" like area, and it's nearly 3,000 feet long!

The sunlight angle and shadowing indicate that it is convex or slightly dome shaped and not concave which would indicate a crater on the edge of a cliff in which case the crater wall would have collapsed over time. There is also a tail structure and possible engine cluster at the rear end of the craft and ground shadowing also reveals that the rim of the craft tapers down and toward the centre thus, it is convex underneath.

All this detail is highly suggestive of an extraterrestrial spacecraft that for whatever reason seem to have come to a gentle rest on an otherwise precarious ledge of the Martian terrain. If indeed it is an ET spacecraft, it appears to have landed or glided across the Martian surface and has been resting in the same location for possibly many millennia. Did the ET beings that piloted the spaceship evacuate their craft ages ago in the distant past because of some irreparable internal damage and then lived out their remaining days on Mars or are they in a state of millennial hibernation sleep?

Once again, this is another area on Mars that requires further up close exploration to determine if this is a natural rock formation or if indeed it is a large abandon spacecraft from eons ago. This is similar to the massive, ancient alien, cigar shape spacecraft found on the backside of the Moon by the **Apollo 15** crew in the **Iszak D Crater** and southwest of **Delporte Crater.**

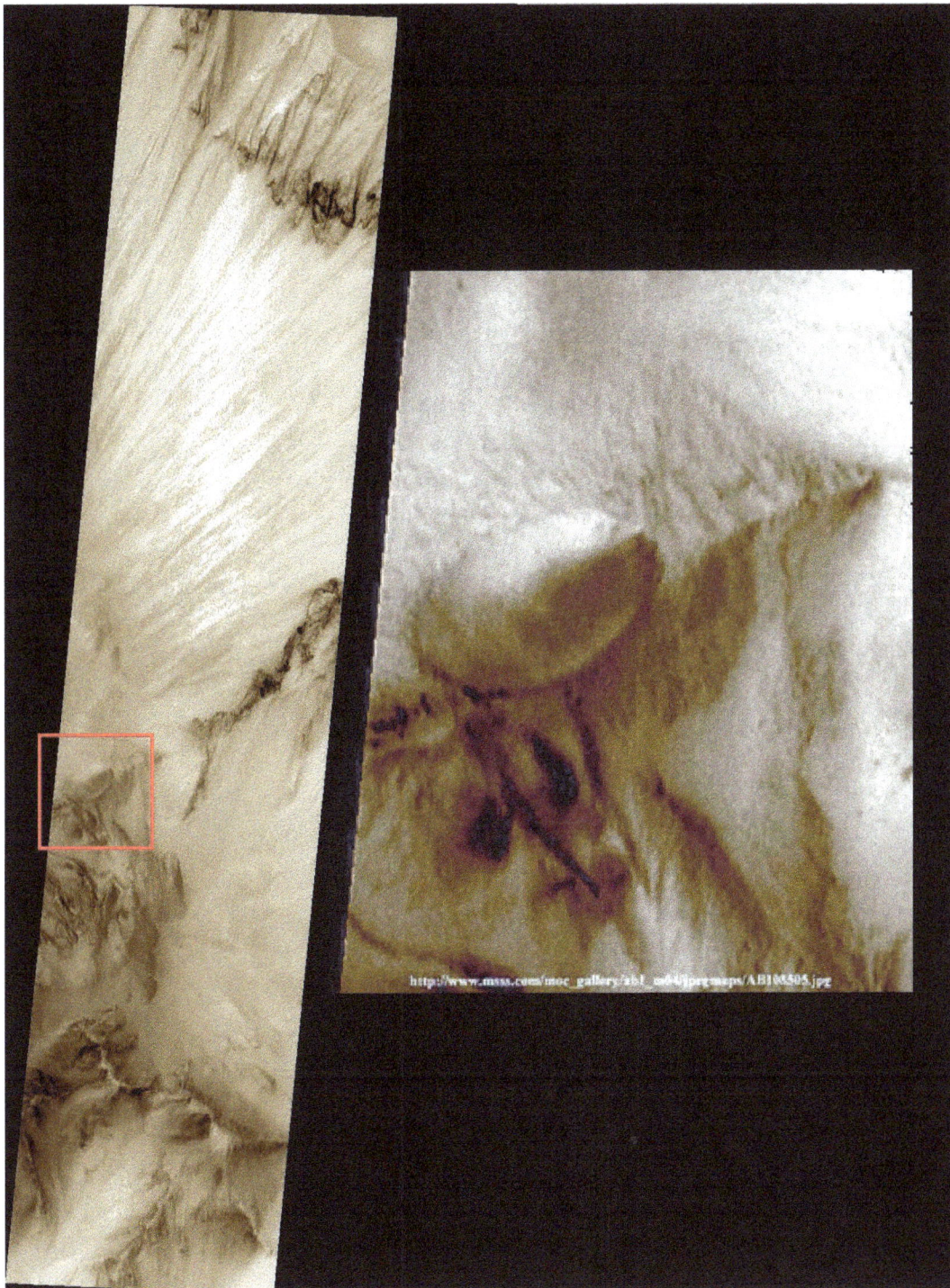

A crash landed Martian spacecraft from MSSS MOC image AB108505 North Wall of East Ophir Chasma (see insert). Enlargement of image showed a curvilinear structure with convex shadowing, sitting precariously at the edge of a cliff. There is a tail structure at right and possible windows around the edge of the craft

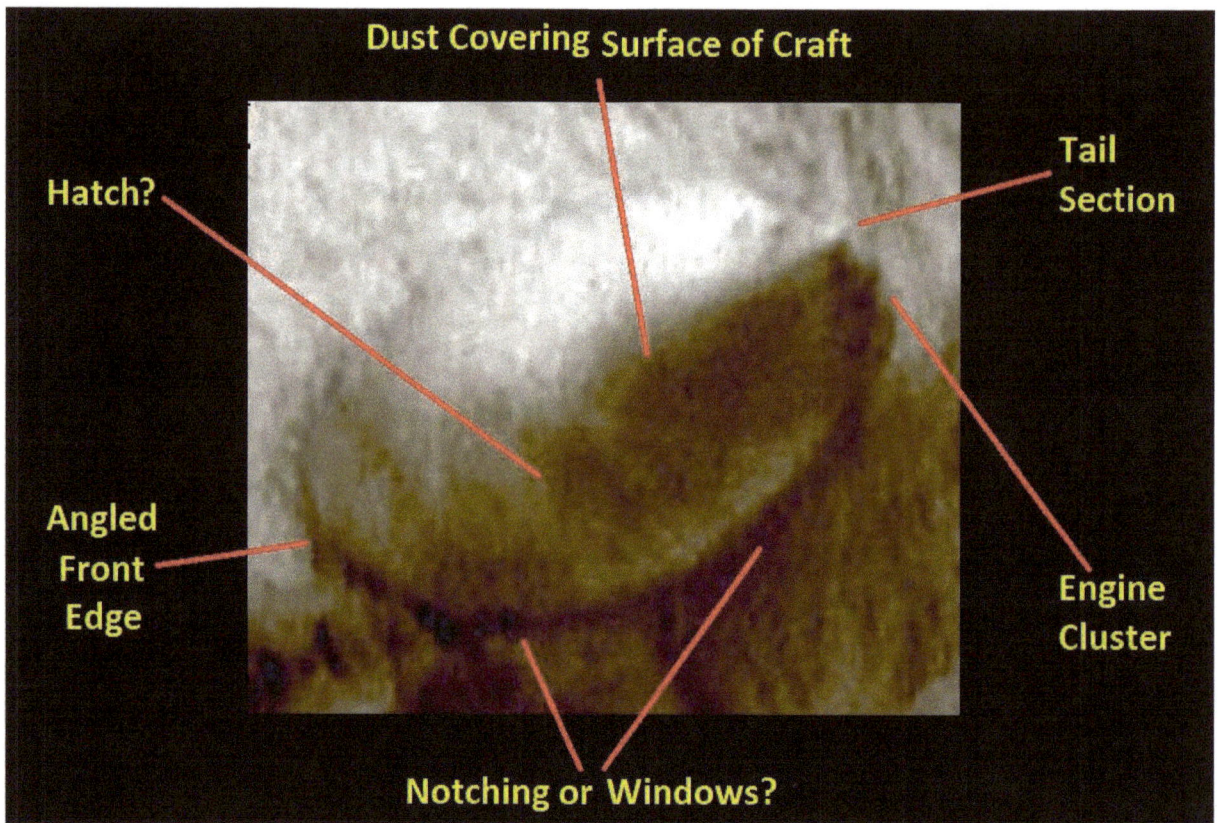

Close up and colour enhanced image of Martian saucer with some details of its structure
http://www.msss.com/moc_gallery/ab1_m04/images/AB108505.html

Another interesting photograph which happens to be one of the last images taken by the Mars rover Spirit before it became inoperable was the photo below of a strange dark object that appears to be a kind of RV camper or maybe it is a landed spacecraft. The size of the object is hard to gauge because there is nothing of a known size close by to compare it to, yet an approximate size would be about 30 to 45 feet in length and about 10 to 15 feet wide and about 15 feet in height at the back portion of the vehicle. The back end part of the craft could be a sort of observation area or crew quarters. Who knows for sure as this is just speculation!

Note the rectilinear structure and layering of the object which seems to have what looks like a "headlight" section at the "front end" of the vehicle. It also seems to have a "windshield" and a smaller window above it and further back and then, the main navigational portion of the vehicle which appears to be detachable from the outpost/crew section of the vehicle.

This concept is nothing new as many vehicle designers in Europe and America use ultra modern fluidic concepts to create trains, trucks, buses, aircraft, ships and just about anything you can think of that travel on wheels, fly or sail the seas. The fact the ETI may have designed something is both conceptually amazing but, not too unexpected. The only differences are the functionality and the method of propulsion generation to propel the vehicle.

172

**One of the last photos taken by Spirit rover on Sol 2114 before it became inoperable shows
an object in the valley area that looks like an RV camper or possibly a spacecraft!
Compare this to the landmaster ATV (insert) from the movie Damnation Alley**

https://areo.info/mer/spirit/2114/#10 and https://www.pinterest.com/pin/469359592392934404/

173

In these colour altered, enhanced and enlarged photo images, there is symmetry of structure and rectangular windows with a superstructure on top of a Martian vehicle. It is not a Mars Rover. © Terry Tibando

The last photo image in this set of vehicles or devices that resemble possible spacecraft is the one found by this author. When the researcher examines the original NASA orange-brown colour image, he sees a landscape of rocks, sand and the typical Martian barrenness to the terrain. However, there are unusual darker objects that have curves, twists, rectilinear structure which jump out to the viewer as possible objects of artificiality.

In the photo image below one sees square spaces and openings, notches, scoop-like indentations, tubes or piping, turret-like structures similar to a tank turret and gun-cannon and what appears to be the propulsive end of the vehicle that resembles a rocket or jet nozzle.

First impression by most people is that this object is distinctly metallic in appearance and mechanical in function and the second realization is that this device appears to have been severely damaged, warped and melted by extreme heat directed at it from an external source. By itself this is unusual and maybe, it represents a one in a billion chance of a downed aircraft or spacecraft that met an unfortunate end due to a mechanical mishap. It happens all the time on Earth and it may have also happened on Mars, except for the fact that very close to this craft *another turret cannon* was also found by the author and the *"fossilized body of humanoid"* probably the pilot of this crashed vehicle! There are even, other carcasses of creatures beside the **"mummified pilot"** that appear to be both aquatic and land-dwelling!

The Martian cadaver and the other creatures, as well as the other piece of mechanical hardware, were almost overlooked because they resembled rock and stone which may be why no one has posted this amazing Curiosity discovery site. The fact that NASA deliberately falsifies the true color of the Martian terrain, aids them in their ability to hide in plain sight, those things that they don't want the public to see or find out about.

This is a jackpot find for any scientist, let alone for an amateur researcher such as this author!! (See photo below and the above images in this subsection on Martian technology).

174

Scrutinizing the raw NASA photo image which is composed of a series of images that have been spliced together, one sees in this immediate area other evidence for mechanical craft and devices which are partially buried and poking through the surface and each object appears to have been subjected to a high thermal melting point yet, the land itself has not suffered this same intense heat as it is not vitrified but does seem to have been pulverized into small rock debris and sand!

(C) Terry Tibando - February 2013

This NASA Curiosity image taken on Sol 107 was found by the author that depicts a severely damaged and weather-worn craft, possibly a spaceship. A humanoid body was also discovered near the craft beside an additional gun turret from the craft
(c) Terry Tibando

This is the sort of technology that would intrigue planetary scientists as it is a positive indication that Mars had a highly advanced civilization that became extinct in the not too distant past, perhaps as long ago as 20,000 years to 3 million years ago. They either became completely extinct as a civilization or they escape their fate by going underground or they survived by going to another planet in this Solar System like Earth. It is also possible that they left this star system for a new star system or returned to an even older star system. Perhaps, one day they may return to our Solar System; there are those who believe that they already have returned.

Finally, the question that we are left with is there an intelligent and sentient life form on Mars today? This has been somewhat answered with the enigmatic Martian surveillance orbs or probes that have been following the NASA Mars rovers' exploits which means they are being operated by an intelligence possibly indigenous to Mars or one that has come from another star system which may have set up shop for the last few millennia.

There may, in fact, be more than one intelligent species living on the Red Planet, perhaps living in peaceful co-existence with each other. We already seen that there is sufficient evidence to indicate that one race of intelligence may be quite tall, giant size in comparison to humans with an average height between 10 and 12 feet tall. They have a similar human appearance with obvious differences judging by the skeletal and fossilized remains we have already discovered or uncovered from NASA's very own raw photographic data.

Joseph McMoneagle, a Fort Meade former member of SRI remote viewing team and under the purview and aegis of the US Army's **Stargate Program** was asked to remote view Mars to determine if he could "see" any Martian beings still living on the planet and to describe them. McGoneagle's perceptions were of people who were very tall and thin: ... *"it's only a shadow, It's as if they were there and they're not, not there anymore."*

Monroe (of **Monroe Institute** fame), his monitor, instructed McGoneagle to go back to a period of time when these people were present. Initially, McGoneagle encountered interference, like "static on a line", where the connection became fragmented and broke up. His monitor advised him to report the raw data and not try to piece things together. *"I just keep seeing very large people,"* said McGoneagle. *"They appear thin and tall, but they're very large. Ah... wearing some kind of strange clothes."*

It is only natural at this point for the reader to be rather skeptical of subjective information derived from remote viewing which at the present time remains only speculative without any physical substantiation of proof. This makes Joe McGoneagle's subjectively acquired information, only proof by association with the evidence we have uncovered thus far, but it is a type of proof, no matter how remote. (Pardon the pun)!

Another type of being whose fossilized remains have been photographed at least a couple times by Curiosity is the small 3 to 5 foot tall ET beings (see photo image near the end of Chapter 80 which shows the being close to some alien technology).

At this point, seeing a live Martian ET being running around on the surface of Mars remains as elusive as ever, but this is not evidence of their non-existence. There is, fortunately, indirect evidence of their presence with the orbs seen monitoring and hovering, off in the distance or very close to the Mars rovers as indicated in the previous photos.

Further photographic evidence of Extraterrestrial spacecraft flying around the Martian skies and over nearby mountain tops which seem to be also monitoring the Mars rover's progress from an aerial advantage point. (See the photo set below). There is also a video of a Martian UFO (Spacecraft) lifting off into the sky toward a mountain as can be seen below.

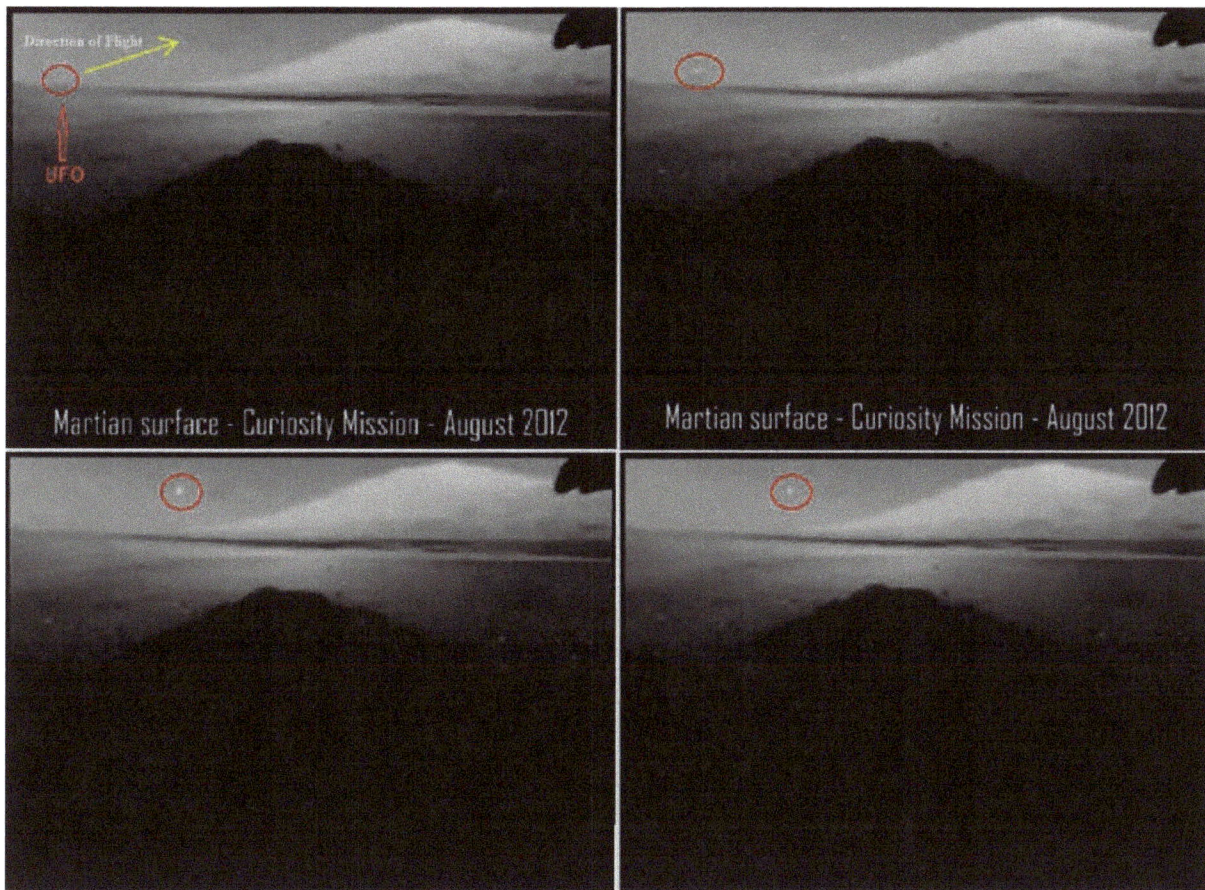

A UFO flies up from the surface and toward the mountain in these stills taken from a video made by Mars rover Curiosity (contrast enhanced)

https://www.youtube.com/watch?v=dykjYBHaCEg

Spherical UFO floats or flies cloud-like above the Martian terrain in front of Curiosity Rover in false colour image (left) and in colour corrected image (right)

(Google Images)

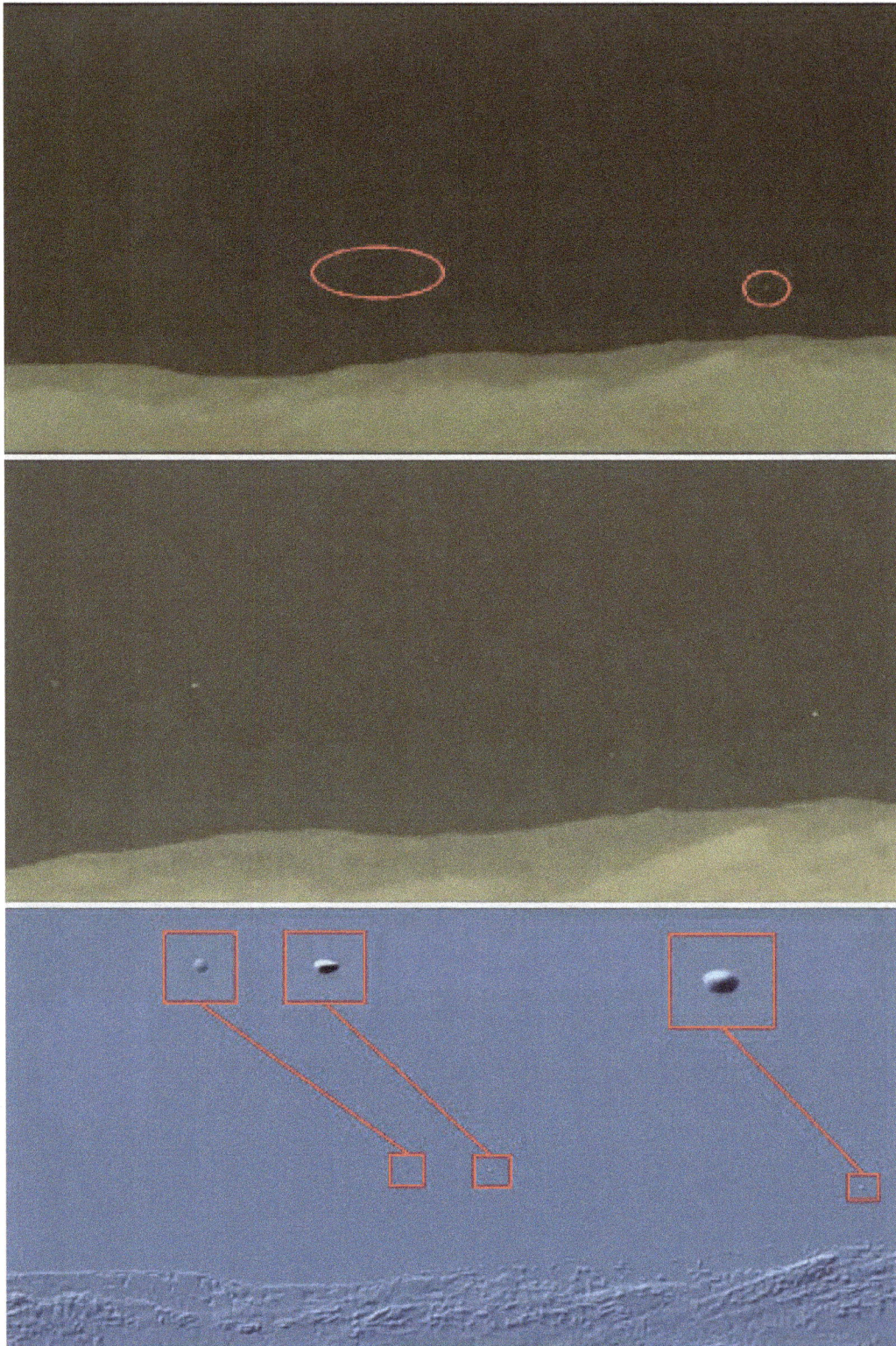

UFOs are seen and photographed by curiosity flying over Martian mountain tops. Bottom, blue high relief photo image shows enlargement of Martian UFOs

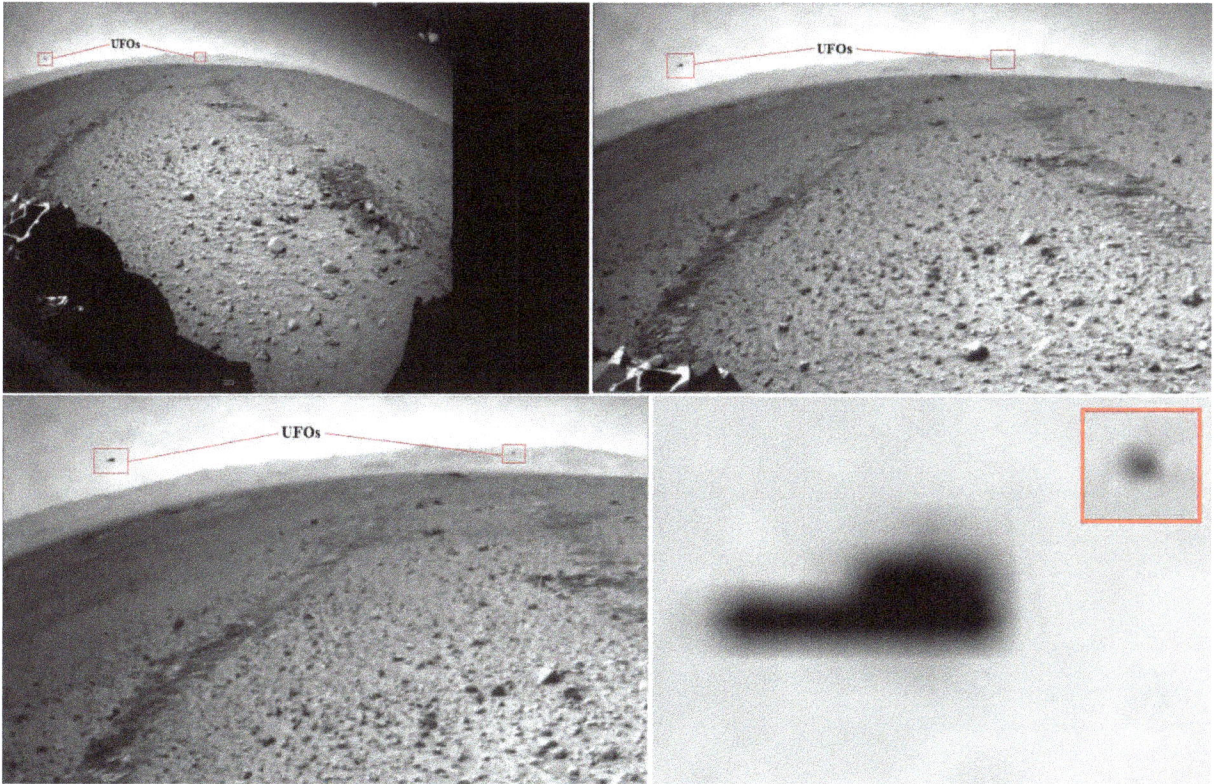

Curiosity's Rear Cam captures two UFOs off in the distance by the mountains. One appears like a shoe (left) and the other is ball shape (right)
(Google Images)

This black and white photo reveals a very bright spherical UFO captured by Curiosity once again hovering over Mount Sharp within Gale Crater
(Google Images)

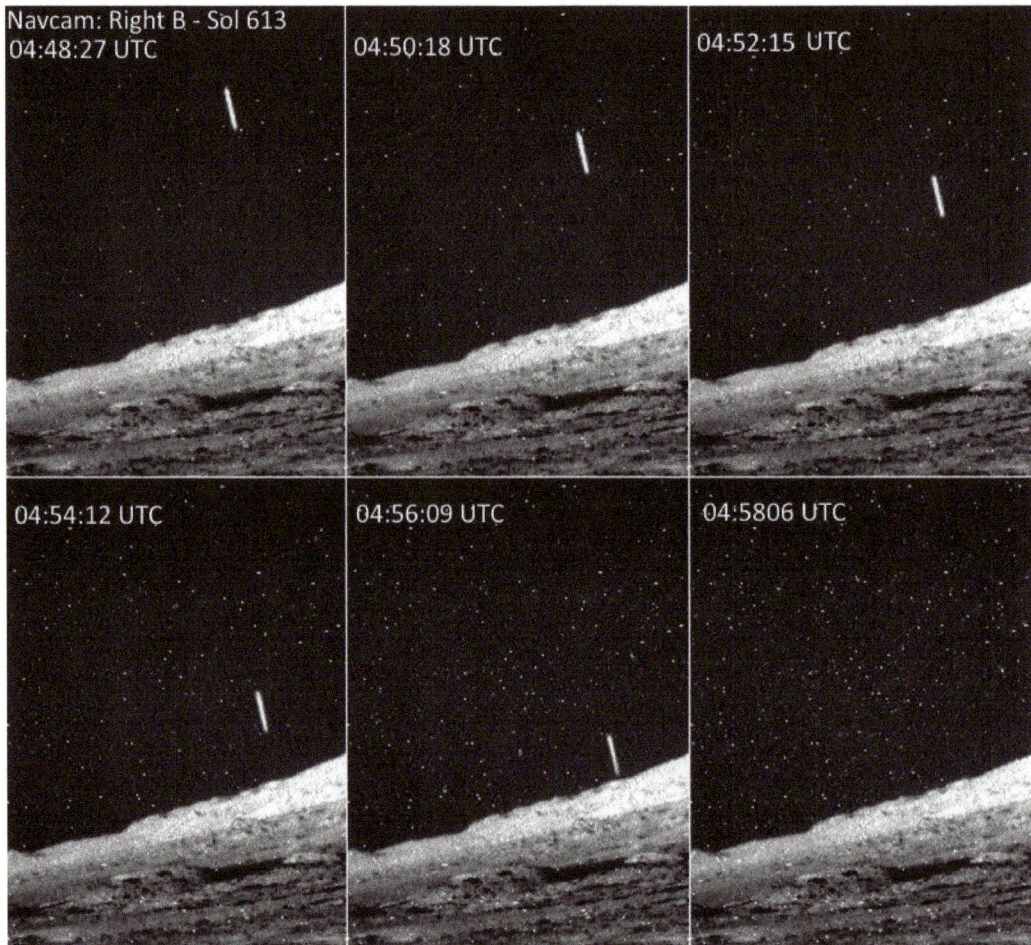

Curiosity captures a large cigar shape UFO heading towards Mars at night

https://mars.jpl.nasa.gov/msl/multimedia/raw/?s=613&camera=NAV_RIGHT

It appears that wherever humans and their space probes go in the Solar System, some high intelligence keeps a watchful eye on us!

http://hight3ch.com/did-nasas-curiosity-rover-capture-a-ufo-on-mars/

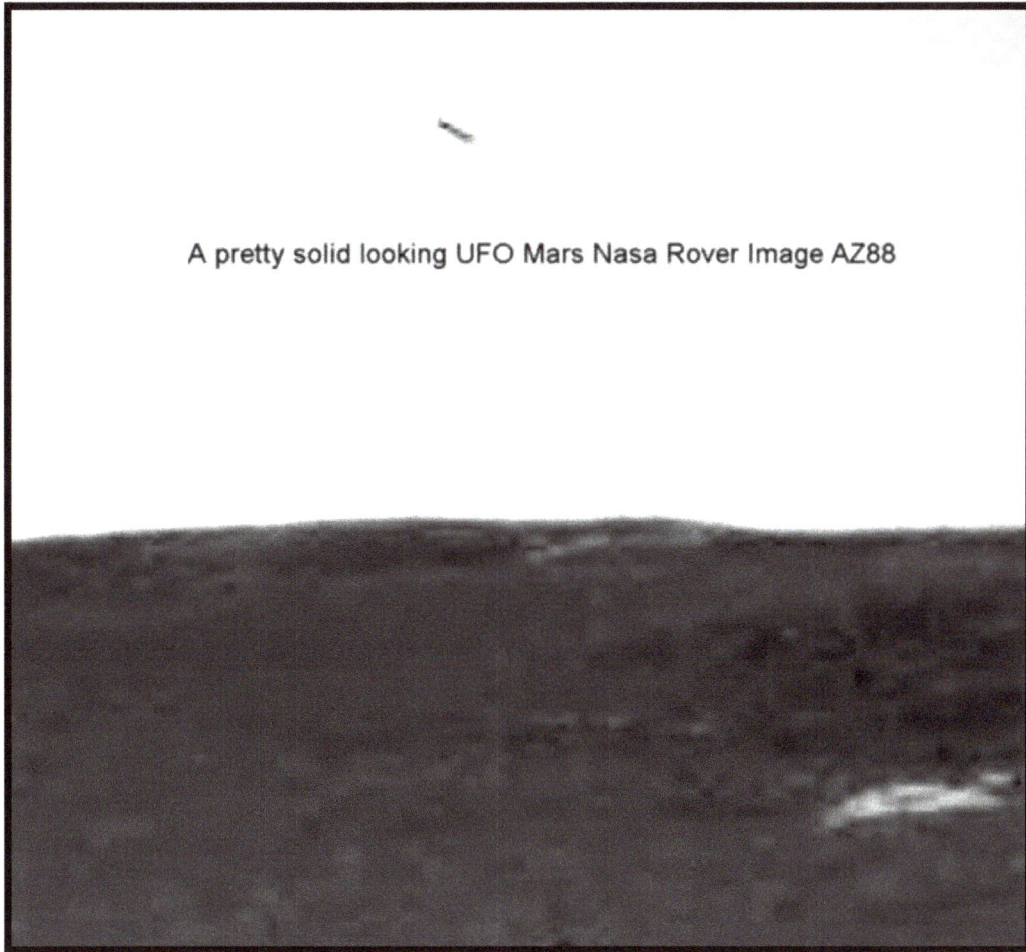

Another UFO taken by Mars Curiosity rover image AZ88
http://historyandmystery.homestead.com/Mars-UFOS.html

The Cabal and the Path Toward a Breakaway Society

This sub section in this chapter may seem to be out of place on our virtual tour of the solar system visiting planets in search of extraterrestrial life, yet many nations have spent a great amount of time and resources to discover what is on Mars. As we have already stated, NASA is not just interested in finding life on Mars but the possibility of finding alien technology that will help America leap further ahead of most nations or to help them achieve their covert agenda on behalf of the **Cabal** of the wealth corporate elite.

The discovery of so much evidence of an ancient civilization and their technology would have almost anyone with an interest in exploration salivating at the chance to go to Mars and explore the Martian terrain for themselves. For this reason we need to consider some of the plans and motives of searching for alien technology.

Alien technology has both frightening and hopeful implications for humanity. **In the wrong hands of secret covert groups, like secret societies operated by wealthy corporate elitists or**

in the hands of rogue military juntas or even fanatical quasi-religious groups or a combination of these groups, who operate with an eschatological modus operandi are quite frankly dangerous to the rest of humanity. In no way should these types of people gain control of any alien technology yet, with this having been stated, it may already be too late!

Remember the warning by the late **US President Eisenhower** in his last public address to the American people, *"...beware of the military industrial complex!"* To be precise, from an important section of his address, **President Eisenhower** specifically stated:

"A vital element in keeping the peace is our military establishment. Our arms must be mighty, ready for instant action so that no potential aggressor may be tempted to risk his own destruction.

Our military organization today bears little relation to that known by any of my predecessors in peacetime, or indeed by the fighting men of World War II or Korea.

Until the latest of our world conflicts, the United States had no armaments industry. American makers of plowshares could, with time and as required, make swords as well. But now we can no longer risk emergency improvisation of national defense; we have been compelled to create a permanent armaments industry of vast proportions. Added to this, three and a half million men and women are directly engaged in the defense establishment. We annually spend on military security more than the net income of all United States corporations.

This conjunction of an immense military establishment and a large arms industry is new in the American experience. The total influence -- economic, political, even spiritual -- is felt in every city, every State house, every office of the Federal government. We recognize the imperative need for this development. Yet, we must not fail to comprehend its grave implications. Our toil, resources, and livelihood are all involved; so is the very structure of our society.

In the councils of government, we must guard against the acquisition of unwarranted influence, whether sought or unsought, by the military-industrial complex. The potential for the disastrous rise of misplaced power exists and will persist.

We must never let the weight of this combination endanger our liberties or democratic processes. We should take nothing for granted. Only an alert and knowledgeable citizenry can compel the proper meshing of the huge industrial and military machinery of defense with our peaceful methods and goals so that security and liberty may prosper together.

Akin to, and largely responsible for the sweeping changes in our industrial-military posture has been the technological revolution during recent decades.

In this revolution, research has become central; it also becomes more formalized, complex, and costly. A steadily increasing share is conducted for, by, or at the direction of, the Federal government.

Today, the solitary inventor, tinkering in his shop, has been overshadowed by task forces of scientists in laboratories and testing fields. In the same fashion, the free university, historically the fountainhead of free ideas and scientific discovery has experienced a revolution in the conduct of research. Partly because of the huge costs involved, a government contract becomes virtually a substitute for intellectual curiosity. For every old blackboard, there are now hundreds of new electronic computers.

The prospect of domination of the nation's scholars by Federal employment, project allocations, and the power of money is ever present and is gravely to be regarded.

Yet, in holding scientific research and discovery in respect, as we should, we must also be alert to the equal and opposite danger that public policy could itself become the captive of a scientific-technological elite.

It is the task of statesmanship to mold, to balance, and to integrate these and other forces, new and old, within the principles of our democratic system -- ever aiming toward the supreme goals of our free society.

Another factor in maintaining balance involves the element of time. As we peer into society's future, we -- you and I, and our government -- must avoid the impulse to live only for today, plundering, for our own ease and convenience, the precious resources of tomorrow. We cannot mortgage the material assets of our grandchildren without risking the loss also of their political and spiritual heritage. We want democracy to survive for all generations to come, not to become the insolvent phantom of tomorrow."

However, even now, in many of the world's wealthiest nations, there are corrupt individuals and rogue agencies within governments and among politicians; within military branches among high-ranking officials; within powerful intelligent agencies and chiefly among the powerful banking cartels; the private industrial sectors and the multi-national corporations who have become the new governments of the world in absentia of true political leadership. They operate transnationally by implementing carefully orchestrated covert agendas on multiple levels of society designed to upset the social equilibrium and destabilize those governments while at the same time sucking dry all financial and natural resources of the planet, without regard for the global populace or the environment. These rogue groups are monitored and influenced by an oligarchy selected from among this group of wealthy industrialists, the political and religious elite who annually gather together in different locations around the world and discuss the geopolitical and social-economic direction of the planet and its ultimate future.

This Cabal continuously maintains their financial leverage over the masses of society and governments of the world by buttressing their position of power and control through the acquisition of natural resources and technology. Corruption is widespread everywhere, thus the Cabal seizes upon this corruptive influence sometimes in very subtle ways and enforces their control through manipulation of governments, their military, their police forces and almost every imaginable social institution in any country, even if those forces

are not aware that such manipulation comes from sources high up, beyond their knowledge or ability to investigate those sources.

One has to wonder what technology is at the disposal of this **Military Industrial Complex**, a.k.a. **Majestic 12, Majic**, et al that enables it to enforce the whims and ultimate direction of this oligarchical cabal.

President Eisenhower said it best, that US was without an armaments industry after the Second World War and later when it found itself involved in other world conflicts. It was big business and opportunists that stepped up to the plate to supply the US Military with what it needed most...weapons of mass destruction! The merging or marriage between an immense military establishment with the large arms industry which grew in proportion to the demands and needs of an ever-burgeoning war industry had never realized before in the American experience.

The real force behind any war was war industrialists who keep the war machine well oiled and the ensured that as long as there was a war huge financial profits would continue to pour into their pockets. Realize that in many ways the downfall of many monarchies in Europe and other parts of the world, particularly in the mid to late 1800s, was due largely in part by the emergence of the mercantile system and the industrial revolution. The balance of power was gradually shifting from the ruling class, the monarchies, and the ecclesiastical order to the common people of every developing industrialized nation. Add to this was the continued international conflicts and wars that grew with each passing century supported with the buildup of massive armies and weapons development. The world stage was set for the outbreak of an immense global conflict or world war.

By the mid-Twentieth Century, **private industrialists** saw the opportunity to expand the industrial empires by ensuring that when there were wars they could always profit regardless of the outcome or who won. Obviously, when there was a lull or cessation of conflict where peace broke out, this meant building and selling armaments to potential enemies often times with the blessing of the US government, as long as the armaments were not recent development but were outdated weaponry. And when there wasn't a war that would be the time to instigate political agitation, turmoil and provoke hostilities in order to get the war factories manufacturing weapons at full production. Thus, the value of war was made lucrative beyond imagination, banking cartels got in on the action, militaries were in full support as their needs for newer and better weapons were always a top priority, politicians could be bribed, universities could be involved to investigate and develop new theories and concepts in the sciences that would lead to cutting edge prototype technology.

Technology was now the number one priority in industry and the military and the solitary garage inventor was being pushed aside or his inventions were either being bought up or stolen outright, sometimes never seeing the light of day again, because of their revolutionary breakthrough nature. Computers were replacing the blackboards of schools, as they had already done in the military industrial complex of America. Huge government contracts to industry were replacing the intellectual curiosity.

All of a sudden other major institutions in society could be corralled to be cooperative with inducements, financial gain, and prestige or with threats of loss of tenure and institutional funding or awarding lucrative projects and programs to other universities and technical institutes that were more cooperative. America was now on a slippery slope downwards and for the most part, the American public was ignorant of the fact that their constitutional rights had been run roughshod over and were becoming increasingly null and void, replaced by the new up and coming corporation governments.

The fact that a few UFOs had been found in US southwest didn't hurt matters any and was a windfall initially for the military and eventually for their new marriage partners, the private industrialists.

If we accept that the US military has been in the business of crash retrievals of downed alien spacecraft since the '40s then, by the early to mid '50s when major scientific breakthroughs from reverse engineering of these ET craft had been understood, they began building prototypes, test flying them and commenced a full secret production of these new **ARV (Alien Reproduction Vehicles).** UFOs and the National Security State, the Cover-up Exposed 173 – 1991 by Richard M. Dolan; 2009; Published by Keyhole Publishing Company; USBN 978-0-9677995-1-3

The MIC must surely have had by the "60s their own fleet of flying saucers as we have stated earlier in a previous section of this book, operating in and out of Earth's atmosphere and probably well into deep space behind the cover of NASA and the numerous misperceived UFO accounts reported by the public.

Alternative 3 and the Breakaway Civilization

Without too much stretch of the imagination, these exotic reverse engineered spacecraft probably had interstellar capabilities for trips to other planets within our Solar System and possibly beyond. It is, therefore, conceivable that manned flights to Mars would be in the realm probability and that Mars could already be occupied by humans, as suggested by **Sgt. Major Robert Dean** and if there is some truth to the **Majestic Documents**, there might be alien colonists there right now and vindicating **Percival Lowell's** claims of life on Mars! From this point forward, it is clear that we are probably dealing with a **"breakaway society"** within the US and most likely, the expansionist ideas of transnational corporations and banking cartels have evolved into a **"breakaway civilization"** from the rest of the world.

If **MJ 12** or **M.I.C.** or the **Cabal** or whatever name that is currently (most recent terminology: **M.I.I.C. (Military Intelligence Industrial Complex)** in used by this secret group, has their own fleet of flying saucers, it also gives rise to the possibility that a British televised mock documentary titled: ***"Science Report: Alternative 3"*** or ***"Alternative 3: End Game of the New World Order"*** or simply as **"Alternative 3"**, may actually have in fact a basis of truth behind it and is a prophetic precursor to the concept of a breakaway civilization. Some explanation is therefore warranted to understand the relevance of the Alternative 3 documentary in today's current global geo-political and socio-economic situation.

Alternative 3 was originally scheduled to air on April 1,1977 (April Fools' Day) but aired in June 1977, which rocked British airwaves with a television program so controversial it was aired once and only once before being banned. It was then shown in Australia where the reception prompted yet another ban. Finally, it was optioned by NBC in the United States only to be roundly rejected by its **Broadcast Standards and Practices** department: it was too dangerous to be aired on mainstream network television and represented a major risk to the network. Since then, the program has lived on in infamy and has been the source of many a raging debate about its legitimacy and ramifications. http://www.paranoiamagazine.com/2013/01/alternative-3-end-game-of-the-new-world-order/

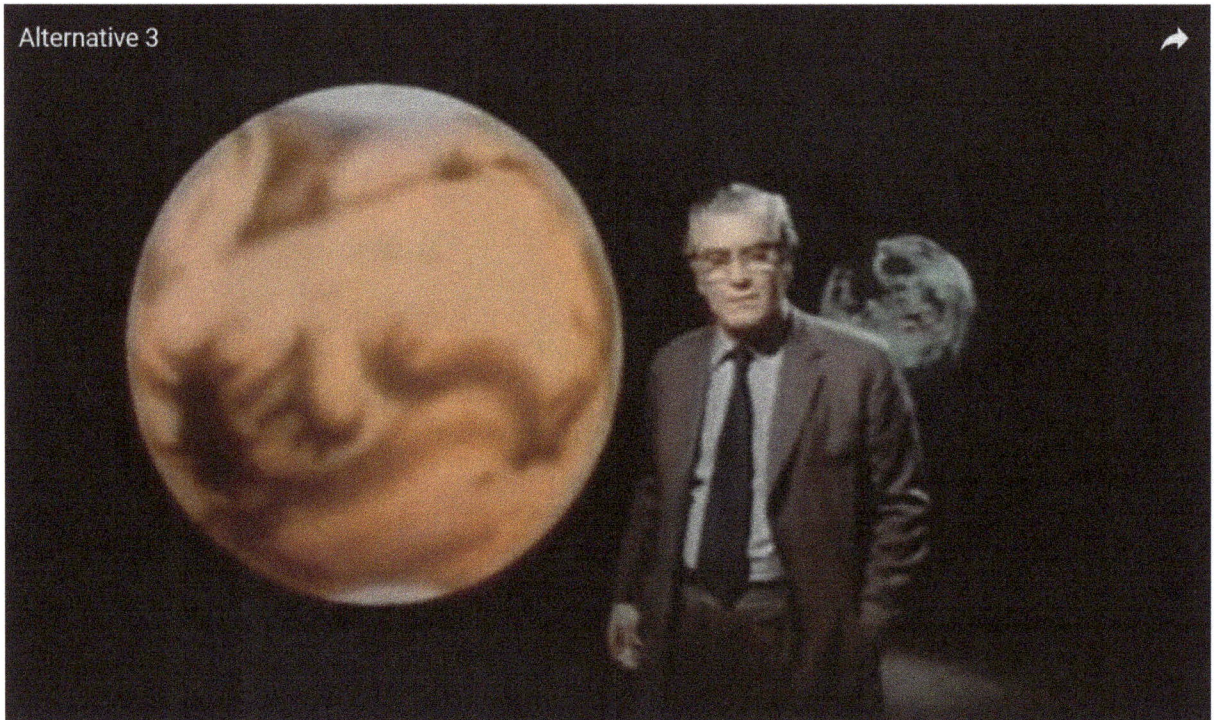

The TV movie Alternative 3 labelled "too dangerous to be aired" by NBC Network in the USA! Could it have been too close to the truth about humanity's future?
https://www.youtube.com/watch?v=gmNFzBVKqvE

The *"mockumentary"* is a grainy 10th generation VHS taped copy of a strange TV show aired in the UK in the late 1970s, the final episode being Alternative 3. It shows a newscaster and a few news reporters interviewing many of the common folk and the political leaders to track down an alleged story that the wealthy corporate elite and the politically powerful were preparing for the evitable demise of civilization on Earth. This elite group of society recognizing that the geopolitical and social chaos in the world, along with all its attendant cold war threats between nations, the pollution and environmental damage placed upon the Earth and its resources saw the evitable collapse and destruction of the human civilization as imminent. In an attempt to save the best of humanity, the wealthy, the politically powerful, the smartest scientists, the most influential religious leaders and a handful the best working class of society, were selected by proxy to become the ultimate survivors of a doomed humanity. They came up with three alternatives to save themselves. https://www.youtube.com/watch?v=jSDBl0FMX0s and https://www.youtube.com/watch?v=gmNFzBVKqyE

What could make a television show like **Alternative 3**, so dangerous to its viewers? In the minds of those in ultimate power and control, it connects all the dots of so many conspiracies in the public consciousness today, in a kind of elegantly rational and simplistic grand unified conspiracy theory.

Syndication of film rights and sales of film through diverse markets in the television industry is a financial cornerstone of business. In the case of Alternative 3, and as a direct result of the loss of all contracts legal, documents and master copies pertaining to *Alternative 3*, purportedly destroyed in a fire ITV can no longer assert authority over the show and has lost all ability to legally protect the show. It is as though Alternative 3 never existed. http://www.paranoiamagazine.com/2013/01/alternative-3-end-game-of-the-new-world-order/

Alternative 3 can be likened to **Orson Welles'** Halloween radio theatre broadcast entitled War of the Worlds, a kind of trick or treat for adults that aired over the radio airwaves on October 30, 1938. Interestingly, back at that time, many people in America had never heard or read of the story originally written by H.G. Wells from Britain and thus, were woefully ignore to its premise, believing that these were real events that were happening at that moment. Yet, as time marched forward, the Earth did have a sort of silent but, peaceful invasion of ET spacecraft commonly known as UFOs. Fact in this case did follow fiction to a degree.

Let's look at each of the three alternatives as well as their real life equivalents to better understand the hoax and the truth behind it. The story of Alternative 3 has many of the same elements of War of the Worlds that blend facts with fiction and it too may have more fact than fiction in its storyline premise.

Alternative 1 *scenario was to use a series of nuclear devices to attempt to blow a hole in the upper atmosphere to vent pollutants and gases into space.*

The 1958 high-altitude nuclear weapons test known as **Operation Argus** or **Floral** was ostensibly to test manipulating the Earth's radiation belts from which this alternative may have been inspired. Additional evidence suggests that the top-secret **High-frequency Active Auroral Research Project (HAARP)** may also be used for such a purpose. The Obama administration is currently investigating attempting to modify the atmosphere as well which include chemtrails of barium and aluminum at high altitudes. It's a modern implementation of Dr. Edward Teller's theory that global warming could be reduced or abated by jettisoning large quantities of metallic debris into the upper atmosphere to deflect sunlight and cool the Earth's surface.

In the storyline of Alternative 3, such projects have failed, so resources and planning have been shifted to Alternative 2. http://www.paranoiamagazine.com/2013/01/alternative-3-end-game-of-the-new-world-order/

Alternative 2 *recommended moving a percentage of the population into underground facilities.*

This project is ongoing, and if Alternative 3 is to be believed then it will probably continue to the very end. The evidence for Alternative 2 is ever- present around us, and the doomsday preparations going on are mind-boggling. We need to look no further than the Nordic Seed Vault, a kind of Noah's Ark for seed species. In the seed vault, *there are no genetically engineered seed varieties*. That means definitely, *no Monsanto seeds!* Other such vaults are either built or being planned around the world.

From the human point of view, one among many examples is **Camp David or Site R**, which can easily maintain fairly large populations for years if not decades. Such facilities exist all over the world. In Norway is the **Sentralanlegget**; in France, a hardened facility in **Taverny**; in the United Kingdom, **Whitehall**—each facility spawned by the cold war and retrofitted today for an impending manmade or natural apocalypse.

Nor should we forget the **Denver Airport**, central to **NORAD** and massive governmental migration, indicates that something is being planned. (For underground facilities, including a worldwide maglev train network, read the excellent work of Richard Sauder.)

As with **Alternative 1**, from a program point of view **Alternative, 2** was deemed a failure, due in part to the inability of various governments to build fast enough facilities that can survive long enough and hold enough people. http://www.paranoiamagazine.com/2013/01/alternative-3-end-game-of-the-new-world-order/

This leaves the ultimate alternative, **Alternative 3**: *to move a selected population off-planet to Mars, utilizing the Moon as a way station—a grandiose and desperate plan, and the most nebulous.*

In the cases of Alternative 1 and 2, there is more than enough evidence that these projects are real but, obfuscated as cold war survivability projects against the real likelihood of a nuclear weapons war or for the building of underground command-and-control bunkers. But in the case of Alternative 3 of moving selected populations off-planet implies a secret space program of some capacity. Given the time frame of this story, absolute proof back then was an impossibility, even with the notion that not all UFOs seen during this time period were alien spacecraft.

Connecting the *"conspiracy dots"*, the evidence for Alternatives 1 and 2 or the modern derivative of such scenarios is straightforward and undeniable at this current time. Though nothing in the clandestine world of *national security* is ever straightforward, the tentacles of Alternative 3 touch almost every modern conspiracy we have seen, read or heard—the grand unified glue of a 50-year conspiracy framework. Again in 1977, there was no "smoking gun" style evidence that included a third scenario for elitist survivability, but there is now abundant evidence throughout this book to prove convincingly that Alternative 3 is not only a real possibility, it may already be an established fact!

Delving directly into Alternative 3, itself is a much more difficult and murky proposition. We get glimpses of elements of the larger project manifesting as little clues left in strange and out-of-the-way places; manifesting in photos, rumors, and minor comments made by notable figures like former **President Dwight D. Eisenhower** (about the military-industrial complex**), Buzz Aldrin** (about a monolith existing on one of the Mars' moons), and **Richard C. Hoagland** (about an artificial moon of Mars). The glimpses are there but difficult to ferret out. http://www.paranoiamagazine.com/2013/01/alternative-3-end-game-of-the-new-world-order/

 In order for the Alternative 3 scenario to work according to our outline, we must have spaceships capable of reaching not only the Moon but Mars. These ships must be capable of carrying large amounts of cargo and personnel, be they "batch consignments" or scientists with families. So we need a vessel powered by the Bell power source and large enough, such as the *Haunebu* electromagnetic disc, two or three of which were in development late in World War Two. The Haunebu was capable of at least Moon travel by the late 1940's.

Inductively, the evidence supports the history of the Nazi secrecy behind the development of wonder weapons - flying saucer craft during WWII; the successful Apollo Moon landings of the 60s and 70s; the capture, retrieval and reversed engineering of crashed ET spacecraft, all are real facts! The validity of these **alien reproduction vehicles (ARVs)** has been established, their prototypes tested in the Earth's atmosphere and in low orbit and deep space. Support facilities and infrastructure for these spacecraft are located underground worldwide with constant expansion and improvements on them taking place. To the healthy doubters and skeptics, there is corollary research weighing heavily in support of the **Alternative 3 Hypothesis**.

Revisiting the photographic images of the Moon that are available for public scrutiny when explored with respect to recently constructed facilities that appear to be out of place with the rest of the lunar terrain have revealed, as some researchers believe, a waypoint en route to Mars. This leaves two categories that need to be examined from among the populace of the planet's many countries and diverse cultures as candidates for inhabiting Mars, *those who are picked as individuals and those who merely form part of a "batch consignment."*

Recently, the **Webbot** team on **Coast to Coast AM** let it slip that a secret **"breakaway" civilization** has achieved **Type I Civilization** status in interstellar propulsion. (***This comes as no surprise to the author as all the benchmarks for a Type 1 Civilization appeared to have been occurring since the '50s and '60s and have been escalating ever since).*** Or take a discussion between researcher **Jim Marrs** and **George Noory** about the UFO development program run by **SS General Dr. Ing Hans Kammler** from the Skoda munitions works near Pilsen in then - Czechoslovakia. The stellar research of **Joseph P. Farrell**, who has specialized in investigating the **Bell (Die Glocke)**, a secret Nazi torsion devices powered by **"Xerum 525,"** takes you from inception to potential completion and dissects in great depth each element of the project. (Bold italics added by author for emphasis).http://www.paranoiamagazine.com/2013/01/alternative-3-end-game-of-the-new-world-order/

As this author has already stated there was once an ancient Martian civilization that appears to have apparently disappeared from Mars yet, there is evidence that Mars is actually undergoing a resurgence of life on its surface, a kind of re-genesis or a **Genesis II scenario**. The real question that needs to be answered: did the former inhabitants, the Martians, survive an ancient planet-wide cataclysm finally re-emerging from their hidden underground cities to live once more, upon the surface of Mars? Is it possible that there are new inhabitants to Mars, who appear to be terraforming the Martian landscape, people who are decidedly more human than Martian who may be part of an advanced vanguard from a breakaway civilization from Earth?

A Breakaway Civilization and the Birth of Type I Civilization

More and more of these covert developments have come to the public's attention in recent times through investigators and researchers who are interviewed on late night talk programs like Coast to Coast or on many of the internet blog radio programs that cover the whole gambit of UFO and paranormal topics in much of the same manner as **Art Bell**. Much of what we know has come from anonymous "deep throats" and whistleblowers within or in association with these black covert organizations who use second-hand informants or middlemen while still retaining their anonymity thus, making the disclosed information unverifiable or falsifiable. This may be its intent all along which then, discredits the UFO researcher and his body of work ends up becoming suspect. When this does occur, then the researcher loses credibility and the general public finds themselves not knowing what to believe in anymore. For the CIA agent/insider, it's mission accomplished. Chaos of the mind can be a useful tool and be more effective than stage crafting an event to convince someone that something is real or one that leads to an inevitable outcome.

Every once and awhile, however, there are true insiders with a smoking gun piece of evidence that can cause rogue groups and organizations to go into a tailspins where spin doctoring is required to bring the intended agenda back online in order to achieve its desired effect.

When people like **Edward Snowden**, formerly with the CIA, go public as a whistleblower revealing sensitive information like the cell phone companies being forced under national security laws to hand over all their customer accounts and their phone conversations to the CIA and the NSA, it causes bells and sirens to go off in the public domain. The perpetrators of such outrageous privacy infringing operations, the CIA and the NSA have become as mad as hell and are in free-fall damage control of informational with now, a desperate mission to get Snowden under their custody and punish him for disclosing what they felt should have been kept secret and secured. Snowden has been called a traitor by **Dick Cheney**, who is more than anyone else, is a traitor to the American people and to humanity. Many millions of people around the world are rallying and standing by Snowden who is currently in hiding in China, as a hero against American intelligence corruption.

Snowden did what any truly moral individual would have done and that is to shine a spotlight on potentially dangerous agendas against the rights and civil liberties of the individual and the public as a whole. The more the public spotlight shines upon these secretive groups and individuals, the less chance there is for them to complete and pull off their hidden agendas against mankind. Exposure of the **Cabal**, the intelligence agencies, and their corruption through **public awareness** is the biggest, most effective tool that can be wielded against them to short circuit their nefarious ways.

190

People like **Dr. Steven M. Greer, Richard C. Hoagland, Dr./Astronaut Edgar Mitchell** and **Richard M. Dolan, Joseph Farrell and Jim Marrs** to name a few are the kind of people who are the true American heroes of UFO/ETI Disclosure and of **Breakaway Civilizations**. These people continue through their herculean efforts to keep the public spotlight on the people and organizations that sequester knowledge and technology that would help mankind overcome its current global problems and crises.

Although the term **"Breakaway Civilization"** was probably first coined by **Richard Dolan** after his exhaustive and extensive research in his books *UFOs and the National Security State Volumes 1 and 2*, the term, however, has been bantered about for decades. Most Ufologists, historians, and sociologists must have been aware of the way certain segments of society were being pulled into a particular direction from the sixties onward, although, it was never really known or understood in its full context until Dolan brought the concept to the public's attention.

Dolan came to the realization that the deep secretive black budget world had actually become a civilization on its own with its own exotic technology, a different view of the cosmos, our place in it, and a different version of human history. This is a new avenue of research that will delve into the **Occult, Secret Societies, Central Banking, UFO secrecy**, the **Military Industrial Complex**, and the **Secret Space Program** comprising this **Breakaway Civilization**.

Who are they? All one needs to do is look at what is happening with the **Wall Street Bailouts (TARP)** and the manipulation of **Libor,** where approximately *$27 trillion has gone missing*. Where did this money go? It is obviously financing this Breakaway Civilization's space ships, mass underground bases, offshore floating cities, and off-world bases. Why have they completely lost the need to communicate with us, and when? That goes back to the Occult and Secret Societies and the role they played throughout Human history. It is also because our passive attitude to their accumulation of power throughout these 100 years. We have done nothing to stop them, even though the signs were everywhere and people have been warning for years.

Now they have *accelerated operations* because people are waking up. However, when their project of total control is finalized, (and it is on the horizon) this information will be suppressed and we will be the last generation to know the truth. We will have invisible "rulers from above and below" running a totalitarian planet with a one world government, one world central bank and a one world religion. http://breakawaycivilization.com/

This concept of a one world government and one world religion is an intriguing and complex pogrom aimed squarely at the human race and it has become a distorted perversion of genuine spiritual world order that is destined to bring about a golden era of civilization for mankind not just for a select few! This topic will be discussed in the next section of this book with a possible solution to the social problems of the world!

Rarely does a month go by without new Martian "faces" and artifacts being discovered by armchair planetary archaeologists. Most of them are of dubious profiles with only vague resemblances to terrestrial counterparts. More than a few demand significant "retouching" before they can be made to look like anything in particular. There are a lot of wannabes in the UFO community who would like to make a name for themselves and be counted as part of the elite team of Ufologists who make real contributions to science on behalf of the rest of us. So who's to say what might be a candidate extraterrestrial artifact and what's not?

And as with most anomalies, the august priests of orthodox science at NASA and JPL maintain their dignified silence. This has become NASA's undoing in recent years as it comes under fire and increasing investigation for its alleged cover-ups and public suspicions of gross incompetency. http://archives.weirdload.com/nasa-shame.html

According to a growing number of scientists and researchers NASA is hiding evidence of ancient technology scattered across planets and moons in the solar system, manipulating photos of artifacts on the Moon, Mars, Phobos, and Titan, ignoring evidence of fossils on Mars, dismissing proof of life on the Red Planet first discovered more than 40 years ago and massaging the data gleaned from satellites and computer modeling of climate change.

Some question why NASA would hide, ignore and censor evidence. Could it only serve a possible agenda to ignite the public's interest in space exploration with new infusions of financial support for missions back to the Moon and beyond to mysterious Mars or is there a more covert and hidden agenda afoot similar to an ala **Alternative 3 scenario**?

Others say they have the answer: the United States of America has had a 'secret space program' since the 1960s. The US, they claim has had bases on the Moon, and maybe Mars too, for almost 40 years or more. I may also be that all these possibilities are all correct and interconnected as fall back scenarios should one or more fail to come to fruition.

The beleaguered space agency has fallen far since its glory days of Apollo. Initially led towards the heavens by its visionary genius, ex-Nazi rocket man **Werner von Braun**, NASA's gone from a rising star to a falling one. It's become politically correct and no longer has a focused mission or a manned space program. And if such a program does exist then, NASA is not revealing what that covert black space program is to the public.

It is hoped, however, that European eyes will view such anomalies with more scientific zeal than their American counterparts yet; events in recent years suggest very strongly that ESA has been influenced to get their space exploration missions, particularly on Mars, more in line with NASA's space programs. The Russian Space Agency, the **Canadian Space Agency** as well as the **Space Agencies of Japan, India** and other developing space-faring countries may also be implicated in certain Extraterrestrial discoveries from being made public. Why is this?

China is probably the only to date that is not playing ball with the US and its allies and for good reason, because they like an ever growing global public don't trust NASA to give them or anyone else honest information on its space exploration discoveries.

NASA faces hard choices and hard days ahead as the world's other space agencies like Europe, Russia, China, Japan, and India , etc. decide not to go along with the US and its hidden covert space programs, deciding instead to gear up in real earnest with their own lunar missions and Mars probes. If and when that day comes, then the proverbial crap will hit the fan for NASA and the US and this may be sooner than we know, in which case the covert agendas by the **Cabal**, **MJ-12** et al will accelerate and become so obvious that even the cave dweller will suddenly realize what has been going for decades. It may, however, be too late to stop a runaway train full of explosive cargo!

In the recent months of 2013, there has been a call for volunteers to go to Mars from among the public which seems contrary to NASA's former policy to not have the public too interested in all things Martian.

The proposed trip is scheduled for the year 2023 and already people are lining up from around the world as candidate hopefuls to go to the Red Planet. The list of volunteers requiring two men and two women initially, do not appear to be too concerned that it will be a one-way trip which will take over 10 months just to get there. Every two years new astronaut crews will be sent to Mars to colonize the planet. If the astronauts don't encounter problems along the way or die unexpectedly, their actual chances of surviving an unknown environment, regardless of what the NASA says, is very slim to nil even, with prior robotic supply ships sent out ahead of time to ensure survivability during this precarious adventure to Mars.

This would be for all intents and purpose a **Kamikaze suicide mission to Mars**, particularly if things go from wrong to worse and become unfixable. If the astronauts die en route to Mars then, the US via NASA would go down in history as being the first nation to put *dead people* on the surface of another planet! To add to mis-adventurous initiative is the fact that NASA is willing to do it over and over again in the hopes of getting it right!
The method of getting people to Mars will be the outdated technology of rocket power even if the spaceship is powered by a small nuclear reactor as its main engine and source of energy generation, these astronauts can never return.

This begs the question why send people in a technology that has never been needed for the last six decades, when NASA and the DOD can send them or anyone else to Mars expediently in comfort and safety in mere hours or minutes on board one of their anti-gravitic, electro-gravitational flying saucers or by entering into teleportation **" jump room"?**

And when these volunteer astronauts landed, they would be greeted by a contingent of welcoming US military officials and scientists who had already arrived on Mars before them, sixty years ago with *a wink and one finger over their smiling lips!*
Nudge, nudge, wink, wink! Say no more!! Know what I mean? Know what I mean? Say no more!!!

Let's hope for the sake of all parties concerned that they seriously reconsider this whole mission to Mars madness with saner heads and avoid the potential of billions of dollars wasted on an adventure that is less than well thought out and technically is not yet fully developed in the world of mainstream science.

CHAPTER 84

PHOBOS AND DEIMOS

Since we began this exploration of our local solar neighborhood we have followed the cosmic trail left by ETI visitors to our planet, it has led us to the Moon and to Mars and beyond Before we leave Mars entirely our next stop is to the two asteroid size moons **Phobos** and **Deimos** orbiting Mars.

These asteroid Martian bodies were discovered in 1877 by **Asaph Hall** and were named after the mythological Greek and Roman characters of **Phobos** (panic/fear) and **Deimos** (terror/dread). In Greek mythology, they accompanied their father Ares, the god of war, into battle. Ares was also known to the Romans as Mars. It is thought possible that Mars may have moons smaller than 50 - 100 meters and a dust ring between Phobos and Deimos may be present but, none have been discovered. http://en.wikipedia.org/wiki/Moons_of_Mars

There has been much speculation as to what these two asteroid moons really are in nature and origin since their discovery due to their unusual albedo ratio. Were they natural lunar bodies or captured asteroids or of artificial manufactured origin that had maneuvered and parked themselves into orbit around Mars.

When Viking 1 and2 went to Mars, they photographed the two moons as it settled into their Martian orbits and NASA concluded from the images that they were natural asteroid bodies that had been captured by Mars into close orbits which are gradually degrading. The USSR, as Russia was known at the time when the USA launched the Viking probes to Mars, thought and concluded differently that these asteroids were more artificial and perhaps were actually massive hollowed spaceships orbiting Mars because of their unusual reflective brightness to the incidence of light.

This theory was first postulated by famous Soviet Astronomer and Astrophysicist **Dr. Iosif Samuilovich Shklovsky**, who calculated that Phobos orbit around Mars is not only faster than it should be, but is decaying. From this Shklovsky concluded that Phobos is actually a mammoth spaceship. This theory was first postulated by famous Soviet Astronomer and Astrophysicist Dr. Iosif Samuilovich Shklovsky, who calculated that Phobos orbit around Mars is not only faster than it should be, but is decaying. From this Shklovsky concluded that Phobos is actually a mammoth spaceship. **"Mars' Moon Phobos May Be The Death Star"; published by Martin J. Clemens; 26 March 2013** and http://martinjclemens.com/mars-moon-phobos-may-be-the-death-star/

Could Phobos be a hollowed-out space station of gigantic proportions?

The Soviet Union decided to launch it own space probes to determine for itself what was nature of the Martian moons. **Phobos2**, as you may recall, was to be the beginning of a joint Soviet/US venture to study Mars and its moon, Phobos. The Phobos program was an unmanned space mission consisting of two probes **Phobos 1 and 2** launched by the Soviet Union to study Mars and its moons Phobos and Deimos. **Phobos 1** launched on July 7, 1988, and Phobos 2 on July 12, 1988, each aboard a **Proton-K rocket**.

The Russian Space Agency Spacecraft Phobos
https://motherboard.vice.com/en_us/article/the-probes-and-cons-of-russia-s-martian-moon-shot

While en route to Mars, Phobos 1 suffered a terminal failure, never reaching its destination. Phobos 2 attained Mars orbit and returned 38 images with a resolution of up to 40 meters, but the communications with the probe were lost prior to the planned Phobos lander's deployment.

Everything was going according to plan with Phobos 2; the spacecraft aligned itself with Phobos. Its ultimate goal to transfer to an orbit that would make it fly almost in tandem with the Martian moonlet and explore the moonlet with highly sophisticated equipment that included two packages of instruments to be placed on the moonlet's surface then, suddenly strange things started happening. On the 28th March 1989 the Soviet mission control center acknowledged sudden communication "problems" with the Phobos 2 and **TASS (Telegraph Agency of the Soviet Union)**, the official Soviet news agency, reported :

"Phobos 2 had failed to communicate with Earth as scheduled after completing an operation yesterday around the Martian moon Phobos. Scientists at mission control have been unable to establish stable radio contact."

The malfunction was a mystery, what could have happened to the Phobos 2 spacecraft whenever

everything seemed to be functioning as expected?
http://www.bibliotecapleyades.net/marte/marte_phobos05.htm

According to **Boris Bolitsky**, science correspondent for **Radio Moscow**, just before radio contact was lost with Phobos 2; several *unusual images* were radioed back to Earth, described by the Russian as *"Quite remarkable features"*.

A report taken from *New Scientist* of 8 April 1989, described the following:

"The features are either on the Martian surface or in the lower atmosphere. The features are between 20 and 25 kilometers wide and do not resemble any known geological formation. They are spindle - shaped and proving to be intriguing and puzzling."

The "anomaly" seen in the Phobos 2 transmission was a thin ellipse with very sharp rather than rounded points (a marquise shape) and the edges, rather than being fuzzy, stood out sharply against a kind of halo on the Martian surface. (See photo below).

Dr. John Becklake of the London Science Museum described it as "something that is between the spacecraft and Mars, because we can see the Martian surface below it," and stressed that *the object was seen by both the optical and the infrared (heat seeking) camera.* So what could have been "interfering" with the photo-shoot of the Red Planet? According to many Ufologists, this image was proof of extraterrestrial contact.

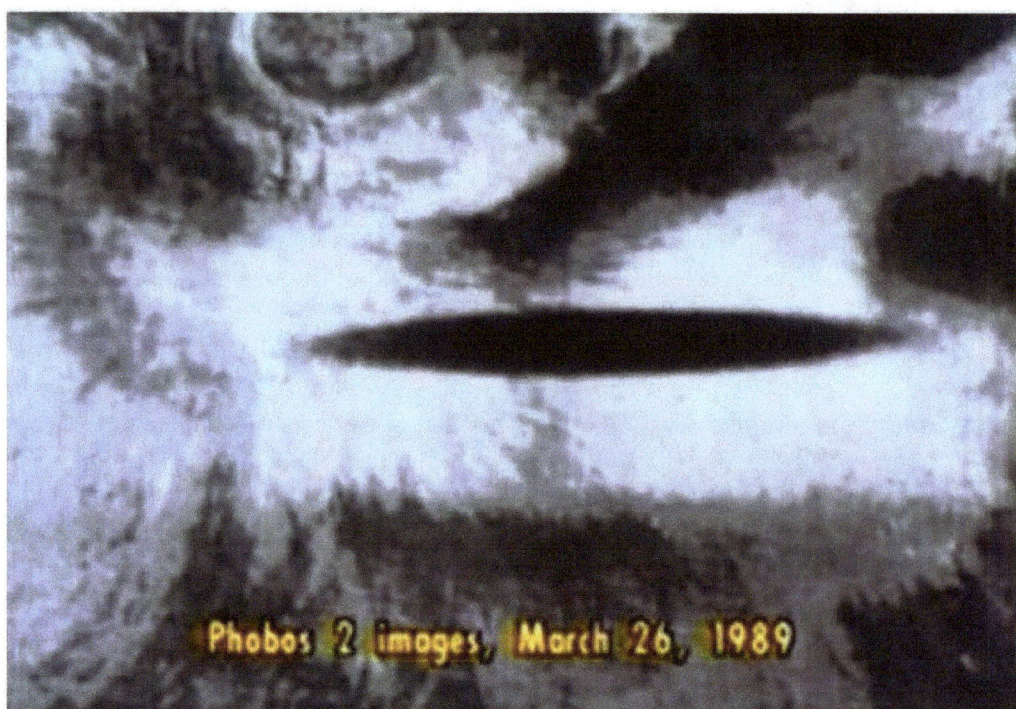

Phobos 2 images, March 26, 1989

Is this 20 km plus size object an ET spacecraft photographed by the Russian Phobos 2 before it stop transmitting back to Earth and disappearing altogether?
http://www.cover-up-newsmagazine-archiv.de/phobos-sonde-filmte-ufo

A Soviet consideration was recognizing the symmetry of the dark "thin ellipse" which ruled out the possibility that the shadow cast onto the Martian was not the shadow from its moonlet Phobos. The shadow of Phobos recorded eighteen years earlier by Mariner 9 was much different, casting a shadow that was a rounded ellipse and fuzzy at the edges, as would be expected by the uneven surface of the moonlet.

A strange as this image appears to be of an immense 20 – 25 km long object flying between the Russian space probe and the surface of Mars, there was one more image taken before the Phobos craft went silent. Russian space officials didn't want to release it to the public as it was something that should not be there, something that apparently hit the spacecraft sending it tumbling out of orbit, either into space or causing it to crash on Mars.

So, what was it that collided or impacted into Phobos 2? Was the space probe shot out of space for seeing or photographing things that it should not have? What is in the last secret frame from Phobos2? This elusive photo was eventually tracked down and in a carefully worded interview with "Aviation Week and Space Technology", the chairman of the Soviet Space Agency referred to the last frame, saying,

"One image appears to include an odd-shaped object between the spacecraft and Mars."
http://www.bibliotecapleyades.net/marte/marte_phobos05.htm

This "highly secret" photo was later given to the Western press by **Colonel Dr**. **Marina Popovich**, a Russian astronaut and pilot who has long been interested in UFO's while attending a UFO conference in 1991. At the conference, Popovich gave to certain investigators some interesting information that she "smuggled" out of the now ex-Soviet Union. Part of the information was what has been called "the first ever leaked accounts of an alien mother ship in the solar system". This statement in hindsight years later proved to be a correct assertion, as photographs of other massive UFOs or mother ships around the Sun and orbiting other planets have been imaged by space probes, as seen elsewhere in this section.

The last transmission from Phobos 2 was a photograph of a gigantic cylindrical spaceship, a huge, approx, 20km long, 1.5km diameter cigar-shaped "mother ship", that was photographed on parked near to the Martian moon Phobos by the Soviet unmanned probe Phobos 2.

After that last frame was radio-transmitted back to Earth, the probe mysteriously disappeared; according to the Russians, it was destroyed - possibly knocked out with an **energy pulse beam (EPB)**. http://www.bibliotecapleyades.net/marte/marte_phobos05.htm

It is interesting that the Russians thought this was a possible a high energy pulse beam weapon which was not well known publicly at the time of the space probe's lost, even though both the Americans and the Russians had been developing and testing such weapons. Even, to the point that an EPB was used to fire upon an Extraterrestrial spacecraft in near orbit around the earth as can be seen in the **video of STS-48** from **Space shuttle Discovery** in September 1991.

The cigar shaped craft in the penultimate frame taken by Phobos 2 is apparently the object casting the oblong shadow on the surface of Mars in the earlier photo. The shadow - spindle- or cigar-shaped - is inconsistent with any possible shadow cast by the moon Phobos, which is an irregular potato shape.

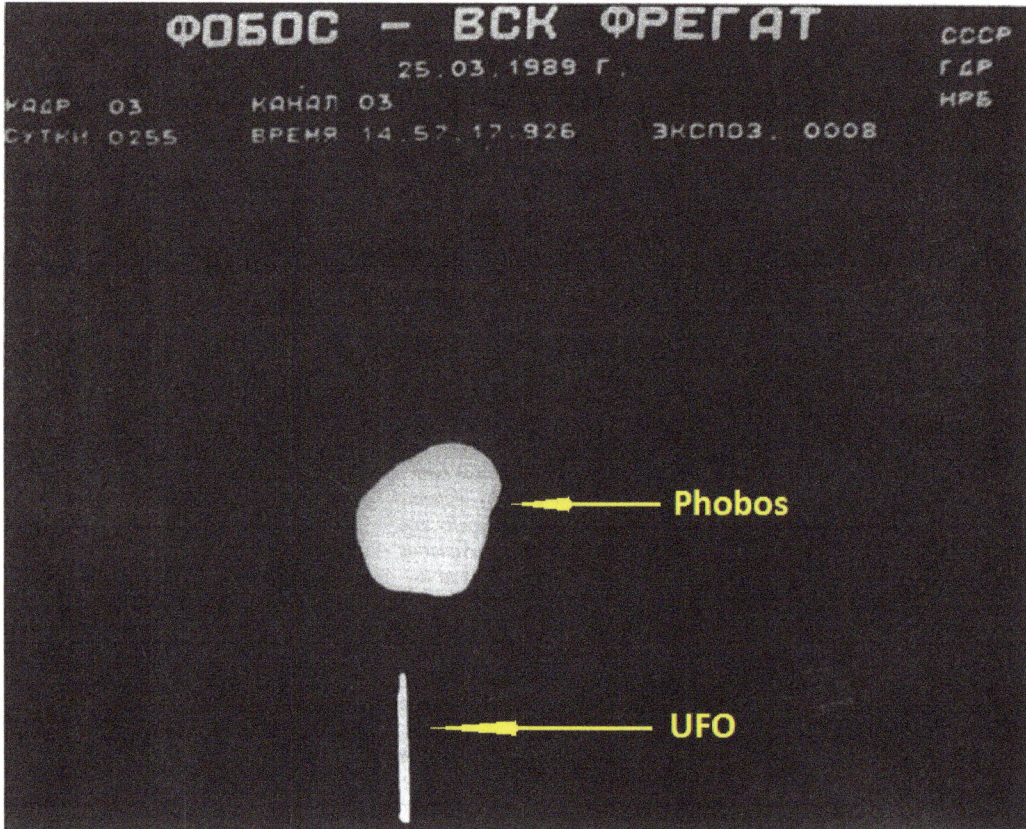

On March 25, 1989, an object 15.5 miles in length was photographed approaching the moonlet of Phobos, it is the same length as the UFO in the above image
http://www.abovetopsecret.com/forum/thread638427/pg2

The last image the Phobos spacecraft managed to return to earth is seen in the above image which shows a huge object just under the moon Phobos. The object seen in the images is several kilometers in length approx 20 km. Strangely, the "thin ellipse" seen in the image above, is also approx 20 km in length according to scientists. Several explanations were given to the image as "plausible explanations" but their central claim is that the object is nothing more than "camera artifacts" and "lens malfunction". Others, such as **cosmonaut Marina Popovich,** who has long been interested in UFO's, claims this object to be an Unidentified Flying Object of Alien origin.
http://www.bibliotecapleyades.net/marte/marte_phobos05.htm

After that last frame was radio-transmitted back to Earth, the probe mysteriously disappeared and according to many Russians, Phobos 2 was destroyed. At the time just before these images were taken of an unidentified object that had apparently come up from the surface of Mars ending the space probe's mission, the Soviets had made a startling discovery. One of their scientific instruments aboard the ill-fated spacecraft was designed to perform specific infra-red type scans of the surface which in turn are transmitted back to Earth. According to one Soviet cosmonaut, what they had revealed was that there was a massive city just beneath the surface of the planet, evident by the heat signatures given off by the crust of the planet. The photo images

198

revealed rectilinear structures, a latticework resembling an aerial view of city blocks similar to Los Angeles about 60 kilometers wide, according to **Dr. John Becklake.**

A Canadian television (probably **CBC**) had aired as a weekly diary news piece on planetary science which presented an infrared scan radiometer image of the Martian surface that showed clearly defined rectangular areas. There were no corresponding surface features taken by regular cameras. This suggests the heat signature of what may be a set of underground cavern or channels that are just too geometrically regular to be formed naturally.

The television sequence thus released focused on two anomalies. The first was a network of straight lines in the area of the Martian equator; some of the lines were short, some longer, some thin, some wide enough to look like rectangular shapes "embossed" in the Martian surface. Arranged in rows parallel to each other, the pattern covered an area of some six hundred square kilometers (more than two hundred thirty square miles).

The television clip was accompanied by a live comment by Dr. John Becklake of England's Science Museum. He described the phenomenon as very puzzling because the pattern seen on the surface of Mars was photographed not with the spacecraft's optical camera but with its infrared camera - a camera that takes pictures of objects using the heat they radiate, and not by the play of light and shadow on them.

In other words, the pattern of parallel lines and rectangles covering an area of almost two hundred fifty square miles was a source of heat radiation.
http://www.bibliotecapleyades.net/marte/marte_phobos05.htm

Could the Phobos 2 probe been destroyed by a particle beam weapon of earthly design emanating from a secret multi-national, US lead space base located at some equatorial region of Mars? If this is the case and many researchers and scientist strongly feel that it is then , this is the first proof that not only is Mars habitable but, that an "**Alternative 3**" breakaway civilization scenario has been taking place since the sixties and this is evidence of that agenda! The destruction of the Russian Phobos probe was a way to ensure secrecy of the underground Martian base built and occupied by humans in a US lead coalition of nations and mega corporations and private industrialists.

In fact, the one thing that was not stated in the above discussion of the Alternative 3 TV documentary is that there is a realistic looking video near the end of the program with alleges an actual landing on Mars by a joint American and Russian astronauts/cosmonauts crew. It depicts blue skies and a red Martian landscape as viewed through a glass portal of the spacecraft (flying saucer or ARV?) as it flies over the terrain and eventually lands. This Martian atmosphere and landscape are exactly as we have illustrated in this section on Mars and this movie is apparently a precursor to those facts.

As the spacecraft lands on Mars, the crew immediately sees movement under the Martian soil close to the spacecraft. We then hear the exclamation from an American astronaut shouting that there is life on Mars! *He continues to say that this day May 23, 1963 (?) would go down in history as the day man landed on Mars and that they discovered life on Mars!*

The news anchorman (actor) concludes the documentary facing into the TV cameras and asks the viewers, is Alternative 3 real and did the Mars landing take place? It is left to you to decide! Could this have been a genuine leak by an insider to get actual information out to the public by disguising it as a mock documentary program or is it just a clever TV entertainment that was originally scheduled for April Fool's Day?

The other possibility conjures up all kinds of speculation with some obvious undertones that the alternative is more ominous in its implications. If this is not proof of a secret US base on Mars but, part of a Martian civilization wishing to not reveal itself as yet. However, by this very action of crippling or destroying a spacecraft with a massively huge weapon or spaceship, would imply that not only is there intelligent life on Mars but, they are more than capable of defending themselves with highly advanced weapons!

If this is what really took place back in 1989 then, NASA, the RSA and all other space agencies and governments need to seriously reconsider all future missions to Mars as they may eventually encounter the people of Mars and they may not take kindly to humans landing unannounced on their planet! We as an intelligent species can ill-afford to perpetuate our outdated precepts and behaviour of war-mongering imperialistic tendencies, upon another planet and its civilization. Our motives for exploration should be altruistic and peaceful by extending the hand of friendship to any and all Extraterrestrial civilizations that we encounter!

There is other evidence that intelligent life currently exists on Mars or very nearby. The last few years the twin moonlets of Mars, Phobos and Deimos have been seriously considered as being massive hollowed spacecraft. The some of Russia's scientists who are more open minded to this hypothesis than their American counterparts or at least less secretive and more vocal about it than NASA and its scientists. Even, Richard Hoagland, NASA's proverbial thorn in the side feels that these asteroid moons are hollow based upon detailed imagery from many photographs taken by the various space agencies. A high albedo surface, pitting, parallel striations, chain-line cratering and other anomalies on Phobos suggest strongly that this moonlet is not your normal asteroid moonlet but, an artificially manufactured structure. In other words, it is a massive, hollow spacecraft deliberately placed in orbit around Mars in a distant ancient time long before human civilization was up and running around!
http://www.enterprisemission.com/Phobos.html and
https://www.youtube.com/watch?v=j4LArNlspjg

Mars and its orbiting asteroid moonlet, Phobos. Is it a hollowed spacecraft as some are lead to believe? Only a manned or unmanned mission to its surface will reveal that answer

http://www.utro.ua/ru/zhizn/nasa_odin_iz_sputnikov_marsa_mozhet_polnostyu_razrushitsya1447224586

If you are **"Star Trekker or Trekkie"** you will recognize parallels to one of the original Gene Roddenberry's Star Trek episodes where the main characters land on an asteroid and discover that it is a hollow world or mammoth spaceship. The world of astrophysics and space exploration in recent years seems to be gripped in a kind of planetary psychosis when it comes to Mars, a *"Mars Madness"* if you will. This madness can be viewed on internet subscriptions like Space.com, the science writers of which are professional spinmeisters in all things astronomical in nature that they wish the public to know. On the opposite scale, radio talk programs like **Coast to Coast** offer alternative news with guest speakers promoting their latest books or videos with the latest informative scientific research and inquiries to the pseudo-scientific and outrageous paranormal claims. Somewhere in all this gobblygook that's promulgated over the digital world of the internet is the common rationale of truth.

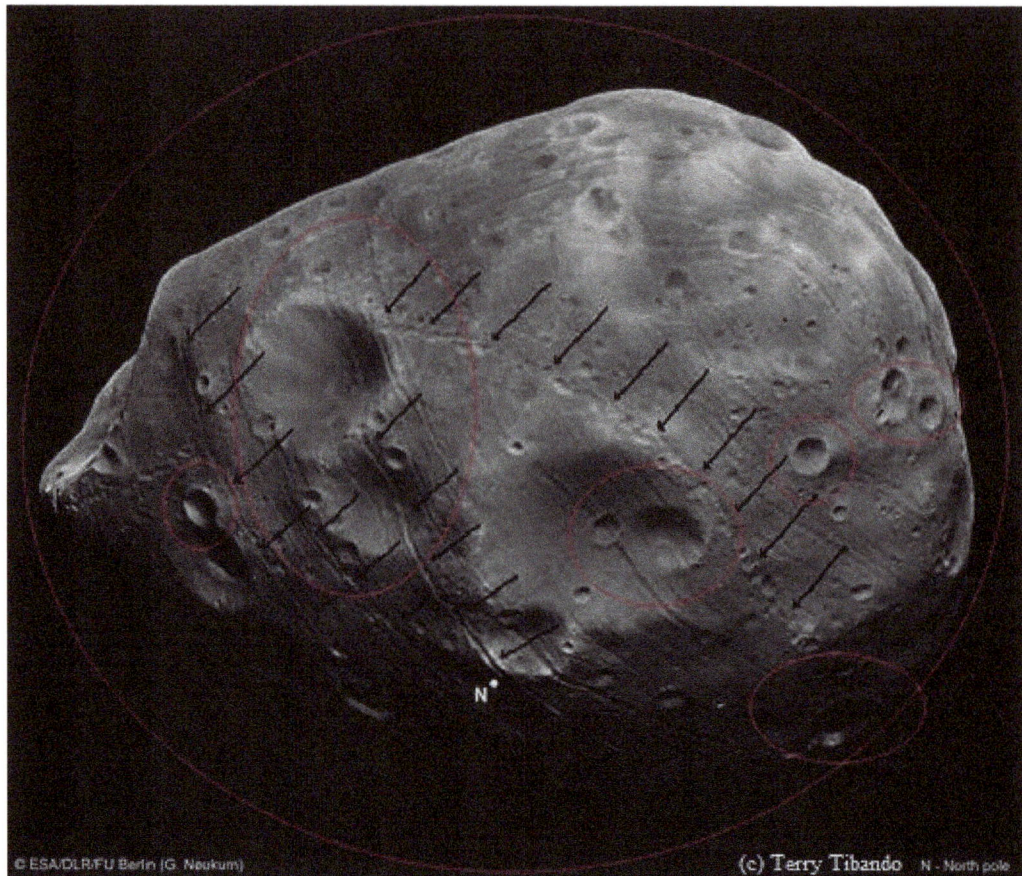

Phobos with black arrows indicating parallel striations, rectilinear geometry and chain cratering like so many dots in a row and red circle areas highlighting major craters with striations running straight through the craters

(c) terry Tibando https://apod.nasa.gov/apod/ap100317.html

Phobos and Deimos in the last few years was specifically targeted in this fervour for all things Mars or Martian. The controversy surrounding these asteroid moonlets has never really waned since these two small satellites were first spotted by Asaph Hall in 1877. Their sudden appearance and their strange orbits have been a source of questions for more than a century. Phobos in particular, aside from the ongoing discoveries on Mars that NASA finds on a daily basis, has been a focal point of many conspiracy theories.

Shklovsky theorized that Phobos, the inner moon of Mars was much, much lighter than previously thought, and was, therefore, subject to tidal forces on Mars, he concluded:

"There's only one way in which the requirements of coherence, constancy of shape of Phobos, and its extremely small average density can be reconciled. We must assume that Phobos is a hollow, empty body, resembling an empty tin can". **"Mars' Moon Phobos May Be The Death Star"; published by Martin J. Clemens; 26 March 2013 and http://www.paranormalpeopleonline.com/mars-moon-phobos-may-be-the-death-star/**

Mars has always elicited a huge amount of interest for millennia and there has also been the technology and the know-how readily available to astronomy for many years prior to the sudden appearance and discovery of both moons in 1877. For well over a hundred years the astrophysical community has realized this but, until that time no one had ever observed these moons orbiting Mars. Why is this? This science oddity supports the concept that Phobos and possibly Deimos are massive spacecraft orbiting Mars. Who built these spacecraft and for what purpose?

In 2010, the **European Space Agency (ESA),** now the second largest space agency in the world, officially and publicly stated its position on the matter of Phobos' artificiality with a very surprising announcement. ESA had concluded and revealed in their study abstract published in the peer-reviewed **Geophysical Research Letters** (a semi-monthly scientific journal published by the **American Geophysical Union**) that Phobos is contrary to what many astrophysicists and astronomers have thought for years, a captured asteroidal satellite.

*"We report independent results from two subgroups of the **Mars Express Radio Science (MaRS)** team who independently analyzed **Mars Express (MEX)** radio tracking data for the purpose of determining consistently the gravitational attraction of the moon Phobos on the MEX spacecraft, and hence the mass of Phobos. New values for the gravitational parameter (GM=0.7127 ± 0.0021 x 10⁻³ km³/s²) and density of Phobos (1876 ± 20 kg/m³) provide meaningful new constraints on the corresponding range of the body's porosity (30% ± 5%), provide a basis for improved interpretation of the internal structure. **We conclude that the interior of Phobos likely contains large voids**. When applied to various hypotheses bearing on the origin of Phobos, **these results are inconsistent with the proposition that Phobos is a captured asteroid.**"* **Casey Kazan, "European Space Agency: Mars Moon Phobos 'Artificial'" and http://realityzone-realityzone.blogspot.ca/2010/06/european-space-agency-mars-moon-phobos.html**

This means that at least one-third and possibly more of Phobos is hollow confirming precisely the same result as the Soviets reported from their own "mysteriously lost" Phobos-2 Mission back in 1989.

This can only be interpreted one way, according to **Casey Kazan**, and that is that Phobos is an artificially constructed spaceship that was built, obviously, by Extraterrestrials.

*"...the **MARSIS radar** reflections officially published on the official ESA Phobos website*

*contained explicit scientific data, from multiple perspectives, which strongly **"supported the idea that this is what radar echoes would look like, coming back from inside 'a huge… geometric… hollow spaceship'."*** **Casey Kazan, "European Space Agency: Mars Moon Phobos 'Artificial'" and http://realityzone-realityzone.blogspot.ca/2010/06/european-space-agency-mars-moon-phobos.html**

The evidence now supports the reality that "the human race via ESA has indeed confirmed its first, clearly artificial, *relic* world…a bona fide **"ancient spaceship"** ... called **Phobos!** http://www.enterprisemission.com/Phobos.html *"For the World is Hollow ... and I Have Touched the Sky!"*

It orbits closer to its primary than any other ***"natural moon"*** in the entire solar system travelling around Mars **every *seven hours, 39 minutes*** under ***"four thousand miles"*** above Mars' reddish sands from ***west to east, three times*** each Martian ***"day."***

According to Hoagland, is Phobos an ancient, Extraterrestrial, very battered, hamburger shape, 15 mile long Spaceship much like the asteroid spaceship of the Star Trek TV episode, "For the World is Hollow ... and I Have Touched the Sky! But, now, from the *Mars Express* ultra-high-resolution imaging, the **REAL** cause of these mysterious "grooves or striations on Phobos" is readily, ***geometrically*** apparent. Phobos as a clearly "artificial moon" is literally coming apart at the seams" because of the increasing, shearing ***gravitational forces*** caused by its slow ***"death spiral"*** into Mars Because of the inevitable "tides of Mars!"

In addition to this, the MARSIS radar imaging experiment aboard the Mars Express has also experimentally confirmed "a Phobos' interior *filled* with 'cavernous, *geometric rooms*, right-angle walls and floors detectable via the semi-regular *'structure of the returning, interior radar echoes* ...' as they were impressed upon the reflected MARSIS signals ....'"

Translation: MARSIS was physically *seeing* (via this radar) a *three-dimensional, totally artificial, interior world ... within* Phobos; and a "reflection void interior geometry" ... which *correlated* eerily with the earlier (lower-resolution) Phobos "interior gravity tracking data ...."
http://www.enterprisemission.com/Phobos.html

In other words, the asteroid moon Phobos has large buildings and other artificial structures contained within it that are geometrical in form and merely random natural rock formations.

One of the interesting anomalies discovered by the Russian Phobos 2 probe before it mysterious end was that it had detected a small atmosphere on the surface of the moonlet! How is that even possible? Phobos has a very weak gravitational field and yet, there it is. The only logical answer would be that the asteroid moonlet is hollow and the atmosphere is the venting of internal gases generated by activity from within!

NASA with all its wisdom and insight has assured other space agencies and everyone from scientist to the general public that the Phobos moon is nothing more than a captured asteroid with a hollow of containing ice that occasionally vents gases much like a geyser on Mars or Earth. Such is the wisdom of NASA!

If the Martian moon Phobos is akin to the asteroid/ spaceship Yolanda from TV's Star Trek in the example stated above and the scientists of ESA and other private researchers are like the crew of the starship Enterprise then, the "Oracle within Yolanda" must ipso facto be NASA, whose corrupt information has set the destination of the asteroid Yolanda and her people toward doom on an inevitable collision course with Daran V. But, we all know what happen in the end in that particular episode, the corrupt Oracle, and its maniacal machinations were short-circuited and reprogrammed with a course correction to a non-lethal destination. Truth, once again, won out!

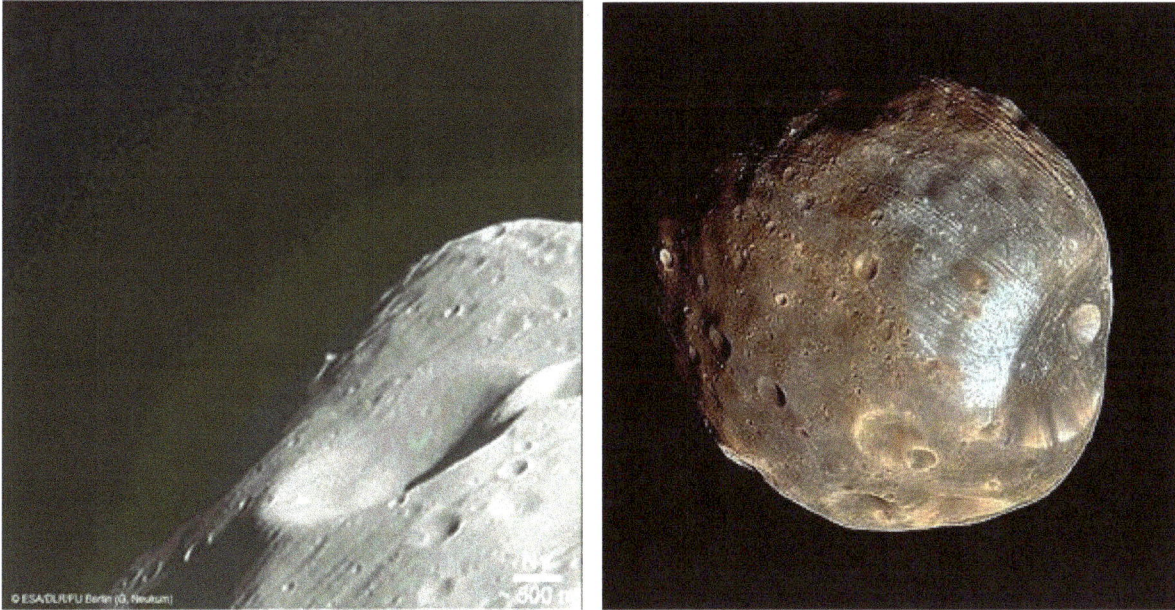

This ESA Image (left) shows Phobos with an atmosphere which was confirmed 20 years earlier by the Russian ill-fated Space probe Phobos 2. The colour image (right) shows Phobos reflective surface around its largest crater, Stickley
http://www.planet.geo.fu-berlin.de/eng/projects/mars/hrsc409-PhobosObservation.php and
http://www.galeriadometeorito.com/2015/04/10-curiosidades-sobre-marte.html

The Monoliths on Phobos and Mars

To top it all off, now there are claims, and photos, of a strange monolithic structure on the surface of Phobos. According to famous **Astronaut Edwin "Buzz" Aldrin** in a C-SPAN interview from several years ago: "We should visit the moon of Mars. There is a monolith there, a very unusual structure on this little potatoe-shaped object that goes around Mars,"
http://www.youtube.com/watch?v=oaiSfn8jlxY

As you might imagine, this theory hasn't gone unnoticed by the skeptics, and Kazan has been accused of perpetuating a hoax originally perpetrated by Shklovsky, though many contend that Shklovsky's original conclusions are valid and should be accepted. Most skeptics, however, ignore a lot of the facts choosing instead to point out evidence out of context. Supporters often cite the many failed attempts at reconnaissance of Phobos and Deimos, what with the disappearance of two probes sent to survey the moons and the technical problems that have plagued other attempts. Mainstream science meaning NASA, however, does not accept theories of giant artificial moons or huge spaceships and to date, the official story is that Phobos is

composed of highly porous *phyllosilicates.*

Highly detailed resolution images of Phobos clearly show that it is covered with monoliths which can be seen in the photograph below. Some are numbered as these are the largest of the monoliths which have been speculated as a kind of "2001: A Space Odyssey" type intelligent monolith or signpost pointing the way to further astronomical phenomena in the galaxy. Another theory is that these are exhaust vents or doorways to the inner hollow world within the asteroid moon as the above atmosphere photo image shows above.

THE MONOLITHS OF PHOBOS

(C) Terry Tibando June 2013

Space scientists and astronauts, like Buzz Aldrin, are intrigued by Phobos' numerous monolith objects on its surface highlighted in red circles with major ones being numbered. Are these monoliths natural rock outcroppings, exhaust vents, communication transmitters or signposts to other galactic destinations?
(c) Terry Tibando and (Credit: NASA JPL)

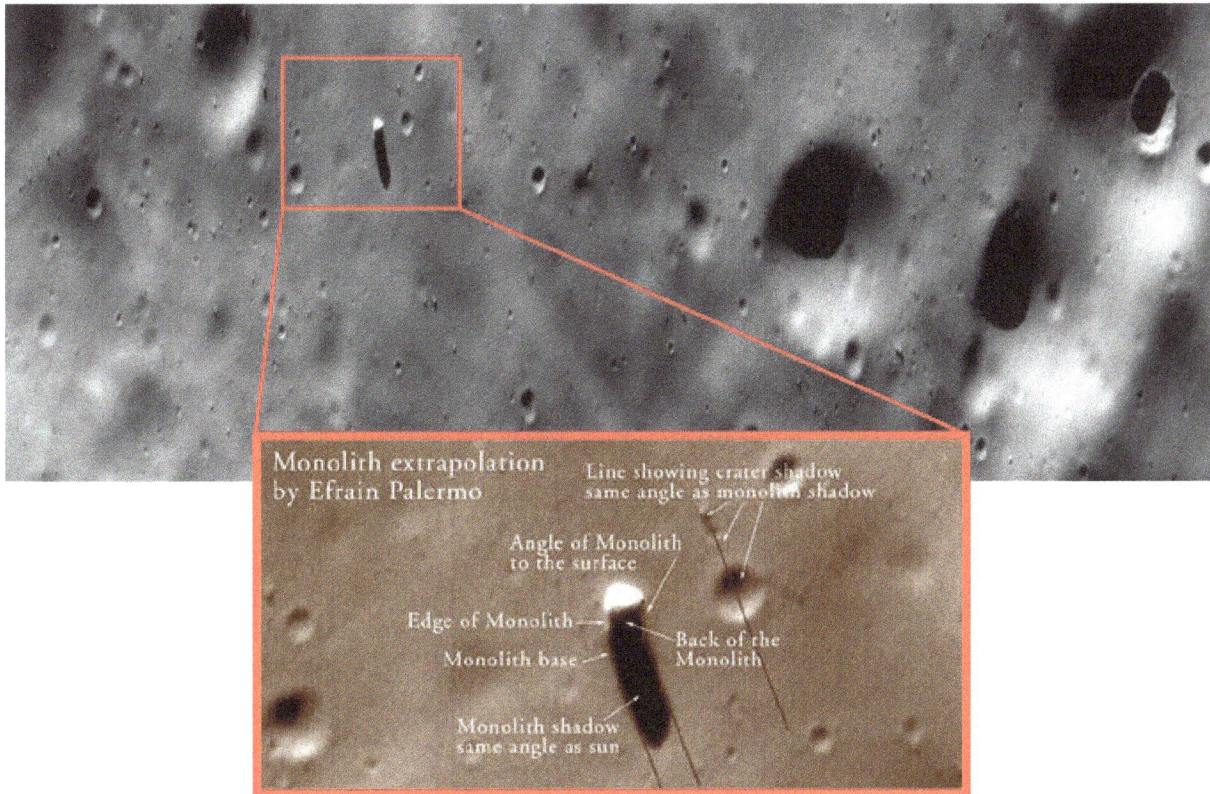

Monolith extrapolation by Efrain Palermo

Line showing crater shadow same angle as monolith shadow

Angle of Monolith to the surface

Edge of Monolith

Monolith base

Back of the Monolith

Monolith shadow same angle as sun

The Phobos monolith (left of center) as taken by the Mars Global Surveyor (MOC Image 55103) in 1998 with Efrain Palermo analysis of the monolith

https://ida.wr.usgs.gov/fullres/divided/orb_0551/55103h.jpg and
http://conspiracioneskilluminati.blogspot.ca/2015/01/los-monolitos-de-marte-cydonia.html

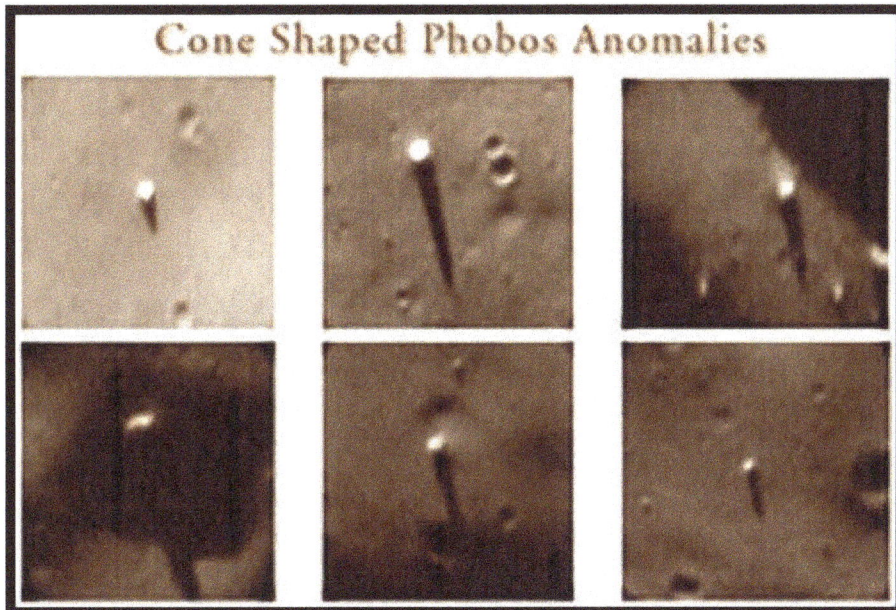

Cone Shaped Phobos Anomalies

Some of the many monoliths found on the surface of Phobos

http://palermoproject.com/Mars_Anomalies/Phobos_Anomalies4a.html

As explored in earlier installments, Mars' moon Phobos features several unexplained outcroppings, shown below.

Perhaps the most interesting of these is the tall feature dubbed the "monolith" by its discoverer, **Efrain Palermo**. Preliminary photoclinometric analysis by **Chris Joseph** shows that the "monolith" may be pyramidal. Since Phobos is a small, rounded body with no water or atmosphere to erode features into interesting shapes, the presence of the pyramidal "monolith" on Phobos is made doubly hard to explain.

However, at the time of this writing, there are many more monoliths discovered which may support the natural geological rock outcropping hypothesis. Yet, we find that there is uniformity in size of these objects and their positions are found on edges of grooves and crater rims or in craters across the moonlet which seems geologically unnatural as they would have been destroyed by impacts from external meteoric debris.

One prosaic option is that the pyramidal outcropping is a chunk of debris that was blasted from Phobos by a meteor strike. But Phobos' gravity is scant, and it's difficult to accept that a shard of debris would fail to achieve escape velocity. We're left with the intriguing option that the monolith is an artifact of unknown function.

Was Phobos once colonized by a technological civilization? A body of evidence suggests this possibility. Before photos of the enigmatic moonlet were relayed to Earth by the Mariner mission, Dr. Iosif Samuilovich Shklovsky, and Dr. Carl Sagan had written about the implications of Phobos' anomalous orbital characteristics. Mainstream astronomers were forced to confront the possibility that Phobos was a hollow artificial satellite.
http://www.mactonnies.com/imperative17.html

In the 1970s, physicist **Gerard K. O'Neill** published "The High Frontier," a blueprint for designing habitats in space which include details of hollowing out a captured asteroid that would allow for human population to survive within using the abundant raw material from the asteroid for self-sustaining microgravity industry.

Richard Hoagland, writing in "The Monuments of Mars," notes that the Martian moonlets seem "fortuitously placed," emphasizing possible artificial origin. Of the two moons, Phobos is the strangest, boasting an enormous crater that may be an opening into an unseen interior. Numerous straight furrows, or grooves, radiate from the crater. While some scientists have ventured that they are cracks caused by tidal stress, their origin is unknown and subject to debate. The Monuments of Mars: A City on the Edge of Forever by Richard C. Hoagland; 1987; published by North Atlantic Books; ISBN 1-55643-118-X

Hoagland goes further to state that close-up detail of the surface of Phobos reveals rectilinear structure that appears to be the metallic surface of the inner spaceship. There are doors or hatch-like entrances in the asteroid ship as well as interior cross beam structures. In other words, we are looking inside of an artificial world and Phobos appears to be coming apart at the seams, its interior structure is now exposed to the ravages of space!
http://www.enterprisemission.com/Phobos2.html

Phobos is the first definitive proof outside of the Earth - Moon environment of a gigantic **Extraterrestrial spacecraft** orbiting around another planet, Mars which is wholly supported by the European Space Agency! Its function may be similar to Isaac Asimov's concept of generational ships that travel between stars or across the galaxy but now, confined to a Martian orbit.

Isaac Asimov championed the idea of "generation arks," enclosed artificial worlds like O'Neill's, designed for long interstellar journeys. Equipped with homeostatic ecologies, Asimov's arks could carry passengers from star to star over durations of hundreds if not thousands of years; the original colonists who set off on journeys aboard the arks would likely not see the voyage through to the end. But future generations, born within the arks' delicate enclosed biospheres, could ultimately finish the journey initiated by their ancestors as per our Star Trek example above,

If the anomalies in Cydonia and elsewhere on Mars are in fact artificial, then perhaps the civilization that built them came from elsewhere, leaving its "ark" to bear mute witness. If so, future exploration of Phobos may ultimately prove as revealing as a manned mission to Cydonia.
http://www.mactonnies.com/imperative17.html

The Monolith on Mars and the Pyramids of Two Planets

Turning our gaze once more upon the Martian landscape, we find an image of what appears to be a mysterious rocky monument on Mars sticking out like the proverbial sore thumb and the excitement level among armchair researchers escalates around the world. A monolith on Mars is photographed by the high-resolution camera onboard the **Mars Reconnaissance Orbiter** and is published on the website ***"Lunar Explorer Italia"***. Is this further proof that there was once life on the Red Planet?

Where the general public sees things that appear artificial in appearance through perhaps, untrained eyes, scientists see every day common occurrences or natural phenomenon in unnatural settings. Their eyes are trained to see things the common public would not see or misinterpret. This would be a fair assessment of science and the public. However, not every situation or circumstance falls into these nice neat parameters of understanding. Quite often it is the public that sees unusual things not normal to the environment and this is where science comes in to make sense of the unusual. This is not a time to identify or make sense of things and then turn around and keep it a secret from the public regardless of the scientific findings.

Thus, scientists at the University of Arizona, who captured the original image, reckon the monolith is just an unremarkable boulder, which could measure up to five meters across.

Spinoza, a spokesman for the **HiRISE** department of the university's Lunar and Planetary Laboratory, gave Mail Online the original image so readers can make up their own minds. He said: *'It would be unwise to refer to it as a "monolith" or "structure" because that implies something artificial like it was put there by someone for example.*
http://www.dailymail.co.uk/sciencetech/article-1204254/Has-mystery-Mars-Monolith-solved.html

On the surface this appears to be good science and honest impartiality on a subject that can explode into controversy, therefore, the opportunity is there for the public to come to their own conclusions which science does not have to acknowledge or even agree with or take an opposing point of view. This is a nice safe position to take that allows an official position to be given at a later time.

Since this tactic by science and in particular by NASA isn't about to change, the public feels to a degree to be in the driver's seat when new discoveries are made by the Mars rovers.

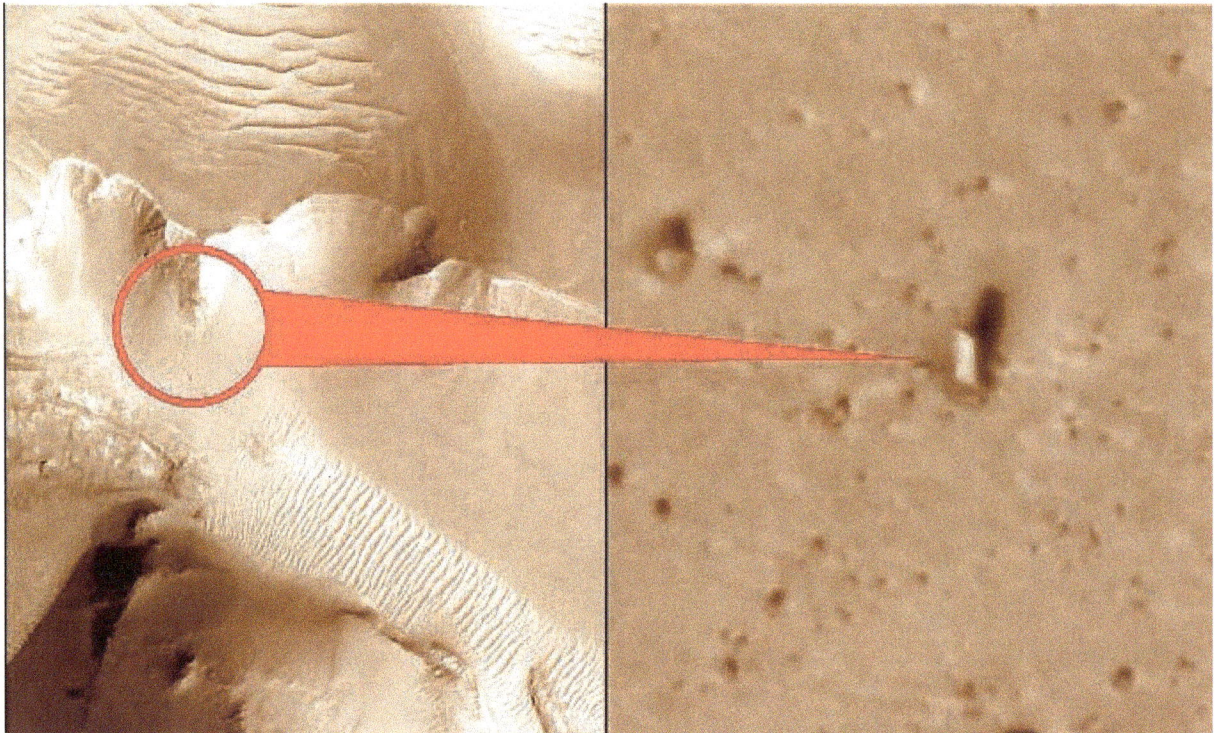

HiRISE PSP_009342_1725 taken on August 2009. The circled area shows where the rectangular feature was discovered
http://www.livescience.com/19636-monolith-mars-2001-space-odyssey.html

The monolith discovered is in reality more likely to be a boulder that has been created by breaking away from the bedrock to create a rectangular-shaped feature.

The image in the above photo resembles the monolith found on Phobos and similar to the black monolith seen at the beginning of the **Stanley Kubrick** film *"2001: A Space Odyssey"* as man evolves from an ape-like creature to become an upright intelligent being in charge of his own destiny.

In defense of NASA and this particular satellite picture, scientist **Alfred McEwen**, the principal investigator from the University of Arizona's HiRISE department, said:

"There are lots of rectangular boulders on Earth and Mars and other planets.

Layering from rock deposition combined with tectonic fractures creates right-angle planes of weakness such that rectangular blocks tend to weather out and separate from the bedrock." http://www.dailymail.co.uk/sciencetech/article-1204254/Has-mystery-Mars-Monolith-solved.html

The black monolith appears at turning points in the film 2001: A Space Odyssey
https://plus.google.com/+ZinzinNaming

Adding fuel to the flames of controversy and throwing doubt upon the official position of NASA and JPL regarding finding any artificiality on Mars, **Buzz Aldrin**, the second man to walk on the Moon alluded to a similar monolith detected on Martian moon Phobos.

Speaking on a U.S. cable television channel C-SPAN and later on other television networks overseas he said point blank:

'We should visit the moons of Mars.

'There's a monolith there - a very unusual structure on this little potato shaped object that goes around Mars once every seven hours.

'When people find out about that they are going to say, 'Who put that there? Who put that there?' Well, the universe put it there, or if you choose God put it there." (Bold italics added for emphasis). https://www.youtube.com/watch?v=Pnt5WKo6SCY

This news announcement went global on all major news networks and viral on the internet causing more than one reporter to ask the question: what does NASA know and what is it covering up with regard to Mars and its moons?
https://www.youtube.com/watch?v=uvkr9ZXnnGs

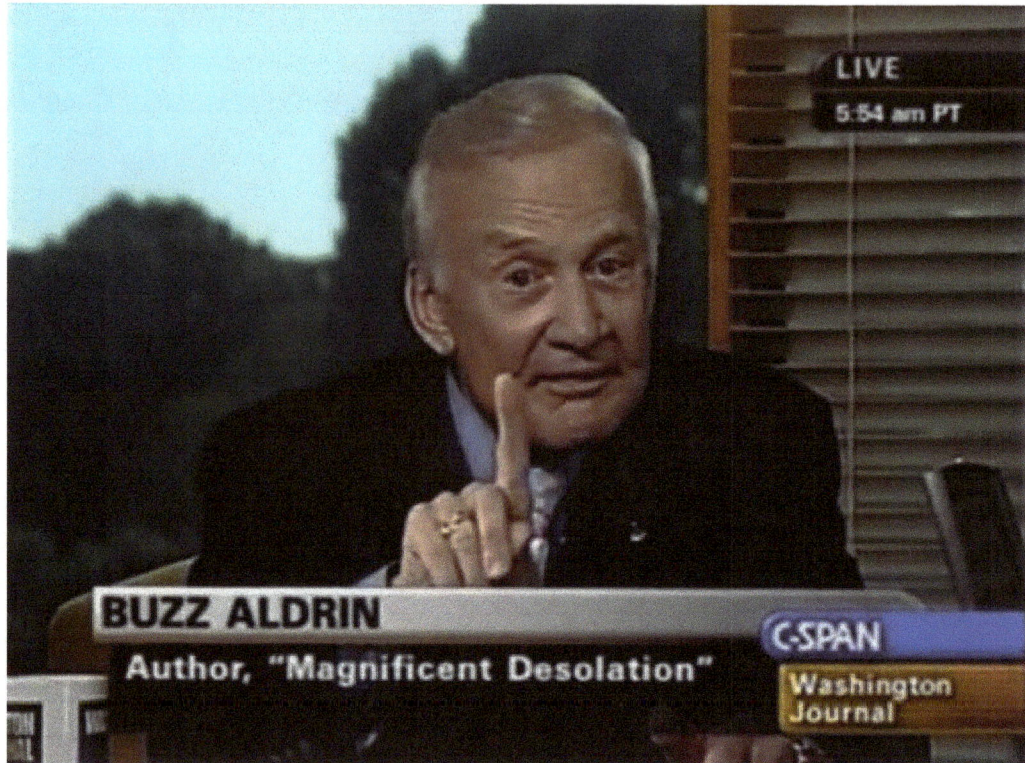

Former astronaut Buzz Aldrin describes how we should go to Mars and to take a closer look at a monolith found on Phobos recently discovered there

https://www.youtube.com/watch?v=bDIXvpjnRws

In 2007 the **Canadian Space Agency** funded a study for an unmanned mission to Phobos known as **PRIME (Phobos Reconnaissance and International Mars Exploration).**

The building-sized monolith is the main proposed landing site but not because scientists suspect UFO activity. They believe the object is a boulder exposed relatively recently in an otherwise featureless area of the asteroid-like moon.

PRIME investigator **Dr. Alan Hildebrand** said it could answer questions about the moon's composition and history. *'If we can get to that object, we likely don't need to go anywhere else,'* he told his science team. (Bold italics added for emphasis).

Now, why do you think that a one-stop landing on Phobos would be the only place you need to go when Mars is a stone throw away? Unlike the position held by NASA that is in constant denial or perpetuates a subterfuge of disinformation, could it be that the **Canadian Space Agency (CSA)** actually agrees with the European Space Agency? Could the CSA perceive Phobos as an artificial moonlet, a hollow world that is, in fact, a mammoth spacecraft on an order of kilometres in size? It certainly appears that way.

In the movie "2001: A Space Odyssey", the black monolith is seen repeatedly at important times in the movie, it imparts knowledge to ape-like creatures with a high pitch sound when it is touched. When a similar object is found on the Moon and astronauts touch the monolith, a high pitch sound is heard once again, imparting even greater knowledge beckoning humanity to travel to Jupiter. Are the monoliths discovered on Phobos and Mars the same kind of divine artifact or merely blocks of rock outcropping?

It is thought that the **Great Pyramids** of **Giza**, Egypt are constructed in precise formation as to emulate perfectly the three stars found in the "Belt" of the Orion constellation. The **Great Sphinx** is likewise aligned at certain times with the rising of the constellation Leo on the horizon. The **Teotihuacan Pyramids** in Mexico are a reproduction of our Solar System indicating all nine planets, including the Sun and the Moon, all correctly positioned as you would find them in our star system. The three major **Aztec Pyramids** in Teotihuacan also align to Orion's Belt, as do the three major **Xi'an Pyramids** in China!

| China: | Mexico: | Egypt: | Constellation: |
| Xi'an Pyramids | Teotihuacan Pyramids | The Great Pyramids | Orion's Belt |

Around the globe in many countries are ancient pyramid structures that are similar to each other and all of them align with the three stars in the belt of the Orion Constellation

On Mars, the **Cydonia City of Pyramids** (aka. Cydonia City) seems to represent a constellation that can be viewed when in that particular area of Mars. In fact, there are seven buildings in the Cydonian city which align perfectly with the seven stars of the **Pleiades** also known as the **"Seven Sisters"** in ancient time. Their Earth counterparts can be found at **Mayan Pyramids of Tikal** in Guatemala and in Britain at **Stonehenge**!

These are other ancient monuments around the world which have astronomical connections or alignments to stars and constellations that are visible at these sacred sites. Why do these sacred sites mirror what is in the heavens or skies above us? Do these ancient monuments act as mnemonic devices on a grand scale to remind mankind of past events or possibly its ancient origins? Were these monuments constructed by star visitors or did they influence and guide humanity in the construction of these edifices? Do these global monuments represent astronomical markers that point to where the star visitors originated from, which may also be our true home?

A young man by the name of **Wayne Herschel** from Cape Town, South Africa, and an author of the book "The Hidden Records" has presented some very original ideas about mankind's past. In Wayne Herschel's book and on his website, **thehiddenrecords.com** he tells of an alternative history of mankind's origins that argues convincingly that our ancestors were from the stars, to be precise from the **Pleiades** in the constellation **Taurus**! Herschel says we are descendants of star visitors who came to our Solar System tens of thousands of years ago. These visitors settled on Mars and Earth and influenced ancient cultures around the globe! Herschel theorizes that these star visitors may actually have replaced/eliminated or interbred with **Neanderthal hominids**, who were the original inhabitants of Earth, before the arrival of the star visitors from the Pleiades, which would explain the missing link in our evolution! All of this would have occurred about before and during the last ice age 20,000 to 12,000 years). It is likely that Neanderthals were genetically enhanced and their evolution was accelerated by the **Pleiadian ETI.**

Herschel states that his theory, for the most part, acknowledges a Divine Creator in the universe and is responsible for all creation through evolutionary processes and where all life forms fit an intelligent design, where all life conforms to a set *"blueprint code of Creation".*

The blueprint code of all life, however, seems to have been interfered with here on Earth as the change in evolution from Ape to Man appears to as a "quantum leap" in evolution. What should have taken another million years or more occurred within an extremely short period of time. A replacement event seems more plausible here, from a visitation… an arrival of our direct ancestors more than anything else and not the direct hand of God.
http://www.oneism.org/orig_right_wit.php

A new concept will be explored that explains the missing link between ape and human.

A few anomalies on humanity that do not fit the pattern of all life on this Earth:

- Modern Human fossils (Homo Sapiens-Sapiens genus) are found in rare and mysterious circumstances predating the abundance of fossil records found globally from approximately 17000 years ago. The findings inspired a book titled Forbidden Archaeology.
- Modern humans suddenly appeared on all continents shortly after Neanderthal became extinct around 17000 years ago.
- There is no genetic link between Neanderthal and Human genus, 'The missing Link' – This is now absolute scientific Fact.
- Humans have approximately a **30% less bone density** and strength than any hominid that walked on this Earth, any ape or any creature. Only bird's bones come close to the frail and light nature of our bone structure.
- Humans have approximately **30% less muscle connectivity** to our bones (i.e. 30% less efficiency due to less mechanical advantage with poor muscle connection) than any hominid that walked on this Earth, any ape or any creature.
http://www.oneism.org/orig_right_wit.php

214

It is as if we originated in a world that had a perfect gravity that would be a lot less than the Earth. For example, if we lived in gravity close to half of what it is today our human ability and efficiency would equal any ape or hominid. Another anomaly is that out skin is not compatible with the harsh strength of our Sun. It is as if we lived in a world that had 'perfect' solar radiation… close to where Mars orbits the Sun! Oh yes… by the way… Mars might have had approximately half the gravity of Earth before it was destroyed by an impact event causing a 2000 km wide impact crater (the largest known impact event in our solar system) where the planet lost its oceans.

New evidence presented by the author of The Hidden Records reasons how humanity is a descendant of those that evolved in another star system. Our ancestors appear to have arrived here replacing the last Earth Hominid… the Neanderthal, around the time of a cataclysmic flood event. http://www.oneism.org/orig_right_wit.php

Herschel stumbled on 'Da Vinci's secret geometry' at Giza in Egypt and realized that all 50 of the pyramids of Lower Egypt appear to represent the brightest stars forming constellations along the length of the Milky Way. But more importantly, there is the proverbial "**X**" that marks the spot and that claims cryptically that human lineage is associated with what appears to be another Sun-like star. The author also proves that Stonehenge has the same cosmic solution and that the pyramid layouts of the Maya, the Incas, the Khmer and many others around the world, all repeat the same star map theme. They all exhibit the same obsession with one nearby Sun-like star in Taurus, a star that the ancients passionately refer to as the star of their gods.

Apart from decoding the meaning of a hidden star map that recurs in pyramid layouts worldwide, The Hidden Records exposes an as yet undiscovered **Da Vinci** secret. Herschel has been researching what he refers to as the Da Vinci codex in the artist's **Vitruvian Man**, which he believes is encrypted as a universal blueprint in all ancient civilizations. This codex is found at Stonehenge and Giza, and in a place not of this world *(Mars).* The new findings could potentially be the ultimate archaeological find... a record of the genesis of humanity. The codex intricately replicates a 'sacred' area in the heavens, isolating one specific star... it is a nearby Sun-like star and it can be confirmed scientifically. Was this the home star of the "gods" who came to our Solar System and to Earth from above?

Herschel's "The Hidden Records" presents the strongest evidence to date that humans have never ever been alone in this vast universe yet, even with his extensive travels to many ancient sites to record and document his discoveries, the news of his work to this day is relatively unknown publicly or even in the UFO community.

In his book, **Herschel** demonstrates that the pyramid ruins of **Cydonia** on Mars and at **Tikal** in Guatemala mirror the Pleiades star systems. In turn, the Mayan pyramids of Tikal match the grouping of pyramids on Mars on their own. How could this be just a coincidence? He further points out that every little anomaly in the Cydonia region has its Earth and stellar counterpart. The important meaning of these star maps (see below) is that they are "**X** which marks the spot". The **X** is the location of these the monuments at Tikal, Stonehenge, and Cydonia, etc. As already suggested, the "home star" of the visiting ETI correlates with the **Face on Mars** structure, only if a star atlas computer program is set to run back in time to an ancient epoch then, the monuments on Mars align perfectly with the stars of the Pleiades!

Tikal (left), Stonehenge /Stone Burrows (right) and the Cydonia Pyramid City (below) all have the same stellar alignments as the Pleiades , proof of an Extraterrestrial Intelligence having visited both the Earth and Mars in the distant past
http://thehiddenrecords.com/mars.htm and http://allgreatworld.blogspot.ca/2011/08/are-ancient-techs-is-aliens-techs.html

The highlight of the Hidden Records book has to be this comparative measurement (below) between all the stars of the Pleiades and all the anomalies in the Mars cluster in Cydonia. There are a spectacular 21 correlations for the small stars matching small anomalies. There are six large magnitude stars correlating with six medium sized anomalies and one primary large star of **Alcyone** correlating with another five-sided anomaly (like the larger D&M five-sided pyramid anomaly which still has one equilateral triangle side perfectly intact).

All in all, 28 spectacular correlations where the Mars presentation still has one major piece of evidence that makes it far beyond just coincidence. It presents a perfectly geometry adjusted layout, altered fractionally not to make it a poor correlation, but ever so slightly adjusted to suggest it is constructed by advanced perfectionist beings... our ancestors. Note the parallel fine lines in yellow, the main shape between the cluster forming a trapezium, and a right angled triangle with two 45 degree angles. **http://thehiddenrecords.com/mars.htm**

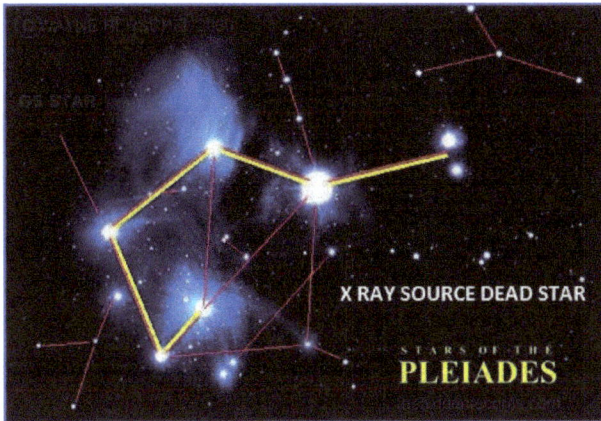

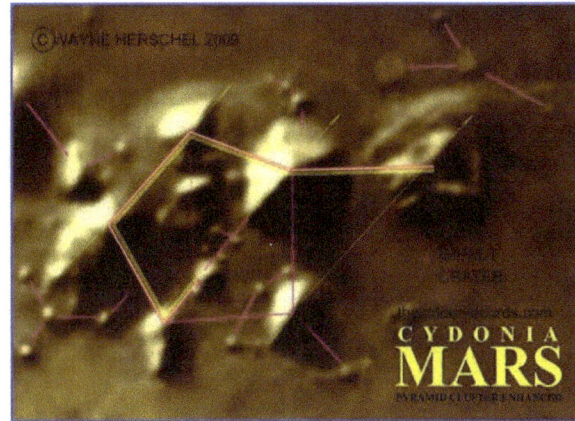

Comparison of the star alignment in the Pleiades in the Constellation of Taurus with the alignment of the pyramid buildings of Cydonia on Mars

http://thehiddenrecords.com/mars.htm

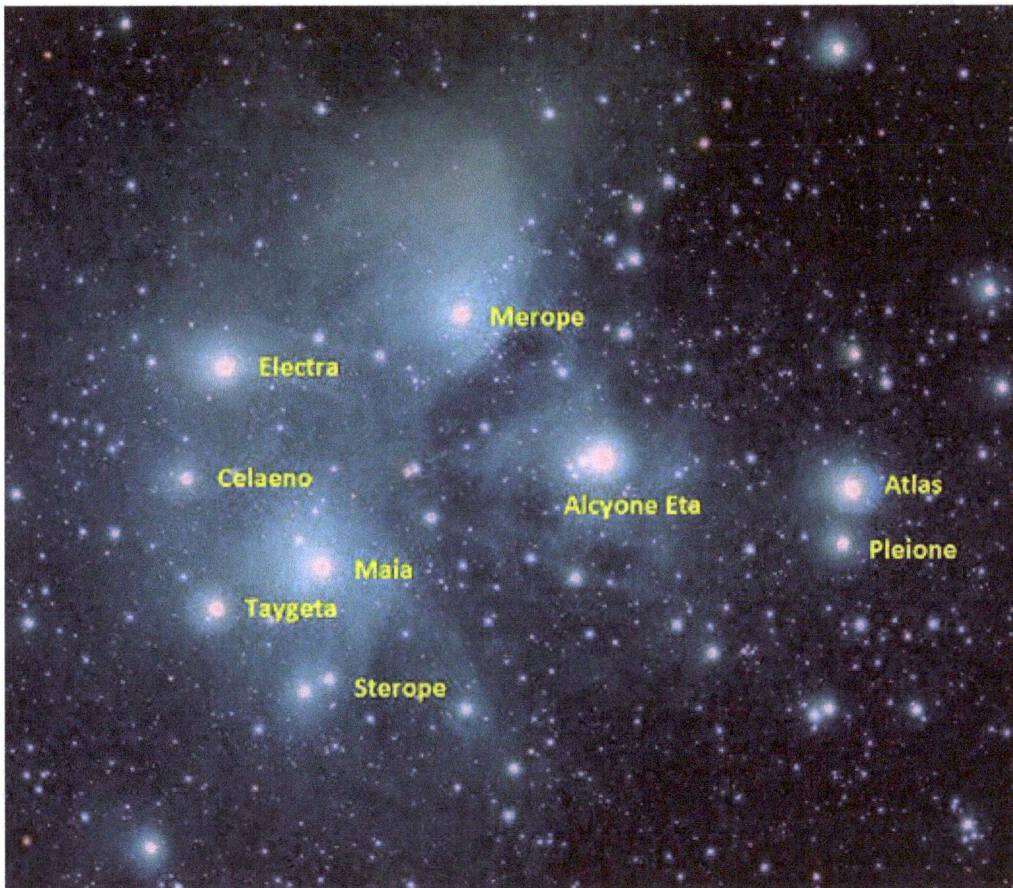

The major stars in the nebula star system of the Pleiades as seen from the Southern Hemisphere

http://cs.astronomy.com/asy/m/starclusters/480502.aspx

Apparently, **Wayne Herschel** has so much evidence that what he has presented in Hidden Records is just the tip of the iceberg and has put out a challenge to other researchers and scientists in particular to peer review his research work. So far no one has come forward, except for the TV program series, **"Ancient Aliens"** and they have used his data without acknowledgement or credit to him for it. Needless to say, his lawyers are looking into the matter.

Author's Rant: I suspect that there are many researchers of planetary anomalies as well as Ufologists who have come to the same conclusions as Herschel but, have not made all the connections and associations as he has done in his investigations. I believe that this young South African is about to become well known globally and that many researchers will be following in his footsteps to confirm for themselves his findings.

Intriguingly, the fact that Herschel theorizes that humanity's roots originate from the Pleiades is either confirmation of the accounts of contactee **Billy Meier** and his communications with the beautiful Pleiadian **Semjase**, a blonde Nordic human-like ET or it is mere coincidence or Herschel's data is seriously flawed thus, a peer review of his evidence definitely would be a step in the right direction.

However, the reality thus far supports Herschel's claims, in fact, there are many accounts where people have reported human type ETs encounters, **Travis Walton's** account being publicly the best-known which has been discussed earlier in this book.

In the early 1960s, **Command Sergeant Robert O. Dean** read an **Above Top Secret Threat Assessment** while working at **NATO** that changed his life. According to Com Sgt. Dean, the report clearly stated that one of the groups of ETs regularly visiting Earth was human. These ETs looked exactly like us and could, therefore, blend in with the rest of earth's population. There was concern among some high-ranking members of NATO that these human ETs could have infiltrated the military organizations and civilian governments of Earth.

Herschel's investigation of archaeo-astronomy and ancient symbolism forces us to carefully re-examine with a fresh perspective and to re-assess with an open mind, the strong possibility that humanity's origins may lay in a nearby corner of the universe and not here on this Earth. The possibility also exists that there may be more than one interstellar location from which humanity's possible origins sprang from in ancient times. As new world are being discovered almost weekly around stars like own Sun, perhaps a search is required in the constellation of Taurus with particular attention focused on the Pleiades star systems, It may be there that we will find Earth-like planets and possibly confirm what this author refers to as the Herschel's **Extraterrestrial Human Origins Hypothesis (EHOH)!** At the very least some of the stars and constellations that we see up in the night sky are almost certainly to be the home of a race of ETI who are currently visiting the Earth.

The evidence thus far indicates that Extraterrestrial Intelligences have been in contact with humanity and have influenced the direction of human civilization, since our early beginnings thus, confirming the fact that we are not alone in the universe and never have been. Regardless, of the type of extraterrestrial intelligences visiting our planet, whether human-like or completely alien in appearance, the evidence is *they are here, now*!

218

CHAPTER 85

THE ASTEROID BELT - DID OUR SOLAR SYSTEM EXPERIENCE AN ANCIENT INTERSTELLAR WAR?

Throughout our search to prove the existence of life on Mars, whether in the distant past or that which is currently on Mars, the overwhelming evidence supports further our current understanding of the UFO/ETI phenomenon currently visiting our Earth. But, before we leave Mars entirely to explore the rest of our mysterious Solar System, there is an obvious yet, nagging question that has not been answered or considered. "What happened to Mars in its distant past that caused so much devastation over the entire planet of Mars, where almost everything was destroyed or pulverized into rock and dust that essentially wiped out almost all life? Was this devastation isolated to Mars or was it interstellar-wide in our **Solar System?**

Significantly prominent theorists, occultists, and astrologers, even physicist and astronomers have postulated that a devastating form of "star wars" may once, have occurred not "in a galaxy far, far away" but, much closer to home - within the boundaries of our very own Solar System! The core of these views stems from the fact that Mars and to some extent the Moon and Venus display evidence of once having flowing rivers and considerably more of an atmosphere than is presently the case! Conditions that would certainly make the existence of intelligent life on these planetary bodies possible! The question is, however, could there have been inter-planetary wars in the solar system?

What we can determine based upon the best scientific astronomical data is the certainty that Mars and the Moon are far from rising up on the planetary scale of evolution actually appear to be decaying! Mars once held life-sustaining atmosphere with running rivers but now look sparse, empty and barren! In fact a closer look at the Martian surface shows that at some stage in its history it was subjected to an enormous and violent upheaval or collision which ripped a canyon over 2500 miles long on its surface and burst the planet's crust with massive volcanoes like proverbial zits on a face, of which **Mons Olympus** is the largest mountain in the Solar System! In addition, photos from the Viking landers showed quite plainly the surface of the planet is extensively strewn with small chunks of rock and artificial debris resulting from some tremendous explosion! Much of the ground shows melting and vitrification from extreme heat but, what could be the source of this planet-wide destruction?

Volcanism on the Mars is evident but, is more of an after effect from something else triggering internal planetary stress, fractures, and upheavals on a massive global scale

Solar wide storms or flares from the sun certainly would have the cataclysmic capability to cause such damage but, in doing so, it would wipe out the other inner planets or cause them to become dead, dry and barren at the very least. To date from all space probes sent to these planets with a possibility of Venus, none have become globally sterile or non-existent.

A self-inflicted global thermal nuclear war is certainly a possibility, even an interplanetary or interstellar war using nuclear, laser and plasma beam weapons could conceivably be the source of devastation. But, who were the Martians fighting with and where did they come from?

The Planet V - Asteroid Belt Origin Hypothesis

The other possibility may lie between Mars and Jupiter in an area known as the **Asteroid Belt** where 700,000 to 1.7 million pieces of rock miles or kilometres in size orbit around the Sun. Some scientists have theorized that a **planet V (or Planet X)** once existed between Mars and Jupiter which literally exploded into many massive fragments creating what is now known as the Asteroid Belt. Its existence was even theorized by mathematicians **Johann Daniel Titius** and **Johann Elert Bode**.

When one thinks about asteroids, one cannot help but think of the comedic routine where an astrophysicist is trying to describe what asteroids are to a student who has incorrectly misunderstood the term thinking the astrophysicist is describing a personal case of anal itch and inflammation requiring suppositories or a soothing crème of Preparation H to cure the problem and thus, bring some needed relief. If only it was that easy but, we are not discussing hemorrhoids but, asteroids and it requires a law of astrophysics to understand how asteroids may have come about.

According to the theory - known as the **Titius–Bode law** (a.k.a. **Bode's law**) the bodies in some orbital systems, including the Sun's, orbit at semi-major axes in a function of planetary sequence. In other words, there exists a definite and repeatable ratio of the distances of the planets to the Sun. The hypothesis correctly predicted the orbits of **Ceres** and **Uranus** but failed as a predictor of **Neptune's** orbit. Six planets were then known but astonishingly the distance of the others as yet undiscovered were correctly deduced by this law which is now relegated to no more than a baffling coincidence!

What is, however, of great interest to us is that according to this law there should be another planet between the orbits of Mars and Jupiter, exactly where the asteroid belt now lies! Planetary theorists believe this was once the location of a sizeable planet that was literally blown apart by an incredible weapon of destruction, or impact with another planetary body! The former would require a weapon considerably more potent than anything we know of today, because although nuclear weapons are quite capable of destroying all surface life, the actual structural existence of the world itself would never be threatened!
http://www.nasca.org.uk/Ancient_Nuc__War/ancient_nuc__war.html

These asteroids blasted into existence from planet X millions of years ago and now travel around the Sun in their individual orbits, but collectively like the rings of Saturn. Sometimes, they hit and jostle with each other, sending one or two asteroids careening off into an elliptical orbit around the Sun with near misses of the Earth or the other planets.

According to late **Zecharia Sitchin**, an etymologist and self-taught in the ancient **Sumerian Cuneiform** language reveals an alternative origin of mankind in many of his controversial "Earth Chronicles" books and in particular Book One: "The 12th Planet". He tells how there was a large Jupiter or Saturn-size 12th planet, (counting from the Sun, the Sun, and Moon are considered as planetary bodies), named **Nibiru** (a.k.a. **Marduk** that has a long elliptical solar orbit of 3600 years. During its orbit inward around the Sun, Nibiru crossed several orbital planes of the outer planets with a near approach to a planet called **Tiamat.** Tiamat was basically a large water planet about the size of Neptune or Uranus with little or no continental land mass and with eleven orbiting moons, the largest known as **Kingu**, a moon on the verge of becoming a life-sustaining planetary body.

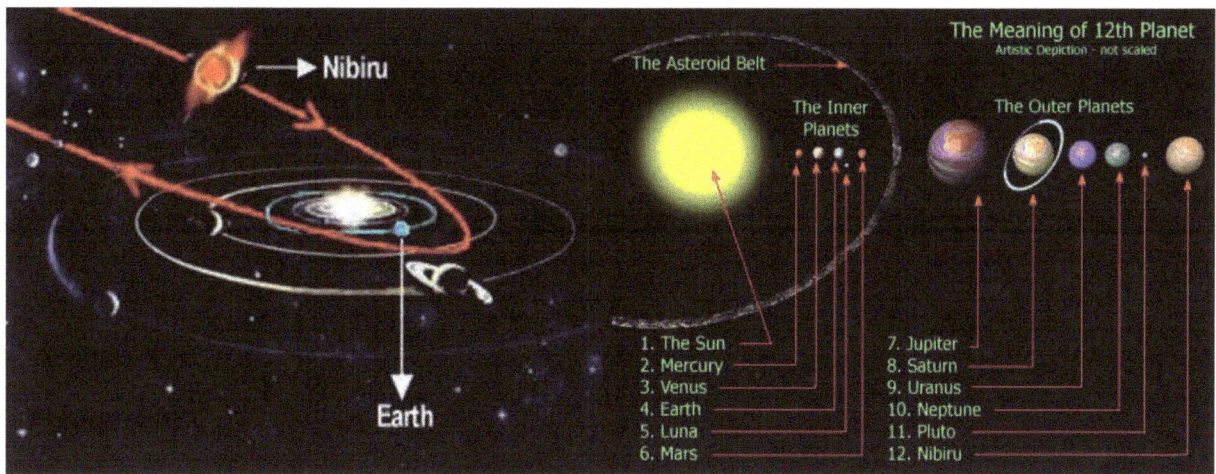

The theorized orbit of Nibiru and its relationship to the other planets in the Solar System
https://centinela66.com/2015/04/19/el-misterioso-planeta-x/

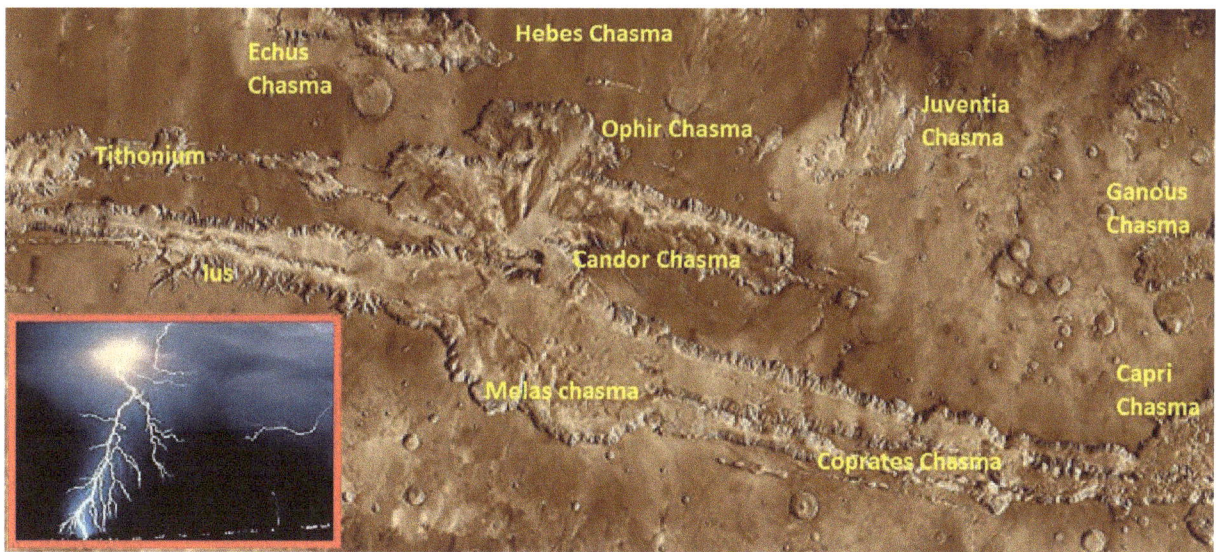

Valles Marineris (with lightning insert) in mosaic of THEMIS infrared images from 2001 Mars Odyssey theorized to have been caused by a massive electrical discharge from a rogue exo-solar planet called by ancient Sumerians as Nibiru or Marduk
https://en.wikipedia.org/wiki/Valles_Marineris

Nibiru did not actually collide with **Tiamat** but some of its own satellites did. The impact from these moons with Tiamat caused it to fracture in two, mortally wounding the planet causing planetary destabilization and destroying some of Tiamat's eleven moons. **The 12th Planet by Zecharia Sitchin; 1976; published by Avon Books, New York, New York; ISBN0-380-39362-X**

221

Nibiru travelled around the Sun but, on its orbit outward from the Solar System it came into near collision with Mars, where strong electrical magnetic fields from each planet discharged sending massive arcs of lightning across space toward each planet. The result was global devastation on Mars killing nearly all life, destabilizing it rotation spin, loss of its gravitational field and removing most of its atmosphere. The **Vallis Marineris**, a long canyon 4,000 km (2,500 mi) long, 200 km (120 mi) wide and up to 7 km (23,000 ft) deep, became the evidence of this interchange of planetary electrical discharge between Mars and Nibiru.
http://en.wikipedia.org/wiki/Valles_Marineris

The Vallis Marineris certainly has the geographical features similar to what you would expect to find on a piece of metal scarred from a high voltage electrical discharge. (See photo above).

Nibiru continued its march of death and destruction through the Solar System as it orbited outward approaching Tiamat once again. This time, the rest of Tiamat's moons were destroyed except for **Kingu** which now became lifeless planetary body from the rain of meteoric bombardment emanating from the destruction of Tiamat and the other moons. Tiamat was dealt a death blow. Half of the planet along with Kingu was dislodged from its orbit and sent careening toward and around the Sun.

Tiamat was cleaved in two with its inner core exposed to the raw elements and vacuum of space and Kingu was severely scarred by numerous massive impacts. Tiamat eventually settled in an orbit between Venus and Mars.

Yes! Earth was once a larger water planet known as Tiamat according to **Sitchin's** translations of the ancient **Gilgamesh text** and other Sumerian literature! Nothing ever really dies but is transformed into something else and this is true even on a planetary level.

Tiamat coalesce and enfolded upon itself and over a period of time reshaped itself into a smaller planet that we call Earth! Its orbit stabilized around the Sun, the remaining waters on the surface separated and land for the first time emerged upon the planet, this fits nicely into Biblical understanding of creation yet, it also explains the age difference between continental land masses and the younger age of the planet's crust found on the ocean floors. It even explains the concept of continental drift of land masses and could explain the concept of the ***"Edge of the World"*** perspective!

Kingu became the Moon as we know it today with a slightly elliptical orbit around the Earth. The other half of Tiamat was pulverized into hundreds of thousands of small and large fragments eventually becoming the now familiar **Asteroid Belt**.

Nibiru

asteroid belt

Sun

Mars

Earth

Mercury Venus

A former planet of the Solar System destroyed by one of Nibiru's moon which created the asteroid belt

Nirbiru strikes Tiamat, splitting it into the newly formed Earth and into the Asteroid Belt

North Wind

Nirbiru

Newly formed Asteroid Belt

Tiamat (New Earth)

Kingu (Moon)

Direction of new Earth

Orbit of Nirbiru

Tiamat's original orbit

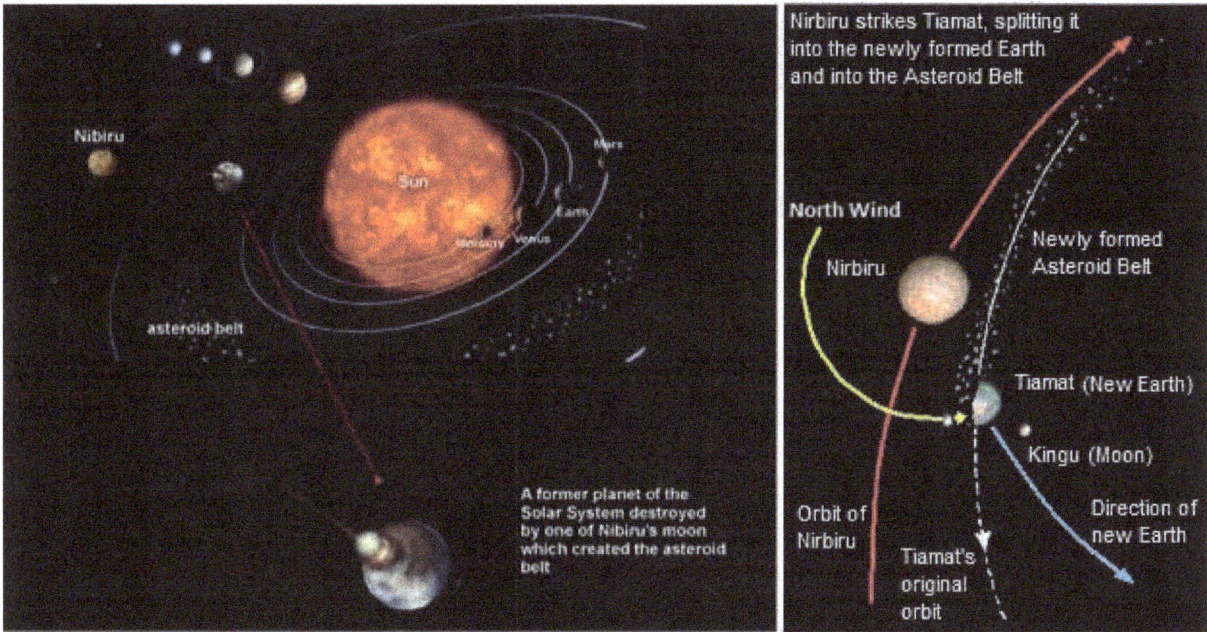

One of Nibiru's moons smashes into Tiamat creating the Asteroid Belt and the Earth

http://www.ancient-code.com/20-things-you-should-know-about-nibiru-planet-x/ and
http://anunnakihero.blogspot.ca/2010_12_01_archive.html

This asteroid origin hypothesis has like everything else, its supporters and detractors and it is viewed as a pseudoscience by mainstream science in much the same manner as the UFO/ ETI phenomenon. It is radical, it is different and it is definitely "out there" yet, it does explain some of the geological oddities on Earth and it explains the **Asteroid Belt** and the surface scarring and damage upon the Moon, Mars and on some of the other planets and their satellites in the Solar System. Could there be some elements of truth in this hypothesis?

Yes, it is possible but, the Earth is far older than a few hundred thousand years or even a few hundred million years. It is billions of years old that is an established fact but, is that the age in which it was originally formed as a supposed larger planet known as **Tiamat** when it orbited between Mars and Jupiter or the age it was formed in its current orbit?

The answer to that question must come from the ground surveys and core samples needed from the Moon, Mars and from some of the asteroids beyond Mars and then compared to see if they are all the same age or of significantly different ages.

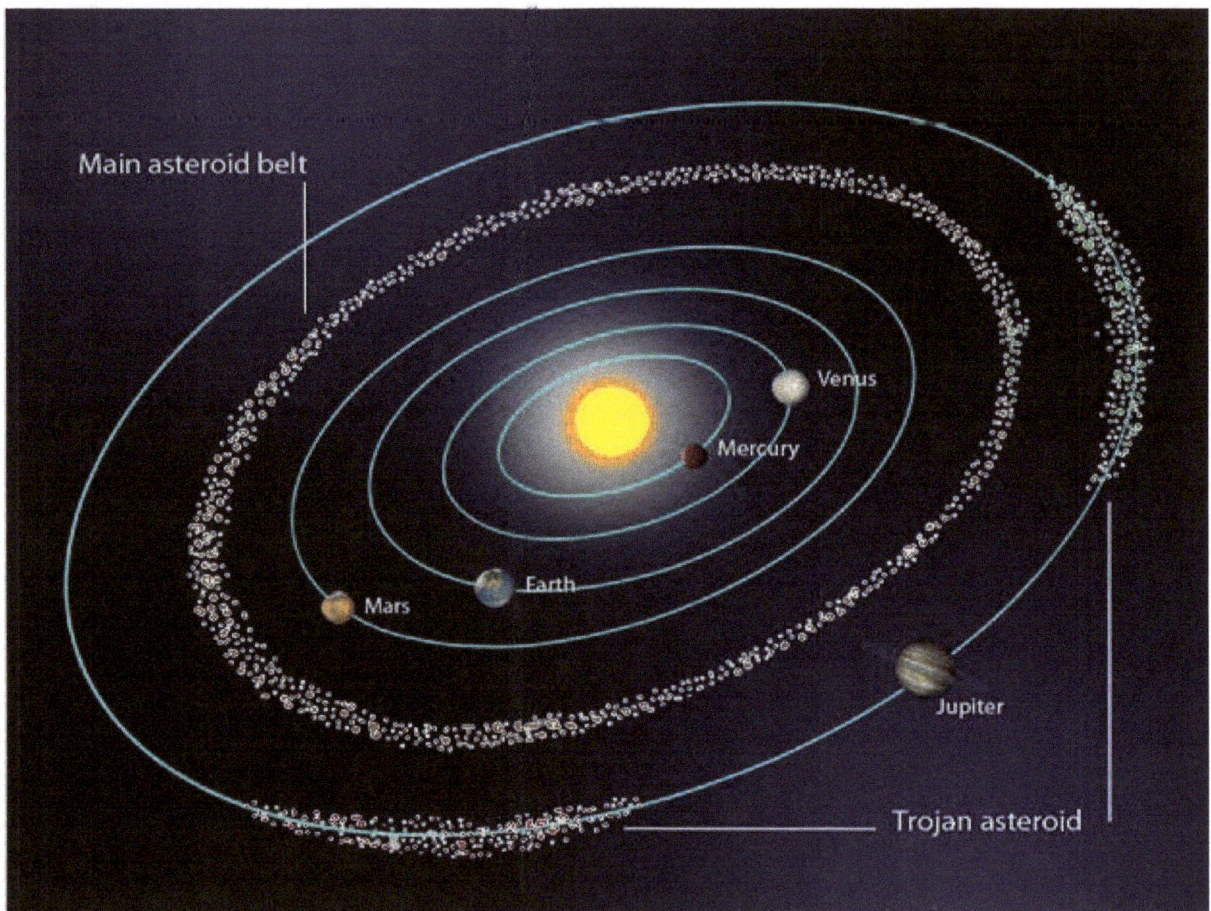

The Main Asteroid Belt and the Trojan Asteroids

Which theory for the origin of the asteroids is the correct one, even scientist do not have any conclusive understanding of the Asteroid Belt's origins. They think that it was merely remnants that came into being when the Solar System first formed but, these remnants never coalesced into forming a planet. It's a convenient theory but, it does not explain the damaged found upon Mars or some of the outer planets' satellites, some like were literally torn apart and then coalesced back together with massive continent size scars shaped like chevrons or like the pieces of a jigsaw puzzle.

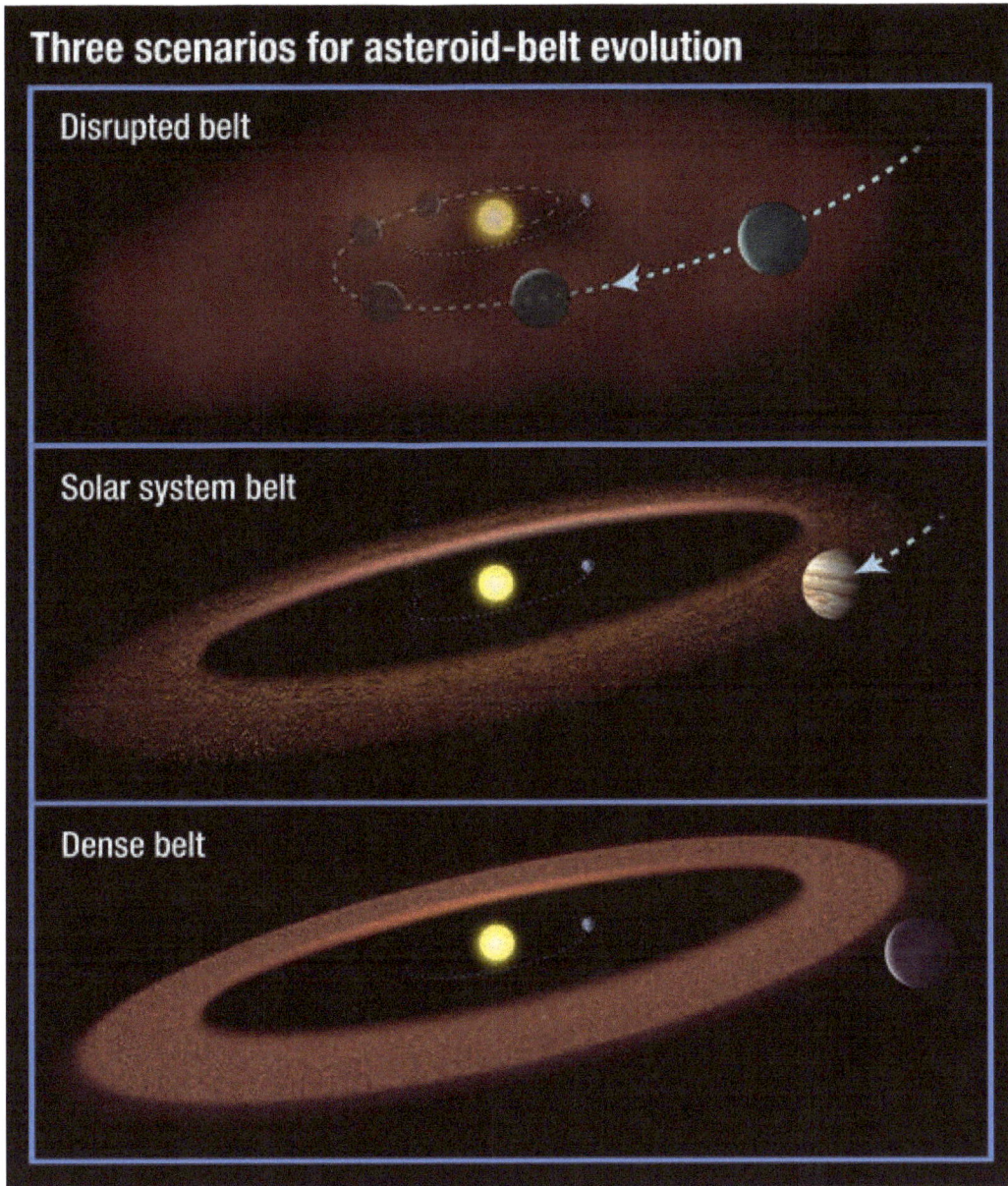

Three scenarios for asteroid-belt evolution

Disrupted belt

Solar system belt

Dense belt

Three possible scenarios for the evolution of the Asteroid Belt
https://physicsforme.com/tag/asteroids/

NASA and ESA theorize that the three most logical scenarios for the evolution of the Asteroid Belts are portrayed with the illustration below. In the top panel, **Disrupted Belt**: a Jupiter-size planet migrates through the asteroid belt, scattering material and inhibiting the formation of life on planets. The second scenario, **Solar System Belt**: shows our solar-system model: a Jupiter-size planet that moves slightly inward but is just outside the asteroid belt. In the third illustration, **Dense Belt**: a large planet does not migrate at all, creating a massive asteroid belt. Material from the hefty asteroid belt would bombard planets, possibly preventing life from evolving.

The Asteroid Belt may once have been a large Neptune-size planet between Mars and Jupiter that exploded from a cataclysmic event in its past
https://www.pinterest.com/pin/528539706250362682/

Where could such an explosion have come from? Was planet X an inhabited planet that fell into one of the above war scenarios resulting in a catastrophic end where the planet literally exploded apart like a scene out of the movies of Star Wars or Star Trek? The Solar System as we know it from our space exploration now looks magnificently tranquil and serene that it's hard to imagine such violence!

Could the universe be in a constant game of galactic billiards as proposed by Zecharia Sitchin? The answer is that surprisingly the evidence supports the astrophysics and dynamics of planetary collision, whereby, even whole galaxies can on frequent occasions collide with each other to form even larger galaxies. If it can happen at the galactic level, it certainly can occur at the planetary and asteroid levels. Could a rogue asteroid or even another planet have come into our Solar System and collided with planet X resulting in an area now inhabited by hundreds of thousands or millions of asteroids as a result of an inevitable galactic billiards game coming to our stellar neighbourhood?

Scientists don't like this theory very much because it's too unpredictable to know when or if it will ever happen to your small corner of the universe! But, the reality is impossible to deny as it happens all the time. The only question remaining, when will it happen again and can it ever be diverted through super science in the future?

It is also conceivable that an alien race blew up this **planet X** in an interstellar war with the human race which in turn caused Mars to be destroyed in the process with planetary fragments impacting upon certain portions of the Martian surface.

There is in ancient mythology of many cultures stories that frequently tell how long ago there were celestial wars, wars in the heavens or in space between the forces of good and evil, wars that took place in the skies over the ancient cities on every continent!

The central core of these mythologies tells the story that there were once inhabitants on the planets Venus, Mars and the Moon (the moon was once thought to be a planet) with advanced societies or civilizations who fought among each other in a interplanetary war similar to the Greek mythological account of the war between the **Titans** and **Olympians** or the great epic aerial battles of ancient Indian mythology as recorded in the books the **Mahabharata** and the **Ramayana.**

These wars and battles nearly resulted in the extinction of the human race. Chaos ruled in the Solar System, a planet was destroyed and the orbit of Venus was forever changed to a position much closer to the sun! Fragments and vast debris from the exploding planet rained down upon Mars and the Moon and outward toward the outer planets in the solar system. The Martian and Lunar atmospheres and landscape were drastically altered, making life on these planetary bodies forever uninhabitable!

The evidence for this ancient interplanetary war was primarily from historical accounts considered to be mere mythology with no actual substance of fact or real evidence. It wasn't until a few archeologists, private landowners and the common folk quite unexpectedly started turning up ancient ruins, gravesites and artifacts literally beneath their feet from around the world which were not only unusual in the archeological sense but, truly out place or unexplainable in their origin.

The little gold airplanes of Mesoamerica, the giant axes of ancient Greece, the giant graves and skeletons found in Sardinia, and the 12,000 year old temple site of **Gobekli Tepe** in Turkey, the vast areas of vitrified ground throughout the middle East, the very ancient and radioactive skeletons found in **Mohenjo-Daro** and **Harappa**, India are just a few examples of real places and artifacts that indicate cultures of antiquity far advanced beyond their time even, by today's standards.

The continent of India is steeped in deep rich antiquity with writings recording incredible ancient nuclear wars! Not only are there physical scars upon the Earth but also, the graphic written accounts of these events in some of the most ancient texts on Earth. In the five thousand-year-old Mahabharata, the epic Indian saga speaks of flying machines called **Vimanas** and even how to build them that were used to launch a powerful weapon of destruction; *"a single projectile charged with all the power of the universe!"*

In one passage of the **Mahabharata,** it describes the destruction unleashed by two warring sets of adversaries, the cousins Kauravas and the Pandavas: *"The Earth shook, scorched by the terrible heat of this weapon. Elephants burst into flames and ran to and for in a frenzy seeking a protection from terror. Over a vast area, other animals crumpled to the ground and died. The waters boiled, and the creatures residing therein also died. From all points of the compass, the arrows of the flame rained continuously!"*

Later we find: *"An incandescent column of smoke and fire, as brilliant as ten thousand suns rose in all its splendour. It was the unknown weapon, the iron thunderbolt.....a gigantic messenger of death!"* The effect of this weapon was that *"The corpses were so burnt that they were no longer recognizable. Hair and nails fell out. Pottery broke without cause. Birds disturbed, circled in the air and were turned white. Foodstuffs were poisoned!"*

**Arjuna and Lord Krishna charge into battle while their army fights
both on the ground as well as in the sky in their flying ships!**
https://www.youtube.com/watch?v=R1azoyQvs-E

Whether or not nuclear weapons have been used in the past, the influence of the Hindu Epics is prominent today, notably, in association with the first detonation of the atomic bomb. **J. Robert Oppenheimer** was a world famous physicist and the head of **Project Manhattan**. Upon witnessing the nuclear test in 1945, Oppenheimer famously recalled verse 32 from chapter 11 of the **Baghavad Gita** quoting **Lord Krishna** (avatar of **Vishnu**) once he has transformed into his celestial form in front of **Arjuna**: *"Now I am become death, destroyer of worlds"*.
http://www.abovetopsecret.com/forum/thread518421/pg1

It takes little imagination to see the similarity between these ancient accounts of India and what we all know these days about the awesome capability of nuclear weapons and their immense destructibility. We only have to look at the Japanese cities of **Hiroshima** and **Nagasak**i that were the first modern cities to be bombed with atomic weapons during the Second World War.

Did the ancient Indians of thousands of years ago know something about flight and nuclear war that we as a collective humanity seem to have forgotten? Was this the first time nuclear war occurred on Earth? Is it even possible, that Neanderthals were originally human but, became genetically mutated or altered from radiation poisoning stemming from a nuclear holocaust?

Nor is the **Mahabharata** the only ancient text to describe such things. Similar accounts of great destruction can also be found in the **Tibetan Stanzas of Dzyan**, the oral traditions, and beliefs of the **Hopi Indians**, and even the Bible speaks of the destruction of **Sodom** and **Gomorrah** in a location now occupied by the mysterious **Dead Sea hollow**, one of the most inhospitable places on earth! http://www.nasca.org.uk/Ancient_Nuc_War/ancient_nuc_war.html

The best physical evidence in support of these ancient wars is the vitrified remains of fortresses, ziggurats, and towers that have been subjected to an unaccountably sharp blast of heat. Charred ruins can be found between the River Ganges in India and the mountains of Rajamahal. "The walls have been glazed, corroded, and split by tremendous heat. Within several of the buildings that remain standing even the surfaces of the stone furniture has been vitrified: melted then crystallized. No natural burning flame or volcanic eruption could have produced heat intense enough to cause this phenomenon. Only the heat released through atomic energy could have done this damage! Also in this same region, a human body was discovered with a radioactivity *"which was fifty times above the normal level"*.

This baffling enigma of vitrified ruins is to be found all across the world. No more so than North America where the strange remains of vitrified rocks and dwelling places defy logical explanation. In South America the Brazilian ruins of **Sete Ciddaes** are enormously revealing, the ruins are melted by apocalyptic energies! Elsewhere, in Mesopotamia sizeable ziggurats - a form of early pyramid - have been found melted to their base in a vitrified mass!

In the **Arabian Desert, blackened stones** litter the sands over a wide area, showing signs of having been subjected to intense radiation. In Israel, the location of the Dead Sea and its mysterious connection with Sodom and Gomorrah bears evidence of an amazing focus of heat that is thought to have gouged out the entire area in a massive explosion. **Vitrified rock** created under intense pressure is a frequent discovery, and in 1952 archaeologists discovered a vitrified area of sand that stretched out over hundreds of square feet! Apparently, deposits like this are similar to those left behind at the White Sands atomic testing site in America!

In fact, wherever we look in the world the baffling enigma of vitrified ruins challenges our intellect! From Peru to Scotland and Scandinavia; to the plateau's of China and India, this indelible evidence attests to some undeniably violent act! Not everyone will be convinced of a nuclear answer but as we have seen the evidence is extremely compelling. It is almost impossible even for established scientists to not find themselves intrigued by the evidence.
http://www.nasca.org.uk/Ancient_Nuc_War/ancient_nuc_war.html

Evidence of an ancient nuclear war on Earth is literally buried beneath our feet, it can be found in areas now considered desolate and barren as well as in areas near modern cities. The question, however at this time, did humanity's origins begin on another nearby star system and did they bring the knowledge of advanced weapons technology to this planet? The growing archeological evidence that is not suppressed or covered up clearly indicates that we share a common ancestry with Mars and that Mars has massive monuments similar to Earth that are signposts pointing to some of the major constellations around the plane of the elliptic of our Solar System. These constellations may be home to current human colonies and outposts, not from recent times but from great ancient times, long forgotten. The evidence is strong that suggest that one or more of the star systems in the Pleiades within the constellation Taurus could very well be where our human ancestry began.

It may never be known what actually occurred on Mars so many epochs ago or what happened to the fifth planet between Mars and Jupiter, now occupied by millions of asteroids. Was this planet destroyed by natural astronomical cataclysm or from interstellar war? Like so many things in the human experience wherever, there is some rudimentary knowledge surrounded by myth and legend, there is a basis of truth. It goes without saying that more archeological research of ancient sites on Earth is required to determine what that truth really is. In addition, planetary exploration of the ancient ruins on Mars and elsewhere in our Solar System is also needed perhaps, in the near future with manned missions to Mars, its moons, as well as to the asteroid belt and the other planets beyond.

If the space agencies of Earth can be trusted to carry forward an honest and open exploration of space with full disclosure of their findings and discoveries to general public then, we will not only be a true space faring civilization but, we will eventually have the answers to many of our questions regarding the history of the Solar System, what happen to Mars and the fifth planet" X" and what lies beyond our star system.

It should be understood, that we cannot paint everyone or every department of NASA or any of the other space agencies with the same brush. Frequently in most departments and agencies, there are situations that become a manifestation of the right hand not knowing what the left hand is doing, particularly if a certain rogue segment of that agency has a covert hidden agenda that is not a part of the original mission statement of that organization. In such situations, there are people within NASA that are operating outside of the agency's mission and mandate following their own agenda or in partnership with some higher authoritative organization and have probably been doing so since its inception.

The fact that other nations have developed their own space programs strongly suggests that they are not accepting everything that NASA is disclosing to them and that they are exploring the Solar System in order to discover the truth for themselves. This has forced NASA and the DOD into spin control by applying political pressure on these space-faring nations to fall into line with NASA's thinking. Ultimately this is a covert way of thinking that serves only the interests of the military industrial complex and those with an eschatological bent designed to enhance their power base and control the masses and the financial resources of the world.

The ultimate covert agenda may very well be as **Richard Dolan** has suggested that there is a **Breakaway Civilization** carrying out an **Alternative 3** scenario which appears to be more blatantly apparent as it steadily plays out with each passing year.

We see many nations self-destructing from within or from outside influences with well-orchestrated, carefully manipulated attacks on the financial, political or religious institutions of certain nations in order to create chaos and strife. The goal is to bring about an international cooperation through a corrupted version of democracy that asks the citizens to give up much of their freedoms, rights, privileges and privacy in order to maintain a heightened sense of national security.

The powers that be, namely the bank cartels, the oils companies and the military industrial complex, et al are arrogantly flexing their muscles and are the ones behind the global chaos in order to take control of this planet and they don't care who they step on or who gets hurt in the process.

NASA and one or two other space agencies are doing their bit to bring about this dark future and at this stage, all we can do is to bring the facts to the public's attention while we hitch a ride on their space exploration endeavours in order to get a bigger understanding of what it is that they have discovered in our Solar System.

Before leaving the **Asteroid Belt**, let's see what else is out there in these orbiting pieces of rock that were once part of a large planet. **Phobos** and **Deimos** now officially confirmed by ESA to be artificial in nature meaning that they are huge ancient hollowed spacecraft can now be used as a benchmark for looking closer at other large asteroid fragments to see if they too, hold signs of artificiality in their appearance.

Asteroid 2867 Steins - A Case for a Very Ancient
Type Two Civilization in Our Solar System

It appears that the diamond-shaped **Steins** asteroid is another candidate for being potentially a large spacecraft because of its shape and surface features. The evening of September 5, 2008, **ESA (the European Space Agency)** successfully flew its unmanned **"Rosetta" spacecraft** within 500-miles of a tiny (~ 3 miles across), newly-discovered (1969) asteroid, known as **"2867 Steins."** http://www.enterprisemission.com/Rosetta/Rosetta-analysis-test.htm

Although the low resolution of the asteroid was the best that ESA could achieve due to a software glitch with the high-resolution imaging data, the shape was distinctly geometric –

a faceted "diamond in the sky" or in this case in space! (See comparison photo images below).

We will see that ESA is not being completely honest with these low res images stating a supposed software glitch at the critical moment when the world was expecting to see high res pictures of the Steins asteroid. It seems that ESA chose instead to follow a page out of NASA's playbook of cover-up lies, denials, misinformation, and obfuscation.

Whenever astronomical bodies or artifacts on a planetary surface display rectilinear geometry the best approach to understanding its potentiality is to observe the noted words of the late Dr. Carl Sagan:

"Intelligent life on Earth first reveals itself through the geometric regularity *of its constructions ...."* **Carl Sagan, "Cosmos" (Random House, 1980)**

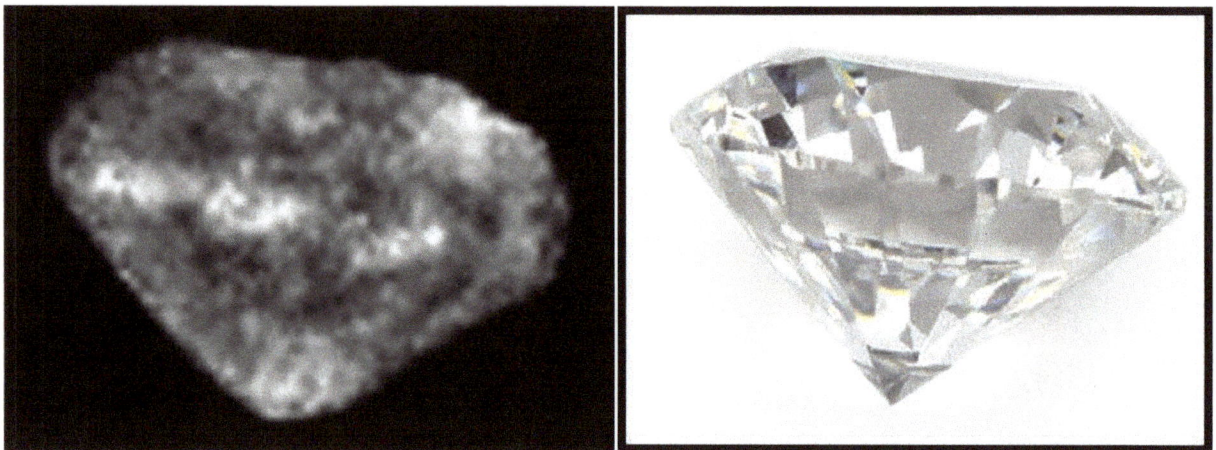

ESA's diamond in the sky, "2867 Steins" compared to a faceted diamond and also to graphic features of a diamond below
http://www.enterprisemission.com/Rosetta/Rosetta-analysis-test.htm

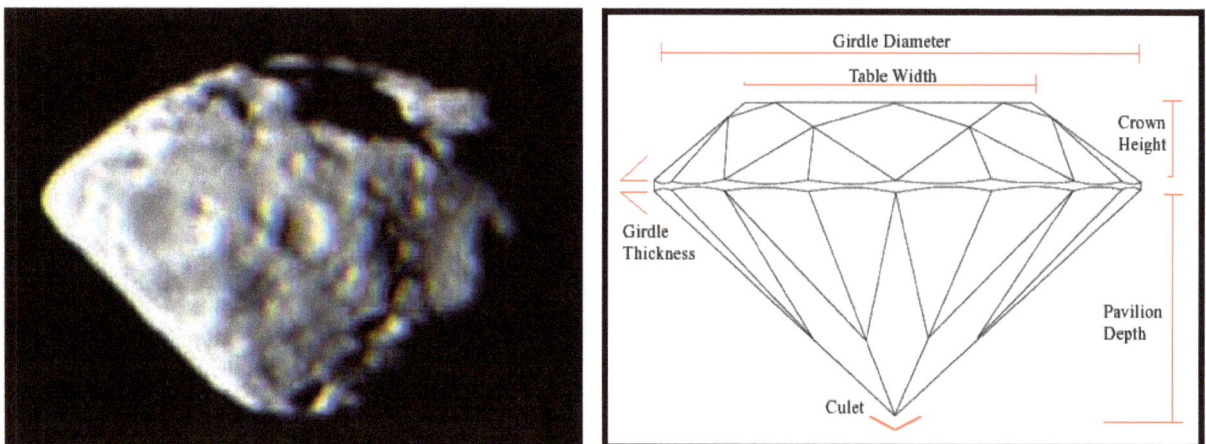

Eastern cratered hemisphere of Steins on left and graphic features of diamond on right
http://www.enterprisemission.com/Rosetta/Rosetta-analysis-test.htm

According to **Richard Hoagland**, *"geometric regularity"* has become the baseline archaeological standard of "intelligence" on Earth, on Mars and is now, the core foundation of other NASA imagery of potential "ancient intelligent ruins" located on *other* planets in this Solar System.

Hoagland posits that the ultimate expression of this "geometric intelligence criterion" was the discovery that *an entire outer solar system object* which appears to follow this essential "geometric intelligence criterion": **Iapetus**

Iapetus, the third largest satellite in the Saturn system was imaged in 2004 by NASA's Cassini Mission; on the best Iapetus images sent back to Earth, the distinct, straight-edged, *highly geometric outline* of this entire ~900-mile-diameter "moon" is striking, blatantly apparent.
http://www.enterprisemission.com/Rosetta/Rosetta-analysis-test.htm

Before we examine Iapetus, the odd moon of Saturn, the geometric regularity of 2867 Stein also needs to be examined to determine what it really could be. ESA's **Wide-angle Camera (WAC)** determined that Steins had some unusual features on its surface. Most notably was the brightness difference between the Western hemisphere and the Eastern hemisphere (more cratered side) of the asteroid which indicated a high albedo difference (reflectivity) of nearly 50%. This averages to a visible 35% albedo which is very high for any asteroid that would normally average from 1% to 20%, making Steins an extremely reflective object due to the striking difference in *hemispherical* reflectivity.

When ESA's WAC photo imaged from the lower bottom perspective, below the "diamond girdle plane" of Steins, other interesting geometric features were revealed on the western hemisphere that displayed "raised vertical flanges" regularly spaced in groups of two just below the girdle plane around the object. (See middle image – red circles).There also appears to be three massive curved buttresses running at 60 degree angles laterally from the culet to the girdle plane of the asteroid supporting the flanges, indicated by the red vertical lines, a feature which ESA seemed to have overlooked or ignored chalking it up as just a typical solar system asteroid. (See below).
http://www.enterprisemission.com/Rosetta/Rosetta-analysis-test.htm

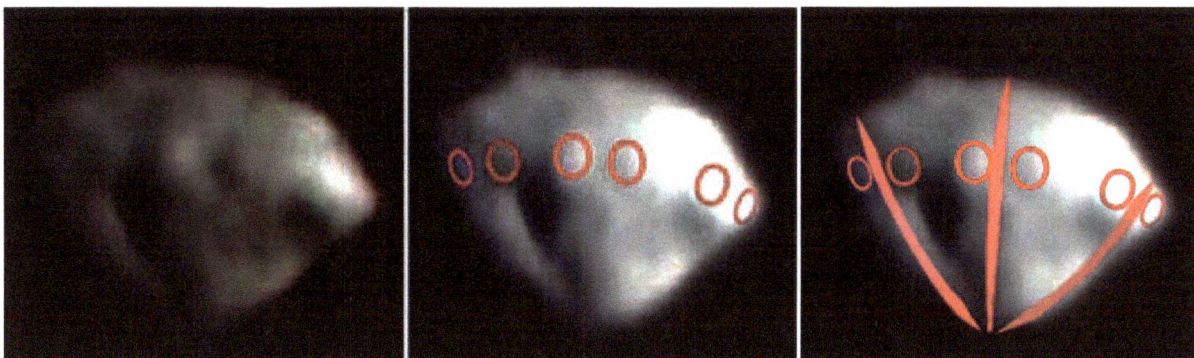

Note the flanges or "bumps" circled in red below the "girdle plane of the diamond" Steins
http://www.enterprisemission.com/Rosetta/Rosetta-analysis-test.htm

Steins shows meteoric battering that much is certain but the concentration of them on one of the asteroid is odd for a supposed natural body of rock but, it is also most revealing.

"... *images beamed from the spacecraft show the diamond-shaped asteroid **'looming like a threat'** from the famous **'Asteroids'** computer game before the deeply pockmarked surface becomes clear. It has provided scientists with their closest look at an asteroid to date. Huge impact craters up to 1.2km wide can be seen covering the surface of the 3-mile diameter asteroid **in a far higher concentration than would be expected for such a small object.***

"'*There is also a chain of seven craters **that we would not expect to see on such a small body,**' said Professor **Uwe Keller**, a principal investigator at the **European Space Agency (ESA),** which is behind the Rosetta mission. 'We normally see craters like this on moons like our own. We have to look at why they are there, but clearly, Steins has a **complex collision history.....**" (Bold italics added for emphasis added). http://www.enterprisemission.com/Rosetta/Rosetta-analysis-test.htm

The largest crater on Steins' surface, the one completely dominating the northern "pole" (below, red arrow) is *so large* (approx. 1 mile across ...), compared to the 3-mile diameter of the Steins itself that, according to **Space.com,** "(the ESA) scientists were amazed that the asteroid *survived the impact*...." (Bold italics added for emphasis).

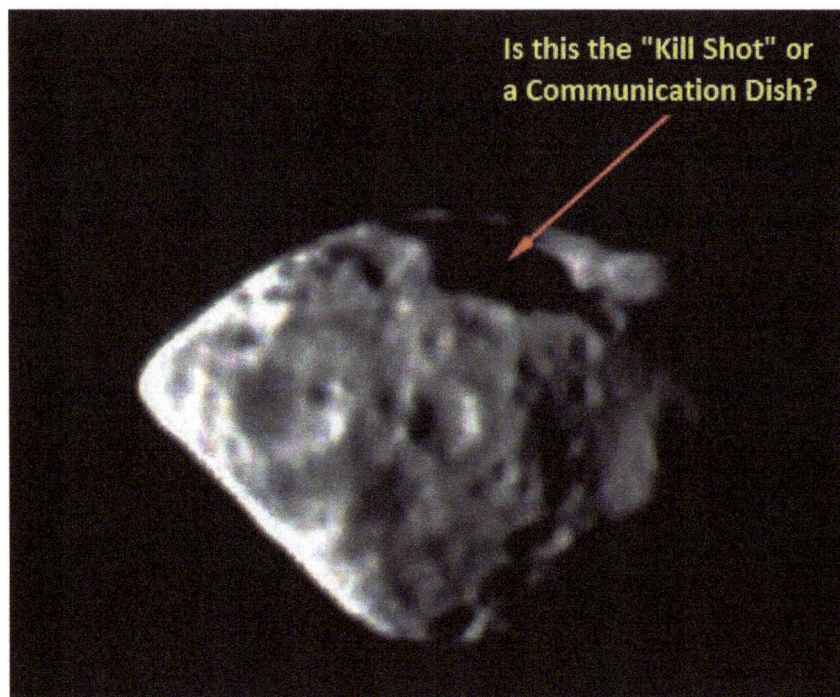

Was the one mile wide impact crater on the "table of the diamond" 2867 Steins caused by a meteor or a fragment of an exploding planet which should have destroyed the asteroid completely or is Steins really a massive spacecraft eroded by incessant micrometeorites over tens of millions of years and the large crater is actually a communication dish?
http://www.enterprisemission.com/Rosetta/Rosetta-analysis-test.htm

What we are seeing on the surface of Steins is repeating geometric regularities on the western hemisphere and cratered impacts on the eastern hemisphere with the overall symmetry similar to a diamond. The cratering impact on the asteroid should have destroyed it leaving only a pile of small floating rock. Of particular interest is the large impact crater on the "table" of the diamond which would have been the **"kill shot"** to the asteroid suggesting that the cratering on the eastern hemisphere, the duller, less reflective side resembles the surface crater strafing found on the Moon. This high difference in albedo between the two hemispheres, one with geometric regularities and the other with impact craters seems to indicate a composition difference. This difference, we will discover is also the same remarkable feature found on Saturn's moon Iapetus affecting half of that moon's surface.

What could be the reason for this possible composition and reflectivity difference for this 3-mile wide object?

The that best explanation for these observations as predicted by the Hoagland's Enterprise Mission team is that "Steins" is, indeed, an *artificial object* ... that has suffered, as Rosetta scientist Uwe Keller termed it, "a complex collision[al] history."

This *Enterprise* "artificial" model predicts that because of those collisions, *one entire half* of this small celestial body has suffered massive stripping of ***a former, highly reflective "outer casing" or "outer hull"***— (Bold italics added for emphasis).

A "casing" which is still (more or less ...) intact across *one* hemisphere and which shows "*layering*", the side that Rosetta captured with its one distant **Narrow-angle Camera (NAC)** image before the fly-by. http://www.enterprisemission.com/Rosetta/Rosetta-analysis-test.htm

In addition to these anomalies, Hoagland sees the striking appearance of several additional geometric features most notably, what appear to be enormous, distinctly separated *"windows"* aligned horizontally, in two parallel rows, just above Steins' highly reflective "girdle" (below, right).

These amazing, familiar-looking artificial features are then joined (further to the east ...) by an even more compelling detail -- what appear to be exposed "structural components" (resembling a giant "box-like structure"...), located above the "girdle" -- at the top of one of those massive "buttresses" seen on Stein's preceding hemisphere.

Could this feature be an exposed, major structural element of "Steins" indicating that its protective covering had been stripped away exposing more of the same major impact damage seen all across this hemisphere? This same type of damage has apparently removed almost all the rest of Stein's brilliantly reflective covering in this entire hemisphere, except for western sections of "the girdle" (below, left) ... and some additional regions around the pointed "culet" at the base of Stein's basic diamond shape. http://www.enterprisemission.com/Rosetta/Rosetta-analysis-test.htm

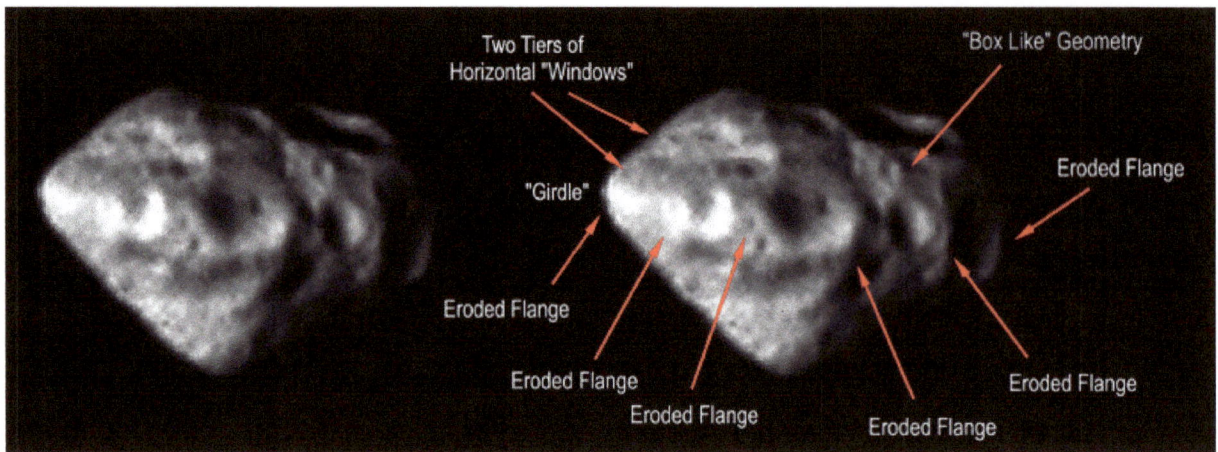

Artificial anomalies circumscribed around the surface of Steins strongly suggest that this is a massive spacecraft and not an asteroid!
http://www.enterprisemission.com/Rosetta/Rosetta-analysis-test.htm

The evidence is suppositional at best based on the photographic image interpretation made by Hoagland but, if the data from **ESA's Rosetta spacecraft** fly-by is correct then it supports Hoagland's hypothesis of artificiality, that 2867 Steins is not a natural solar system object but is, "in fact, a 3-mile-wide *manufactured artifact*, currently of unknown surface composition, about the size of a small town.'"!

Based on telescopic visual and IR spectral data gathered prior to Rosetta's flyby, Stein's external composition may actually be *"refined metal";* this is based, in part, on its unique spectrum classified as **"E"** according to planetary catalogues *making Steins among the rarest of all asteroid types* ... a spectral classification shared by only about *30 other objects in the entire solar system* out of over three hundred thousand asteroids known! (Bold italics added for emphasis). http://www.enterprisemission.com/Rosetta/Rosetta-analysis-test.htm

Historically, **"E-class"** asteroids have also shown unique responses to reflected *radar* signals.

The **Deep Space Network (DSN)** like that found at the **Arecibo Observatory** observed two medium-size **E-class asteroids** similar to 2867 Steins, **"44 Nysa"** and **"434 Hungaria"** indicating some "highly polarized" radio waves reflected back to Earth could actually be coming, not from their surfaces ... but from multiple reflections *deep inside!*

These two **E-type objects** are also about 20 percent *more efficient* at reflecting radar signals overall compared to any other type of asteroid!

These combined observations are thus totally consistent with E-type asteroids being **a)** covered with some type of *highly efficient, electrically* **conductive** *"casing"* and **b)** also being, essentially ...*Hollow!*

In other words, the data suggest that some of the radar echoes are literally returning from deep

236

within -- "scattered" multiple times (thus, *polarized* ...) by countless reflections between the flat, geometrically-spaced walls of ***innumerable interior corridors and rooms!***

Based on the preceding Earth-based visual and radar observations of these E-class objects -- combined with Rosetta's history-making fly-by of the first example seen in close proximity -- Steins could, indeed, now be the best-known representative of *a totally new class* of solar system "object"--

One of only a few dozen "E-type" ***ancient space platforms*** ... still orbiting the Sun ...
http://www.enterprisemission.com/Rosetta/Rosetta-analysis-test.htm

Correct interpretation is everything if the hypothesis for a former, ancient Extraterrestrial civilization in our Solar System is to hold up under close scrutiny. The images and data presented here from the Steins' fly-by are verifiable by referring to the appropriate ESA Rosetta websites. However, the correct scientific interpretation is only as good as the quality and quantity of scientific data made available from ESA or any other space agency given the political and psychological atmosphere extant at this current time

Steins has at first cursory examination all the superficial features of an asteroid but, Steins **IS NOT** an **ASTEROID!**

Hoagland points out the "extraordinarily ***specific*** topological ***symmetry*** (a "diamond"), coupled with its remarkable, repeating reinforcing that perception (regularly-spaced vertical protrusions in its surface, ***but only in the 'diamond girdle' plane*** ... the "buttresses" ... and remnants of complex, surface structural remains ....), all argue compellingly for Steins as artificial. The "asteroid's" marked overall ***asymmetry*** (in terms of the light unevenly reflected by the two opposing hemispheres ...) -- on an object otherwise remarkably ... anonymously ... ***symmetric*** -- completes the "artificial" picture ...." http://www.enterprisemission.com/Rosetta/Rosetta-analysis-test.htm

Yet, amid the self-congratulatory handshakes and pats on the back going on among the ESA scientists of the Rosetta Mission from the first post-fly-by in September 2008, it was clear that not everything was successful. At least that was the impression given at the news press announcement when ESA stated that their "the highest-resolution NAC Steins images have been LOST ... nine minutes ***before*** Closest Approach." Was ESA keeping something back on the data they had collected from Rosetta? How could a *"malfunction"* occur at the most critical point in the mission?

We have seen these types of *"malfunctions"* occur with other spacecraft and probes whether from the **Russian Space Agency** e.g. **Phobos I and II**, with NASA's **Pathfinder Mission** and most currently with ESA and the **Rosetta Mission**. There are countless other missions to the various planetary bodies with too many incidences of spacecraft malfunctions occurring at the most critical and in opportune times where important raw data has been irretrievably lost. Or has it?

Is it possible that everything went according to plan and that ESA saw something on 2867 Stein

that reeked of artificiality and they were able to get high-resolution images to confirm their suspicions but those images have never been released to the public? Was there *critical proof* here that supported Hoagland's **"ancient, inhabited solar system"** model?
http://www.enterprisemission.com/Rosetta/Rosetta-analysis-test.htm

In Richard Hoagland's mind, it was no mere coincidence as to "the nature and timing of the bizarre "camera failure" announcement. Well into the 'morning after' press briefing," ESA casually blamed the malfunction on a software "safe mode error" that the Narrow-Angle Camera went into at **nine minutes** before the minimum distance from Steins. And that the images taken just before cut-off would NOT be any better than the BEST images taken by the wide-angle camera at Closest Approach! Is this a plausible technical excuse or just mere coincidence?"

Instantly, Hoagland wondered what the Rosetta team had seen on those 5X higher resolution images at Closest Approach but, couldn't tell the public.

It would appear that they were deliberately suppressing those key images! According to Hoagland, ESA was forced to invent such an obvious "bold-faced lie" to cover up the fact that startling NAC close-ups of Steins **were already on the ground!**

So do we have "smoking gun" evidence from the imaging data, we collected thus far to determine what "2867 Steins" is, what happened to it to create a distinctly "two-faced" object and exactly when did it happen?

The overwhelming clue to all these questions is oddly enough, the dramatic "hemispherical asymmetry" of Steins which is tied to Hoagland's and Bara's Enterprise Mission of a **"reconstruction of an ancient solar system chronology"** along with **Tom Van Flanderns'** the **Exploded Planet Hypothesis (EPH).**

One of the major predictions of the "exploded planet hypothesis" (EPH), is that "normal" solar system events came to an abrupt and catastrophic crescendo ~**65 million years ago** with the sudden, literal *explosion* of one of the Sun's major inner planets ... the one NOT orbiting where the current asteroid belt now lives!

Tom Van Flandern theorizes that if a planet exploded in the ancient past in the area now occupied by the asteroids, there would be "visible scarring and other effects of such an inconceivable catastrophe on other bodies in the solar system".

" ... a major [planetary] explosion would send *a blast wave through the solar system,* blackening [and *cratering*] exposed, airless surfaces in its path. Most such solar system surfaces are indeed blackened [and heavily cratered], even for icy satellites. But a few cases have such slow rotation that only *a little over half of the moon* gets blackened. Saturn's moon Iapetus is one such case because its rotation period is nearly 80 days long ... One side is icy bright; the other is coal black. The difference in albedo is a factor of five ...

Thomas Van Flandern in 2007

"Perhaps the most basic explosion indicator is that all fragments of significant mass will trap smaller nearby debris from the explosion into *satellite orbits.* So, explosions tend to form [*systems* of] asteroids and comets with multiple nuclei of *all* sizes. Collisions, by contrast, normally cannot produce fragments in orbits because any debris orbits must lead either to escape or to re-collision with the surface. Moreover, collisions tend to cause existing satellites to escape, leading to asteroid "families" (many of which are seen). Our prediction that asteroids and comets would often be found to have satellites *has been confirmed* in recent years. The first spacecraft finding (by Galileo) was of moon **Dactyl** orbiting **asteroid Ida** in 1993. More recently, Hubble imagery found that **Comet Hale-Bopp** has at least one, and possibly three or more, secondary nuclei. http://www.metaresearch.org/solar%20system/eph/eph2000.asp

This period of 65 million years ago coincides with the extinction of the dinosaurs on Earth. Would this explosion of a 5th planet explain the sudden massive wipeout of most reptilian life on Earth with the impacts of massive meteorites during that time period?

The **extraterrestrial impact theory** stems from the discovery that a layer of rock dated precisely to the extinction event is rich in the metal **Iridium**. This layer is found all over the world, on land and in the oceans. Iridium is rare on Earth, ***but it's found in meteorites*** at the same concentration as in this layer. This led scientists to postulate that the iridium was scattered worldwide when a comet or asteroid struck somewhere on Earth and then vaporized. A 110-mile-wide (180-kilometer-wide) crater carved out of Mexico's Yucatán Peninsula, called

239

Chicxulub has since been found and dated to *65 million years ago*. *Many scientists believe the fallout from the impact killed the dinosaurs*.
http://science.nationalgeographic.com/science/prehistoric-world/dinosaur-extinction/

Did Planet V explode becoming the Asteroid Belt from gravitational instability or from internal stress forces and loss of its magnetic field or was it external forces originating from a collision with another planetary body? Perhaps, it was destroyed with weapons of mass destruction in an interstellar war!
(Google Images)

Van Flandern extrapolated that when a planet explodes for whatever reason, that it becomes a solar system-wide catastrophic event affecting most of the other solar system objects with outward rushing debris. This debris field would be composed of fine dust to extremely large miles-wide planetary fragments which depending on the satellite or planetary body's rotation period would result in debris impacting perhaps, only on one hemisphere side of that solar system body.

Our Moon seems to be more heavily impacted on the "dark side" hemisphere than on the light side or near side hemisphere that face towards Earth. Mountains on the Moon have been eroded to near smooth rounded features with little jaggedness remaining due to the constant bombardment from fine micrometeorites over a period of nearly sixty-five million years or more. The same effect is apparent on Mercury as well as on many of the satellites of the outer planets; in particular Saturn's moon Iapetus shows a high albedo difference with one side of the satellite more reflective while the other side is darker and less reflective indicating that due to a slow rotation of this moon, only one side was affected by the outward blast of debris moving towards it.

240

Now, if **Steins** is a massive spacecraft as we believe it to be then, its damage on its eastern hemisphere was due to its 6-hour rotation while being relatively close to the "blast point" or "ground zero" when Planet V exploded. Not too close to be utterly destroyed or too distant to be unaffected but, close enough to receive damage from meteoric bombardment. The fact that it was damaged on its entire eastern flank means that the whirling dust and debris cloud travelling at a minimum 5 miles per second was approaching directly from the source of ***"an explosion"*** ... ***behind*** Steins, as confirmed by the Rosetta WAC images meaning that Steins **"moving away"** from the planet when it exploded.

Reconstruction of the probable sequence of events follows:

"Subjected to a relatively brief (but intense) rain of "planetary crustal fragments" from the horrific explosion of the distant Planet V, combined with the effects of an expanding ***micrometeorite*** cloud of condensed planetary mantle rock -- all rushing past at several miles per second -- Steins had its highly reflective "casing" completely scoured off *one-half* of its exterior (the side facing the explosion ...), immediately followed by the larger, solid fragments impacting directly that same side ... with visibly catastrophic results leaving the unmistakable signature of a **LOT** of highly destructive impacts ... but only on ONE-HALF of Steins--

 "... in a far higher concentration than would be expected for such a small object ...."

 That latter, critical observation **CANNOT** be explained by any "steady-state" model of impacting debris ... spread across the lifetime of the solar system!

It can **ONLY** be satisfied if Steins was, indeed, exposed to "a sudden, short-duration rain of *high-speed blast-wave- objects*... causing out-of-proportion destructive impacts from the distant, cataclysmic destruction of *an entire planet!"*
http://www.enterprisemission.com/Rosetta/Rosetta-analysis-test.htm

This leaves one remaining question about 2867 Steins based upon the evidence from Van Flandern's **EPH Model** from a previously determined existence of an ancient, approximate 65 million year old solar system civilization. Why was Steins built in the first place? Why build a ***three-mile-wide, space platform*** shaped like a perfectly cut *"**diamond in the sky**"* which was then, almost destroyed. (Bold italics added for emphasis).

The answer is: "Steins was one of 30 known **'Arks of Salvation'** from a dying world! "They were spacecraft designed to house million or tens of millions of inhabitants to escape an impending catastrophe on their home world, the mysterious fifth planet that once existed between Mars and Jupiter!

This conclusion was not reached in any cavalier sense of the word but, from through logical assessment of the evidence determined through known astronomical facts and then extrapolating then most probable hypothesis that supports that evidence.

Almost every current model for the formation of 2867 Steins, including the ESAs assumes that this object is similar to the hundreds of thousands of other catalogued asteroids that is as

241

Hoagland puts it, simply "a chip off the old block." Steins, as one of many asteroids, through incessant collisions over billions of years of solar system evolution would have been inexorably *"ground down"* by relentless external impacts, like a rock being polished in a rock tumbler, into their current configurations.

Therefore, Steins' unique diamond-cut geometry is an oddity but, purely accidental, the intriguing result of our own Solar System's *"rock tumbler"* due to those 'innumerable random interplanetary collisions'.

However, when all is said and done, Steins' deliberate, intelligently *manufactured* design is not a random occurrence but, a *carefully-planned* bona fide **"artificial construct"** ....'diamond in the sky'

The fact that Steins' deliberate geometric design is diamond shape, that is neither too shallow nor too deep a cut but, the "ideal cut" serves a purpose. Where the Saturn's moon, Iapetus was constructed for *radar-stealth,* the opposite is true for 2867 Steins; it was created to be *radar-visible* to a high degree of prominence! It was meant to be detected and not hidden like Iapetus!

In other words, every aspect of Steins' geometry from the technical *radar-return* perspective shouts out to *any* active radar system, "**HERE I AM**"!

Which, when put together with **WHERE** Steins was discovered orbiting (10 degrees above, but within, the Main Belt Asteroids ...), and how close this is to the proposed orbital location of the original "exploding **Planet V**" leads directly to Steins' potential **PURPOSE**. As a *"city-sized" rescue craft* from the same ancient solar system catastrophe, we've been discussing! http://www.enterprisemission.com/Rosetta/Rosetta-analysis-test.htm

In other words, as mentioned above Steins is a kind of an "**Ark of Deliverance or Salvation**".

If you look even casually at Steins' remarkably faithful "diamond-cut" geometry (below, left), and compare it with the classic *reasons* for that specific cut (below, right), you reach a very interesting conclusion: that diamonds are cut in such a way as to display the most brilliance of light possible!

In the case of Steins, the geometric "diamond principle" of reflection and refraction of *lightwaves* still applies with *microwaves*, *if* the interior and exterior manufacturing geometry also *remains the same*.

And, if the material used in this construction interacts with these *radar wavelengths* in the same fashion as "light" does (within the crystalline geometry of a *real* diamond ...). The overall effect will then *also* be precisely "diamond-like."

Thus creating an "artificial, miles-wide diamond in the sky" -- which would literally "sparkle" (on a radar screen) with the reflected/refracted *microwave* energy being beamed at it ... just as a real diamond does when hit by light (see below)!

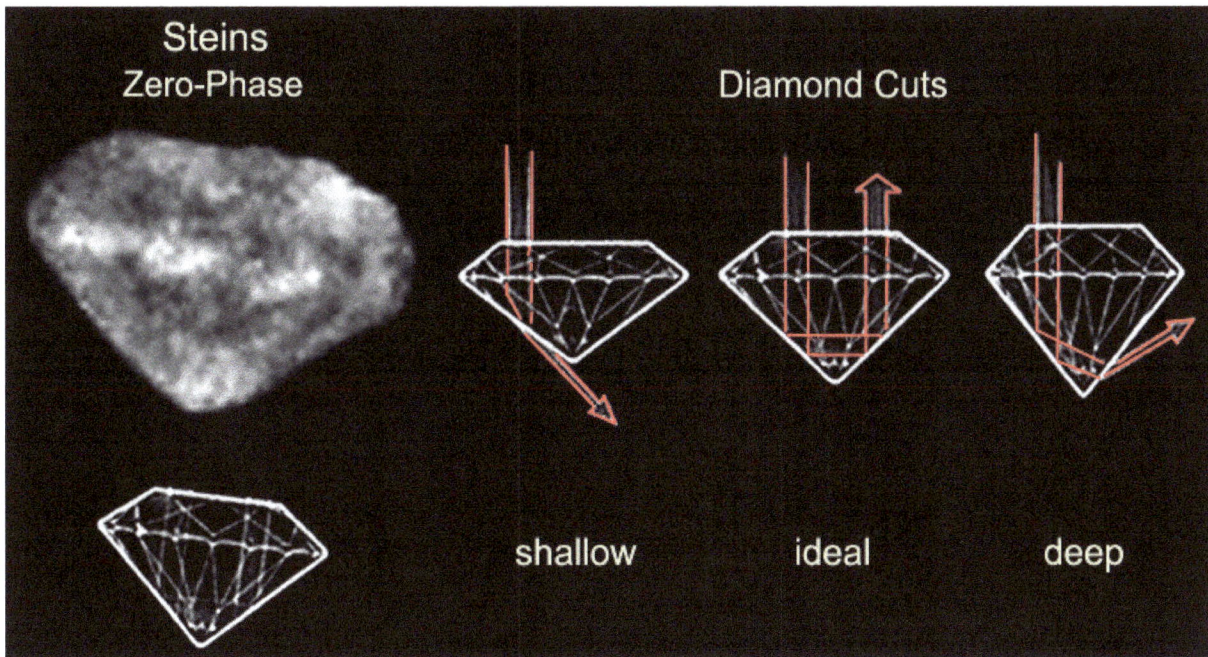

Steins
Zero-Phase

Diamond Cuts

shallow   ideal   deep

The "Asteroid" 2867 Steins' diamond shape is no accident of nature but, a deliberate design of intelligence to utilize high radar visibility to make it easily detected. Steins is, therefore, a city-wide rescue spacecraft, an Ark of Deliverance!
http://www.enterprisemission.com/Rosetta/Rosetta-analysis-test.htm

Recall, all of those other *30 plus the E-class asteroids* currently being investigated because of those *"highly anomalous radar signatures",* what would those current radar observations now reveal in light of the nature and origin of 2867 Steins..?!

This, of course, is exactly the type of "passive beacon" property you'd *want* to build into any space *fleet* of "rescue platforms" (designed to house literally millions of refugees from a dying planet) ... platforms that you would desperately need a *foolproof* system to relocate, anywhere in the solar system ... *after* the catastrophe!

The proper "diamond cut" technology was successfully used by the Apollo 11 Moon mission experiment with the deployment of an array of *optical, diamond-shaped prisms* on the Moon by the astronauts for the **"Moonbounce Laser Experiments"(MLE)** that has continued to function for last five decades, since their departure from the Moon.

Steins seems to embody the same *"retroreflection"* principle in its diamond-shaped "space platform," is the *one parameter* that would have ensured its quick location (if it survived ...) *(which it appears it may not have, unless in survivors were rescued soon after the planetary explosion or still in a hibernation sleep or dead!)* regardless of its damaged condition or orbit *after* the Planet V explosion. (Bold italics added by author for emphasis).
http://www.enterprisemission.com/Rosetta/Rosetta-analysis-test.htm

Remember that massive almost a mile wide "impact crater covering the entire northern rotational pole of Steins on an object only about *three times* wider than the crater itself? Well, ESA scientists have wondered, "Why didn't that disproportionate collision **totally destroy** this tiny object!?

The answer lies in the acceptance of the hypothesis that Steins is a *"city-sized, space platform"* with geometry specifically built-in to return *"microwave scanning beams"* (radar) directly to their sources. This yawning *"polar crater"* is **NOT** just another *"impact crater"* ... but a specific, integral part of Steins' carefully designed, geometrically precise *"diamond structure"*.

A carefully tailored opening (now, of course, badly eroded by mega-years of micrometeorite abrasion ...), originally designed to allow microwaves to penetrate deeply, *directly* into Steins' interior, so those same microwaves would then be reflected *directly back* to the radar stations searching for Stein's location after the explosion of Planet V!

Even, the smaller craters around and below the diamond girdle of Steins may also, not be impact craters but, radar or communication transmitter/receiver dishes. The official ESA analysis noted their curious abundance *"... in a far higher concentration than would be expected for such a small object ...."*

Perhaps, they are cavernous airlocks, spacecraft docking bays, or fuel storage tanks, whatever their purpose, their voluminous spaces are now totally, catastrophically exposed to the elements of space but, are now in a death-state preservation, their tell-tale geometric spacing, waiting for some future manned exploration mission to the asteroids to uncover their actual purpose. http://www.enterprisemission.com/Rosetta/Rosetta-analysis-test.htm

It is highly conceivable, in fact, probable that one day, when we do start to explore among the asteroids, we may actually come across asteroids with ruins upon them or even unusual artifacts floating in space. There may even be pieces of buildings, objects of metal construction, possibly even humanoid bodies still floating in among the asteroids that are literally tens of millions of years old, perfectly preserved by the vacuum of space, which at first glance may appear to be nothing more than rocks but, in fact are the last remaining vestiges of evidence from a former but, extremely ancient civilization!

By now, if it is not already obvious to the reader, you are starting to get the big picture about the potential ancient history of our Solar System and the profound implications it will have on all of humanity? The history books of every culture, from every society on Earth, will have to be re-written in light of the revelation that must come sooner or later to the discovery of a former, very ancient, highly advanced civilization that once existed in our Solar System. What we thought we knew and what we were taught about our past will undoubtedly require a re-examination of our history as a species. A lot of preconceived ideas, notions, theories, scholarly papers, books and the big reputations founded upon them will have to be set aside, if not relegated to the scrap pile of history of outdated ideas, useless doctrines and misleading precepts as well as the worn-out shibboleths that were built upon the false prejudicial standards of a select elite few.

The Solar system seems to be filled with objects, ruins, artificial structures and planetary constructions that beg us to re-examine and re-evaluate our current understanding of this particular corner of the universe. We are finding that as we explore outwards and beyond our

244

stellar neighbourhood that not all planetary bodies appear to be of natural origin. In fact, many of the megalithic projects like "designer-worlds and moons", massive miles-wide spacecraft and mega-size building structures were the creations of very ancient, highly advanced intelligence. They were in all probability an intelligence on an order of a type two civilization, which once inhabited our Solar System.

They were us! They are us!! But, for reasons not completely understood, they met with a catastrophic epochal end.

It is unlikely that any Type Two Civilizations would be wiped or reduced in population from a natural planetary disaster or from having a star go supernova; as such civilizations have the advanced technological capability to control, manipulate and even create such planetary bodies. Such civilizations would be constantly monitoring for such potential catastrophes in much the same way we monitor for earthquakes and volcanic activity on Earth but, with greater technical sophistication.

Such knowledge and has been hidden from and forbidden to the masses of society by a select powerful and privileged few, who are determined to control the informational flow of such knowledge that will benefit only the few and not the many.

There is growing evidence to suspect that the fly-by of "2867 Steins" was no mere chance or coincidence but based upon prior knowledge right from the star as a prime objective of ESA's clandestine Rosetta Mission! The original public announced mission to "Comet Wirtanen" was never intended but, used as a ruse to go to 2867 Steins, because of its unusual E-class characteristics as an asteroid which suggested artificiality. Comet Wirtanen may only have been a secondary objective as a backup choice if Steins was not achievable.

Somehow, ESA knew what they may find in its fly-by of Steins because they already possessed the knowledge from some other source.

Perhaps, as this author has already suggested earlier in the section that ESA was revealing too much of its discoveries into the public domain that NASA took it upon itself to corral ESA into NASA's **Brookings Institute Report** type of way of thinking by placing political pressure on the nations that support the European Space Agency as well as the agency, itself.

"Play ball with us or you'll find your space probes shot down or your countries in political and financial chaos!" Such threats or similar tactics of intimidation and perhaps, promises of alien technology being shared with them might produce the results NASA would hope for. Whatever, the political pressures NASA used, it appears to have worked!

As **Hoagland** and **Barra** have stated in their search for evidence to support the model of "an ancient, solar-system-wide, extraterrestrial civilization ..." was met by a "completely unexpected *social and cultural* environment surrounding this most elusive topic." There has been evidence of some arcane, *ritual* reasons within NASA namely, secret **"Masonic rituals"** as presented in their book: "Dark Mission: the Secret History of NASA" which are embodied in essentially every NASA launch, every NASA planetary fly-by or NASA mission landing that support this idea.

The solicited cooperation of nations with space programs by NASA including ESA appears now to have a definite *hidden agenda* "to locate and secretly investigate *ancient, extraterrestrial ruins in the solar system ...*" This search is driven as Hoagland has stated on numerous occasions by some kind of bizarre "ritual plan"; they have standing orders to *conceal* this **Prime Objective** (and *how* they carry it out ...) with a "plausible, scientific cover-mission." The latter, of course, is designed primarily to keep their respective taxpayers (and even most scientists involved ...) *completely in the dark* about the real reasons for these "competing" interplanetary missions' increasingly being conducted by, the Chinese, the Japanese, the Russians, the Indians, the Europeans ...and their *true objectives*. http://www.enterprisemission.com/Rosetta/Rosetta-analysis-test.htm

As this author has also, stated previously that NASA's true planetary missions and space explorations is the *quest and acquisition of highly advanced Extraterrestrial technology* and now these other nations appear to be doing the same!

Is it possible that ESA got incredibly lucky by collating all the existing Earth-based spectral observations, photometry, radar scientific data on the "E-class" asteroids and realized that Steins was a candidate of this unique asteroid type, which was serendipitously just happened to be on the travel route to an eventual *secondary* comet target (**Comet 67P/Churyumov-Gerasimenko**), in 2014 as publicized.

It is somewhat far-fetched but, not out of the realms of possibility. It is more likely, that some of ESA's planners and scientists had access to *"secret documents"* from other sources **"in the know"**... NASA, maybe), regarding the *true, artificial nature* of "2867 Steins" -- and thus, as Hoagland says, based their entire 10-year "secret mission" on the early rendezvous with Steins with the much-publicized "later comet rendezvous" actually being just the "cover story"

And as Hoagland points out the dead giveaway for this type of reasoning is the fact that the mission's overall, official *name* is directly taken from NASA's own covert ritual **"Masonic Playbook"... Rosetta!**

Rosetta as every student of history well knows takes its name from the **Rosetta Stone** which gave archaeologists the tools to decipher **Egyptian hieroglyphics.** ESA project scientists hope their orbiter-lander mission is the astronomical equivalent of its namesake, giving astronomers the tools they need to decipher the nature of comets.

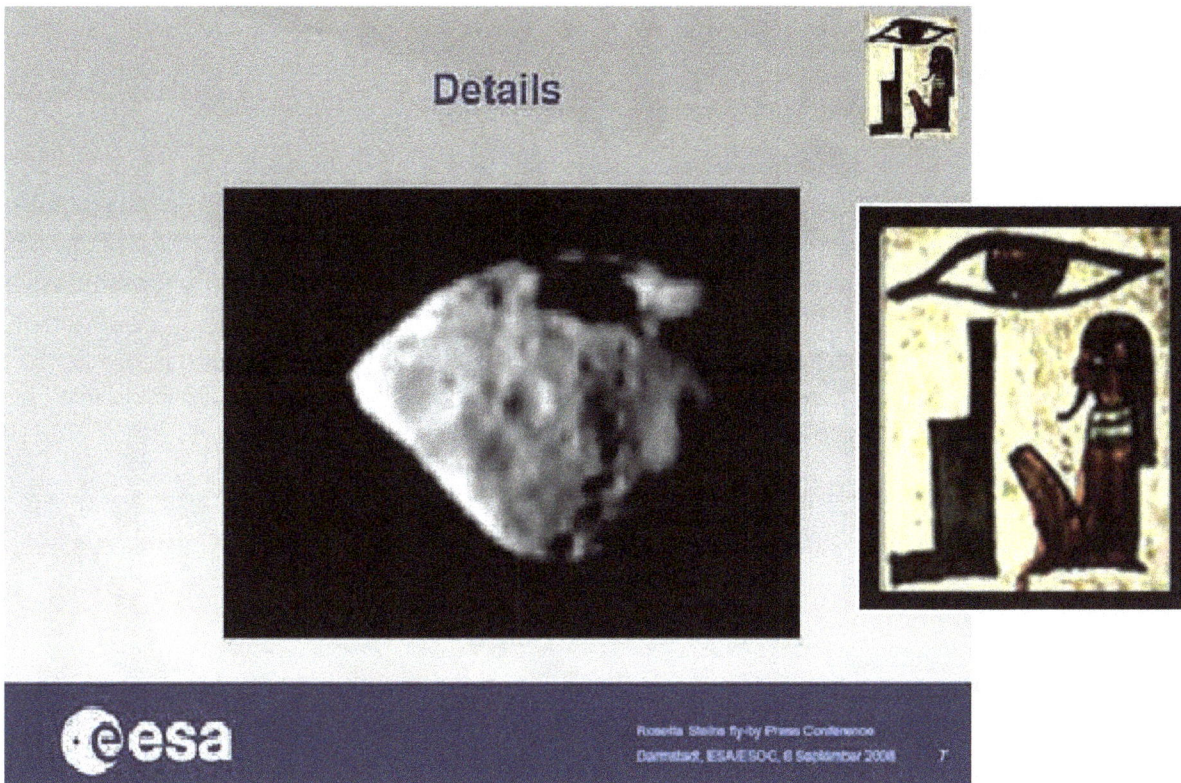

**Note the Egyptian hieroglyph in the upper right-hand corner and to the right,
the symbol for Osiris on the photo image is from ESA's Rosetta website**
http://www.enterprisemission.com/Rosetta/Rosetta-analysis-test.htm

ESA has even named it key camera system aboard its comet-chasing spacecraft, "OSIRIS" in keeping with the Egyptian iconography. There is a logo on every officially released Steins "OSIRIS image" literally spelling out (in Ancient Egyptian!) the name of this most important of Egyptian "gods".

So, why is "Osiris" important to any potentially *secret* Rosetta Mission to an ancient space platform? It is because it is a way for certain scientists and administrative staff within NASA and now, ESA to pay homage to the ancient memory of Egyptian and Sumerian demi-gods Extraterrestrial astronauts) of old, who may have one time walked upon the Earth and aided humanity to arise to become a viable thriving species.

According to Hoagland and Bara, Orion (Osiris), the Apollo 16's Lunar Module was purposefully not sent crashing back to the Moon as had been the case with previous LMs but, returned home with the astronaut signifying the literal 'stellar embodiment' of the sacred Egyptian god of the **Dead *and Resurrection* ... Osiris**"

If the ESA Rosetta Mission's *secret* objective was "to fly-by one of the long-dead, ancient solar system's surviving *artificial space platforms*" thereby, *resurrecting* its memory then, the naming choice was perfect!

247

But, could ESA be looking at Steins as more than just a spiritual acknowledgement of death and resurrection of Osiris, could ESA also be viewing it for some future physical resurrection? In other words, could ESA be considering possible future salvage attempts to revitalize the operation of the space platform, with repairs or reverse engineering? Nothing is out of the realm of possibility given that American has a secret black space program utilizing reverse engineered alien technology then, why wouldn't ESA have developed a similar black space program that could become operational within the next decade, if not sooner?

How is it possible that ESA knew before they began the mission that 2867 Steins was in fact, an *ancient spacecraft!?* ESA may have had access to ancient solar system information e.g. ancient Sumerian texts like the **Gilgamesh accounts of creation** and the early rise of the Mesopotamian civilization. Perhaps, there were other carefully preserved texts with actual images*!*

Whatever form this information took, wherever it may have originated from, the history was dutifully and faithfully written down to preserve actual "ancient solar system events" including incredibly useful *navigation* data with exact ephemerides (locations) of some of the surviving ancient spacecraft still orbiting the Sun masquerading as "mere asteroids"!

And somewhere in Europe or Washington, tucked away in the archives of ESA or NASA are actual enlarged, high-resolution images of 2867 Steins taken "from Rosetta's *fully-operational* Narrow-Angle Camera, photographs *five times closer* than the ones we have now ... photographs that show astonishing, *individual artifacts* from Steins' ancient, manufactured origins ... artifacts lying in ruin on its blasted surface even! http://www.enterprisemission.com/Rosetta/Rosetta-analysis-test.htm

No doubt other asteroids, particularly those of the **E-Class** type, due to their high albedo surfaces will once again catch the attention of space scientists and the interest of the public. It will most probably be this class of asteroids that we will discover that they are in reality ancient space platforms and not asteroids as in the case of 2867 Steins. The question at that time will be, are any of these 65 million year old space platforms still operational and are they still manned with the inhabitants of the former Planet V or have these inhabitants already resettled on another planet like …Earth …?!

CHAPTER 86

VESTA AND CERES, THE DWARF PLANETS
IN THE ASTROID BELT

Vesta (Virgin Goddess of Home and Hearth)

If **Asteroid 2867 Steins** was not enough to convince researchers and debunkers of its artificiality and the possibility that it is a mammoth size spacecraft, then perhaps, Vesta and most certainly that Ceres may be prove that an Extraterrestrial Intelligence once existed in this Solar System as these asteroids/planetoids were possible sentry outposts occupied and operated by ETI.

Vesta, minor-planet designation **4 Vesta**, is one of the largest asteroids in the Solar System, with a mean diameter of 525 kilometres (326 mi). It was discovered by **Heinrich Wilhelm Olbers** on 29 March 1807 and is named after Vesta, the virgin goddess of home and hearth from Roman mythology.

Vesta is the second-most-massive object in the asteroid belt after the dwarf planet **Ceres**, and it contributes an estimated 9% of the mass of the asteroid belt. The less-massive Pallas is slightly larger, making Vesta third in volume. Vesta is the last remaining rocky protoplanet (with a differentiated interior) of the kind that formed the terrestrial planets.

Could it have been a small moon hundreds of millions of years ago that when **Planet V** exploded that this moon became so battered by meteoric bombardment from fragment debris emanating from Planet V that its appearance became unrecognizable leaving it in its current large asteroid form or is it a large remnant of a planet that once existed where the Asteroid Belt is now?

Vesta is the brightest asteroid visible from Earth. Its maximum distance from the Sun is slightly greater than the minimum distance of Ceres from the Sun, though its orbit lies entirely within that of Ceres.

NASA's *Dawn* **spacecraft** entered orbit around Vesta on 16 July 2011 for a one-year exploration and left orbit on 5 September 2012 heading for Ceres. Researchers continue to examine data collected by *Dawn* for additional insights into the formation and history of Vesta.
http://en.wikipedia.org/wiki/4_Vesta

In 1802, Heinrich Olbers discovered Pallas the year after the discovery of Ceres. *He proposed that the two objects were the remnants of a destroyed planet.* He sent a letter with his proposal to the English astronomer **William Herschel**, suggesting that a search near the locations where the orbits of Ceres and Pallas intersected might reveal more fragments. These orbital intersections were located in the constellations of Cetus and Virgo. Olbers commenced his search in 1802, and on 29 March 1807, he discovered **Vesta** in the constellation Virgo — a coincidence, because *Ceres, Pallas, and Vesta are not fragments of a larger body.* Because the asteroid **Juno** had been discovered in 1804, this made Vesta the fourth object to be identified in the region that is now known as the **Asteroid Belt**. The discovery was announced in a letter addressed to

German astronomer **Johann H. Schröter** dated 31 March. Because Olbers already had credit for discovering a planet (**Pallas**; at the time, the asteroids were considered to be planets), he gave the honor of naming his new discovery to German mathematician **Carl Friedrich Gauss**, whose orbital calculations had enabled astronomers to confirm the existence of Ceres, the first asteroid, and who had computed the orbit of the new planet in the remarkably short time of 10 hours. Gauss decided on the Roman virgin goddess of home and hearth, Vesta.
http://en.wikipedia.org/wiki/4_Vesta

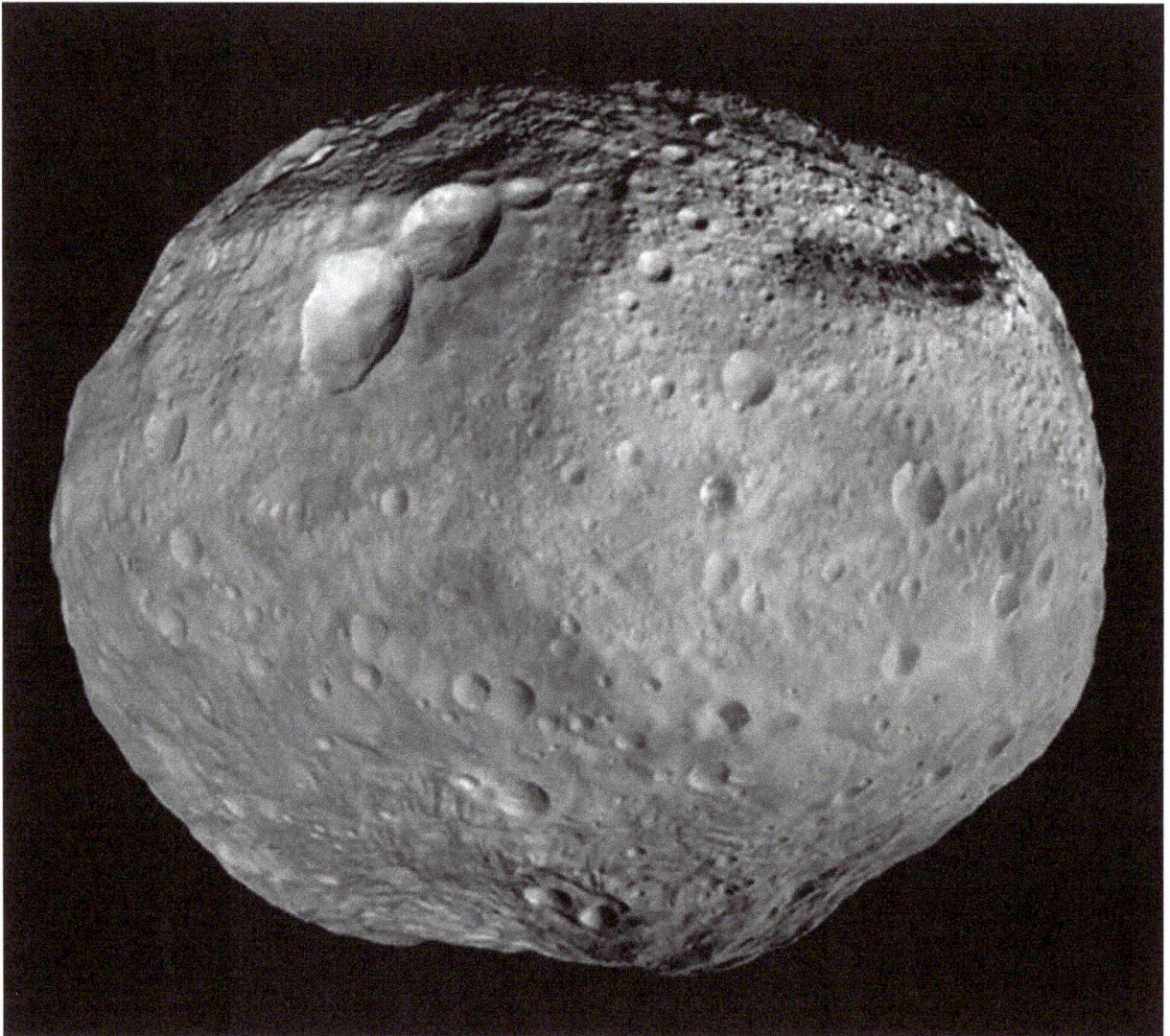

Vesta is a dwarf planet but, its appearance is more like a large asteroid and similar to the Martian moon Phobos only many times large
https://en.wikipedia.org/wiki/File:Eros,_Vesta_and_Ceres_size_comparison.jpg

Ceres and **Vesta** are the two most massive residents of the asteroid belt. Vesta is a rocky body, while Ceres is believed to contain large quantities of ice. The profound differences in geology between these two protoplanets that formed and evolved so close to each other form a bridge from the rocky bodies of the inner solar system to the icy bodies, all of which lay beyond in the outer solar system. At present, most of what we now know about Vesta and Ceres comes from

ground-based and Earth-orbiting telescopes like NASA's Hubble Space Telescope. The telescopes pick up sunlight reflected from the surface in the ultraviolet, visible and near-infrared, and by emitted radiation in the far-infrared and microwave regions.
http://www.nasa.gov/mission_pages/dawn/ceresvesta/#.VRjOUOH7OVo

Vesta is considered to be an asteroid even though its size qualifies it as a small protoplanet that was probably formed during the formation of the Solar System and the planets about 3.5 to 4.5 billion years ago. Heinrich Olbers hypothesized that Vesta, Ceres, Pallas may be remnants of a destroyed planet, even though, it is currently considered not to be as such. This does not preclude the theory that the balance of asteroids between Mars and Jupiter are remnants from a destroyed planet. In fact, billions or even hundreds of millions of years could have been enough time for the spherical shape of **Vesta** to have been formed through multiple collisions with other asteroids in the **Asteroid Belt**. This means that Vesta could have been a remnant of a destroyed planet that is now the Asteroid Belt.

This is just a theory, however; these larger asteroids or protoplanets may, in fact, be the escaped former moons of a planet that was destroyed in some cataclysm in the ancient distant past which has now become the Asteroid Belt.

In other words, Vesta and especially, **Ceres** which is spherical in shape and probably a frozen partially icy dwarf planetoid that once orbited a former planet that once existed between Mars and Jupiter. In fact, this author is going out on a limb by hypothesizing that is the case with Ceres which lends support to **Tom Van Flanden's Exploded Planet Hypothesis** model which explains the origin of the Asteroid Belt. Essentially, when the mysterious **Planet V** which orbited between Mars and Jupiter for billions of years, until in recent astronomical times, 50,000,000 to 70,000,000 years ago according to the Van Flanden model, it suffered a planet-wide cataclysm that blew the planet apart into millions of fragments that became the Asteroid Belt. In this destruction, the moons of Planet V, one of which may have been Ceres, somehow managed to escape destruction except for cratering but, remained in near orbit to its original orbit in relation to its mother planet's position.

Could this explain the shape, the positions and the geological attributes of the dwarf planets of **Ceres, Vesta, Pallas** and other large asteroids as possible moons of a larger, former planet that once existed between Mars and Jupiter? Only NASA's *Dawn* space probe will uncover those answers when it finally enters into a steady orbit around Ceres which will occur sometime in April 2015.

Ceres (Goddess of Agriculture)

As the ion propulsion spacecraft, *Dawn* departed from the asteroid/planetoid Vesta, NASA aimed it at its next target in the **Asteroid Belt**, Ceres a spherical planetoid or dwarf planet. Ceres was discovered on January 1, 1801, by **Giuseppe Piazzi** of Italy, it was the first asteroid/dwarf planet discovered in the Solar System. Its size is estimated to be 975 by 909 kilometers (606 by 565 miles). Its rotation is once every 9 hours, 4.5 minutes. For a planetoid to be in among the **Asteroid Belt**, it seems to go against what is understood in astronomy and planetary sciences as asteroids by rule are considered to be fragments or remnants of a possible planet and they are irregular in shape not spherical like Ceres with planetary qualities.

The object is known by astronomers as "1 Ceres" because it was the very first minor planet discovered. As big across as Texas, Ceres' nearly spherical body has a differentiated interior - meaning that, like Earth, it has denser material at the core and lighter minerals near the surface. Astronomers believe that water ice may be buried under Ceres' crust because its density is less than that of the Earth's crust, and because the dust-covered surface bears spectral evidence of water-bearing minerals. Ceres could even boast frost-covered polar caps.

This image was taken by NASA's Dawn spacecraft of dwarf planet Ceres on Feb. 19, 2015, from a distance of nearly 29,000 miles (46,000 kilometers). It shows that the brightest spot on Ceres has a dimmer companion, which lies in the same crater basin.
https://en.wikipedia.org/wiki/File:Eros,_Vesta_and_Ceres_size_comparison.jpg
Credit: NASA/JPL-Caltech/UCLA/MPS/DLR/IDA

Astronomers estimate that if Ceres were composed of 25 percent water, it may have more water than all the fresh water on Earth. Ceres' water, unlike Earth's, is expected to be in the form of water ice located in its mantle.

http://www.nasa.gov/mission_pages/dawn/ceresvesta/#.VRjbO-H7OVp

Two unusual "bright spots" which remain bright as the planetoid
Ceres rotates as the red arrows indicate
https://www.nasa.gov/feature/jpl/the-case-of-the-missing-ceres-craters

The Dawn spacecraft is now in a slow insertion trajectory that will put it in orbit around the dwarf planet, however, when it was approximately 29,000 miles (46,000 kilometers) from Ceres, it took some photographic images of the planetoid which were unusual, to say the least, which has NASA scientists mystified. Ceres shows two very bright spots on its surface both within a large single crater, one bright spot being larger with a dimmer companion. (See images below of the bright spots as they remain bright and visible during Ceres' rotation ruling out the possibility of a **cryovolcanism**.

A volcano, in its broadest and simplest sense, is an opening in the surface of a body through which material is ejected. The most well-known type of volcano is the type that occurs on Earth which erupts molten rock, gasses and pyroclastic material; although this is not the only type of volcano on Earth—sedimentary volcanoes erupting sand and mud do exist.

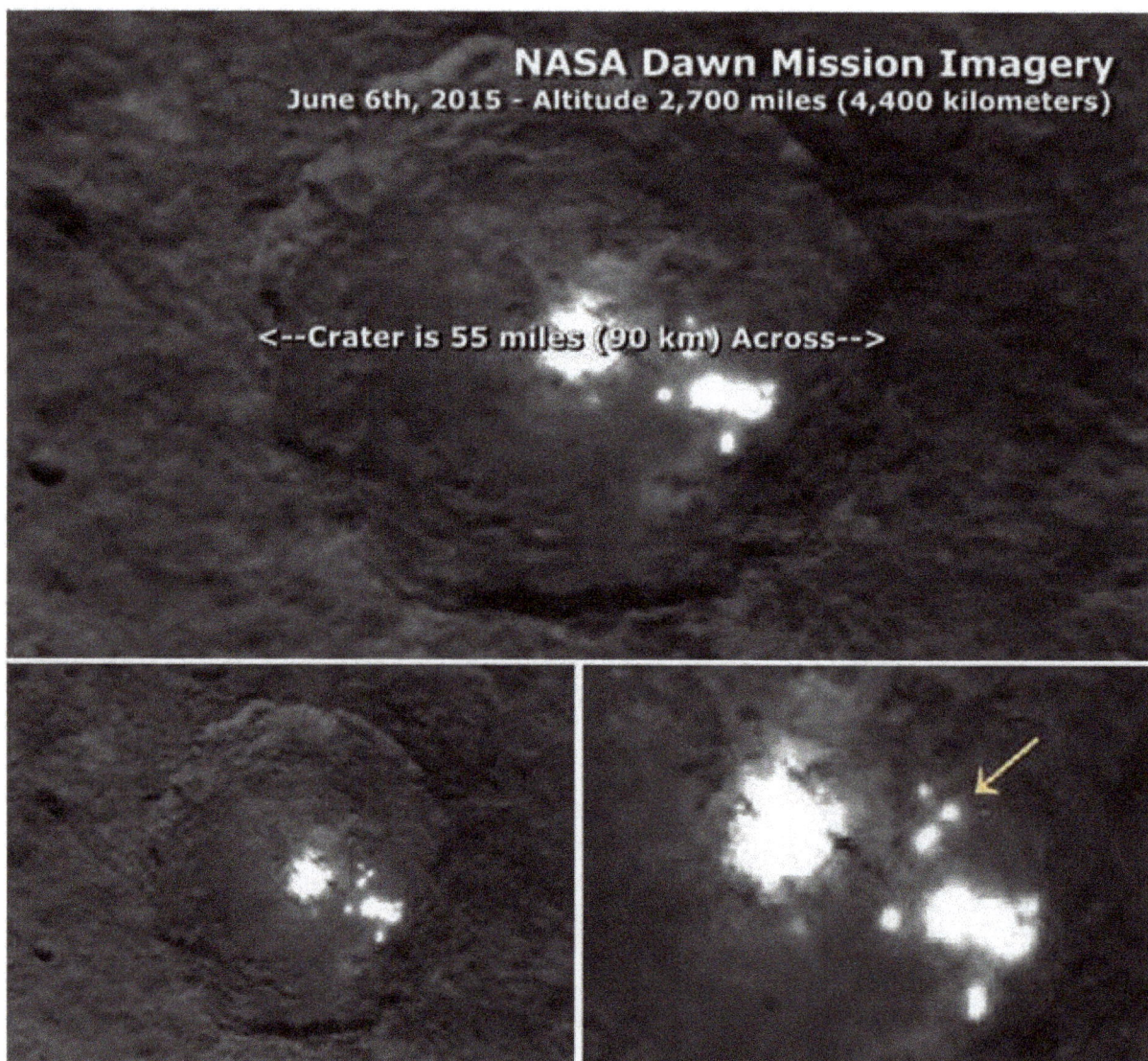

High resolution indicates that the two bright spots in Occator Crater on Ceres photographed on June 6, 2015, are actually ten or more bright spots
https://www.nasa.gov/jpl/bright-spots-shine-in-newest-dawn-ceres-images

A cryo-volcano is an opening on the surface of an icy body, like the moons of our outer planets that, instead of erupting molten rock erupts chemicals with a low boiling point otherwise known as volatiles such as water and methane. http://hagablog.co.uk/demos/enceladus/volcanism/

Some of NASA's scientists think that this could be the case with the two bright surface spots in which the vapor from **ice volcanoes** is reflecting sunlight and thus the cause for its very high albedo. However, this brightness continues to show even with the rotation of Ceres into the darkness which is not possible without sunlight. This means that *the light emanating from the surface is probably not natural in origin* when compared to other bright reflections from other crater walls which do disappear losing their sunlit reflection off it surfaces as the planetoid rotates.

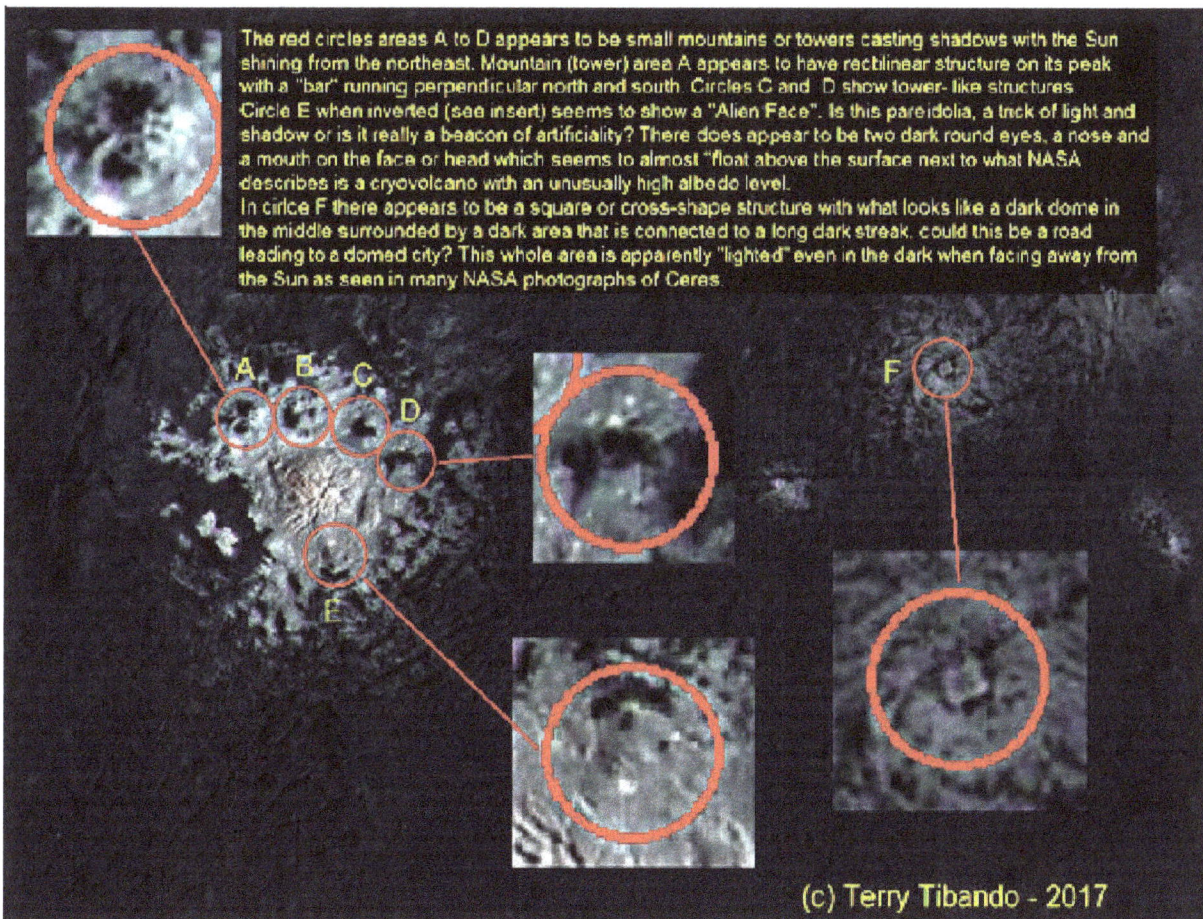

The red circles areas A to D appears to be small mountains or towers casting shadows with the Sun shining from the northeast. Mountain (tower) area A appears to have rectilinear structure on its peak with a "bar" running perpendicular north and south. Circles C and D show tower-like structures. Circle E when inverted (see insert) seems to show a "Alien Face". Is this pareidolia, a trick of light and shadow or is it really a beacon of artificiality? There does appear to be two dark round eyes, a nose and a mouth on the face or head which seems to almost "float above the surface next to what NASA describes is a cryovolcano with an unusually high albedo level.
In cirlce F there appears to be a square or cross-shape structure with what looks like a dark dome in the middle surrounded by a dark area that is connected to a long dark streak. could this be a road leading to a domed city? This whole area is apparently "lighted" even in the dark when facing away from the Sun as seen in many NASA photographs of Ceres.

(c) Terry Tibando - 2017

A close-up photo of Ceres' bright spots which appear to be artificial in origin. There are four small mountains or towers, one with geometric structure and there even appears to be an "Alien Face" with eyes, a mouth and nose, a strong sign of artificiality!
https://www.nasa.gov/feature/jpl/dawn-identifies-age-of-ceres-brightest-area (c) Terry Tibando

Stubbornly, NASA refuses to think that the strange bright spots can be anything than volcanic ice erupting or bubbling up to the surface from a deep subterranean ocean. According to Chris

Russell, principal investigator for the **Dawn Mission**, the origin of the bright spots still remains the same: *"Reflection from ice is the leading candidate in my mind"*.

The problem with this theory is that the bright spots besides being highly reflective, even on the dark side of Ceres (away from the sun) still remains very bright as if the source of the lights may be artificial in origin! (See above photo of the artificial structures found by this author).

There is even a large three-mile-high mountain, probably the only mountain on the surface of Ceres that stands out like a huge pimple, a possible **cryovolcano** of ice which disproves the theory that cryovolcanism is the cause of the high reflectivity for the bright spots, this mountain is only slightly reflective in comparison to the rest of the Ceres terrain!

Crater walls with high sunligh reflectivity which diminishes as Ceres rotates

Note that the sunlight reflectivity off the crater walls has diminished

(c) Terry Tibando - 2013

The above photos illustrate the change in albedo brightness off the crater walls of Ceres as it rotates on its axis
(c) Terry Tibando

Further investigation by the **Dawn spacecraft** is necessary to determine the nature of these bright spots which has an albedo brighter than the natural sunlight seen on some of the crater walls. Are these bright spots of light sign of artificiality?

Yet, tenaciously some NASA scientists hang on to this theory like a Pit Bull with a bone in its mouth as they speculate on what may be causing the two bright spots.

"This may be pointing to a volcano-like origin of the spots, but we will have to wait for better resolution before we can make such geologic interpretations," said **Chris Russell**, principal investigator for the Dawn mission, based at the University of California, Los Angeles.

Dawn will enter orbit around Ceres on March 6. As scientists receive better and better views of the dwarf planet over the next 16 months, they hope to gain a deeper understanding of its origin and evolution by studying its surface. The intriguing bright spots and other interesting features of this captivating world will come into sharper focus.
https://www.youtube.com/watch?v=jSDBl0FMX0s

"The brightest spot continues to be too small to resolve with our camera, but despite its size, it is brighter than anything else on Ceres. This is truly unexpected and still a mystery to us," said Andreas Nathues, lead investigator for the framing camera team at the Max Planck Institute for Solar System Research, Gottingen, Germany.
http://dawn.jpl.nasa.gov/feature_stories/Bright_Spot_Ceres_Dimmer_Companion.asp

Author's Rant: All Ufologists and Planetary Anomaly Researchers should anticipate that NASA's findings will ensure that their high resolution images of these two bright spots will "reflect" an appropriate answer in the traditional manner that we have all come to expect. Namely, from NASA's years of public disclosure to such things that are mysterious and smack of artificiality, they will conclude that they are either cryovolcanoes or very highly reflective geological mineral or rock that is *"worthy of further investigation"* after it has been sanitized of any possibility of showing anything strangely alien.

But, given NASA's inclination toward a natural cause explanation for the high reflectivity within the crater, there are fortunately other strange crater anomalies to offset any possibility that Ceres is anything but, a natural planetary body resident within the Asteroid Belt. These less than natural looking craters have a polygonal symmetry to them which portend to similar anomalies that as we will discover, are to be found on other moons further out in the Solar System.

Ceres shows at least two very large hexagon-shaped craters, each with six sides; one very large flat plain crater with one small central crater within it and four small craters almost equidistant to each other radiating out from the central crater in an almost perfect Celtic Cross shape, all of which is south of the equatorial region. Is there a hyperdimensional tetrahedral geometry in play on this small dwarf planet of Ceres?

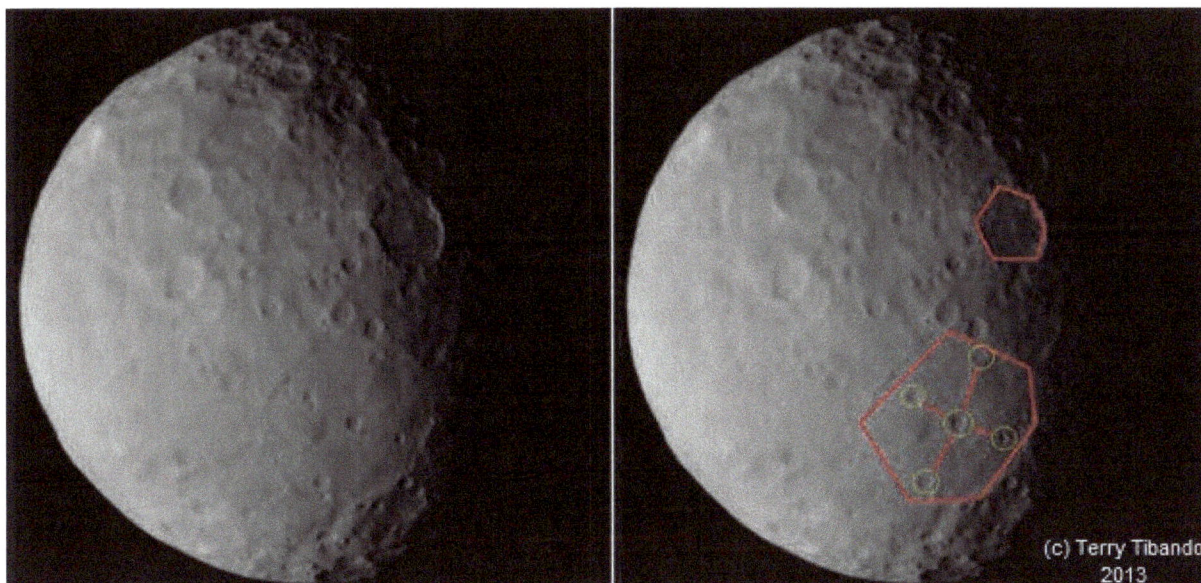

Ceres has at least two craters with polygonal shapes, with the larger crater basin displaying an almost perfect Celtic cross layout of four small craters with a slightly larger central crater
(c) Terry Tibando

This large hexagon, flat plain crater with its **Celtic Cross** design within its circumference jumps almost off the planet's surface begging the viewer to further investigate the possibility of a hyperdimensional physics operating on this planetoid that may have been originally created by a higher intelligence as a calling card to other travelling star-faring intelligences passing near our Solar System. (See images above and below).

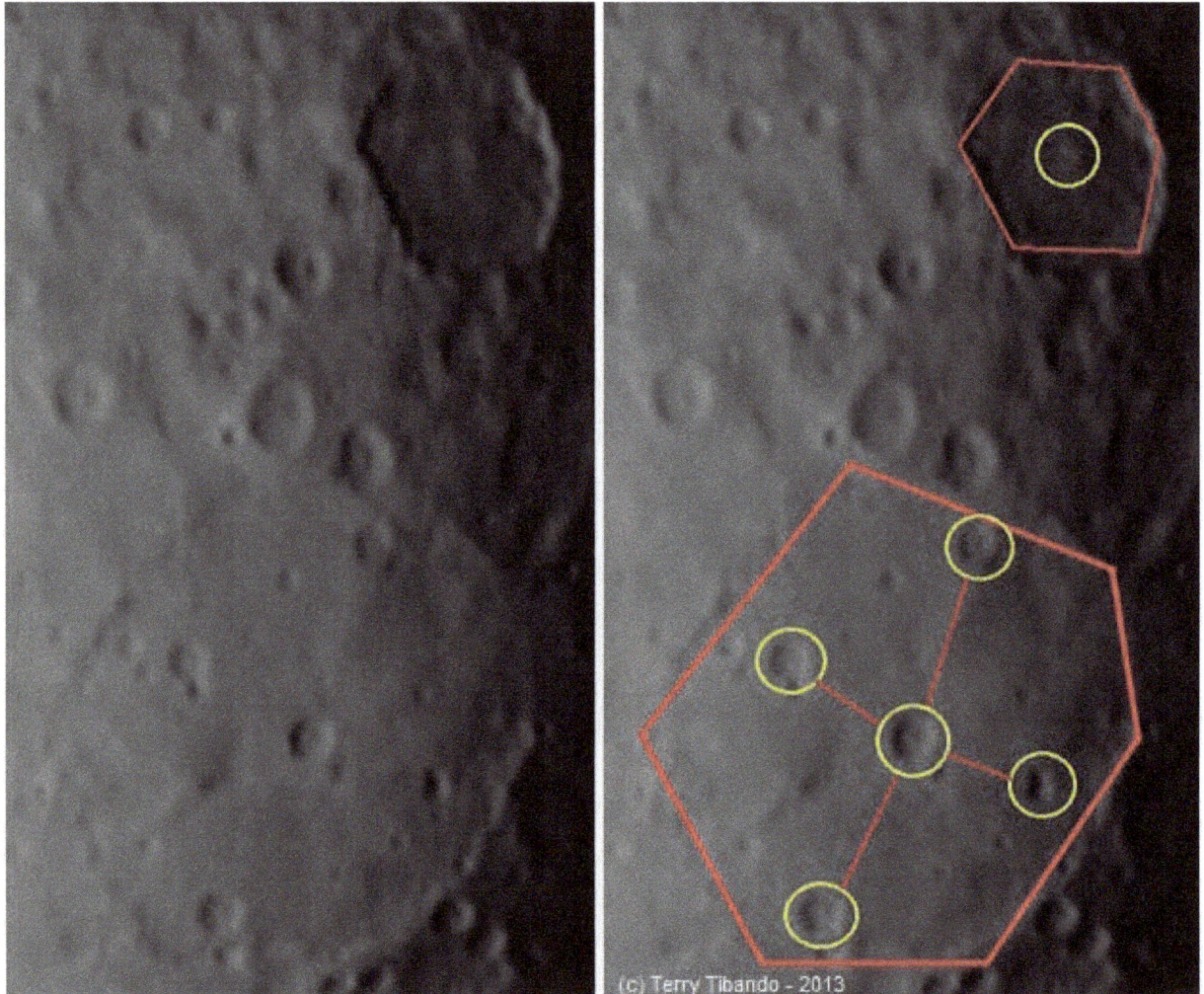

Enlargement of the two crater regions on Ceres with the Celtic Cross of small craters within the much larger hexagon crater, Kerwan Basin indicating a possible hyperdimensional physics in operation created by an ancient higher intelligence that once existed in our Solar System
(c) Terry Tibando

To the north of the equatorial region, there is a smaller hexagon crater with a central mountain within its centre that also stands out against the rest of Ceres planetary terrain, acting like a redundant calling card feature to ensure that any intelligence travelling by this planetoid will not miss the obvious artificial shape of these craters. Perhaps this dwarf planet is a sentry outpost for higher intelligences in our Solar System used in distant bygone ages, millions of years ago or possibly it may still be active. Only when *Dawn* is in near orbit around Ceres with its high definitions cameras mapping out the surface terrain will we know the answers to these questions, assuming NASA is more forthcoming in its public disclosure of its discoveries about our solar system.

Ceres is an asteroid moon with many unusual polygon craters on its surface
(c) Terry Tibando

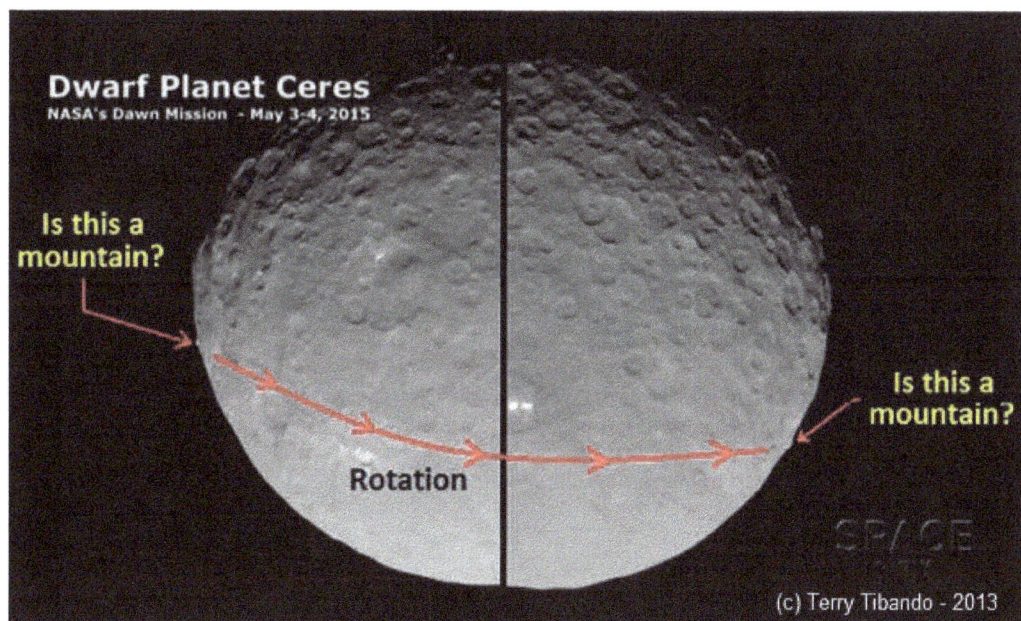

The Dawn spacecraft captures what appears to be a large mountain on Ceres
(c) Terry Tibando

259

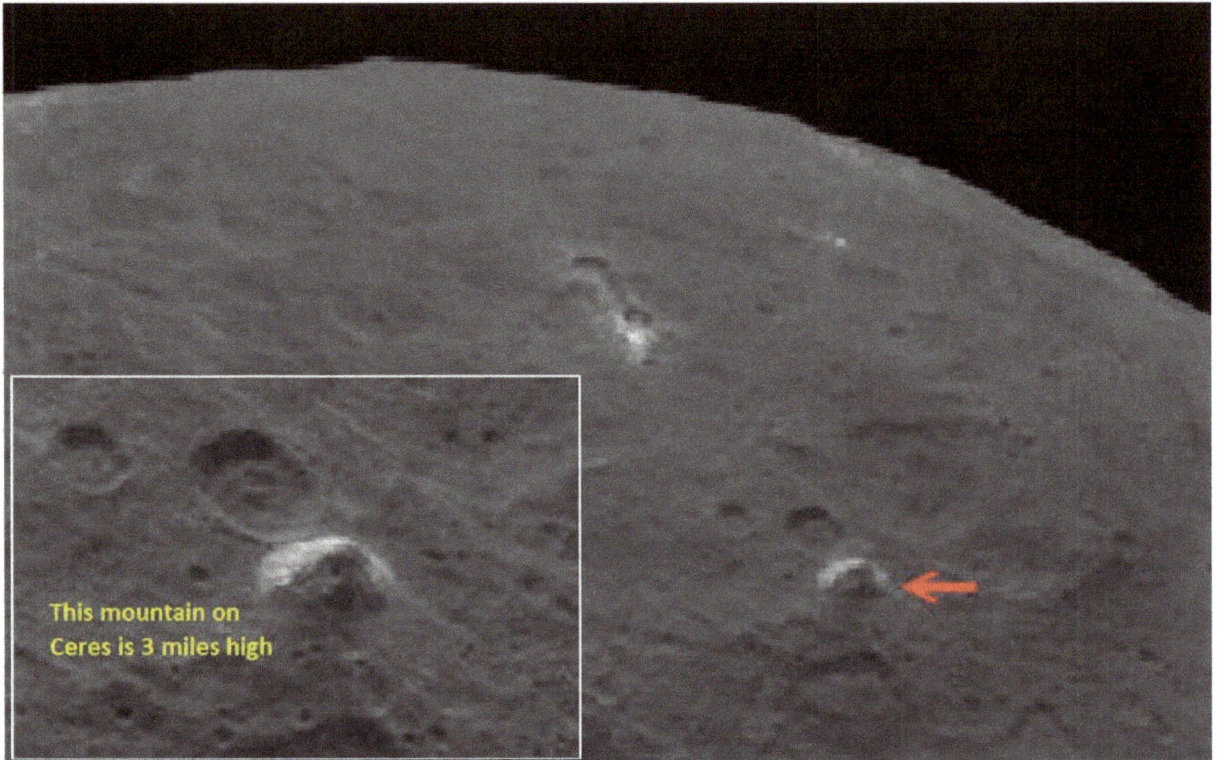

The three mile high mountain Ahuna Mons on Ceres stands out like a big sit on the crater terrain. It appears to be the only mountain on Ceres of questionable cryo-volcanic origin

http://www.cite-sciences.fr/fr/ressources/science-actualites/detail/news/un-volcan-de-glace-sur-ceres/?tx_news_pi1%5Bcontroller%5D=News&tx_news_pi1%5Baction%5D=detail&cHash=d32fbdb76aebba015e97ff5721b03cea

Ceres' lonely mountain, Ahuna Mons (left, is seen in the simulated perspective view (right). The elevation has been exaggerated by a factor of two. The view was made using enhanced-color images from NASA's Dawn mission and is not as reflective as portrayed in the illustration above but is more like the photo on the left.

https://news.agu.org/press-release/new-research-shows-ceres-may-have-vanishing-ice-volcanoes/
Credits: NASA/JPL-Caltech/UCLA/MPS/DLR/IDA/PSI

CHAPTER 87

JUPITER ("KING OF THE GODS")

Let's explore now, further outward toward Jupiter and her many moons and asteroid-like satellites. Jupiter is like a mini solar system unto itself and in fact it has been consider both in the science of astrophysics and astronomy as well as in science fiction as a dwarf **"brown star"** that never had enough nuclear fusion energy to ignite when the solar system began, to become a true radiant star of its own, as portrayed in **Arthur C. Clark's** book sequel "2010: Odyssey Two" and in the movie: 2010 (also known as 2010: The Year We Make Contact). http://en.wikipedia.org/wiki/2010_%28film%29

Jupiter, sixth planet from the Sun, largest of the gas giants and the largest planet in the Solar System with 67 moons and satellites, it is like a mini solar system unto itself
https://www.glavny.tv/news/19639

Some theories suggest that if Jupiter acquired more mass, the planet would lessen in size. The reason for this is that the interior of the planet would be much more compressed as the gravitational force of the planet would increase as mass increases. This would not happen if the increase in mass is insignificant but, a large enough increase would cause the planet to decrease in volume and diameter.

Jupiter's colourful south polar region is as turbulent as its equatorial region

Because of these theories, it is implied that Jupiter's current size is the largest that such a planet with that kind of composition can become. Hypothetically speaking if Jupiter does increase in mass the shrinking of the planet will continue until stellar ignition occurs. Some astronomers believe that Jupiter was a "failed star." This is an assumption though, as it is still unclear whether the formation of stars is the same as the formation of planets. Jupiter produces as much heat as it receives from the Sun. It seems, our Solar System, may have missed the opportunity to become a binary star system at this time, this may change at some future time, whether this occurs by natural means or by deliberate artificial means. http://planetfacts.org/mass-and-density-of-jupiter/

Jupiter is the fifth planet from the Sun and as all grade school students know is the largest planet in our solar System and the largest of the gas giant planets followed by Saturn, Uranus and Neptune; each a mini solar system unto itself with their many moons and satellites. Jupiter has a faint ring system similar to Saturn and in fact all of the Jovian planets possess ring systems (Jupiter, Saturn, Uranus and Neptune).

Jupiter's mass one-thousandth that of the Sun but, is two and a half times the mass of all the other planets in the Solar System combined. Its radius is 69,911 km; surface area is 61,418,738,571 km². Jupiter's mass is 1.898E27 kg (317.8 Earth mass). The distance from Sun is

778,500,000 km. Some of the major Jovian moons are Europa, Io, Ganymede, Callisto, Amalthea, Adrastea, Metis, plus sixty other satellites

The most prominent feature on Jupiter is the **Great Red Spot** just slightly below the south equatorial belt, situated at latitude of 19.5° that hyperdimensional physical apex or node spoken of by **Richard C. Hoagland** that exists on most planets and the Sun! The Red Spot is in fact a super storm cell as are most of the red and white oval spots which materialize, sometime merge with other spots or simply disappear again on their own.

The Great Red Spot of Jupiter. To give a sense of Jupiter's scale, the white oval storm directly below the Great Red Spot is approximately the same diameter as Earth.
http://www.taringa.net/posts/ciencia-educacion/15279323/El-intrigante-Mar-interior-de-Jupiter.html

The Jovian atmosphere shows a wide range of active phenomena, including band instabilities, vortices (cyclones and anticyclones), storms and lightning. The vortices reveal themselves as large red, white or brown spots (ovals). The largest two spots are the **Great Red Spot (GRS)** and **Oval BA**, which is also red. These two and most of the other large spots are anticyclonic. Smaller anticyclones tend to be white. Vortices are thought to be relatively shallow structures

with depths not exceeding several hundred kilometers. Located in the southern hemisphere, the GRS is the largest known vortex in the Solar System. It could engulf several Earths and has existed for at least three hundred years. Oval BA, south of GRS, is a red spot a third the size of GRS that formed in 2000 from the merging of three white ovals.
https://en.wikipedia.org/wiki/Atmosphere_of_Jupiter

This view of Jupiter's **Great Red Spot** (as seen below) and its surroundings was obtained by **Voyager 1** on February 25, 1979, when the spacecraft was 9.2 million km (5.7 million mi) from Jupiter. Cloud details as small as 160 km (99 mi) (100 mi) across can be seen here. The colorful, wavy cloud pattern to the left of the Red Spot is a region of extraordinarily complex and variable wave motion. To give a sense of Jupiter's scale, the white oval storm directly below the Great Red Spot is approximately the same diameter as Earth.

Jupiter at this point in time does not appear to have any life on or in or above its surface because of the extreme atmospheric pressures present in its atmosphere. However, possible extremophile life can't be ruled out in such harsh conditions as we are currently discovering here on Earth. Life has arisen in the depths of the oceans and in sulfur tubes at the bottom of the ocean floors or in the darkness of the earth hundreds of feet below ground in the soil and rock itself and even in the frigid ice continent of Antarctica. Life as stated before is where you find it!

The **major moons of Jupiter** are **Io, Europa, Ganymede** and **Callisto** which are larger than out Moon (Luna) with the exception of Europa which is smaller than the Moon. There are sixty-seven (!) moons or satellites orbiting around Jupiter and most of all of them are asteroid in shape probably because they captured by Jupiter's powerful magnetic field when **Planet V** exploded 65 million years ago leaving an asteroid belt between Mars and Jupiter. Many large asteroid moons/satellites have erratic orbits around Jupiter and it is also true of the asteroid moons of Saturn, Uranus and Neptune. It is quit probable that these asteroid moons that orbit the outer gas giant planets were once a part of **Planet V** as per **Tom Van Vandern's Exploded Hypothesis**.

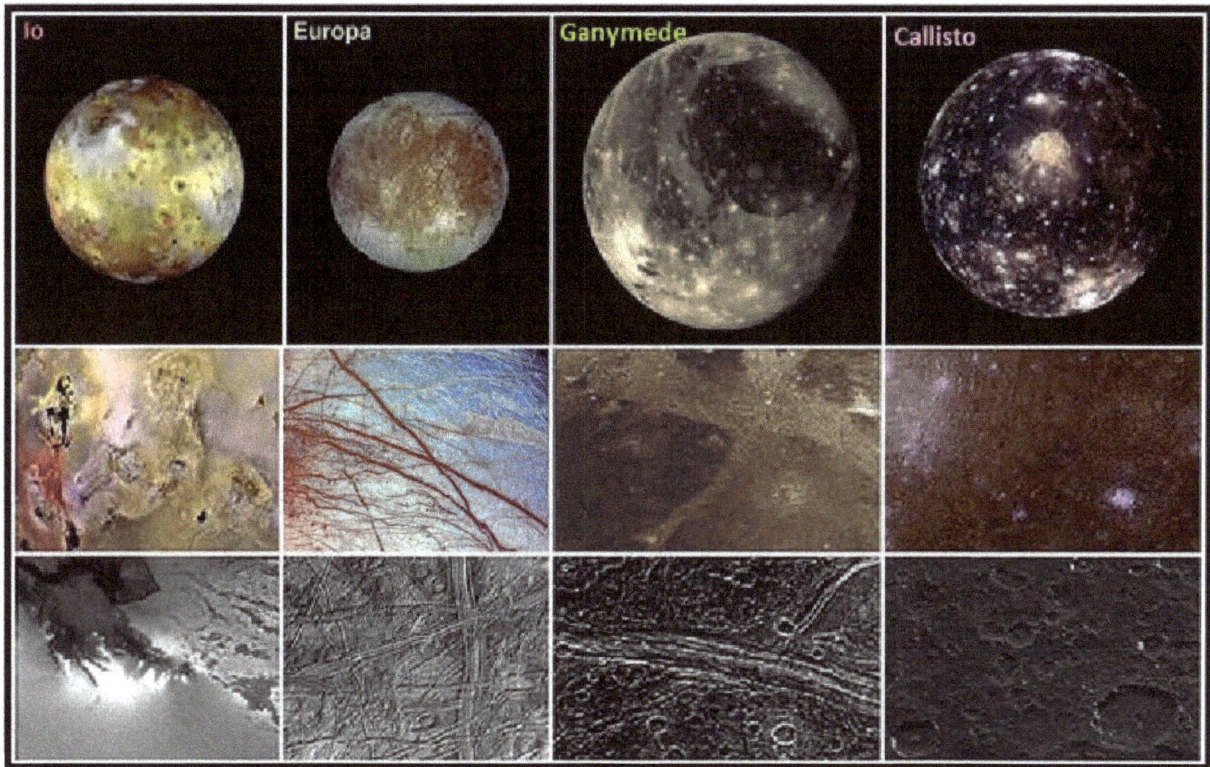

The major moons of Jupiter showing some of their surface detail below each moon
https://www.universetoday.com/52061/moons-of-jupiter/

The major moons of Jupiter or the **Galilean moons** as they are also known by, from their discoverer, **Galileo Galilei** in January 7 – 13, 1610 are some of the most interesting satellites in our Solar System because of their individual uniqueness.

Io

Io is the innermost of the four Galilean moons of the planet Jupiter and, with a diameter of 3,642 kilometres (2,263 mi), the fourth-largest moon in the Solar System. It was named after the mythological character of Io, a priestess of **Hera** who became one of the lovers of **Zeus**.

With over 400 active volcanoes, Io is the most geologically active object in the Solar System. This extreme geologic activity is the result of tidal heating from friction generated within Io's interior as it is pulled between Jupiter and the other Galilean satellites—Europa, Ganymede and Callisto. Several volcanoes produce plumes of sulfur and sulfur dioxide that climb as high as 500 km (300 mi) above the surface. Io's surface is also dotted with more than 100 mountains that have been uplifted by extensive compression at the base of the moon's silicate crust. Some of these peaks are taller than Earth's **Mount Everest**. Unlike most satellites in the outer Solar System, which are mostly composed of water ice, Io is primarily composed of silicate rock surrounding a molten iron or iron sulfide core. Most of **Io's** surface is characterized by extensive plains coated with sulfur and sulfur dioxide frost.

Io's volcanism is responsible for many of the satellite's unique features. Its volcanic plumes and lava flows produce large surface changes and paint the surface in various shades of yellow, red, white, black, and green, largely due to allotropes and compounds of sulfur. Numerous extensive lava flows, several more than 500 km (300 mi) in length, also mark the surface. The materials produced by this volcanism provide material for Io's thin, patchy atmosphere and Jupiter's extensive magnetosphere. Io's volcanic ejecta also produce a large plasma torus around Jupiter.

Io is a moon which appears to turn itself inside out over a period of decades or within several hundred years minimum, thus, it surface features are constantly changing and therefore, it is highly unlikely that life can establish itself on the surface of Io. If there is a physical manifestation of a place like Hell, then Io would certainly be that place. It certainly was the inspiration for the volcanic planet in **George Lucas'** "Star Wars, Episode Three".

The only land feature on the surface of Io that may be artificial, although, it is very improbable as it may be another classic example of pareidolia, nevertheless, it is a land mass resembling a Mayan or Egyptian king or priest located north of the volcano, **Babbar Patera** in the **Lycea Planum** region. (See image below and its enlargement).
https://en.wikipedia.org/wiki/Atmosphere_of_Jupiter

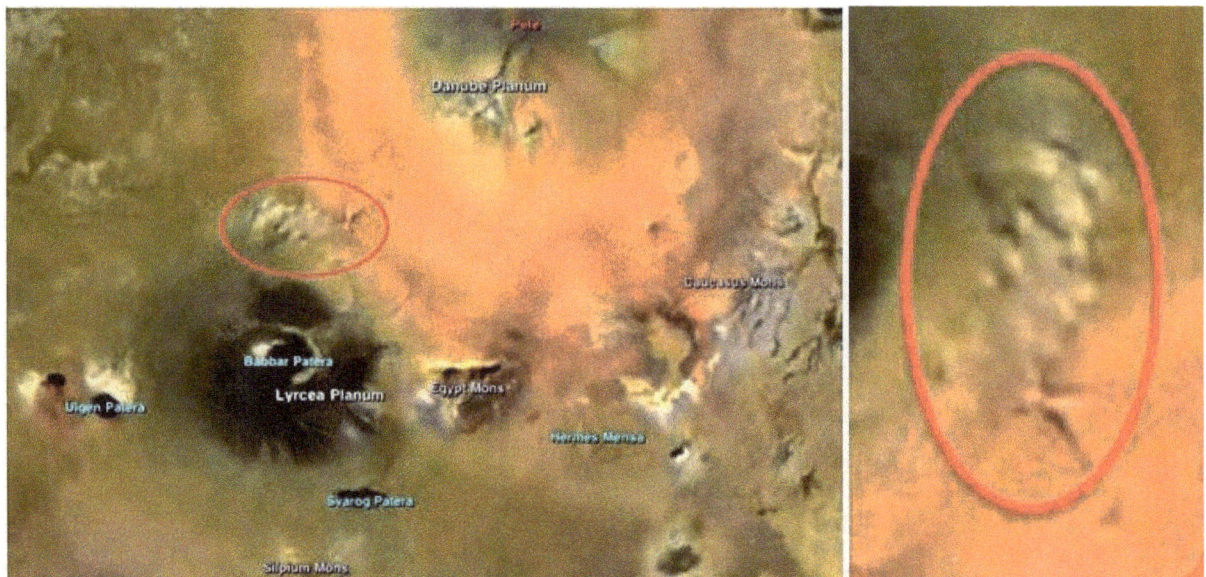

A large "face" - like structure of a royal or priestly visage of a Mayan or Egyptian discovered on July 18, 2011 with an iPad App. located just above the Babbar Patera volcano on the Jovian moon of IO
https://genesisxresearch.wordpress.com/2012/01/31/alien-faces-43/

If there is a Hell in our Solar System, Io, the closest moon to Jupiter would fit that description, as it continuously turns itself inside out with volcanic activity. One of its volcano's is seen erupting taken by the Galileo spacecraft

http://www.space.com/36316-the-most-amazing-space-stories-of-the-week.html

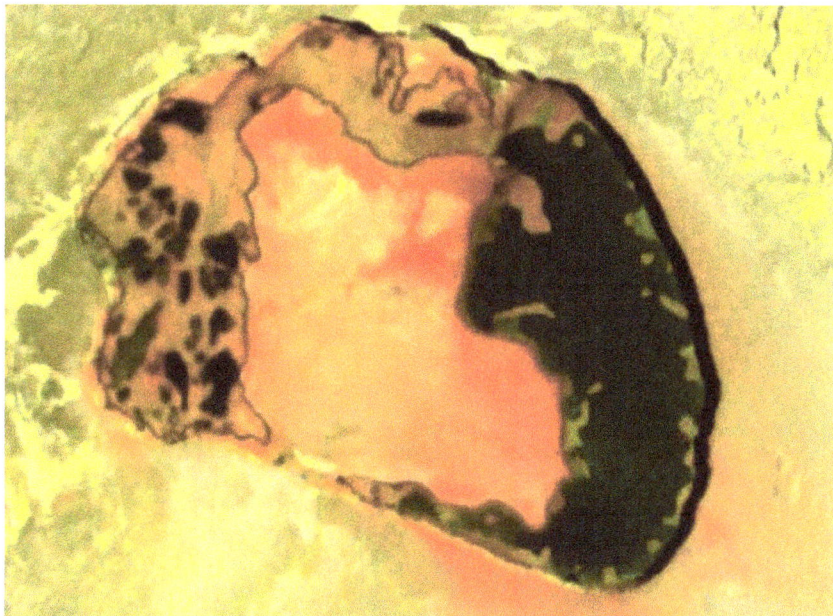

The colourful volcanic crater, Tupan Patera on Jupiter's moon Io imaged on 10 Dec 2001 by NASA's Galileo spacecraft shows lava interacting with sulfur-rich materials.

http://www.thelivingmoon.com/43ancients/02files/IO_Images_01.html

Europa

Europa is the sixth closest moon of the planet Jupiter, and the smallest of its four Galilean satellites, but still the sixth-largest moon in the Solar System. Europa was discovered in 1610 by Galileo Galilei and possibly independently by **Simon Marius** around the same time. Progressively more in-depth observation of Europa has occurred over the centuries by Earth-bound telescopes, and by space probe flybys starting in the 1970s.

Slightly smaller than Earth's Moon, Europa is primarily made of silicate rock and probably has an iron core. It has a tenuous atmosphere composed primarily of oxygen. Its surface is composed of water ice and is one of the smoothest in the Solar System. This surface is striated by cracks and streaks, while cratering is relatively infrequent. The apparent youth and smoothness of the surface have led to the hypothesis that a water ocean exists beneath it, which could conceivably serve as an abode for *extraterrestrial life*. This hypothesis proposes that heat energy from tidal flexing causes the ocean to remain liquid and drives geological activity similar to plate tectonics.

Realistic-color *Galileo* mosaic of Europa's anti-Jovian hemisphere showing numerous lineae https://en.wikipedia.org/wiki/Europa_(moon)

Shortly after the first NASA unmanned Voyager mission to Jupiter, in March, 1979, **Richard C. Hoagland** published in "**Star & Sky magazine**" a radical new theory regarding implications stemming from Voyager's historic fly-by and data return from one of the "Galilean moons": …Europa.

Hoagland proposed that a planet-wide ocean still exists under the tens-of-miles-thick sulphur-tinged ice now completely covering Europa. Further, that in that extremely ancient ocean -- the only other planetary "near-by" liquid water that may have persisted from the beginnings of the solar system (other than on Earth)

Like most of what Hoagland writes about or publicly states, it is usually discredited, debunked or simply ignored yet, at least 90% of the time, his pronouncements and predictions, whether written or stated are typically proven correct over time. Well, it seems NASA and ESA have reluctantly agreed with Hoagland's theory that Europa has an ocean miles deep below it thick ice-covered crust and they are planning to send unmanned probes to this Jovian moon to explore for the possibility of sea life!

Arthur C. Clark, science fiction writer and inventor of the concept of artificial orbiting satellites and, Dr. Robert Jastrow, one of the founders of NASA, and former Director of its Goddard Institute for Space Studies are in full agreement with Hoagland that this time he got it right!

In the acknowledgments to "2010," Clark would write:

*"The fascinating idea that there might be life on Europa, beneath ice-covered oceans kept liquid by the same Jovian tidal forces that heat Io, was first proposed by Richard C. Hoagland in the magazine Star & Sky (The Europa Enigma,' January, 1980). This quite brilliant concept has been taken seriously by a number of astronomers (notably NASA's Institute of Space Studies, Dr. Robert Jastrow), and may provide one of the best motives for the projected **Galileo Mission**..."*

Now, twenty years later, even NASA's most ardent Hoagland haters are being forced to admit that Hoagland's overarching model of Europa -- which included over 20 years ago some kind of highly evolved, *living organisms* potentially swimming in that ocean -- is almost certainly correct. So, having failed to disprove his two-decade-old theory (in fact, just the contrary!), some critics have now desperately launched a full blown, rear-guard attempt to remove Hoagland from his historically preeminent position on this issue, as it has become clearer with each Galileo fly-by of Europa over the past several years that his startlingly original scientific model of Europa is inexorably being confirmed.

These desperate, last-ditch efforts have even included recent charges of outright fraud -- bizarre claims that Hoagland has attempted to take false credit for **Cassen, Reynolds and Peale's** original published work on the European tidal stresses. In fact, from the beginning, Hoagland specifically cited their preceding pioneering work, using it as the credited foundation for his own meticulously worked out biological model of what might be occurring in that ocean, in "The Europa Enigma" itself. Which, decades later, he would update on the "*Enterprise"* web site.
http://www.enterprisemission.com/europa2000.htm

The ***Galileo Mission***, launched in 1989, provided the bulk of current data on Europa. Although only fly-by missions have visited the moon, the intriguing characteristics of Europa have led to several ambitious exploration proposals. The next mission to Europa is the European Space Agency's **Jupiter Icy Moon Explorer (JUICE)**, due to launch in 2022.
http://en.wikipedia.org/wiki/Europa_%28moon%29

Europa has emerged as one of the top locations in the Solar System in terms of potential habitability and the possibility of hosting extraterrestrial life. Life could exist in its under-ice ocean, perhaps subsisting in an environment similar to Earth's deep-ocean hydrothermal vents. Life in such an ocean could possibly be similar to microbial life on Earth in the deep ocean. So far, there is no evidence that life exists on Europa, but the likely presence of liquid water has spurred calls to send a probe there.

Life on Europa could exist clustered around hydrothermal vents on the ocean floor, or below the ocean floor, where **endoliths** are known to inhabit on Earth. Alternatively, it could exist clinging to the lower surface of the moon's ice layer, much like algae and bacteria in Earth's polar regions, or float freely in Europa's ocean.

In September 2009, planetary scientist **Richard Greenberg** calculated that cosmic rays impacting on Europa's surface convert some water ice into free oxygen (O_2) which could then be absorbed into the ocean below as water wells up to fill cracks. Via this process, Greenberg estimates that Europa's ocean could eventually achieve an oxygen concentration greater than that of Earth's oceans within just a few million years. This would enable Europa to support not merely anaerobic microbial life but potentially larger, aerobic organisms such as fish.

In 2006, **Robert T. Pappalardo**, an assistant professor in the Laboratory for Atmospheric and Space Physics at the University of Colorado in Boulder said, *"We've spent quite a bit of time and effort trying to understand if Mars was once a habitable environment. Europa today, probably, is a habitable environment. We need to confirm this … but Europa, potentially, has all the ingredients for life … and not just four billion years ago … but today"*.

in November 2011, a team of researchers presented evidence in the journal "Nature" suggesting the existence of vast lakes of liquid water entirely encased in the moon's icy outer shell and distinct from a liquid ocean thought to exist farther down beneath the ice shell. If confirmed, the lakes could be yet another potential habitat for life.

A paper published in March 2013 suggests that hydrogen peroxide is abundant across much of the surface of Jupiter's moon Europa. The authors argue that if the peroxide on the surface of Europa mixes into the ocean below, it could be an important energy supply for simple forms of life, if life were to exist there. The scientists think hydrogen peroxide is an important factor for the habitability of the global liquid water ocean under Europa's icy crust because hydrogen peroxide decays to oxygen when mixed into liquid water.
http://en.wikipedia.org/wiki/Europa_%28moon%29

Top left image shows two possible models of Europa. The right image depicts possible sulphur steam vents on Europa's ocean floor and bottom image shows Red Tube Worm life found on Earth's ocean floors which may exist in Europa's ocean

https://en.wikipedia.org/wiki/Europa_(moon) **and** http://www.star2.com/travel/asia-oceania/2015/08/29/head-to-the-deep-where-no-one-can-hear-you-scream/

In a paper published in the magazine Star and Sky, **"The Europa Enigma,"** Hoagland went on to propose that volcanic rifts in the ocean floor (much like those found on Earth) would almost certainly contribute further to the "organic soup" required for life and that with 4 billion years of evolution below the ice, very complex life forms might be found on present day Europa. He was the first to propose that dark fissures in the surface crust of Europa were caused by the seepage of organic materials from this deep seated ocean.

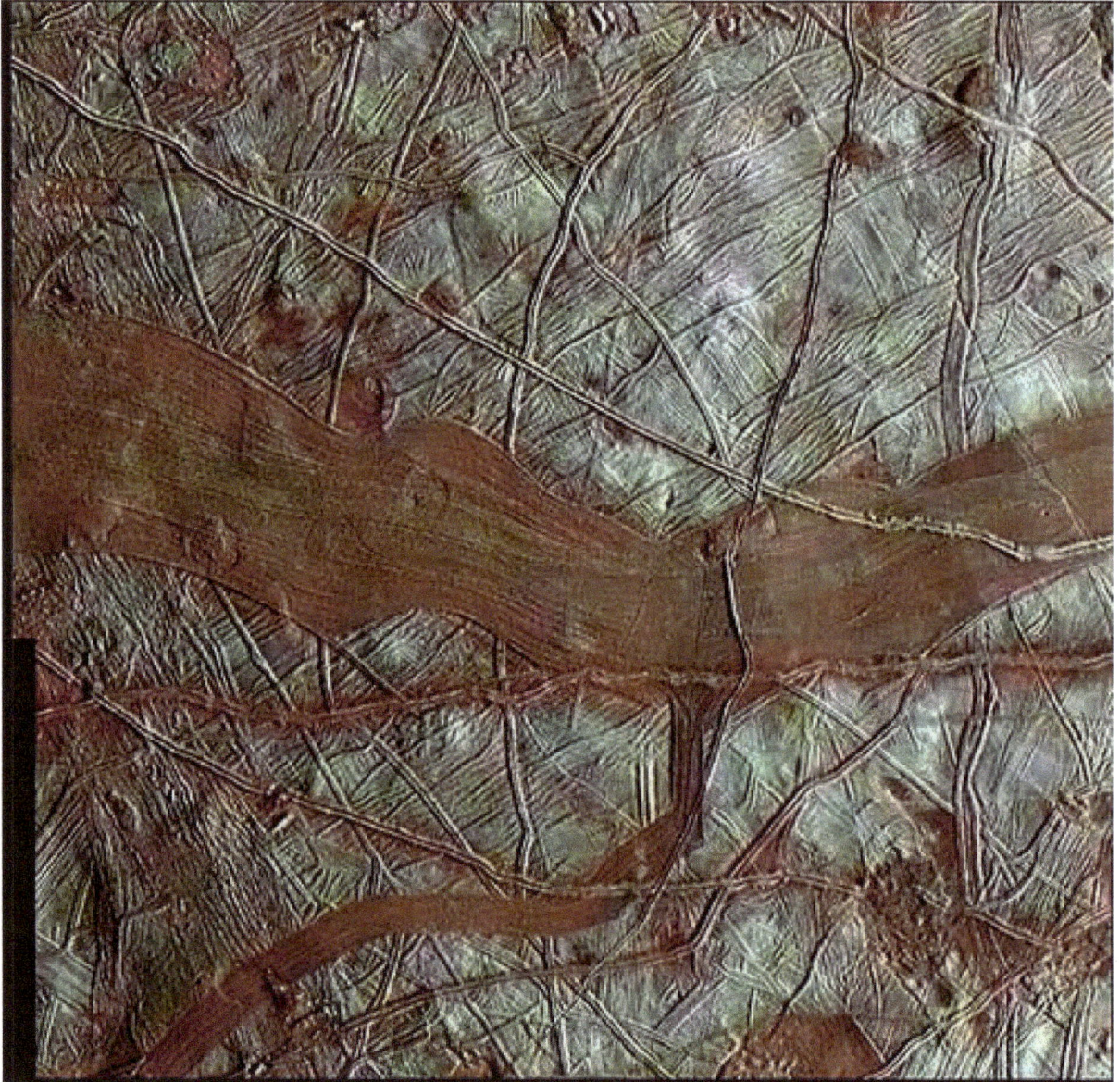

The dark material staining the surface of Europa's ice and the dark fissures in the surface crust may be caused by the seepage of organic materials from the deep ocean below
https://spaceplace.nasa.gov/europa/en/

As we get closer to NASA's in-depth testing of the rest of Hoagland's Europa theories (via new Europa missions themselves), including the content of that dark material staining the surface of Europa's ice, expect these disinformation efforts to dramatically increase. While Europa will wait for no man in her rush to reveal her secrets to these new robotic surrogates from Earth, there are plenty of consequences for those that would seek to obscure the truth. By all means, as these people see it, Hoagland must *not* be allowed to receive appropriate credit for his successful

Jovian (and other ...) scientific predictions. ***For if that happened, some in the media (and inside NASA!) might just take seriously his multi-disciplinary work on another interesting piece of Solar System real estate ...*** (Bold italics added by author for emphasis).
http://en.wikipedia.org/wiki/Europa_%28moon%29

Ganymede

Ganymede is the largest moon in the Solar System. It is the seventh moon and third Galilean satellite outward from Jupiter. Completing an orbit in roughly seven days, Ganymede participates in a 1:2:4 orbital resonance with the moons Europa and Io, respectively. It has a diameter of 5,268 km (3,273 mi), 8% larger than that of the planet Mercury, but has only 45% of the latter's mass. Its diameter is 2% larger than that of Saturn's Titan, the second largest moon. It also has the highest mass of all planetary satellites, with 2.02 times the mass of the Earth's moon.

Ganymede's lighter surfaces, such as in recent impacts, grooved terrain and the whitish north polar cap at upper right, are enriched in water ice.
https://en.wikipedia.org/wiki/Ganymede_(moon)

Ganymede is composed of approximately equal amounts of silicate rock and water ice. It is a fully differentiated body with an iron-rich, liquid core. Like Europa, a saltwater ocean is believed to exist nearly 200 km below Ganymede's surface, sandwiched between layers of ice. Its surface is composed of two main types of terrain. Dark regions, saturated with impact craters and dated to four billion years ago, cover about a third of the satellite. Lighter regions, crosscut by extensive grooves and ridges and only slightly less ancient, cover the remainder. The cause of the light terrain's disrupted geology is not fully known, but was likely the result of tectonic activity brought about by tidal heating.

Ganymede is the only satellite in the Solar System known to possess a magnetosphere, likely created through convection within the liquid iron core. The satellite has a thin oxygen atmosphere that includes O, O_2, and possibly O_3 (ozone). Atomic hydrogen is a minor atmospheric constituent. Whether the satellite has an ionosphere associated with its atmosphere is unresolved.

Beginning with **Pioneer 10**, spacecraft have been able to examine Ganymede closely. The **Voyager probes** refined measurements of its size, while the *Galileo* craft discovered its underground ocean and magnetic field. The next planned mission to the Jovian system is the European Space Agency's **Jupiter Icy Moon Explorer (JUICE),** due to launch in 2022. After flybys of all three icy Galilean moons the probe is planned to enter orbit around Ganymede.

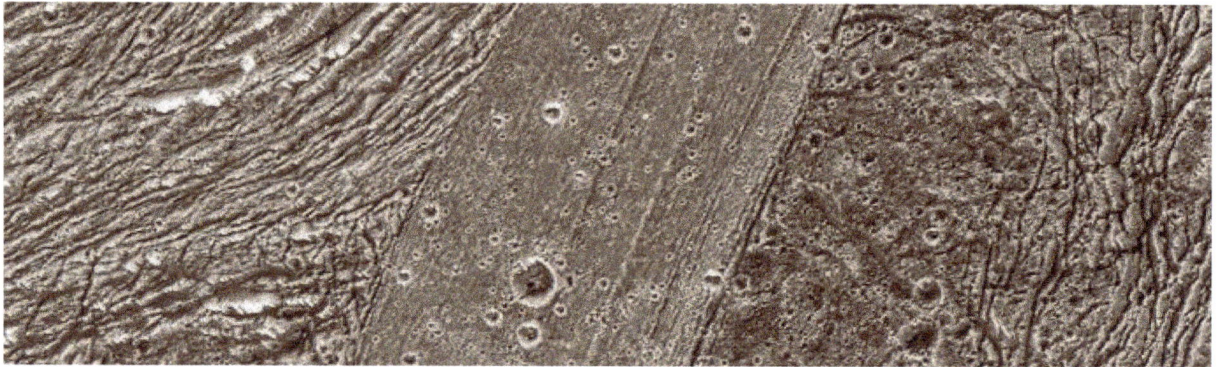

Galileo close-up image of the Nicholson Regio and Arbela Sulcus area of Ganymede shows many of the diverse terrain types: striations, rippling and cratering
http://www.seasky.org/solar-system/jupiter-ganymede.html

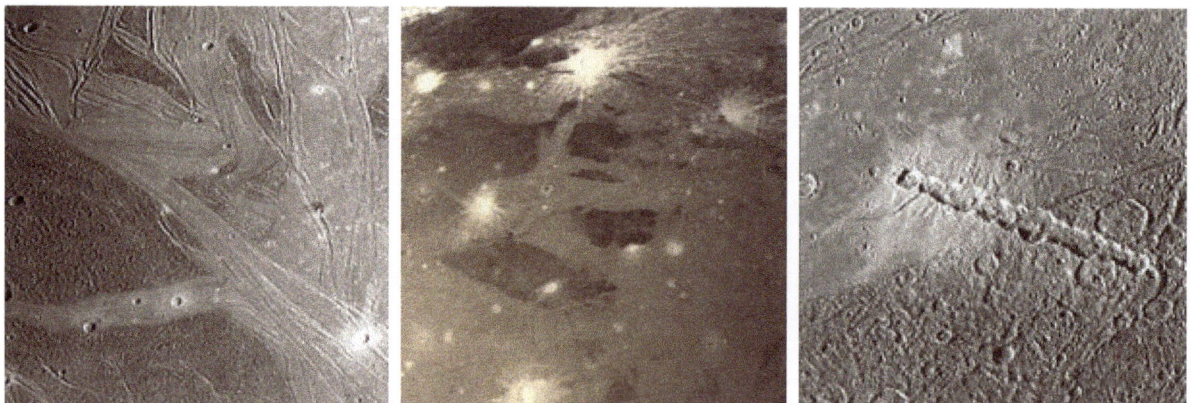

Galileo close-up of Marus Regio and Nippur Sulcus (left), Voyager 1 close-up of Ganymede showing differences in terrain (middle) and Galileo view of Enki Catena Crater formed by a series of meteorites (right)
http://www.seasky.org/solar-system/jupiter-ganymede.html

Until robotic space probes land on Ganymede and drill through the ice mantle to search with submersible probes for sea life, it is assumed at this current time that Ganymede has no life *"on it"*. Life, deep down on the ocean floor may have developed like it has on Earth and as it may have formed on Europa, only time will tell.
https://en.wikipedia.org/wiki/Ganymede_%28moon%29

Callisto

Callisto is the third-largest moon in the Solar System and the second largest in the Jovian system, after Ganymede. Callisto has about 99% the diameter of the planet Mercury but only about a third of its mass. It is the fourth Galilean moon of Jupiter by distance, with an orbital radius of about 1,880,000 km. It does not form part of the orbital resonance that affects three inner Galilean satellites—Io, Europa and Ganymede—and thus does not experience appreciable tidal heating. Callisto's rotation is tidally locked to its revolution around Jupiter, so that the same hemisphere always faces inward; Jupiter appears to stand still in Callisto's sky. Callisto is less affected by Jupiter's magnetosphere than the other inner satellites because it orbits farther away. Callisto is composed of approximately equal amounts of rock and ices, with compounds detected spectroscopically on the surface that include water ice, carbon dioxide, silicates, and organic compounds. Investigation by the **Galileo spacecraft** revealed that Callisto may have a small silicate core and possibly a subsurface ocean of liquid water at depths greater than 100 km. The surface of Callisto is heavily cratered and extremely old. It does not show any signatures of subsurface processes such as plate tectonics or volcanism, and is thought to have evolved predominantly under the influence of impacts. Prominent surface features include multi-ring structures, variously shaped impact craters and chains of craters (*catenae*) and associated scarps, ridges and deposits. At a small scale, the surface is varied and consists of small, bright frost deposits at the tops of elevations, surrounded by a low-lying, smooth blanket of dark material. The absolute ages of the landforms are not known.

Callisto is surrounded by an extremely thin atmosphere composed of carbon dioxide and probably molecular oxygen, as well as by a rather intense ionosphere. Callisto is thought to have formed by slow accretion from the disk of the gas and dust that surrounded Jupiter after its formation.

The likely presence of an ocean within Callisto leaves open the possibility that it could harbor life. However, conditions are thought to be less favorable than on nearby Europa. Various space probes from **Pioneers 10** and **11** to **Galileo** and **Cassini** have studied the moon. Because of its low radiation levels, Callisto has long been considered the most suitable place for a human base for future exploration of the Jovian system.
https://en.wikipedia.org/wiki/Callisto_%28moon%29

Once again, with future planned robotic space probes that will be sent to Callisto for exploration of the terrain and the ocean areas below as is planned for Europa and Ganymede, there is no way to determine at this time, whether life is extant or even possible on Callisto. We will have to abide our time and be patient for that future day to unfold.

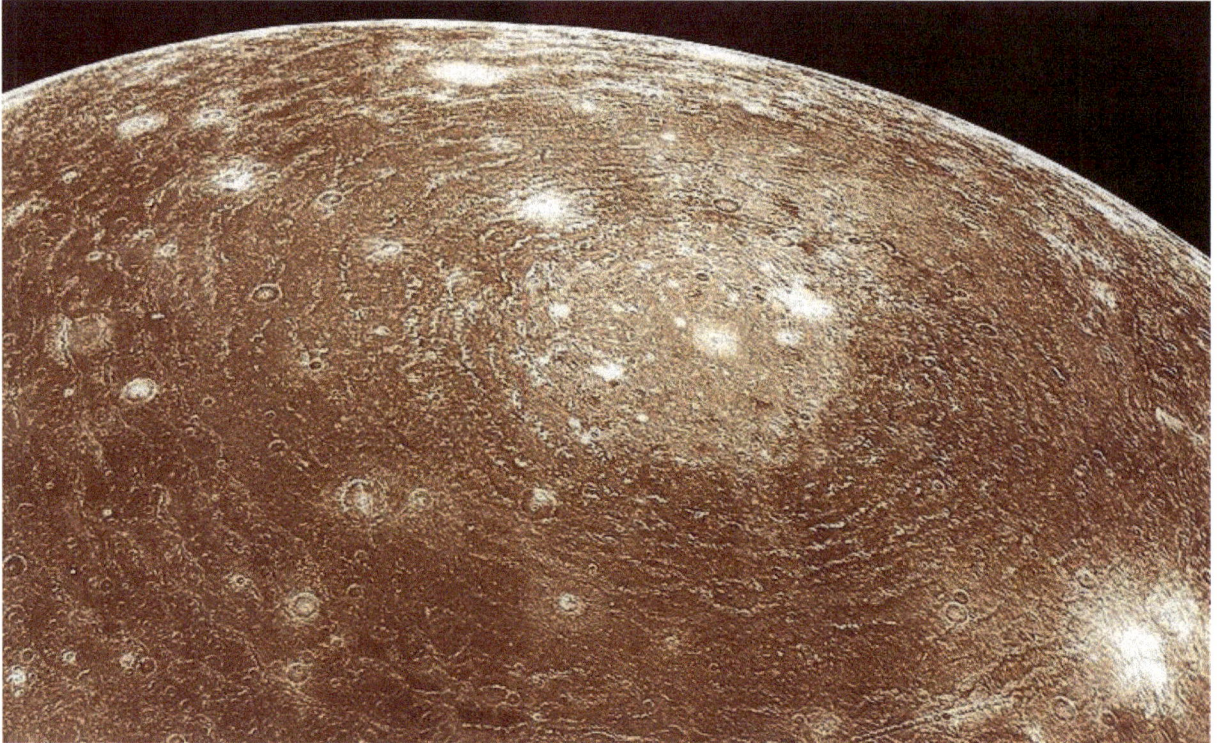

Voyager 1 image of Valhalla, a multi-ring impact structure 3800 km in diameter forms a part of Callisto's surface features
https://en.wikipedia.org/wiki/Callisto_(moon)

Before leaving the Jovian System, even though life it tentative at this time within this mini solar-like system of numerous satellites, there does appear to be life flying around or near Jupiter and it moons in spacecraft that have not been sent from any space agency of Earth.

On a Canadian website called "Keep Your Eyes On The Skies" are some photos that show unidentified flying objects orbiting around or near Jupiter and also around Callisto, Europa and Ganymede. These objects were captured by the Hubble Space Telescope on January 2010 yet, their source as originating from NASA has still to be confirmed; however, they indicate that these **alien spacecraft** are massive in size. Such images, confirm that there are many Extraterrestrial spacecraft miles in size traversing between planets and to the Sun in our Solar System. http://slushpup62.blogspot.ca/2010/01/nasa-pics-show-ufos-orbiting-jupiters.html

276

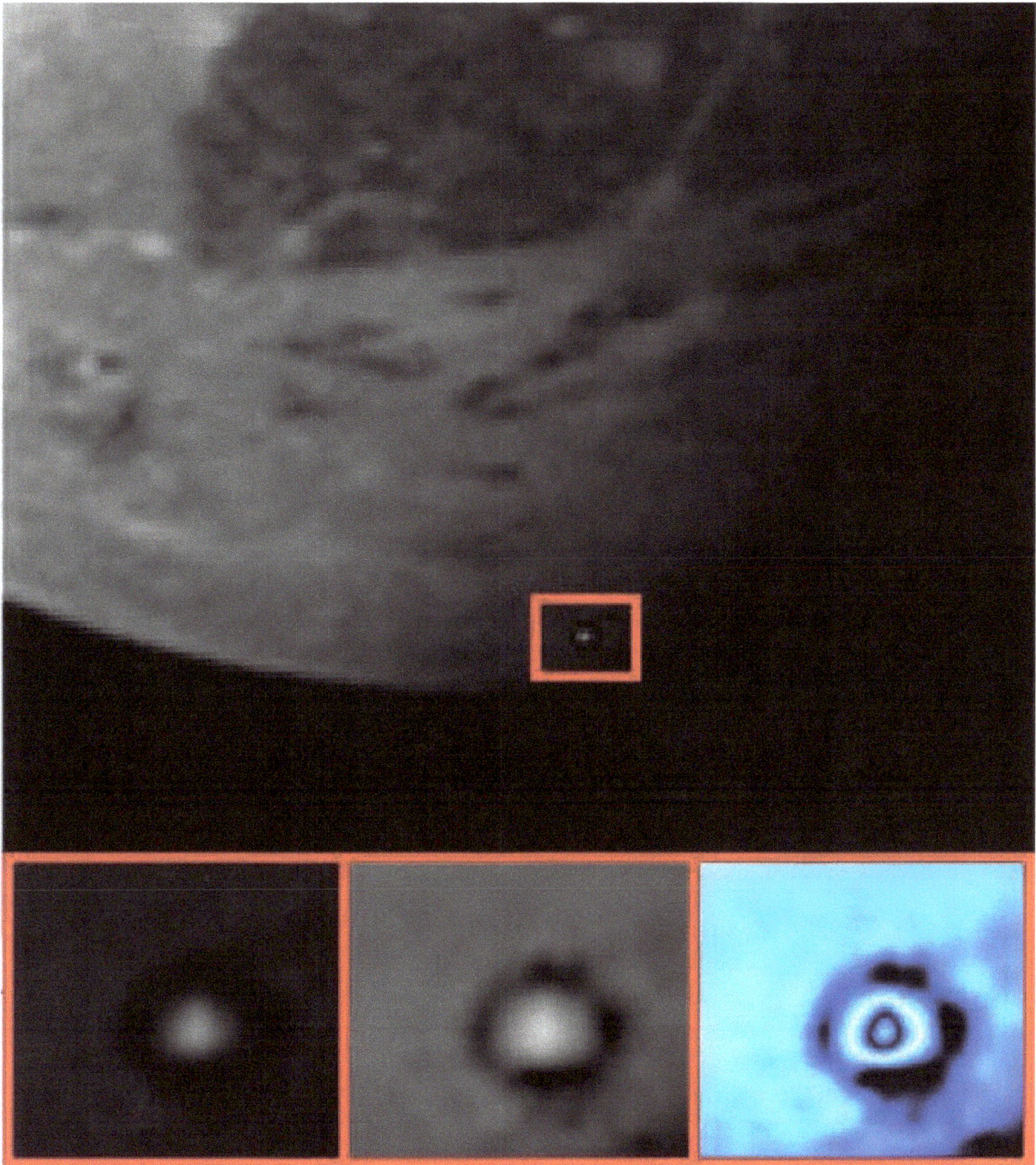

A large UFO was photographed near Jupiter's moon Callisto
(Google Images)

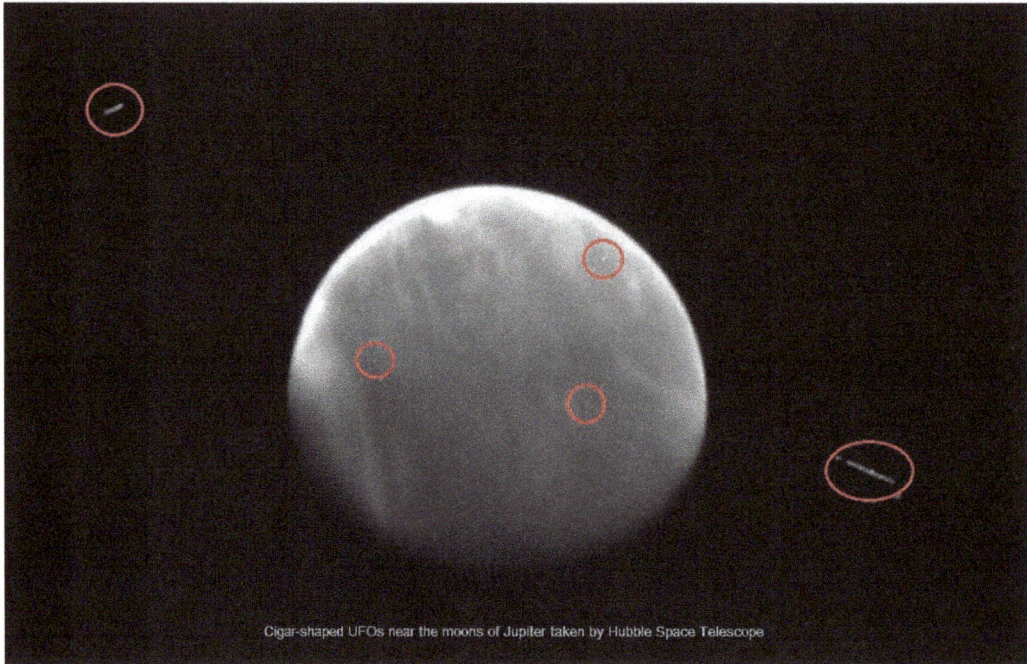

Cigar-shaped UFOs near the moons of Jupiter taken by Hubble Space Telescope

**Massive cigar and global UFOs were captured by Hubble Space Telescope
in January 2010 around Jupiter's moon, Ganymede (images above and below)**
(Google Images)

Cigar-shaped UFOs near the moons of Jupiter taken by Hubble Space Telescope

Massive Cigar shape UFO near Jupiter's moon Ganymede
(Google Images)

278

Could ETs be trying to communicate with us? According to UFO researcher **Rupert Matthews**, an NASA astronomer and insider who worked closely with SETI told him back in the 1980's that NASA and SETI astronomers had detected an artificial alien radio signal originating from Ganymede, one of Jupiter's largest moons.

Such stories are spurious at best and probably sources of disinformation nonetheless, it has only been in the last several decades that astronomers have confirmed strong evidence for liquid oceans underneath all three of Jupiter's largest moons: Europa, Ganymede and Callisto. And as everyone knows, water is the ingredient that must be present for life to develop, as we know it, so while the scenario could be viewed as highly unlikely, it is certainly not impossible.

Unfortunately, the signal appears to have been only detected for a brief period of time and has not been detected since, so it could be said that these aliens might have been just visiting the local area and are not permanent residents of the Jovian moon. As to the signal itself, this is what Rupert Matthews reports in his book "Alien Encounters: True Stories of Aliens, UFO's and other E.T. Phenomenon" which was published in 2008. NASA and SETI astronomers picked up on a radio signal that seemed to have a coded message sometime in the mid 1980's. The signal was coming from Jupiter's moon Ganymede and was definitely artificial in nature. The signal was run through multiple computers and no sense could be made as to the meaning of it. **President Ronald Reagan** was notified and an international conference was assembled to determine how to make an attempt to contact whatever beings were sending the message. Eventually it was decided to send a signal to the exact location that the message was coming from on Ganymede. This message would be sent in **Morse Code** which was considered to be the easiest for an alien civilization to decode.

It would read *"We have received your signal, but we do not understand it. Please resend your signal using this language and transmission code."* Immediately after this message was sent, the mysterious signal from Ganymede ceased its transmission so everyone assumed that we would receive a new signal in Morse Code in a quick period of time. But day after day and then week after week no signal came and officials began to give up hope of actually receiving a response. Finally after more than a month, a signal was detected coming from the original radio source on Ganymede and incredibly it was in Morse Code. Eagerly NASA scientists translated our first direct contact with an alien civilization. And it said *"We were not talking to you."*

If there response wasn't so funny, it would be a crime, if it was determined to be a hoax from within NASA, itself considering the implications and repercussions of fallout up the chain of command to the US President, himself!

Was this a polite but, assertive slap in the face response from ETI with fore knowledge of our human species that we are not yet mature enough to engage in any meaningful dialogue? Did we eavesdrop on an interstellar conversation with ETs communicating with another Extraterrestrial civilization and that they basically told us to hang up the phone, their call was private and not meant for us? Ironically, this wasn't the first time that we've eavesdropped on a **"WOW"! signal** from ETIs as SETI will attest.

Nevertheless, as **Richard C. Hoagland** would often say: "Stay tuned!"

CHAPTER 88

SATURN ("GOD OF PLENTY, WEALTH, AGRICULTURE AND FEASTING")

Saturn is the sixth planet from the Sun and the second largest planet in the Solar System, after Jupiter and is considered the most beautiful planet in the Solar System. Saturn is a gas giant with an average radius about nine times that of Earth. While only one-eighth the average density of Earth, with its larger volume Saturn is just over 95 times more massive than Earth.

How does the rest of the Solar System stack up for extraterrestrial activity, in light of the fact that **UFOs** or extraterrestrial spacecraft have been sighted and photographed from the Sun to all the way out around Jupiter? Is ETI activity present around Saturn and it moons and the other outer gas giant planets? Apparently so!

Saturn and its rings as viewed from the south polar region of the planet where an aurora can be seen
https://commons.wikimedia.org/wiki/Saturn_(planet)

Saturn's interior is probably composed of a core of iron, nickel and rock (silicon and oxygen compounds), surrounded by a deep layer of metallic hydrogen, an intermediate layer of liquid hydrogen and liquid helium and an outer gaseous layer. The planet exhibits a pale yellow hue due to ammonia crystals in its upper atmosphere. Electrical current within the metallic hydrogen layer is thought to give rise to Saturn's planetary magnetic field, which is slightly weaker than Earth's and around one-twentieth the strength of Jupiter's. The outer atmosphere is generally bland and lacking in contrast, although long-lived features can appear. Wind speeds on Saturn can reach 1,800 km/h (1,100 mph), faster than on Jupiter, but not as fast as those on Neptune.

The most distinguishing feature about Saturn is its prominent ring system that consists of nine continuous main rings and three discontinuous arcs, composed mostly of ice particles with a smaller amount of rocky debris and dust. Sixty-two known moons orbit the planet; fifty-three are officially named. This does not include the hundreds of "moonlets" within the rings. Titan, Saturn's largest and the Solar System's second largest moon is larger than the planet Mercury and is the only moon in the Solar System to retain a substantial atmosphere.
http://en.wikipedia.org/wiki/Saturn

A size comparison of the Earth with Saturn and an insert showing where the Earth is in the Solar System as viewed from the orbit of Saturn. Wave everyone!
https://www.universetoday.com/24161/saturn-compared-to-earth/ and
https://www.nasa.gov/image-feature/cassini-earth-and-saturn-the-day-earth-smiled

Other interesting features of Saturn are its ***hexagonal swirling clouds in the North and South Polar regions***, the ***swirling cloud band around body*** of the planet similar to Jupiter also, located at the ***mysterious latitude of 19.5° degrees***, all outward signs of hyperdimensional physics compliments of ala Richard C. Hoagland's hypothesis. (Bold italics added for emphasis).

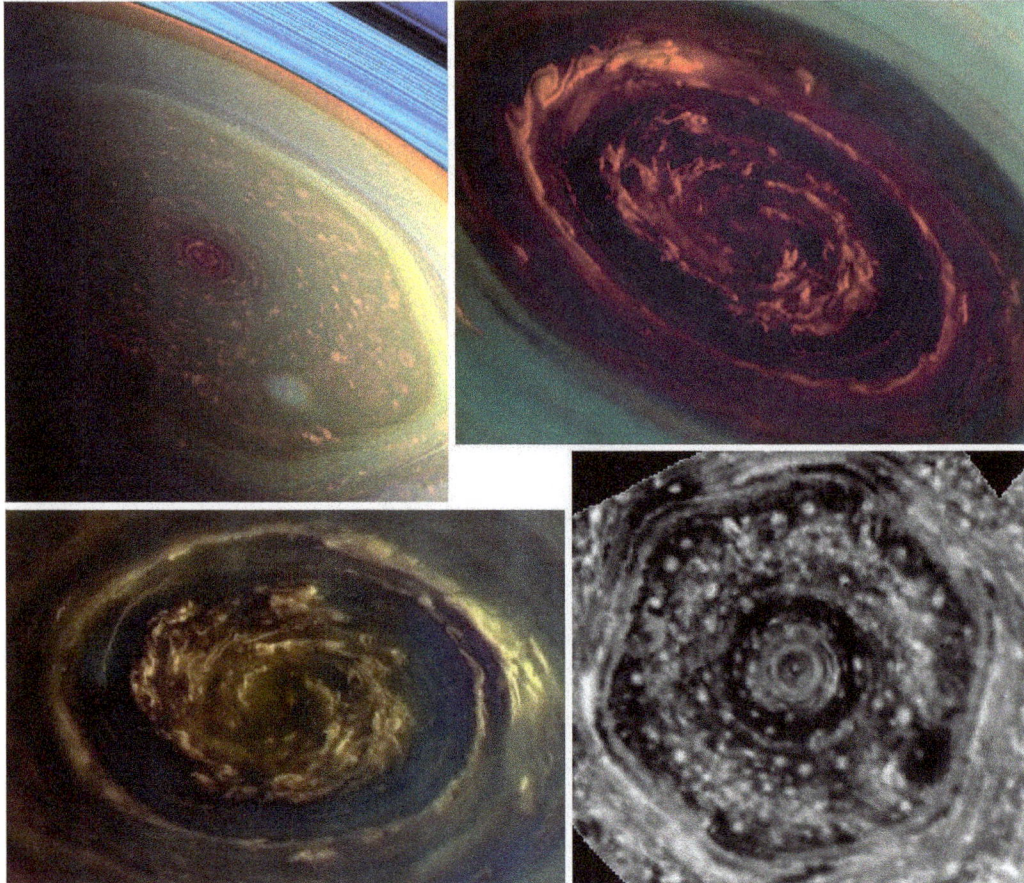

The North Pole hexagonal region of Saturn (top L. to R.) in infrared colour perspective, close-up of false colour view, (bottom L. to R.) close-up true colour image, black and white overhead view showing the vortex of Saturn's colossal hexagon storm
https://saturn.jpl.nasa.gov/science/saturn/ and https://commons.wikimedia.org/wiki/Saturn_(planet) and
https://www.youtube.com/watch?v=UqobxWXUA5k and https://www.youtube.com/watch?v=UqobxWXUA5k

NASA's **Cassini spacecraft** provided scientists the first close-up, visible-light views of a behemoth hurricane swirling around Saturn's North Pole. High-resolution pictures and video show the hurricane's eye is about 1,250 miles (2,000 kilometers) wide, 20 times larger than the average hurricane eye on Earth. Thin, bright clouds at the outer edge of the hurricane are traveling 330 mph. The hurricane swirls inside ***a large, mysterious, six-sided weather pattern known as the hexagon.*** http://www.dailygalaxy.com/my_weblog/2013/04/the-vortex-of-saturns-colossal-hexagon-storm-.html

Truly, one of the most bizarre things seen in the solar system captured in the above images was taken on 27 November by NASA's Cassini spacecraft. Cassini's camera zoomed into Saturn's polar hexagon storm's eye from a distance of about 250,000 miles (400,000 kilometers). The spacecraft observed in infrared wavelengths, which can peer through the top layer of clouds to reveal the complex texture beneath. NASA's Cassini spacecraft has been traveling the Saturnian system in a set of inclined, or tilted, orbits that give mission scientists a vertigo-inducing view of Saturn's polar regions, yielding spectacular images of roiling storm clouds and the swirling vortex at the center of Saturn's famed north polar hexagon.

These phenomena mimic what Cassini found at Saturn's South Pole a number of years ago. In 2007, the Cassini team saw a huge hexagon-shaped structure two-thirds as wide as Earth about 25,000 kilometers across stretching over Saturn's North Pole while planet was in the depths of its 15-year-long winter, and it was impossible to (know) what was going on within the storm. But, with the change of the Saturnian seasons, the sun started to creep over the planet's North Pole and spring lifted the gloom in 2009, and the team was able to view and study the massive storm at the hexagon's core.

The hexagon was originally discovered in images taken by the Voyager spacecraft in the early 1980s. It encircled Saturn at about 77 degrees north latitude and was estimated to have a diameter wider than two Earths. The jet stream is believed to whip along the hexagon at around 100 meters per second (220 miles per hour). (See pictorial explanation below).

Cassini later obtained higher-resolution pictures of the hexagon – which tells scientists it's a remarkably stable wave in one of the jet streams that remains 30 years later – but scientists are still not sure what forces maintain the object.
http://www.dailygalaxy.com/my_weblog/2012/11/saturns-hexagon-one-of-the-most-bizarre-things-seen-in-the-solar-system.html

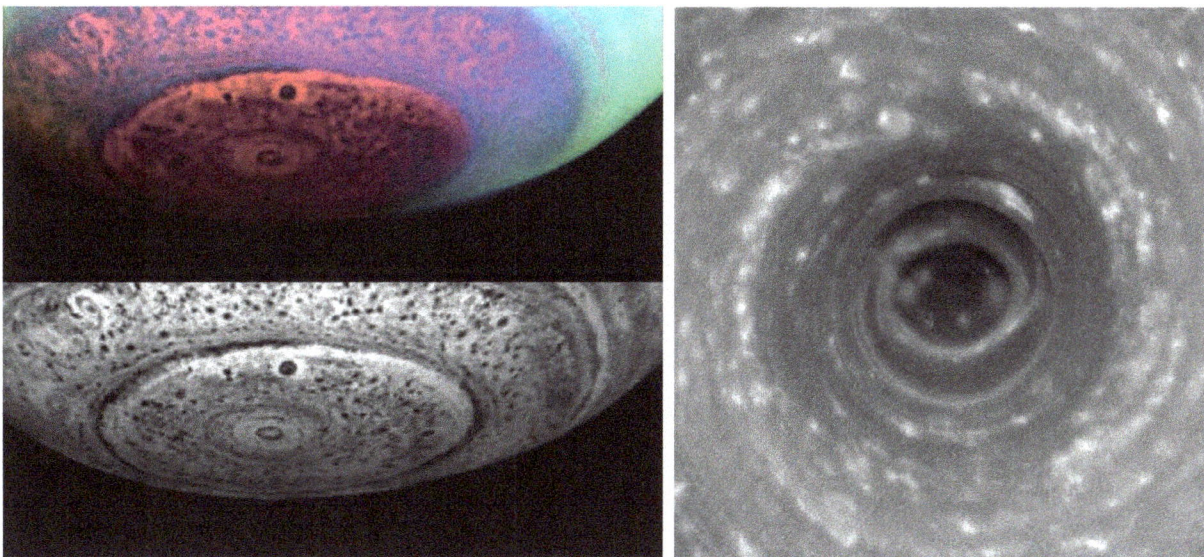

**South Polar region in infrared and black and white images
and close-up of the eye of the Saturnian super storm**
https://www.universetoday.com/19632/violent-polar-cyclones-on-saturn-imaged-in-unprecidented-detail-by-cassini/ and
http://www.abovetopsecret.com/forum/thread785735/pg1

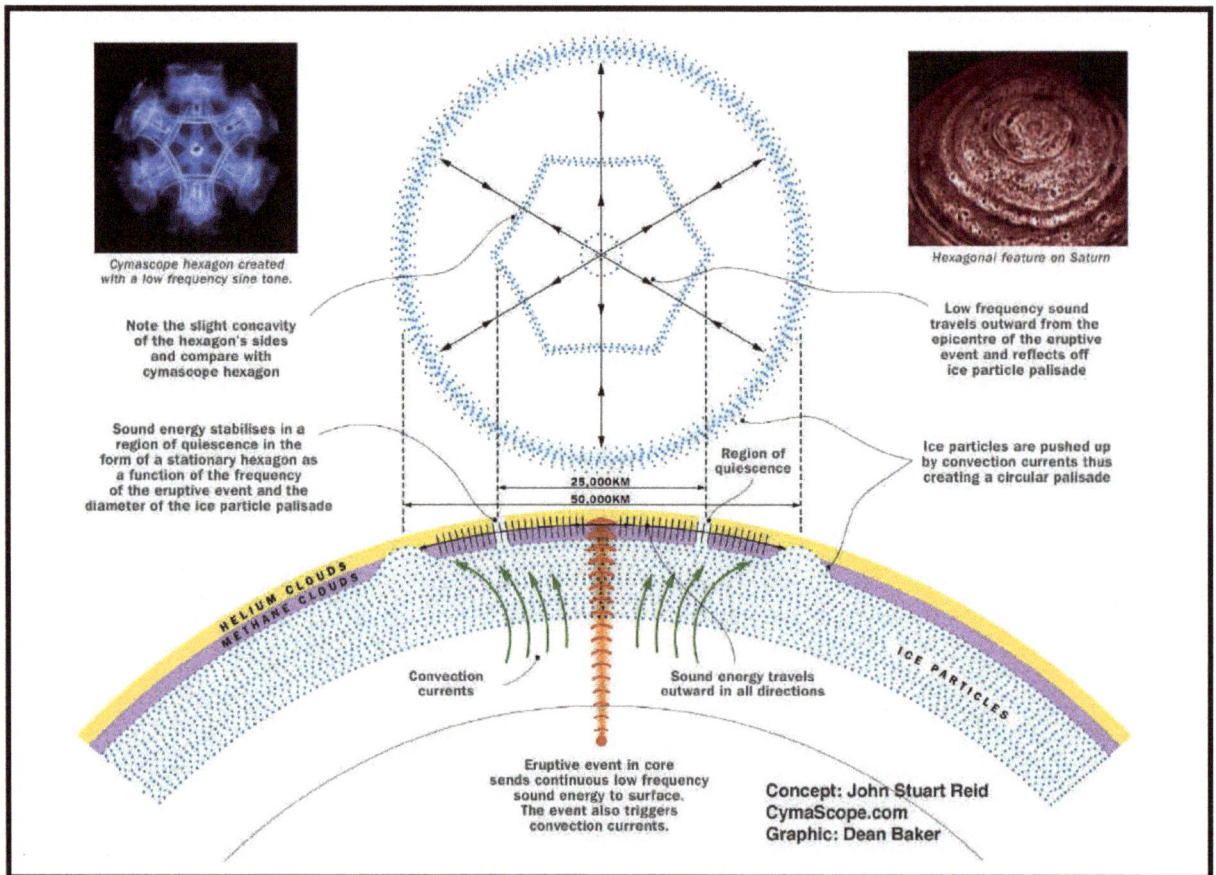

John Reid's theory shows how convection currents of ice particles, sound wave energy and eruptive core events produce the violent hexagonal super storms at the polar regions

http://www.tjmitchell.com/stuart/saturnrosslyn.html

From a distance, in visible light, Saturn's atmosphere looks more boring than Jupiter; Saturn has cloud bands in its atmosphere, but they're pale orange and faded. This orange color is because Saturn has more sulfur in its atmosphere. In addition to the sulfur in Saturn's upper atmosphere, there are also quantities of nitrogen and oxygen. These atoms mix together into complex molecules we have here on Earth; you might know it as *"smog"*. Under different wavelengths of light, like the color-enhanced images returned by NASA's Cassini spacecraft, Saturn's atmosphere looks much more spectacular.

http://www.universetoday.com/24029/atmosphere-of-saturn/#ixzz2alFWBoht

Saturn is about 120,000 km across. It takes 29.46 years to go around the Sun. Like Jupiter, it spins very rapidly - the day lasts for 10 hours and 39 minutes. It has a similar structure to Jupiter. It has a solid core, which is surrounded by a shell of solid hydrogen, which is in turn surrounded by a shell of liquid hydrogen, and then the giant shell of atmosphere. This atmosphere is made of hydrogen and helium gases, and ammonia, with small amounts of other gases. Like Jupiter, Saturn seems to be a bubbling cauldron of liquid and gas. Like Jupiter, the atmosphere of Saturn is NOT in chemical balance, with some unknown process making trace amounts of various gases.

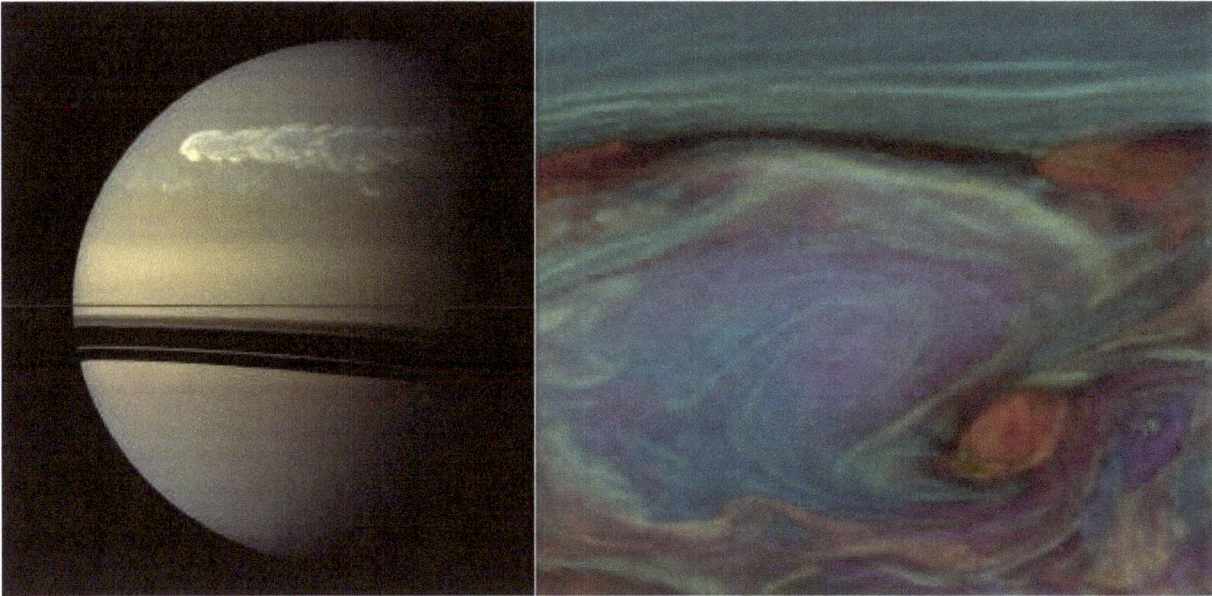

The Great White Spot changes into a giant raging storm in the northern hemisphere of Saturn in these two images from the Cassini spacecraft. The storm clouds probably are made of water ice covered by crystallized ammonia.

NASA's Cassini spacecraft captures a composite near-true-color view of the huge storm churning through the atmosphere in Saturn's northern hemisphere.

Like Jupiter, Saturn gives out more heat than it gets from the Sun. But the heat is made in a different way. On Saturn, the heat comes from the condensing of helium as it sinks in the atmosphere. In the same way that steam gives off heat as it turns from gas into liquid, so helium gives off heat. This heat is the power supply for the weather of Saturn.

Saturn has fierce winds which travel at some 1,700 km/hr near the equator - 3.5 times faster than the winds on Jupiter. For some unknown reason, the winds travel around Saturn in only 10 hours and 15 minutes, some 24 minutes faster than Saturn rotates. While Jupiter has the Great Red Spot, Saturn has the **Small Red Spot** - about 6,000 km across. And yes, it is located at that mysterious hyperdimensional latitude of 19.5° degrees!

But Saturn has a surprise - the **Great White Spot**. It appears about every 30 years - fairly close to the 29.46 year of Saturn. It has previously appeared in 1876, 1903, 1933, and 1960, and it last appeared on September 24, 1990. As before, it appeared in the mid-summer of the northern hemisphere of Saturn. It very rapidly grew to an oval big enough to swallow 3 Earths, and then stayed the same size for about 3 weeks. Then the 1,700 km/hr winds began to change the Great White Spot, and it became even larger as it grew a long tail that began to stretch around the planet. This cloud of crystals of ammonia ice reached some 240 km above the cloud tops. (See images above). http://www.abc.net.au/science/space/planets/saturn.htm

**The Small Red Spot in the southern hemisphere of Saturn
similar to the Great Red Spot on Jupiter**
https://www.jpl.nasa.gov/spaceimages/search_grid.php?sort=id&category=saturn

Perhaps, the most mysterious feature of Saturn is also its most beautiful feature, the rings of water ice which orbit around the planet in a carefully choreographed ballet movement. These rings composed of trillions of icy particles that range in size from 1 centimeter to 10 meters seem to continually coalesce into aggregates or clumps and then mysteriously disperse, only to repeat the dance over again and again.

Saturn's rings are named alphabetically in the order they were discovered. The main rings are, working outward from the planet, **Ring C, B and A**, with the **Cassini Division**, the largest gap, separating Rings B and A. Several fainter rings were discovered more recently. The **D Ring** is exceedingly faint and closest to the planet. The narrow **F Ring** is just outside the **A Ring**. Beyond that are two far fainter rings named **G Ring** and **E Ring**. The rings show a tremendous amount of structure on all scales, some related to perturbations by Saturn's moons, but much unexplained. http://en.wikipedia.org/wiki/Rings_of_Saturn

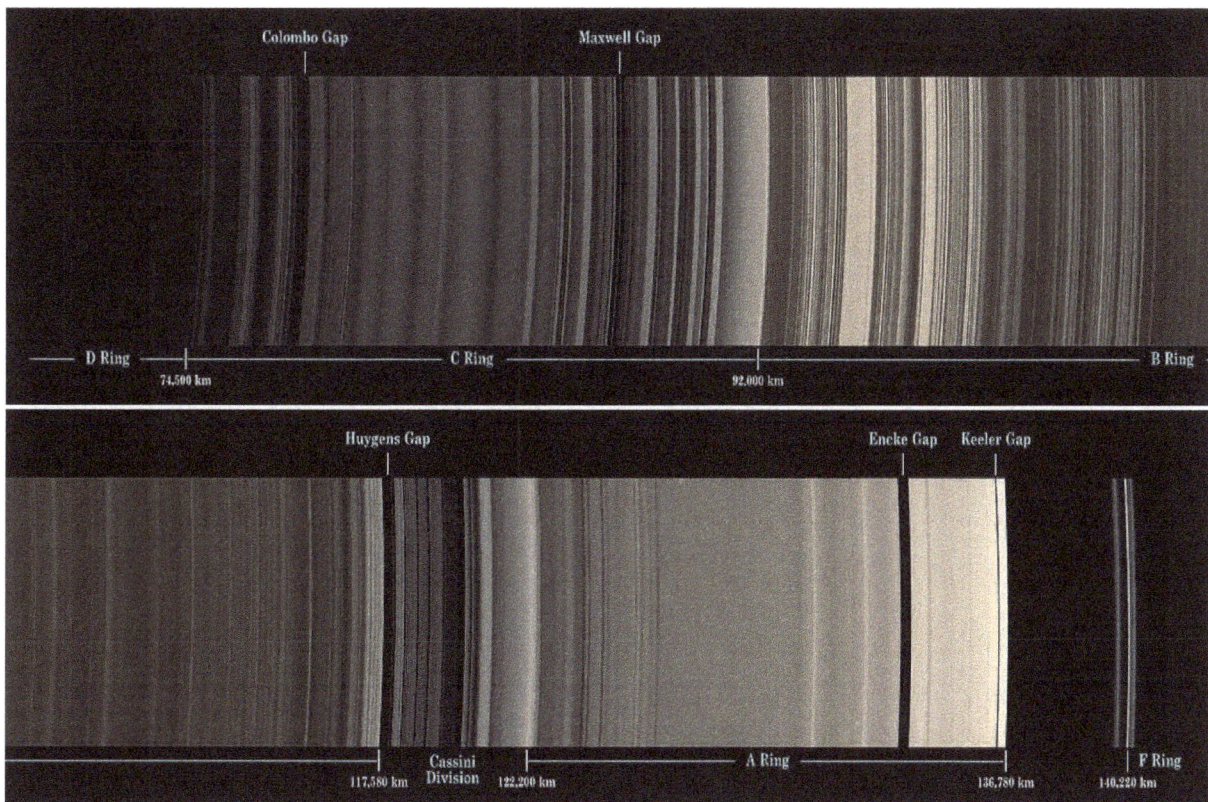

Natural-color mosaic of *Cassini* narrow-angle camera images of the unilluminated side of Saturn's D, C, B, A and F rings (Image strips are one continuous image left to right) taken on May 9, 2007.

https://commons.wikimedia.org/wiki/Saturn_(planet)

Hiding among Saturn's ring strands are small moons, long suspected by astronomers as being the satellite bodies producing some of the unusual structure observed in the F ring. The shepherd moons, Prometheus and Pandora are the main moons of Saturn causing the strange behavior of Saturn's F ring however, these moons cannot explain all observed features. NASA's Cassini space probe has detected lengthy objects that may be either solid moons or just loose clumps of particles within the ring; this has created a dilemma of interpretation among NASA, ESA and **Italian space Agency** scientists. The **Cassini narrow-angle camera (NAC)** images shows bright clump-like features at different locations within the F ring. (See images below).

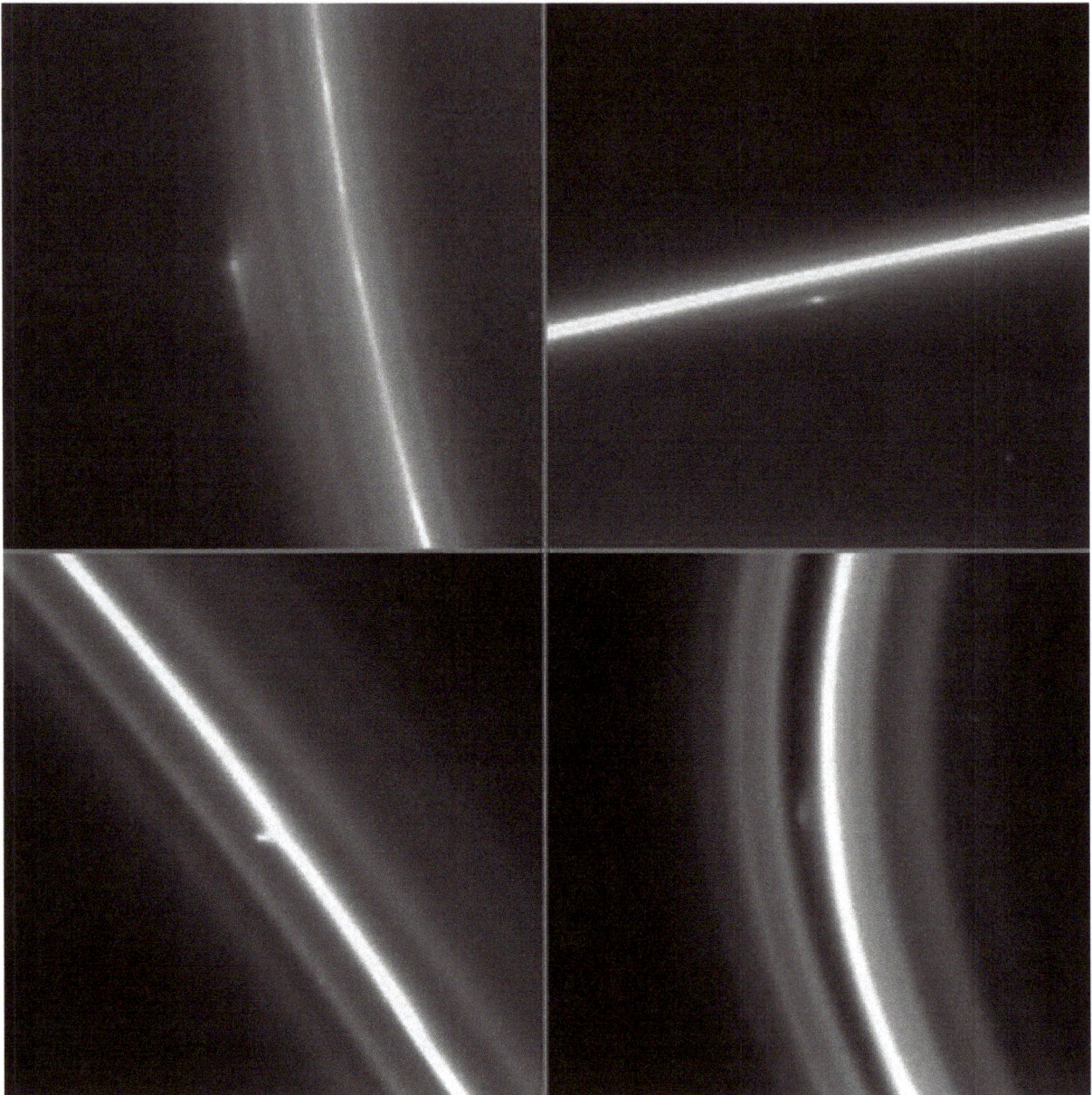

Do the 2000 kilometer long bright spots in each image above represent Saturnian moons, clumps of ice particles or something of artificial design?
http://nexusilluminati.blogspot.ca/2012/04/gigantic-motherships-and-other-bizzare.html

Two objects in particular, provisionally named S/2004 S3 and S/2004 S6, in the above images have been repeatedly observed by Cassini over the past year and a half. These objects do not conform to known astronomical observations and measurements as the orbits for these two objects have not yet been precisely determined, in part because perturbations from other *nearby moons make the orbits of objects in this region complicated.* It also appears from previous observations that *S6 appears to have an orbit that crosses that of the main F ring.*

This unexpected behavior currently is a subject of great interest to ring scientists. Thus, scientists cannot be completely confident at the present time if they in fact have observed new sightings of S3 and S6, or additional transient clumps.

In addition to these uncertainties and oddities of satellite behaviour, the objects in the outer ringlets of the F ring and within the F ring's inner ringlets are *approximately 2,000 kilometers (1,200 miles)* long.

The lower left image shows a feature that may be S3. S3 has been found to have an orbital path that is tightly aligned with that of the main F ring. The lower right image was taken on April 13, 2005; the *object does not appear* to be either S3 or S6.

http://spacespin.org/article.php?story=cassini_new_details_saturn_rings

Could these structures be clumps of ice particles within the F ring which are often transient by nature, appearing and then disappearing within months or actual moons that disturb the material around them or the by-product of interaction between the F ring and larger moons such as **Prometheus**? Or, has NASA and ESA scientists already established that these anomalies have a more artificial explanation which is being kept from the public because of the true nature of the rings is not what we think it is?

CHAPTER 89

"PREPARE YOUR MINDS FOR A WHOLE NEW SCALE OF PHYSICAL SCIENTIFIC VALUES, GENTLEMEN!"

"Prepare your minds for a whole new scale of physical scientific values, gentlemen!" is a line from the movie: **"*Forbidden Planet*"** (a precursor to the TV series **Star Trek** and its spin- off movies) spoken by Dr. Edward Morbius (Canadian actor, **Walter Pigeon**) to Earth spaceship Captain, John J. Adams, (another fellow Canadian actor, **Leslie Nielsen**) and the ship's doctor, Lieutenant "Doc" Ostrow (played by American actor, **Warren Stevens**).

Morbius takes the star captain and his senior officers on a tour of the Krell planet's interior via a subterranean shuttle pod to a vast cube-shaped underground complex, 20 miles (30 km) on each side, still functioning and powered by 9,200 thermonuclear reactors. It is indeed a whole new order of physical dimension and size that anyone can scarcely imagine in their wildest dreams! http://en.wikipedia.org/wiki/Forbidden_Planet

The often quoted adage that *"science is stranger than fiction"* has never been truer, than now. It appears that a page or perhaps, a chapter may be taken from the movie: "Forbidden Planet" with regard to size.

Before we begin to look at the photographs and comprehend the significance of the extreme large scale anomalies that orbit around Saturn, we must understand the nature and reality of the *"new scales of physical scientific values"* of size in the universe. What is about to be revealed, may seem so incredible to our basic understanding of science that it will force us to re-evaluate everything that we thought we knew about the universe and our place in it. Our current perspective of our evolution and development on Earth as an intelligent species, as an emerging global civilization and our exploration of space has only just begun!

So far, in our virtual tour of the Solar System we have found evidence of an Extraterrestrial presence out in Earth orbit, on the Moon and on Mars and out around Jupiter and it moons. The Extraterrestrial presence has even been imaged by robotic cameras on board Mars orbiting spacecraft like Viking and Pathfinder and their landing probes as well as the **Mars roving vehicles** of **Pathfinder**, **Sojourn**, **Spirit**, **Opportunity** and **Curiosity** travelling across the Martian terrain. If life on Mars does not currently exist, we can be absolutely confident that the photographic evidence taken by **NASA**, **ESA** and **RSA** has proven beyond any reasonable doubt that an ancient and a highly advanced intelligence had established a global civilization on Mars in the distant past. The archeological evidence on Earth is steadily accumulating that this ancient intelligence has had direct contact with our own terrestrial ancestors in ancient times and may very well have influenced the course of our evolution and societal development globally.

But the question that needs to be asked at this point, how far out in the solar System does life exist and is that life sentient and intelligent? The answer comes from deep space probes like **Voyager 1 and 2 spacecraft** launched in 1977 with their 1980 photographic fly-by of Saturn on their way out of the Solar System and now, with the current orbiting **Cassini-Huygens space probe** around Saturn since 2004 (launched from Earth in 1997), display images of such extraordinary evidence of artificiality, as to boggle the minds of space scientists and UFO researchers. http://voyager.jpl.nasa.gov/

Disclaimer: "Cameras, being well-established scientific instruments, provide direct data of the "Seeing-Is-Believing" variety. Despite their straight-forward characteristic, actual photographs probably will not establish conviction for everyone." Ringmakers of Saturn© Norman R. Bergrun; 1986 by The Pentland Press Ltd Kippielaw by Haddington East Lothian EH41 4PY Scotland; Library of Congress Catalogue Card Number 86-81530; ISBN 0 946270 33 3

If we accept the photographic data captured by the Cassini-Huygens space probe, it appears that Saturn as well as the outer gas giant planets, all currently have *super, massively behemoth spacecraft in orbit around them that are so gigantic in size as to be even larger than the Earth!*

And as incredibly mind-blowing as that information may sound, what is even more amazing is that these same planet-size spacecraft are actually *constructing the rings around Saturn and the other outer planets!*

Admittedly, no one in the science community has come out to acknowledge this incredible discovery, let alone anyone from any of the space agencies, as it is suspected that they are still trying to get their heads around the incredulousness of this behemoth scientific conundrum. One thing that science does not like to admit to is when they are wrong about an accepted theory or concept or where many reputations have been made and recognized over the years based upon unacknowledged, poorly researched or outdated science. This doesn't happen frequently but, it does happen often enough to the embarrassment of those who have accepted the bad science without question. This embarrassment sometimes becomes so cumbersome as to force a deliberate cover-up of the scientific reality of the truth or to force a complete revision in scientific understanding with the possibility of loss of reputation, tenure, government grants and positions, etc. There is the remote possibility to forgive all pass transgressions of poor scientific research and the cover-ups of real factual scientific data and move forward and build upon the real accurate data that science discovered.
Fortunately, someone did look beyond the status quo thinking of the scientific community to question the raw data and ask the important questions of why is this happening, how come, what's causing it, where did it originate, when did it occur and will it continue?

Norman R. Bergrun, the Ringmaster of Cosmic Size

Norman R. Bergrun is one of the those rare individuals of genius caliber, who decided to come out of the confines of military and intelligence circles to shed some light on the rings of Saturn and other solar system bodies in a revealing, tell-all book called the "Ringmasters of Saturn". His credentials are impeccable, first class in the highest degree, his education is full of degrees, and his career is full of awards, citations, serving in positions as director in many military and aerospace companies!

Dr. Norman Bergrun is an alumnus of **Ames Research Laboratory**, **NACA (National Advisory Committee for Aeronautics)** predecessor of **Ames Research Center**, NASA where he worked twelve years as a research scientist. At Ames, he pioneered the setting of design criteria for airplane thermal ice-prevention and the developing of roll stability laws for airplanes, missiles and rockets. During his time at NACA Ames Research Laboratory, Bergrun took the photos of icing conditions experienced by the Curtiss Wright C-46 test airplane in natural icing environments. Photo laboratory professionals tutored him prior to the first flight. On returning to home base, he specified photo development requirements and analyzed the images obtained.

He joined **Lockheed Missiles and Space Company** (now **Lockheed Martin**) where he was manager of the planning and analysis of flight tests for the **Navy Polaris Underwater Launch Missile System**. Bergrun established a closed area to receive and catalogue all photos and films obtained by test-range cameras. Its purpose was to provide assistance in the analyses of telemeter tracking data. During his thirteen years at Lockheed, he scrutinized miles of photo-coverage film of Polaris flight tests...successfully correlating these data with inception of malfunctions recorded by telemetry and thereby pin-pointing where hardware improvements were needed. He also served as a senior scientist having responsible analysis cognizance of special space-satellite applications. After a short tour of duty with Nielsen Engineering and Research, in 1971 he founded **Bergrun Engineering and Research**, parent of Bergrun Research founded in 1999 especially for World Wide Web activities.

Norman R. Bergrun
http://ufosonline.blogspot.ca/2013/11/dr-norman-bergrun-enorme-nave.html

An Associate Fellow of the **American Institute of Aeronautics and Astronautics (AIAA),** he is active as a leader in Congressional Visits Day events on Capitol Hill. As Deputy Director-at-Large for the AIAA western region, he overlooks section activities in seven western states. Other memberships include The Planetary Society, The Association for the Advancement of Science, The Aviation Hall of Fame, the National Society of Professional Engineers, the Federation of American Scientists and the Scientific Faculty, International Biographical Centre, Cambridge, England.

Bergrun holds a BSME degree from Cornell University, an LLB from LaSalle University Extension, a DSc (Hon) from World University and a California Professional Engineer (PE) License. He also has engaged in graduate aerospace studies at Stanford University. He is a founder of the California Society of Professional Engineers Education Foundation, is author of two books "Tomorrow's Technology Today" and "Ringmakers of Saturn" and has published over 100 papers. Two recent manuscripts, "Lunar Life Forms: Revelations of Apollo 14" and "Mars Puts on a Good Face: The Masquerade" have been registered with the Library of Congress, Washington, D. C. He has lectured in the United States, Canada, England and Europe.

Credited with numerous awards and citations including the California Society of Professional Engineers *Archimedes Engineering Achievement Award,* and Special Service Citations for contributions to the AIAA National Public Policy and to the Regional Sections Activity Committees, he is listed in Marquis "Who's Who in the World", "Who's Who in America", "Who's Who in Science and Engineering", and other reference works.

Soon after publication of "Ringmakers of Saturn" in 1986, approximately four trips annually have been made from the West Coast to the NASA Goddard Space Flight Center. During this time, beginning with the early missions, many thousands of images have been examined of Mars, the Solar System Planets and Satellites, the Moon, the Sun as well as Hubble Telescope and Shuttle images. This process is an ongoing activity of Bergrun Research.
http://www.ringmakersofsaturn.com/About%20Bergrun.htm

Bottom line here, besides his impressive career in the aerospace industry and the many distinctions awarded him, when it comes to photo analysis and interpretation, he knows what he's talking about, he is an expert par excellence!

Bergrun speaks about the new *"physical scientific values"* as mentioned above with some basic understanding of astronomy and with some pictures which speak louder and clearer than words, in order to appreciate the size differential that occurs within the known universe. In this regard, **Norman Bergrun** helps steel our minds for what he is about to unveil to us, he introduces us to the concepts of quasars. Quasars are star-like radio sources that are the largest and brightest single objects known. An example is quasar 3C-273 estimated to be a light year in size and able to produce an energy equivalent to 10 trillion suns and is located so remotely that it takes 3 billion years for its light and energy source to reach the Earth. The universe is a very big place and immensity in all forms abounds. Another example is the size relation of the Earth to other planets, as we have already seen in comparison to Jupiter and Saturn and in comparison to our own Sun. (See photos below).

Size relationship of Earth to other planets in the Solar Systemand the Sun
https://www.youtube.com/watch?v=59Ee-z-syBM and https://www.quora.com/If-one-was-to-compare-the-size-of-the-Earth-with-the-size-of-the-rest-of-the-observable-universe-how-would-it-compare

The Earth compared in size to the Sun and a solar flare (insert). Approximately one million Earths would be required to fill the volume of the Sun
https://www.universetoday.com/65583/is-the-earth-bigger-than-the-sun/

The universe presents a panoply of interstellar and galactic objects of such immensitythat the sheer magnitude of their size is mind numbing to even comprehend. We can understand that our sun when compared side by side with the red super giant star of VY Canis Majoris, that it would not even be visible! (See photo below). Yet, there is a theory held by both mystics, a few scientists and some astronomers that instead of a **black hole** at the centre of the galaxy, there exists an incrediblly brilliant galactic sun within each galaxy of such magnitude as to be 1000 to 5000 lightyears in size and possibly larger, by which all other stars pale into insignificance!

In this size comparison chart of stars, the sun is visible in image 3, it is a pixel in size in image 4 (red circle) and in images 5 and 6, it's so small as to be invisible! VY Canis Majoris is so huge that flying around it in a jet at 900 mph would take 1100 years!

https://en.wikipedia.org/wiki/File:Star-sizes.jpg

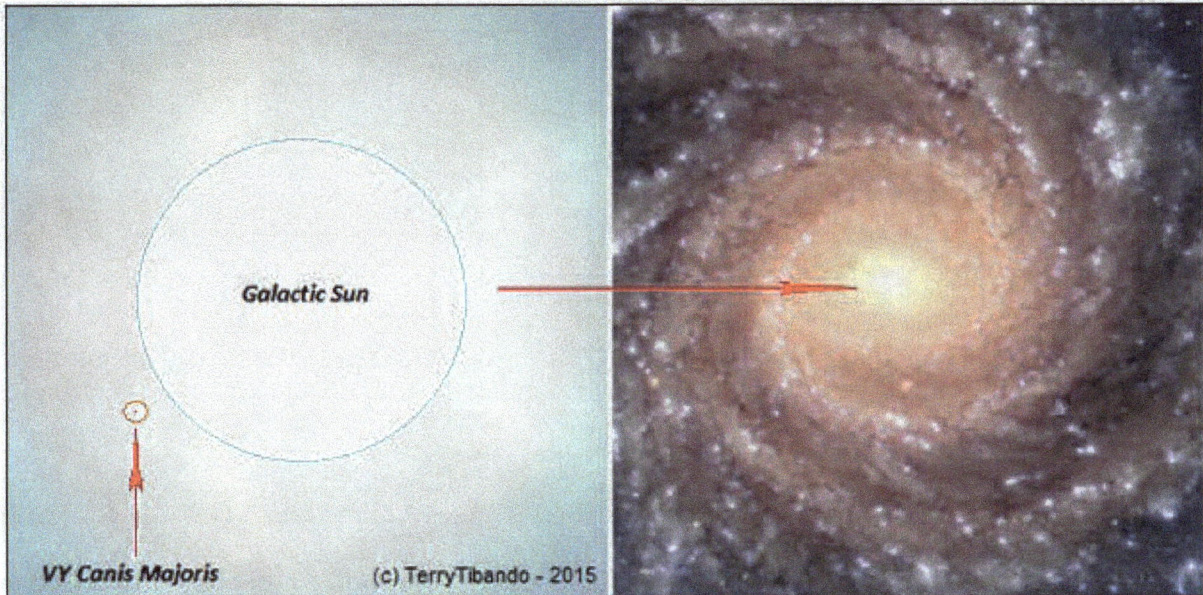

Is there a galactic sun at at the heart of the Milky Way Galaxy, instead of a black hole as most astronomers think? VY Canis Majoris circled in red is a pixel in size, but in reality it would be invisible by comparison!
(c) Terry Tibando

However, most astronomers and cosmologist state that each galaxy has a powerful black hole at its centre that pulls all matter, including stars towards a centre of gravitational nothingness that not even light can escape. So, how can a black hole of nothingness generate light of such intensity as to flood its immediate surrounding stellar neighbourhood with blinding brilliance? Even, if the galactic bulge at the centre which measures 20,000 ly in circumference is a stellar mass of billions of stars, would not the gravitional pull from each star via with each other to the point that many stars would be absorbed into a single star to form one galactic super star?

The Milky Way Galaxy is 100,000 lightyears across and is part of a galactic cluster
http://www.solstation.com/x-objects/andromeda.htm and https://en.wikipedia.org/wiki/Galaxy_cluster

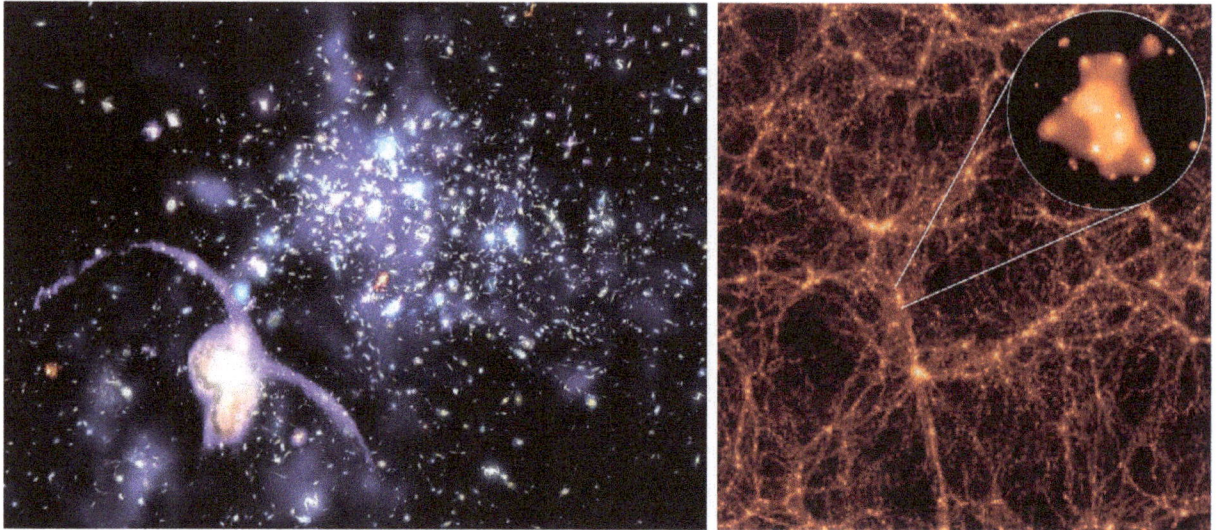

A galactic cloud (left) is composed of many galaxy clusters (insert in right image) billions of lightyears across and these galactic clouds in turn comprise a massive network of galaxies

http://chandra.harvard.edu/press/05_releases/press_040805.html

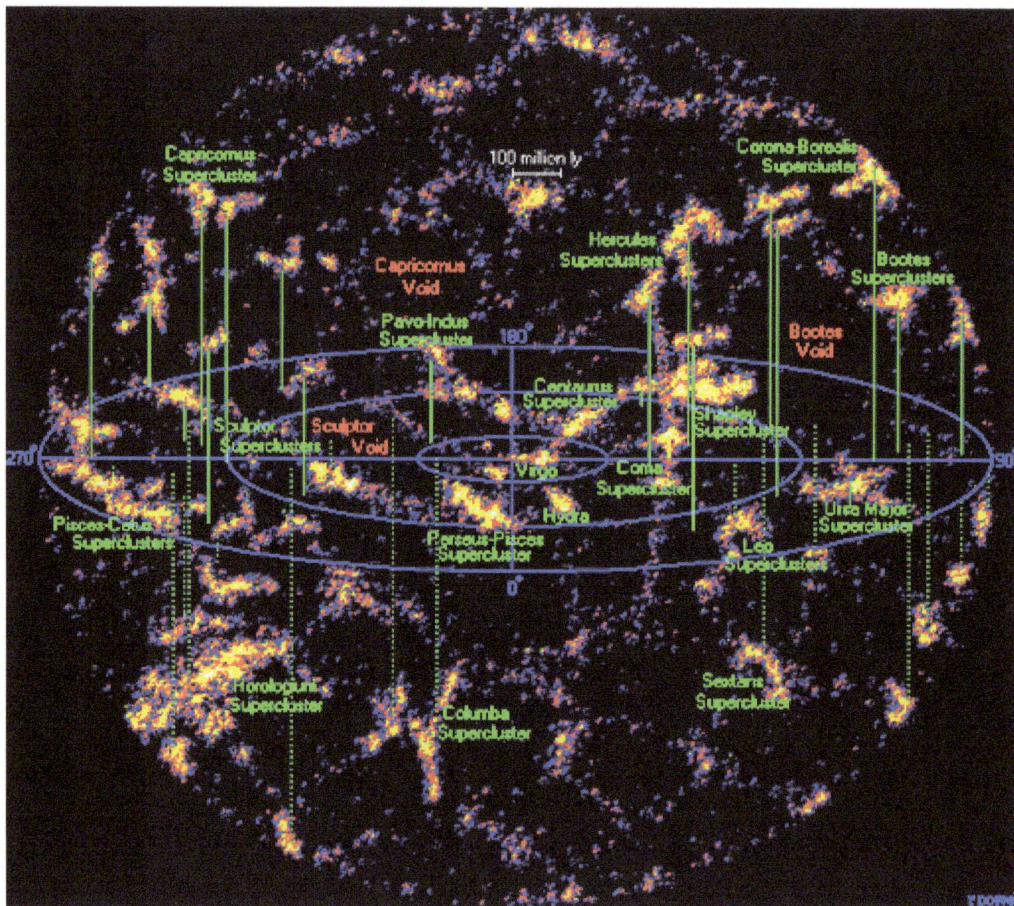

The known universe is a network of galactic superclusters or clouds

https://www.wired.com/2011/12/universe-size/

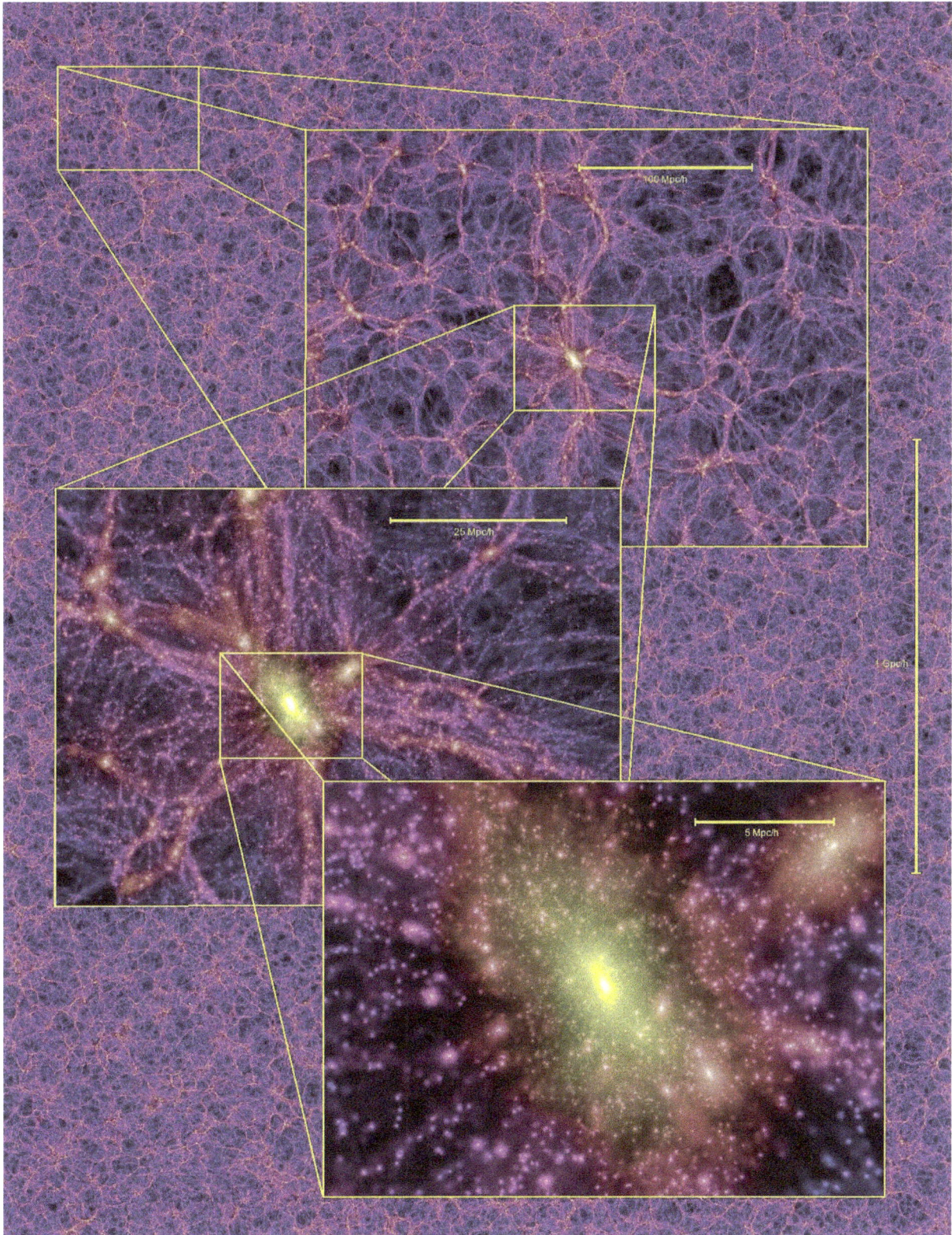

A massive network of galactic superclusters and tendrils interconnected to each other, megaparsecs in size or hundreds of billions of lightyears across!

299

Galactic super stars may or may not exist, this has yet to be proven but, there is no doubt of the immense size of our galaxy, the Milky Way however, it too shrinks in size when compared to the galactic clusters that abound in the universe of which it is part of one such cluster, the Virgo cluster. Many galactic clusters form into ssuperclusters or galactic clouds which all appear to be interconnected by threads or tendrils of other galaxies. These clouds of galaxies in turn seem to be a part of an even larger mass of galaxies that are **Megaparsecs (Mpc/h)** in size which are in turn is part of an universe globe that is **Gigaparsecs (Gpc³)** in volume. The whole universe when compared to the human brain is amazingly similar. The galactic superclusters and the tendril connections between clusters resembles the neurons found in the human brain.

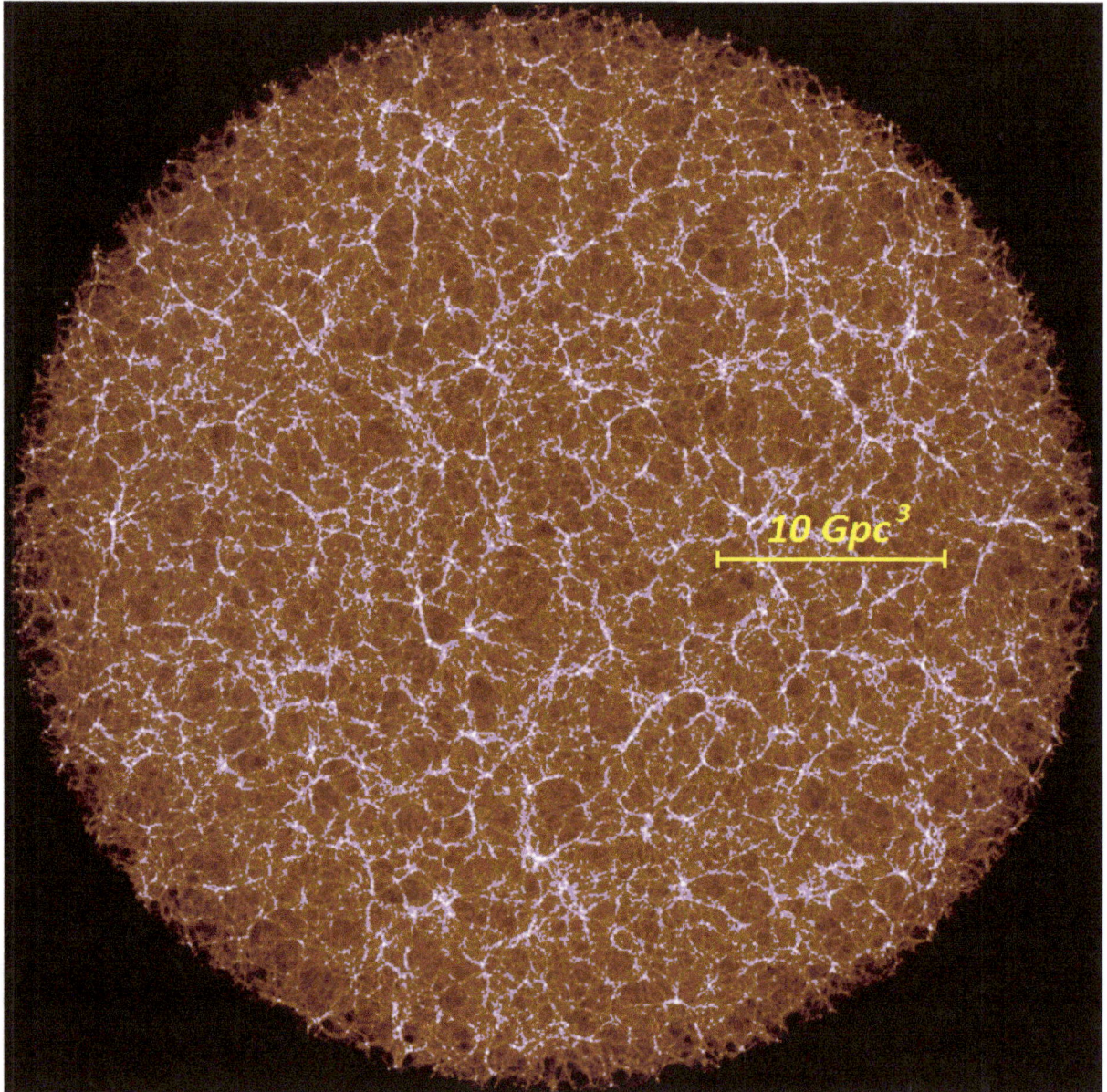

Is this the body of the known universe as imagined in this computer simulation or are there many more such universes beyond this one?
(Google Images)

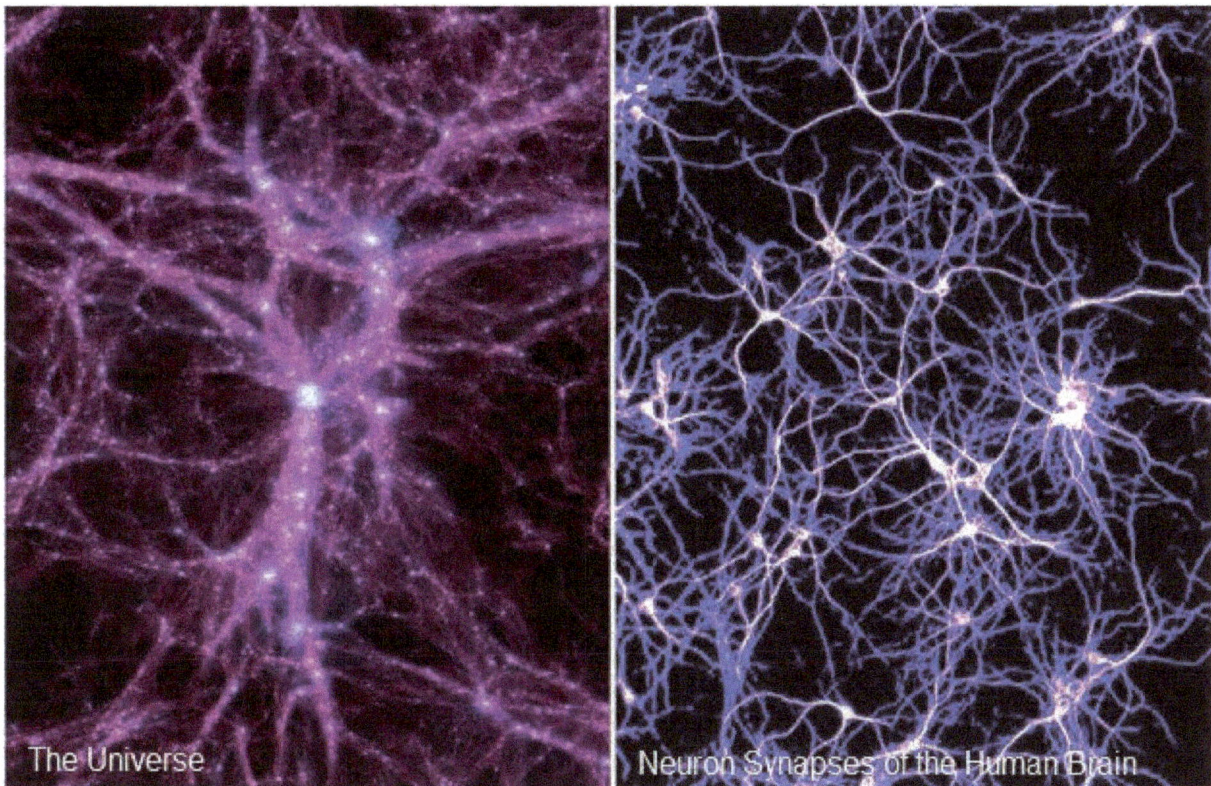

The reality of nature finds expression in similarity within the macroverse and the microverse, where the neuron synapses of the human brain (right) are mimicked by the intergalactic connections of the universe on a scale so unimaginably immense as to be unfathomable
https://pics-about-space.com/superclusters-of-galaxies-filaments?p=1 and

Pondering the immensity of the universe and how small we humans are in the scheme of things, one cannot feel awed and perhaps, overwhelmed by this reality, that we question our existence and our purpose within it. But, we should never feel insignificant or helpless by the nature of this reality, rather we should recall the sublime words of **Baha'u'llah** from the **Baha'i Writings**:

"Dost thou reckon thyself only a puny form
When within thee a universe is folded?"
The Seven Valleys and the four Valleys by Baha'u'llah; translated by Ali-Kuli Khan and Marzieh Gail in 1945; Baha'i Publishing Trust; Wilmette, Illinois

O SON OF SPIRIT! *Noble have I created thee, yet thou hast abased thyself.*
Rise then unto that for which thou wast created.
The Hidden Words of Bahá'u'lláh by Bahá'u'lláh; translated by Shoghi Effendi; 1970; Baha'i PublishingTrust; Wilmette, Illinois

We are more than the sum total of our atoms and molecules as physical beings on this Earth. We are rational intellect, spirit, and soul, we are indivisible and therefore, infinite. Thus, we can understand the universe that we live in, starting with the universe within each of us!

301

Such is the enigma of size yet, even these galactic central suns and the galaxies in which they reside are merely one of billions upon billions of galaxies within a galactic supercluster or cloud and there are myriads upon myriads of galactic clouds with tendrils of galaxies that connect to other galactic clusters in a network that looks like the neuron synapses of the human brain!

This imagery conjures visions of a universe so immense that it may actually be a part of some larger unfathomable organism of which we are merely a small part of that organism!!!

God possibly? Maybe, and then, maybe not! For we are taught by all religions that God is a spirit and not a physical being! But, how wonderfully amazing to see that not only do human brain cells resembles the universe and vice versa but, so does the internet, as does all plant life, animal and aquatic life on Earth as well! We are in reality all connected to each other and to everything else in the universe!

If we can accept the immensity of things in the universe that occur naturally then, we should be able to get our minds around the concept of artificial structures and spacecraft constructed by intelligent beings, that are only tens of thousands of miles in length.

The point being made here is that **Bergrun** is trying to get us to understand the values and magnitudes of size, so that one can then, have an open responsiveness and appreciation for the beings that have created **immense spacecraft** that are literally **thousands of miles in length and width**. These tubular or cigar shape spacecraft are estimated to be, based upon detailed photographic analysis and measurements ranging from **one thousand miles to as large as twenty thousand miles in length!**

We must also give serious consideration to the fact that spacecraft of such immense size were constructed by intelligent beings, who are possibly **giants in stature!** Traditional and biblical literature are filled with legends, myths, and stories of a race of giant beings that once walked upon the Earth. Perhaps, these legends derive from an extraterrestrial source which may account for the megalithic structures found on this planet and on Mars.

According to **Bergrun** from his analysis of NASA's raw photographic data sent back by the **Voyager I and II spacecraft**, these immense spacecraft are responsible for the construction of the rings around Saturn and some of the other outer planets. He states that these spacecraft are not to be confused with the natural satellites and moons of Saturn but, are truly anomalous by nature and movement. The spherical moons of Mimas, Enceladus, Tethys, Dione, Rhea, lie within the radial expanse of E ring with Rhea just outside of the E ring. The other spherical moons of Titan Iapetus and Phoebe are outside the ring system with all the other asteroid or irregular shape satellites occupying between the A to G rings and two of these satellites straddle the F ring as **"Shepherding Satellites"**.

The shepherd moons are within the rings, specifically within the gaps. These shepherd satellites are either "transient clumps of dust" or little "icy moonlets". In other words, NASA scientists and astronomers have no real idea as to exactly what they are, however, they've provisionally named it **S/2004 S 3** and **S/2004 S 6**. The "known" shepherd moons and inner satellites are named **Pan, Daphnis, Atlas, Prometheus** and **Pandora** are considered to be asteroid type moons.

Two major gaps in the ring system are the **Enke Division** found in the A ring and the Cassini division between the **A ring** and the **B ring**. The divisions between rings like the Enke and

Cassini Divisions vary their location and are thus, not consistent in their location or width, even the thickness in the ring plane show variance. This variability in the ring system geometry is hard for astronomers to accept as they are not aware of the physical mechanisms that produce recurrent changes. This aspect is key in understanding what alternative mechanisms are in operation to cause these changes.

This brief understanding in simple basic astronomy is necessary when we try to explain the anomalous features in and around the planets of the Solar system, particularly Saturn so that the reader will understand what is artificial and not a natural formation of the planet.

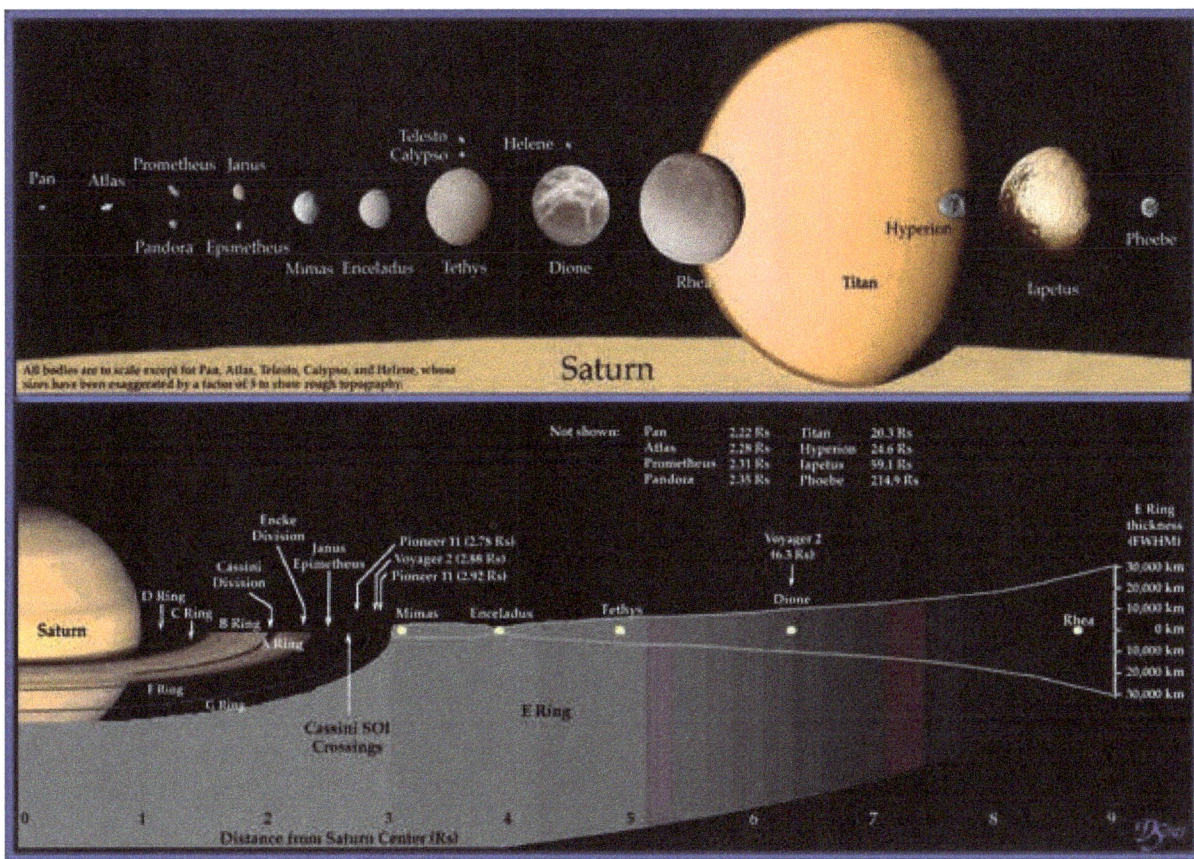

Saturn and its Moons, Satellites and Ring Structure
http://astro.hopkinsschools.org/course_documents/solar_system/outergasplanets/saturn/saturn_rings/saturn_rings.htm

Luminous sources at Saturn have been observed, notably by Herschel, Knight, and Ainslie. In one instance, a fiery source moved suddenly away from the A-ring outer edge. In another unrelated instance a bright, elongated source pursuing a straight-line course entered the A-ring outer edge, traversed the Cassini division, and exited the opposite A-ring outer edge. After these dramatic events, luminous sources did not become a specific subject of inquiry as might be expected - that is, until this analysis many years later." **Ringmakers of Saturn© Norman R. Bergrun; 1986 by The Pentland Press Ltd Kippielaw by Haddington East Lothian EH41 4PY Scotland; Library of Congress Catalogue Card Number 86-81530; ISBN 0 946270 33 3**

Below are two similar images composed from a number of Voyager 1 images of Saturn taken at different times; besides the obvious polar tilt of the planet and the minor colour differences, the image on the left appears to be cropped yet, it is not. What would cause this piece of the outer **A** ring of Saturn to be missing if the photo image taken by the Voyager spacecraft is not cropped? When the image on the left is lightened, it definitely shows an image that is unaltered as it was originally photographed. (See images below).

NASA Press Release Photo
August 13, 1981

Original NASA/JPL P-23876C Image composed from a
number of Voyager 1 images

These two images are are composites of other images. The left image is an "uncropped" photo with a small section of the outer A ring missing and an anomalous orange spot on the outer ring. The right image was also taken by Voyager 1 that shows a complete ring system but, according to Bergrun, the ring system in the left image is being "constructed"!
http://www.thelivingmoon.com/46_mike_singh/03files/Ships_Saturn.html

There are also a number of luminous sources that appear in the Voyager raw photo data, one of these light sources, in particular, is located in the A ring as seen circled in yellow within the red box in the colour lightened photograph below (Chapter 88). This light source is not a satellite or moon of Saturn as it has an internal luminosity and not reflected luminosity and is approximately the size of the Earth.

CHAPTER 90

GIGANTIC ANOMALIES ORBITING SATURN

There are also a number of luminous sources that appear in the **Voyager raw photo data**, one of these light sources in particular which is orange-red in colour is located in the A-ring at the left edge as seen circled in yellow within the red box in the photograph below. This light source is not a satellite or moon of Saturn as it has an internal luminosity and not a product of reflection from the sun or from Saturn and is approximately the size of the Moon. The blue orb in the photo is another anomalous light source. (See below enlarged photo image).

Light source

Missing Ring Section

This is a colour lightened image from the photograph above enlarged to show the missing section of the A ring and the near Earth-size luminous light source
http://www.dailygalaxy.com/my_weblog/2013/05/saturns-naturally-occurring-radio-signal-new-insights.html

The break in the outer A-ring is pproximaely the width of Saturn's diameter or 10 Earth diameters across in distance, this makes the break over 20,000 miles in length. **Norman Bergrun** in his book shows a enlarged photo image of this break "Ringmakers of Saturn" and through photographicmetric analysis (possibly from a different photo image than the one shown, see below), states that there is efflux being generated along the length of a slender body which is exhausting at both ends creating Saturn's A-ring and that this cylindrical body is orbiting in a clockwise motion around Saturn depositing a wide trail of material in its travel!

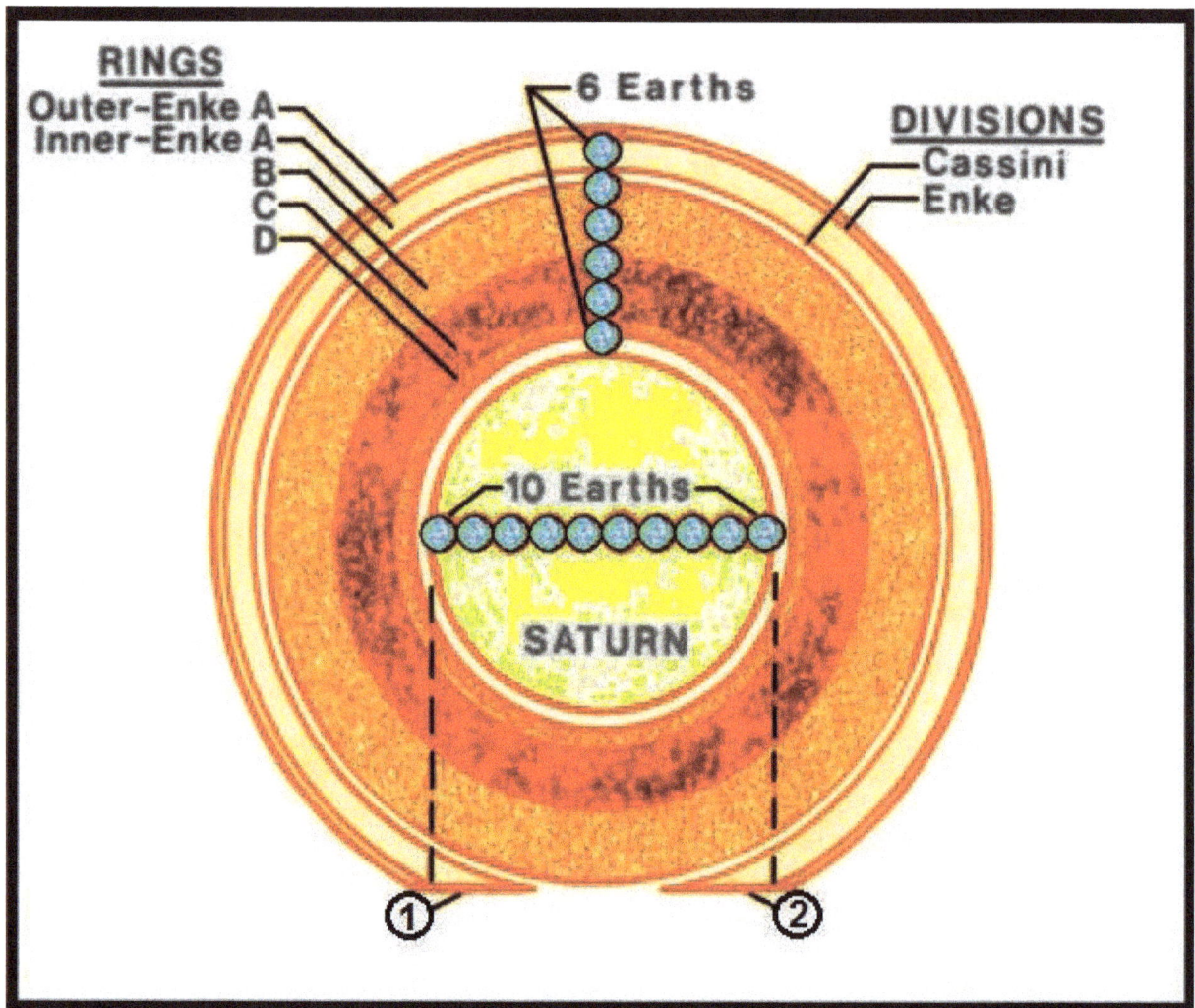

The size of the Earth in comparison to Saturn and its rings provides the estimated size (fineness ratio) of the space vehicles (1 and 2) that form the rings of Saturn

https://www.slideshare.net/dormsfornorm/ringmakers-of-saturn-by-norman-r-bergrun

In October 1999, Dr Bergrun was interviewed by **MUFON´s Don Ecker** regarding his conclusion that *'Huge Artificial Machines'* are operating in our solar system. Here is part what he had to say with regard to the apparent cropped rings of Saturn:

*Figure 4 is obtained from figure 2 by brightening the entire ring system. By performing this operation, all of what is going on can be seen in one image. A blue arc appears connecting ring-system termination on the left-hand side with that on the right-hand side. That this can occur, indicates a flow between the two sides. The other observation is that a discontinuity of the ring system now is readily apparent as indicated by the change in colors from a yellow-green with purplish overtones (upper ring system) to blue (terminal arc of the ring system). Separating the abruptly different colors are slender, long objects (called **ElectroMagnetic Vehicles**, or **(EMVs)**, in Ringmakers of Saturn). These can clearly account for the presence of the blue arc. (Note: The brightened ring-system image is fairly unique. Different persons should be able to obtain similar*

306

images; but the chance of every person making them identical to one another is somewhat remote).

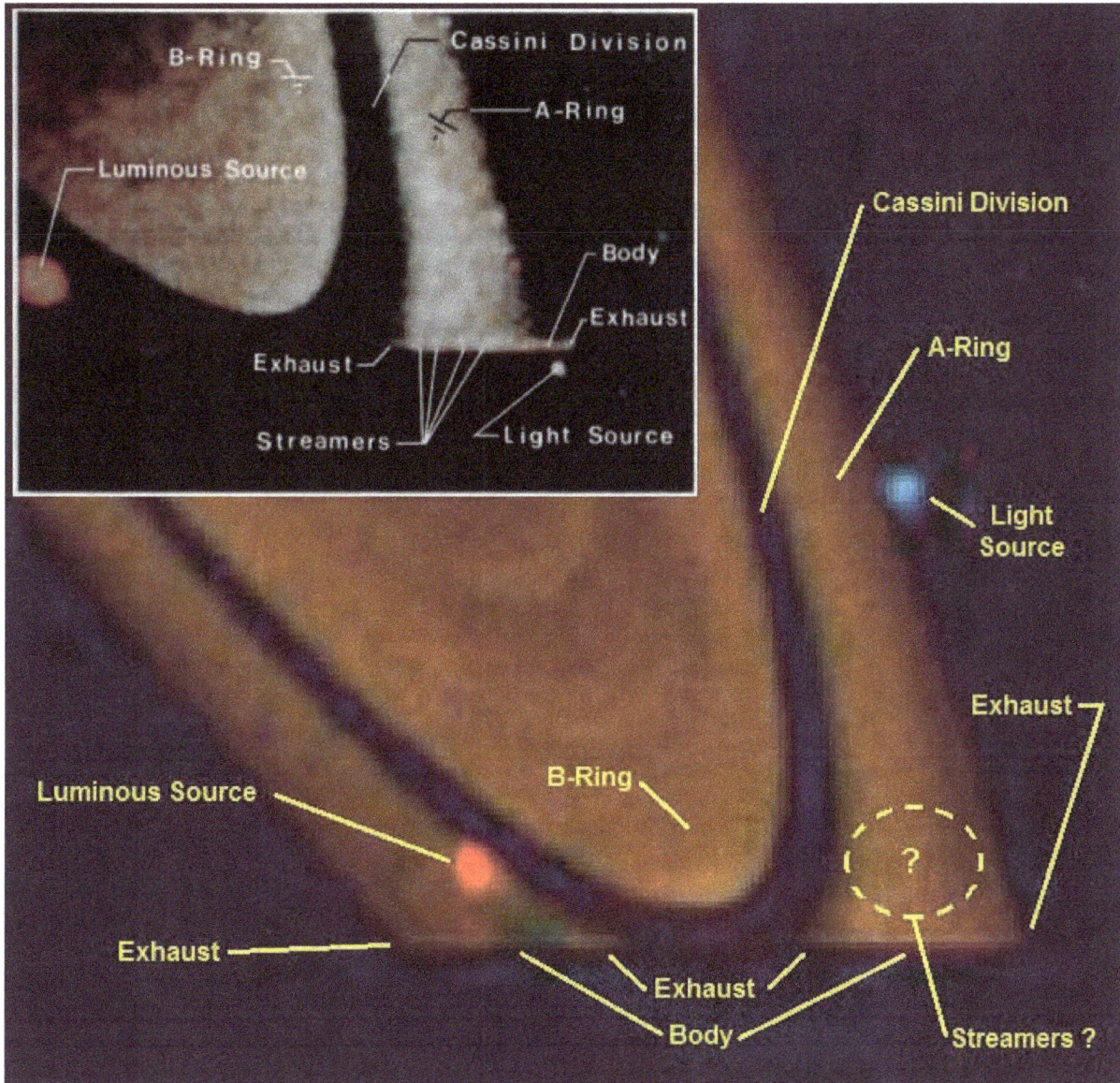

The enlarged image and the insert from Bergrun's book "Ringmakers of Saturn" are clearly not the same photographs, but used for comparison purposes. Each photo however, does show unexplainable anomalies not batural to the Saturnian system

http://www.abovetopsecret.com/forum/thread587701/pg7 and
http://www.dailygalaxy.com/my_weblog/2013/05/saturns-naturally-occurring-radio-signal-new-insights.html

In figure 4, it is interesting to note that a fireball appears on the left-hand side of the ring system (this same thing also was observed and pointed out in Ringmakers of Saturn). However, no fireball appears on the right-hand side of the ring system. In Ringmakers of Saturn, this fireball was scaled to be larger than Earth's moon, indicating the presence of a huge amount of energy.

The orange color of the fireball versus the absence of a fiery color on the right-hand side is an indicator that the left-hand side is at a higher temperature than the right-hand side.

*Ringmakers of Saturn made out the case that the efflux from **EMVs** is electrical in character, indicating that electrical potentials (voltages) are present. The temperature difference between the two sides of the ring-system in figure 4 can be translated, then, to mean that a potential difference exists between the left- and right-hand sides of the ring system at the discontinuity and that, inferentially, the flow of the efflux portrayed by the blue arc is from left to right.*

Figure 2: Saturn Cropped from NASA P23870 (As published)

Voyager 1

Atypical ring

Voyager 1 (from NASA P23870)

Interaction at ring tip demonstrates that no imaging malfunction exists, and that the rings proper are not cut off by image framing.

Figure 4: Figure 2 with brightness applied selectively to ring system only.

Saturn's cropped A-ring is enlarged (right) where a blue arc appears connecting the ring. A discontinuity of the ring system is apparent indicating a flow between the two sides

http://www.alienhub.com/threads/gigantic-motherships-and-other-bizzare-anomalies-near-saturn.21727/page-4

In conclusion, that two samples of the same event have been recorded in itself says that the event does not owe to chance, but rather is real. That activity appears which shows a completion of the ring system compounds the physical reality of the event.

Bergrun points to the fact that this trail of material in the form of streamers is recognizable as the A-ring due to the absence of the **Enke Division** which is being generated by efflux emanating from nearly the entire length of the cylindrical body. "These streamers pass over toward the right, proceed above (and below) body and contribute to the A-ring trail." Bergrun also suggests that there is "exhaust flames from both ends of the body" with a somewhat bulbous appearance of the streamers as" they pass over the the body" indicating a possible "circular cross –section for the body". There is also a secondary light source below and to the right of the cylinder that appears to have a diameter greater than the cylinder which is "attached to the body with interconnecting emmissions turning to an arange-red arc along the top edge". Ringmakers of Saturn© Norman R. Bergrun; 1986 by The Pentland Press Ltd Kippielaw by Haddington East Lothian EH41 4PY Scotland; Library of Congress Catalogue Card Number 86-81530; ISBN 0 946270 33 3

The net effect of this emission is to move the craft in a Newtonian cause and effect direction or action-reaction dynamic to complete the A-ring. This motion indicates an intelligent design with intent and hence justifies calling the slender body a vehicle or spacecraft. The size of this spacecraft is based upon a fineness ratio derived from the apparent length to thickness ratio which is 13 to 1; this comes from scaling with Saturn's diameter and the length of the A-ring inner edge to the inner edge of the **Enke Division.** Thus, **Bergrun** determines the vehicle size as 36,200 km or 22,500 mi in length and a width of 2785 km or 1730 mi. truly, this is an incredible size however; there is not one, but two such spacecraft with the second craft 5600 km or 3480 mi. in length with a similar width to the first craft! This second craft is located to the left directly opposite to the first craft between the outer A-ring and the inner edge of the Enke Division.

In the next photo image beneath the light source can be seen a wire-like arm which curves upward into the foreground toward the left. At about 1/3 of its length from the bottom, the arm has a bulge in it. This bulge appears to be a doughnut-shaped formation, or a toroid through which the arm passes. Presence of a toroid indicates that the arm is acting as a conductor carrying electricity. Such an indication is given because physically a circular conductor of concentric circles (i.e., circles with a common center). Magentizable matter caught in such a field will align itself concentrically with the conductor and collectively assume a toroidal shape. Without this arm, maintenance of the luminous source probably would be impossible.

With the scaling ratio as previously determined of 13 to 1, as the two vehicles approach each other in what would appear to be a collision (which it is not), the A-ring is then completely formed. Based upon the apparent fineness ratio of the two vehicles thickness, the A-ring thickness would aslo be the same which is 2785 km or1730 mi. which would become thinner at the ring's edges, the further from the vehicle.

Historically Bergrun says that the formation of Saturn's A-ring confirms the earlier observations by **Herschel, Knight** and **Ainslie** of a variable light source or a luminous source as bright as a star. These obsevations fit well with the current photgraphic data and micro-photometric analysis including the chordal path of the Knight-Ainslie moving source is the same chordal element defined by the location and orientation of the two vehicles. Ringmakers of Saturn© Norman R. Bergrun; 1986 by The Pentland Press Ltd Kippielaw by Haddington East Lothian EH41 4PY Scotland; Library of Congress Catalogue Card Number 86-81530; ISBN 0 946270 33 3

The variance in width of the Cassini and Enke Divisions and gaps as well as the A-ring is neither accidental nor repeatable due in part the orbit radius of the vehicles, variability of the efflux emissions along the length of the vehicles as well as the trialing flux can alter the radial location of the inner edge of the A-ring. All these possibilities can account for the differences in astronomical measurements from observers over the years.

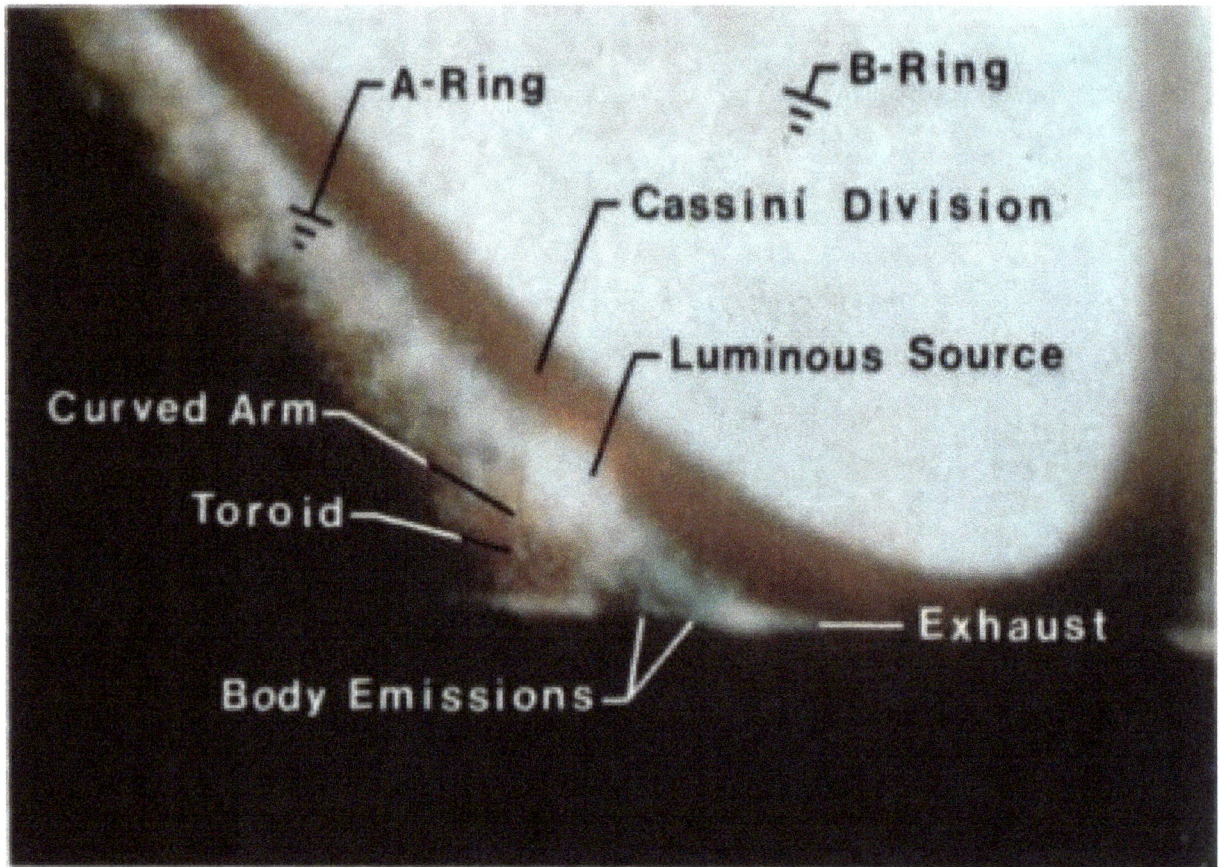

A slender vehicle form an A-ring trial which includes a luminous source also, note the thin curved arm running vertically upward from the vehicle forming a toroid cloud

Electro Magnetic Vehicles (EMVs) and Saturnian UFOs

Bergrun points out that from examination of numerous photographs of the rings of Saturn that there are not only vehicles producing the A-ring but, another vehicle with similar dimensions but smaller also located at the outer edge of the **Enke Gap** creating the **Enke Division**. The process of the creation of the Enke Division was also determine to be the same as the development of the A-ring yet, there were minor differences in spacecraft operation that created the Enke Division, particularly in the emission patterns. These differences would account for the Enke Gap being "located almost anywhere or not at all" thus, "the difficulty of early observers in pin-pointing a single radial location for the Enke Division". There are other such vehicles which Bergrun has taken to call **Electro Magnetic Vehicles (EMVs)** because of the electromagnetic properties producing the pinched plasma formations known as the rings of Saturn which are located in every ring from the A-ring to the F-ring! These EMVs are ever present within the ring system that they created, continuously recreating or maintaining the ring system! **Ringmakers of Saturn© Norman R. Bergrun; 1986 by The Pentland Press Ltd Kippielaw by Haddington East Lothian EH41 4PY Scotland; Library of Congress Catalogue Card Number 86-81530; ISBN 0 946270 33 3**

**Norman Bergrun demonstrates a mock-up model of the Electro Magnetic
Vehicle (EMV) which creates and maintains the rings around Saturn**
https://cosmicrevelationsblog.wordpress.com/2016/05/03/norman-bergrun-and-the-ringmakers-of-saturn/

It is conceivable that any of the **Saturnian moons, Mimas, Encedadus, Tethys and Dione** as well as **Rhea** may actually have one or more of these powerful slender EMVs shadowing them. In the photo above and the enlargement below, **Voyager 1** photographed an orange luminous globe within the **Cassini Division**, larger than the Earth's Moon that is not a Saturnian moon or a shepherding satellite, nor an asteroid.

Cassini was launched in October 1997 with the **European Space Agency's Huygens probe**, its six instruments were designed to study Titan, Saturn's largest moon. It landed on Titan's surface on Jan. 14, 2005, and returned spectacular results. Cassini completed its initial four-year mission to explore the Saturn System in June 2008 and the first extended mission, called the Cassini Equinox Mission, in September 2010.

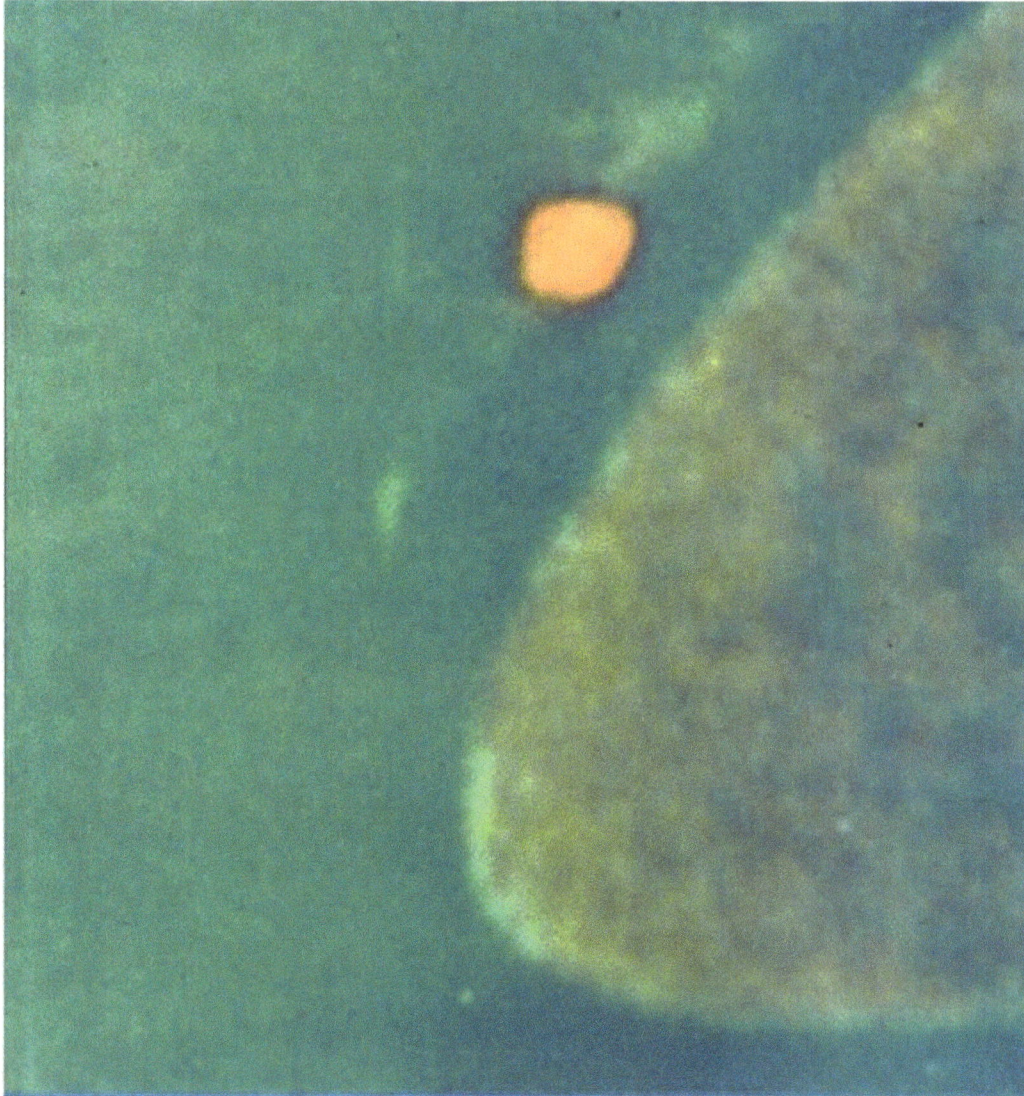

**This glowing unidentified flying object orbiting around in the rings
of Saturn is estimated to be larger than our Moon!**

Now, intriguingly NASA has decided to extend its **Cassini-Huygens mission** seeking to make exciting new discoveries in a second extended mission called the **Cassini Solstice Mission**.

The mission's extension, which goes through September 2017, is named for the Saturnian summer solstice occurring in May 2017. The northern summer solstice marks the beginning of summer in the northern hemisphere and winter in the southern hemisphere. Since Cassini arrived at Saturn just after the planet's northern winter solstice, the extension will allow for the first study of a complete seasonal period.

Among the most important targets of the mission are the moons Titan and Enceladus, two of Saturn's most mysterious moons as well as some of Saturn's other icy moons. Towards the end

of the mission, **Cassini** will make closer studies of the planet and ***"its rings."***
http://saturn.jpl.nasa.gov/mission/introduction/

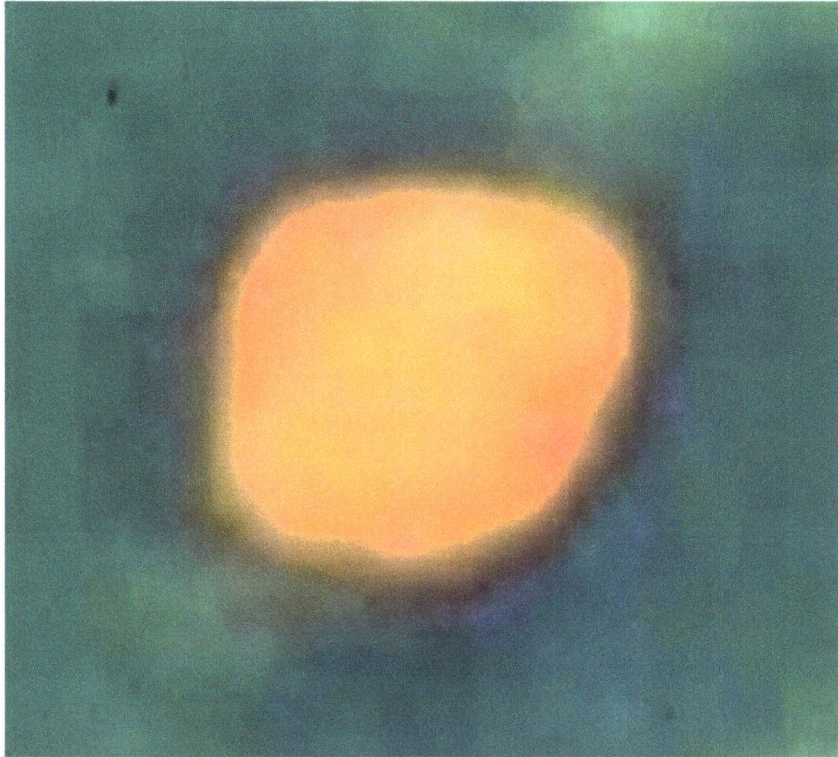

**Close up of planet-size UFO near Saturn's rings taken
during the Voyager 1 fly-by in November 1980**

https://weveneverbeenalone.tumblr.com/post/105968886574/the-ringmakers-of-saturn-pdf-norman-r-bergrun

It stretches the imagination to believe that giant spaceships are creating the rings of Saturn which
obviously flies in complete contradiction to the traditional astronomical views as taught in
universities and as studied by astronomers which conclude that the rings are natural debris bodies
of large ice chunks and rock in various orbits around Saturn that have been there since the
planet's formation in the solar System. Yet, the photographic evidence speaks volumes that
gigantic craft exist around Saturn as evidence through micro-photographic analysis by Bergrun.
These immense space vehicles thousands of miles in length and width are according Bergrun,
what he calls **Electro Magnetic Vehicles (EMVs)** creating the many rings of Saturn and
continually maintaining them.

As with all such evidence, it has spawned its detractors, naysayers, debunkers and skeptics who
say that with recent advancement in photometric equipment, cameras and photo lab analysis and
interpretation that past evidence compared with today's technology does not support such
outrageous conclusions as giant spaceships creating the rings of Saturn. The science of
astronomy as far as the detractors, skeptics and debunkers are concern is safe as we know it from
such fallacious nonsense. Is it really?

Just, when we start to believe that the scientists and the skeptics may be right, more evidence surfaces and the pendulum swings back to supporting the UFO/ETI theorists as evidenced by the photos below. In the immediate photos images below, the rings of Saturn in black and white and in colour, we see strange, very long black streaks crossing over the rings of Saturn. These images are reminiscent of the large bright object photographed by Apollo astronaut Neil Armstrong on his way to the Moon and it's similar to the Russian Grunt space probe, **Phobos 2** that was sent to Mars to photograph the Martian moons **Phobos** and **Deimos**, just before it became disabled and forever lost.

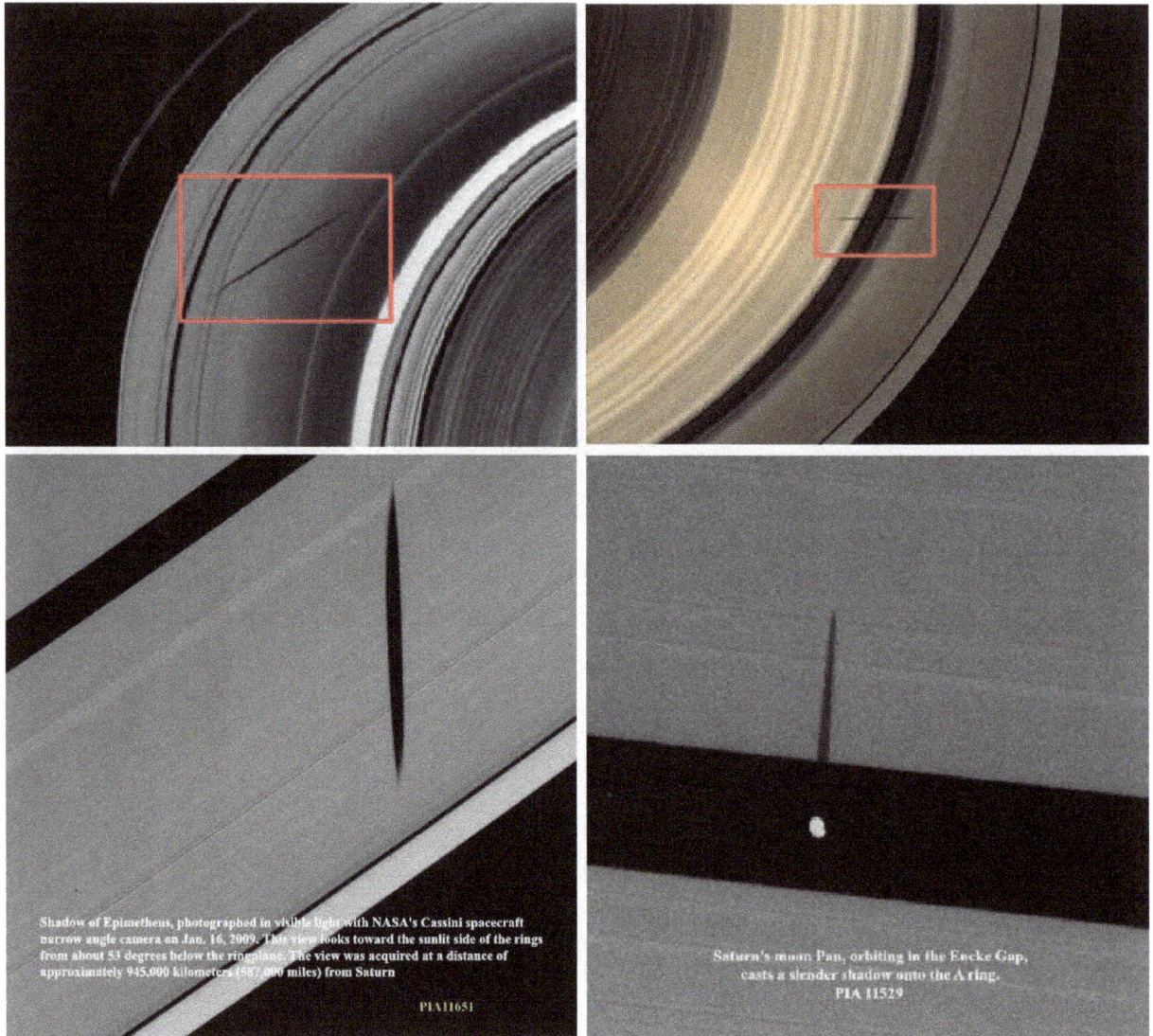

Shadow of Epimetheus, photographed in visible light with NASA's Cassini spacecraft narrow angle camera on Jan. 16, 2009. This view looks toward the sunlit side of the rings from about 53 degrees below the ringplane. The view was acquired at a distance of approximately 945,000 kilometers (587,000 miles) from Saturn

PIA11651

Saturn's moon Pan, orbiting in the Encke Gap, casts a slender shadow onto the A ring.
PIA 11529

These black streak objects seen above the rings of Saturn are thousands of miles in length. In the bottom left image, NASA states this tapered and pointed end shadow is cast by the Saturn moon, Epimetheus yet, where is this moon? In the bottom right is the clearly visible moon Pan in the Enke Gap casting a shadow over the rings! Now compare the remarkable similarity to those objects in the images below taken by Apollo 11 astronauts and by the Russian Mars probe Phobos 2
(Google Images)

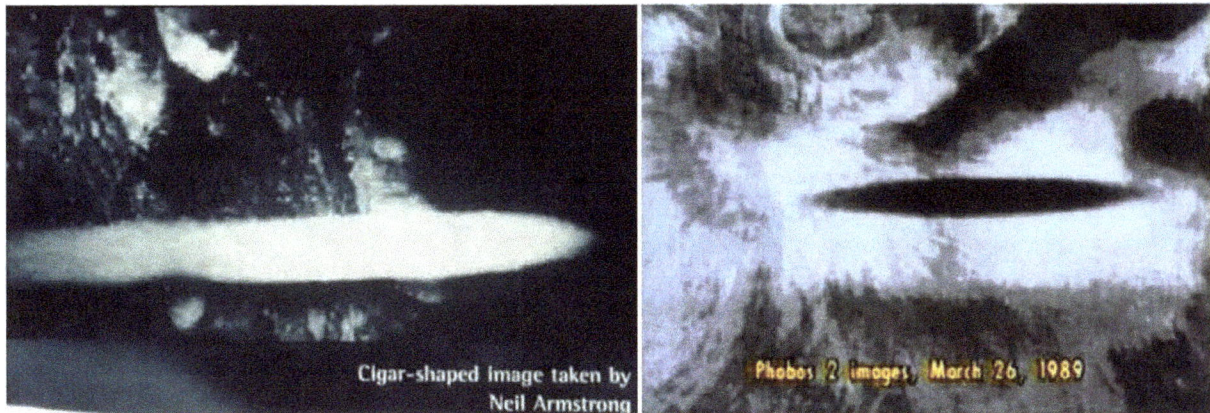

A comparison of astronaut Neil Armstrong 's cigar-shaped UFO over the Moon (left) and the Russian Phobos 2 photo image of a black streak travelling up from the surface of Mars before the space probe was disabled and lost

http://heavy.com/news/2015/11/astronaut-ufo-sightings-videos-pictures-photos-news-encounters-interview-testimonies-edgar-mitchell-gordon-cooper-apollo/4/ and http://www.ufocasebook.com/phobos2.html

In the image above, we see a round object most likely the moon, **Epimetheus** in **Cassini Division** between the rings casting a shadow of considerable length across the A-ring indicating that its position is just above the equatorial plane of the rings. Is this a moon or asteroid satellite of Saturn as most astronomers would surmise or anomalous object of artificial origin? This is odd indeed but, even stranger are the shadows cast by immense structures that appear to sit on the edge of the rings or that seem to be a part of Saturn's ring system. They are near moon size but, very narrow in proportion to their height, much like a pyramid with a peak but, flatten on both sides as if purposely constructed to sit on the very edge of the rings to allow the best view of the Saturnian ring system and the **Cassini Gap**.

Are these photos more evidence that gigantic spacecraft are orbiting saturn and creating the ring system as posited by **Norman Bergrun**? It would certainly appear to validate his hypothesis that an extra solar intelligence has entered our system or possibly a former intelligence from this Solar System has returned to pick up where it had left off. They definitely are busy constructing and maintaining the ring system around Saturn as can be seen from the raw photographic data from NASA and ESA.

NASA has no official statements as to what these apparent anomalies may be other than to say that they are a part of the Saturnian system meaning, natural orbiting bodies or no comment s at all. Regardless, these anomalies appear on the internet on a regular basis gleaned by UFO enthusiasts and researchers; some are mistaken identifications and pure speculations from raw NASA photographic data. Individually, these photo images portend something anomalous yet, unsubstantiated but, collectively as compiled with all the information presented in this book, the evidence becomes overwhelming that not only is the Earth being visited by Extraterrestrial Intelligences but, they have an affinity to our solar System and a real interest in certain planets like our Earth, Mars and Saturn.

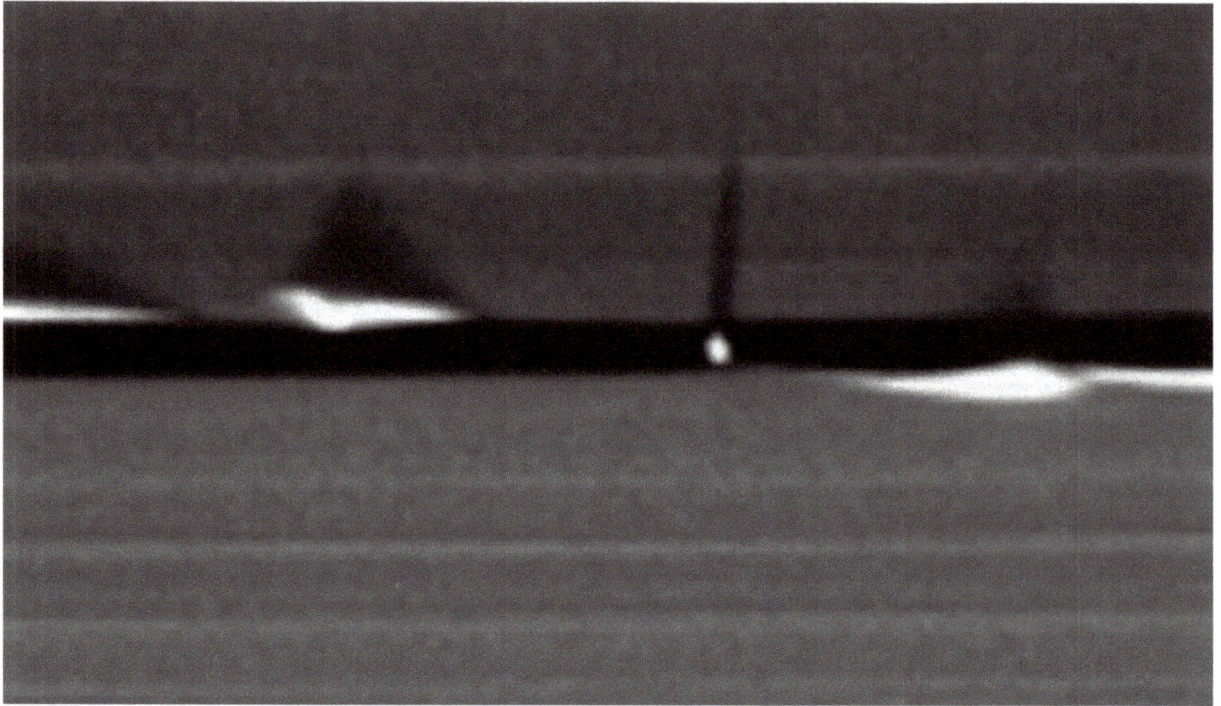

What is the round object in the Cassini Division and the mountainous like structures on the A-ring and B-ring that cast such immense shadows across the ring plane system?

http://ciclops.org/view_event/110/Towering_Edge_Waves_Pop_Into_View

As pointed out earlier in this section, NASA was going to terminate the Cassini mission but has decided instead to increase the mission duration of the **Cassini-Huygens space probe**. It seems that NASA's on-going curiosity for Saturn hasn't been satiated, there is still much that NASA is learning and still needs to know about this planetary system. Could it be that NASA continued **Cassini mission** has more to do with the work of **Bergrun's** photographic analysis of the ring of Saturn being constructed by gigantic spacecraft and that some of its moons are artificial in nature?]

Another interesting set of photos are the images below taken by the Hubble Space Telescope using the Infrared setting on its camera to photograph the ring system of Saturn which shows initially, two objects A and B that are cylindrical in shape in a *"clockwise"* orbit around Saturn within the A-ring. There is also a long streak ithe first image #1 (counting from left to right and down as they were photographed by the **Hubble telescope** chronologically) which does not appear in the other images of this set. Is this a meteorite or one of those long cigar-shaped spacecraft known as **Electro Magnetic Vehicles (EMVs)**?

In images #7 and #8 can be seen a cigar-shaped object travelling outside of the A-ring in a *"counter clockwise"* orbit. The other globular objects in the photo set are not anomalous objects but some of the moons orbiting around Saturn.

https://www.youtube.com/watch?time_continue=257&v=ibT4SFNcGcY

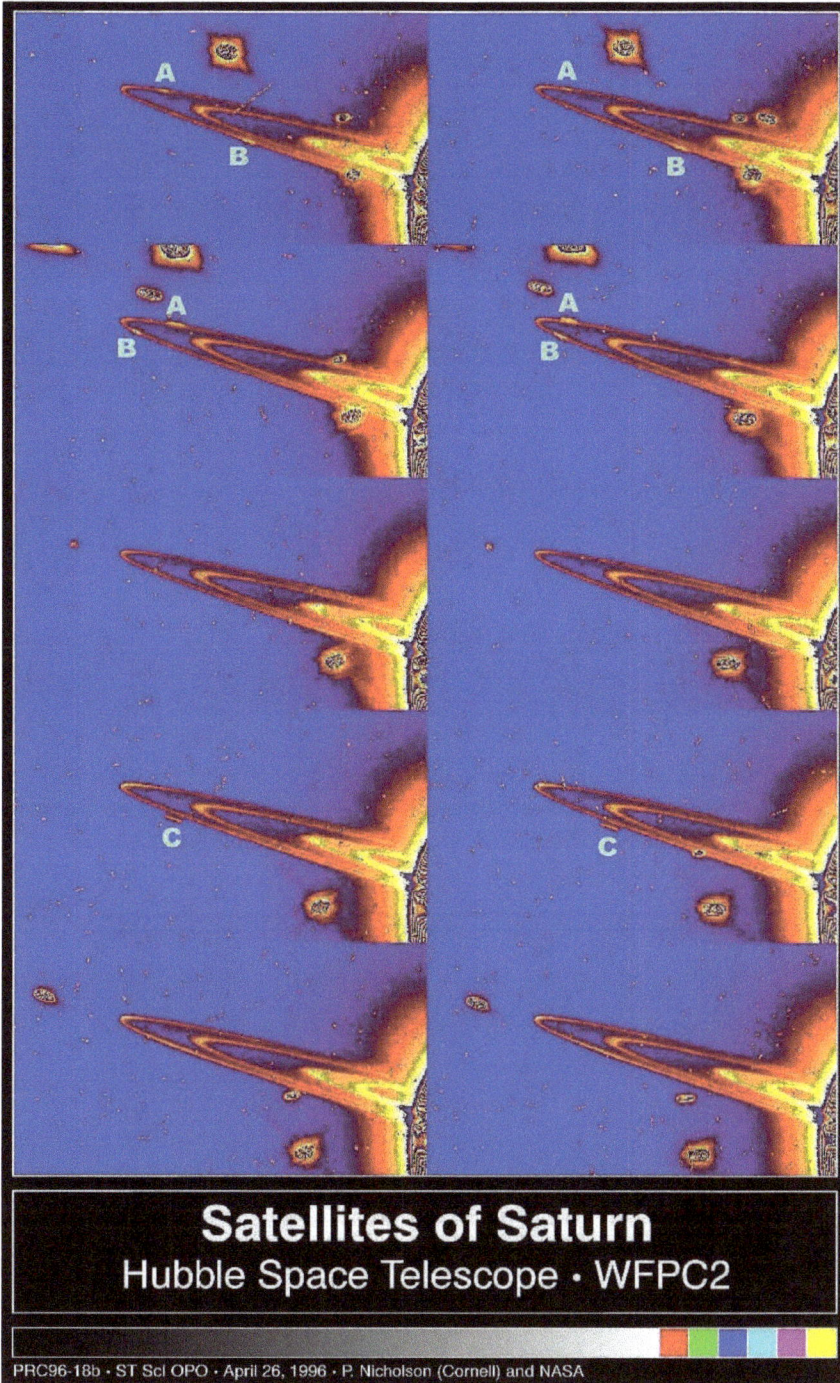

Satellites of Saturn
Hubble Space Telescope · WFPC2

PRC96-18b · ST ScI OPO · April 26, 1996 · P. Nicholson (Cornell) and NASA

In these Infrared images of Saturn's rings there are cylindrical objects
A, B, C (possibly EMVs) orbiting around in the A-ring.
Note object C is outside of Saturn's A-ring

http://www.thelivingmoon.com/46_mike_singh/03files/Ships_Saturn.html

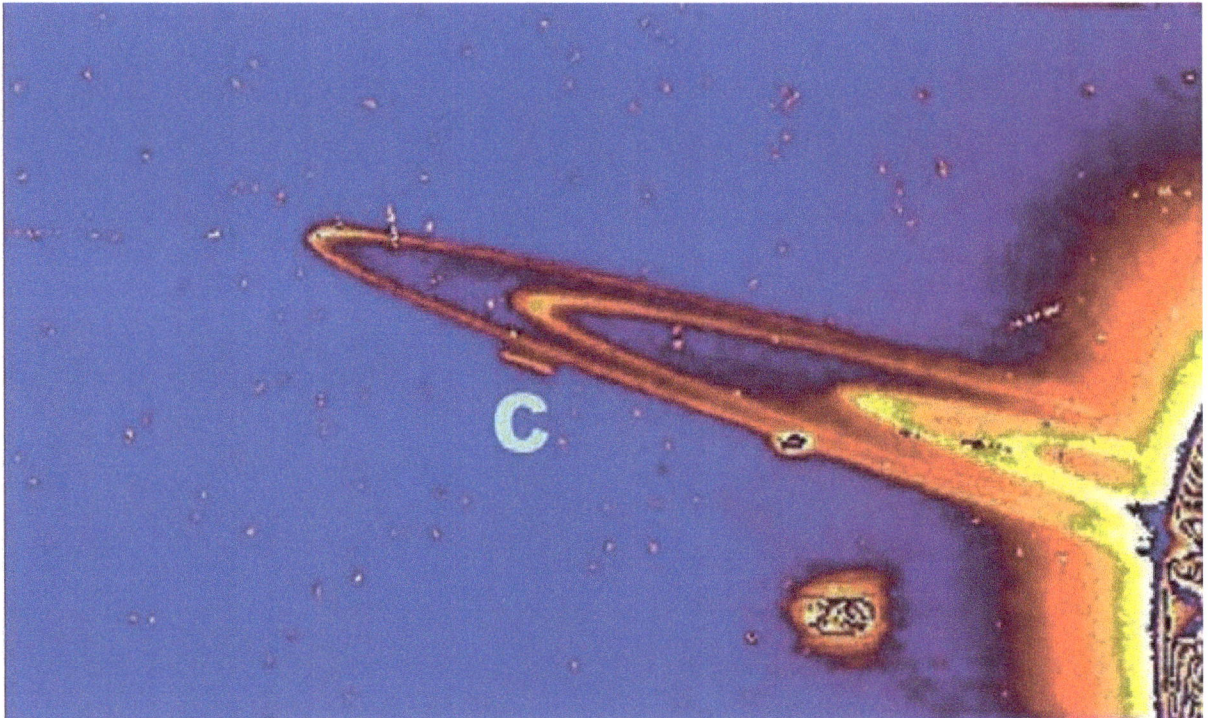

**Close up of the cylindrical object C orbiting around outside of the A-ring of Saturn.
The large orange blobs near or well outside of the ring system are Saturnian moons**

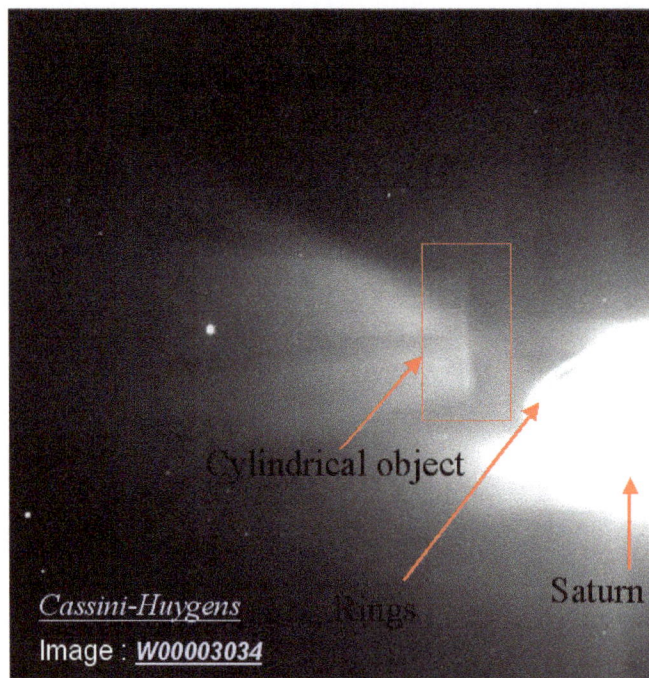

Cassini-Huygens spacecraft photographs another cylindrical object in the rings of Saturn

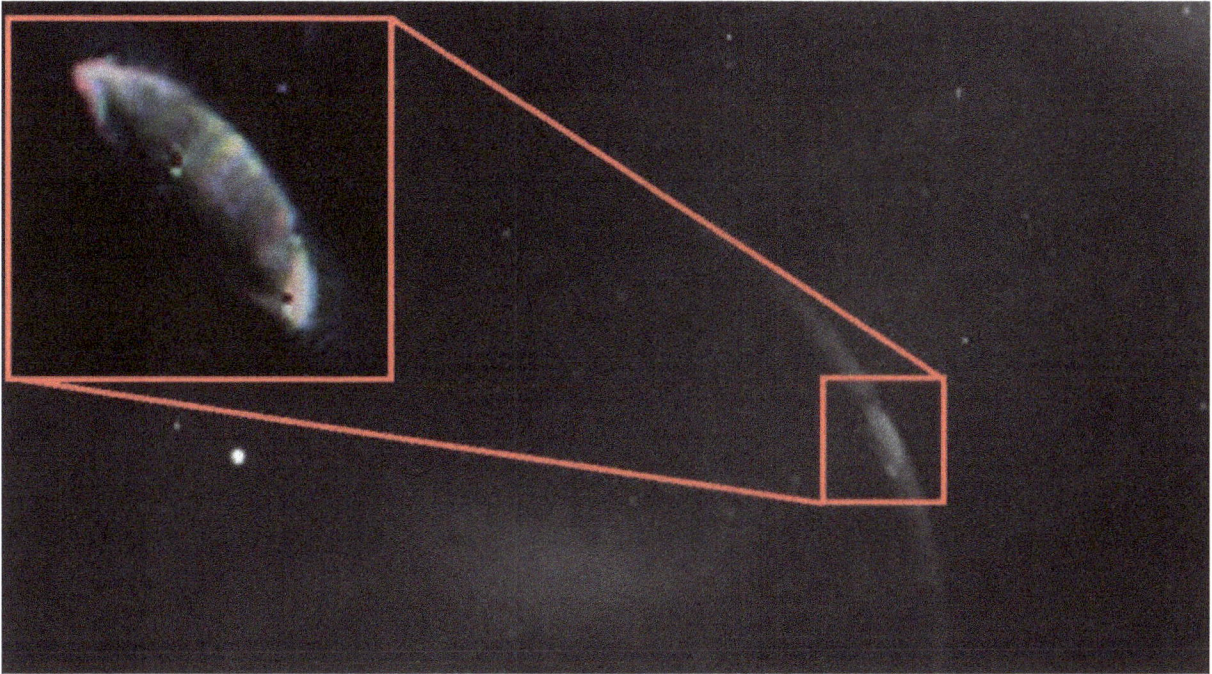

This tapered cylindrical object is not a Saturnian asteroid satellite, it shows artificiality with symmetrical ring structures along its length and luminosity on both ends

http://www.alienhub.com/threads/gigantic-motherships-and-other-bizzare-anomalies-near-saturn.21727/ and
http://nexusilluminati.blogspot.ca/2012/04/gigantic-motherships-and-other-bizzare.html

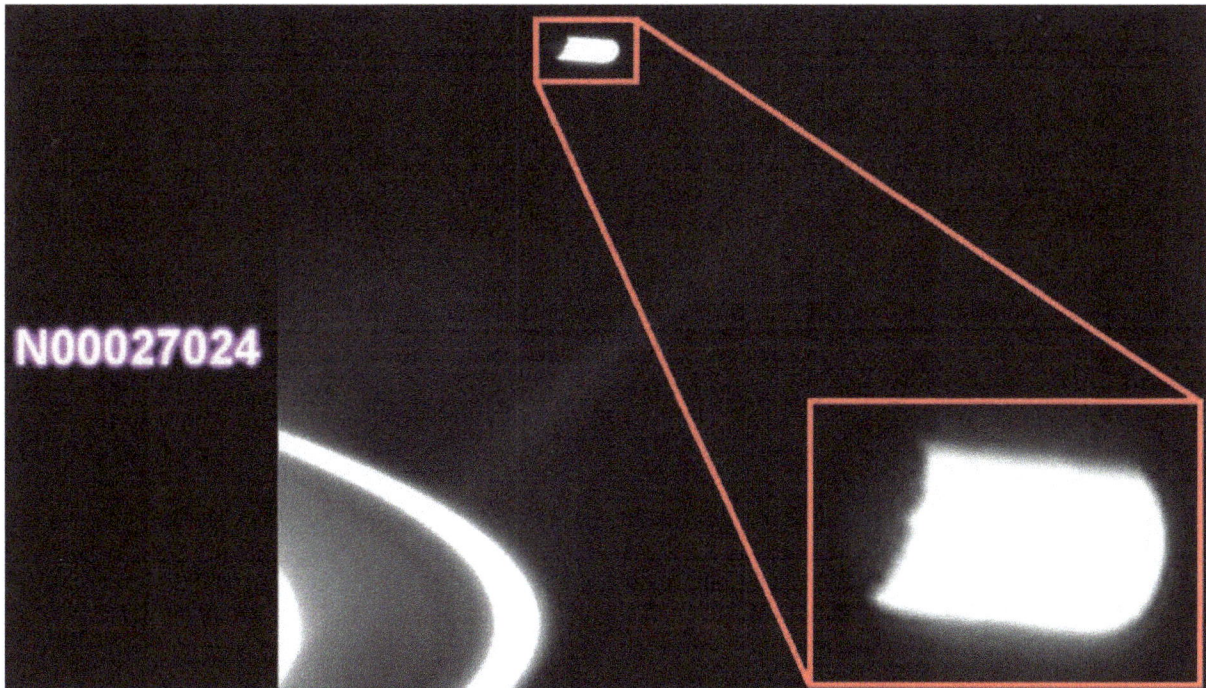

This immense object is reminiscent of the Canadian Avro Y2 Flying Saucer and the Roswell UFO photographed just before the famous Roswell saucer crash in 1947

http://ununiverso.altervista.org/blog/bob-dean-enormi-astronavi-aliene-sono-state-fotografate-nei-pressi-di-saturno/

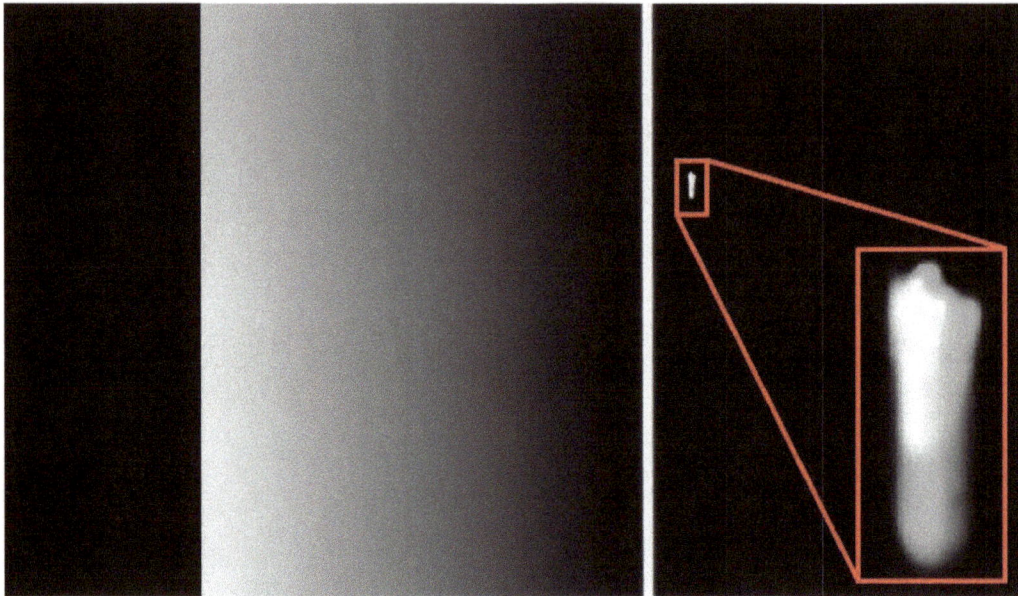

Another object beside Saturn's rings, possibly the same object in the previous image but, seen on an oblique angle. Note the rounded "nub-like" tail similar to the craft above

http://www.whatsupinthesky.com/index.php/photos/rings/3899-n00047648

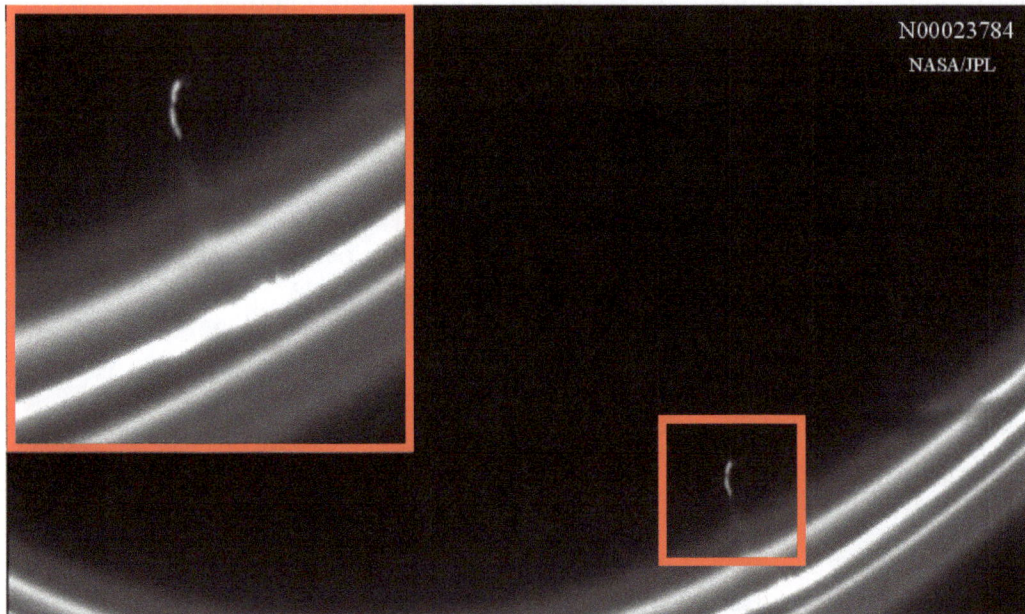

N00023784

NASA/JPL

Cassini took this photo image of a curved cylindrical object above the Saturnian rings, NASA believes this to be the Saturn moon Prometheus

http://www.abovetopsecret.com/forum/thread437208/pg9

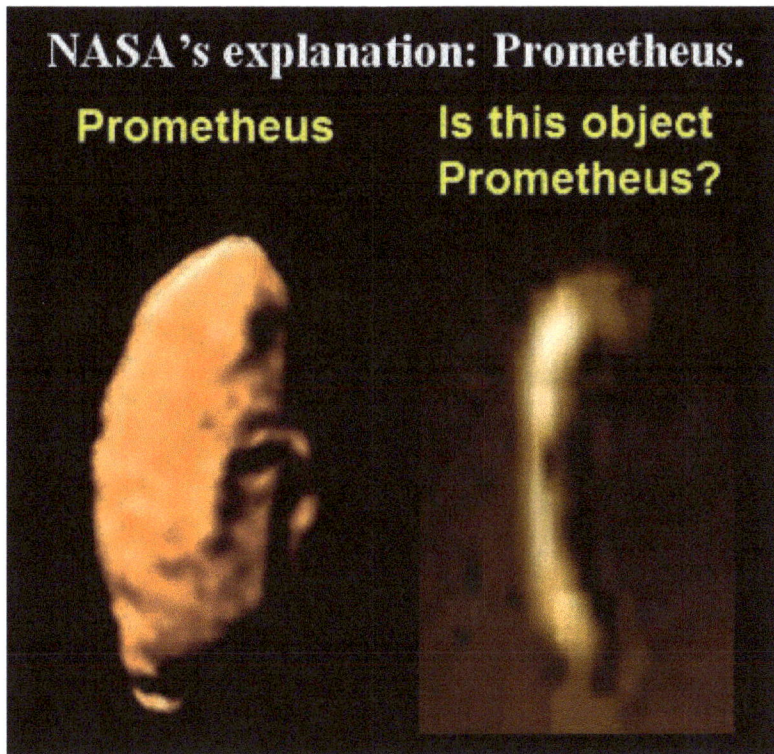

This is an enlarged colour image of Prometheus compared to the object imaged by Cassini, the objects are definitely not the same; the right one looks to be artificial in appearance

This object moving up through the ring system of Saturn displaces a cloud of ring matter

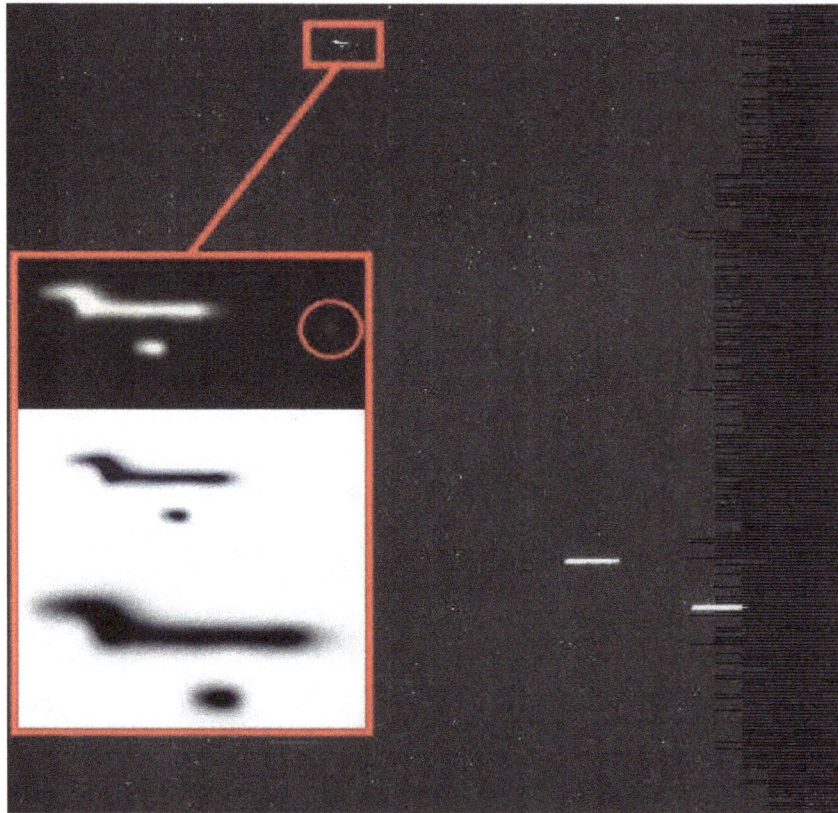

This long cylindrical object with a large tail structure appears to be paced
by a bright disc object below and possibly a fainter one ahead of it.
Note also the missing pixels on the right side of image

Compare this image of a Japanese duplicate of an American NASA photo
taken by astronauts aboard the Apollo 13 command capsule

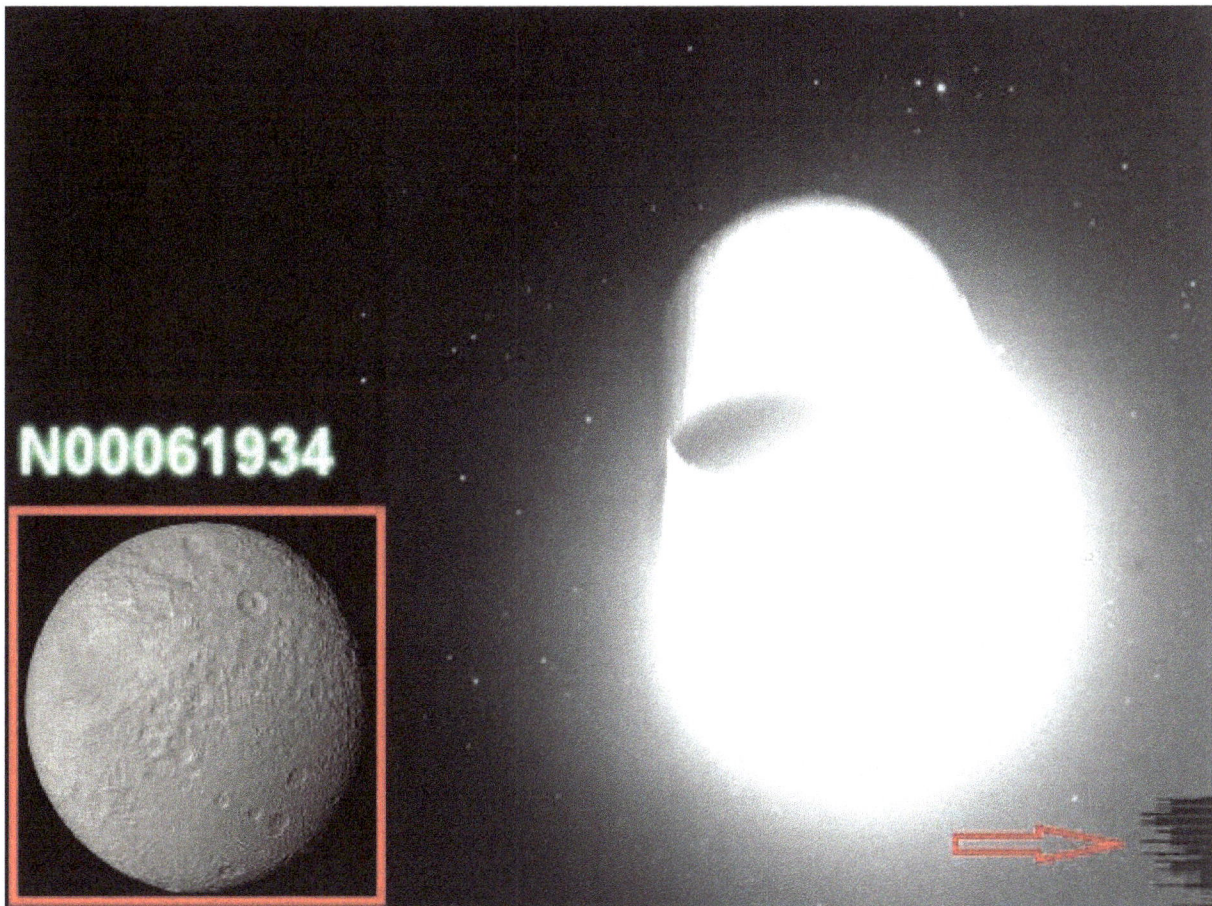

NASA says that this "one eyed light bulb"- like object is Tethys but, the inset in left corner is how Tethys really appears. Note arrow pointing to missing pixels in image

http://www.thelivingmoon.com/46_mike_singh/03files/Ships_Saturn.html

In July 25, 2004, NASA's Cassini spacecraft using its radio and plasma wave detectors while passing over Saturn's rings upon its arrival to the planet detected intense radio emissions coming from Saturn. Some people have speculated that it may be a transmission of alien communication to fellow ETs or an interstellar symphony. Again, NASA is in the dark about these sounds and states, "A most intriguing file, we do not know what to make of it..." and neither does SETI know either. It is the sort of thing that SETI would be searching for and you would expect a typical "WOW" response to be written somewhere in the margins off the discovery. Alas, no so!

The tones are emitted as radio waves which the University of Iowa research team headed by Don Gurnett reduced the frequencies by a factor of five in order to make them audible to the human ear. Gurnett and his team were "completely astonished" when they heard the musical notes. When the same radio signal is raised in pitch by twelve tones, a speech–like pattern emerges! Gurnett's explanation for the short tones is that each tone is produced due to the impact of a meteoroid on the icy chunks of Saturn's rings. http://saturn.jplnasa.gov/news/press-release-details.cfm?newswsID=589

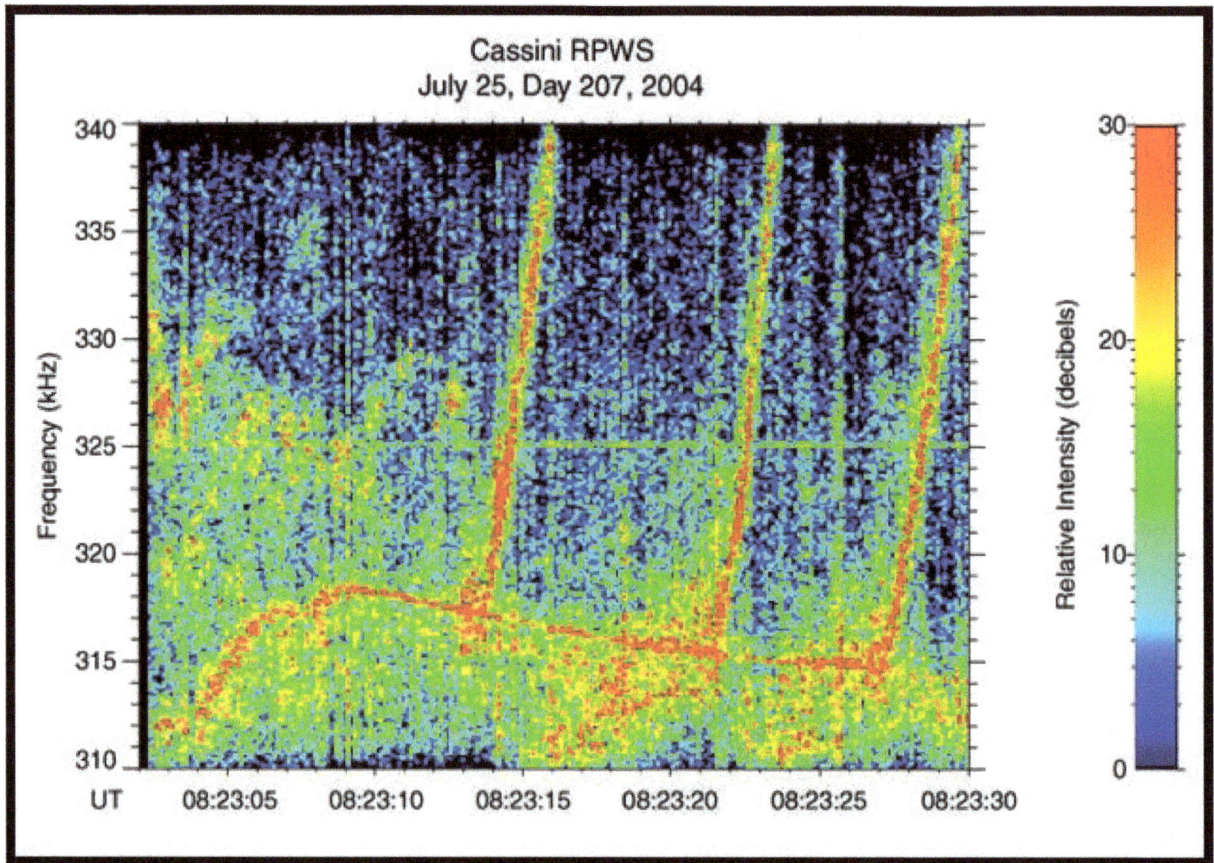

Cassini RPWS
July 25, Day 207, 2004

**The strange sound wave pattern as transmitted from Cassini spacecraft back to Earth.
Is this an interstellar alien symphony or alien speech or communication?**
http://www.youtube.com/watch?feature=player_embedded&v=pGeWBiLVn8g
http://nexusilluminati.blogspot.ca/2012/04/gigantic-motherships-and-other-bizzare.html

Recently, September 05, 2008 another discovery was made of an additional ring or partial ring (arc of shepherding satellite debris) further out from the main rings around Saturn

 NASA's Cassini spacecraft has found two new, partial rings around Saturn that each accompany a small moon, shedding light on what determines whether a partial or complete ring forms with the moon.

The partial rings, called ring arcs, extend ahead of and behind the small Saturnian moons **Anthe** and Methone in their orbits.

Both Anthe and Methone orbit Saturn in locations called resonances, where the gravity of the nearby larger moon Mimas disturbs their orbits. Mimas provides a regular gravitational tug on each moon, which causes the moons to skip forward and backward within an arc-shaped region along their orbital paths.

Scientists believe that the faint ring arcs likely consist of material knocked off the small moons by micrometeoroid impacts. The material doesn't spread all the way around Saturn to form a complete ring because the interactions of the moons with Mimas confine the material to a narrow region along the moons' orbits.

The material that orbits with Pallene, Janus and Epimetheus, however, isn't subject to the same powerful resonant forces and is free to spread out around the planet, forming a complete ring. http://www.space.com/5800-partial-rings-discovered-saturn.html

In this Cassini image of Anthe's ring arc, the moon is moving downward and to the right. Most of the visible material in the arc lies ahead of Anthe in its orbit. However, over time the moon drifts slowly back and forth with respect to the arc.
https://www.nasa.gov/mission_pages/cassini/multimedia/pia11101.html

In 7 October 2009, Saturn's ring system has just got a lot larger, with the discovery of a faint ring that stretches out millions of kilometres into space. But this new ring, which follows the orbit of one of Saturn's moons, **Phoebe,** is unlike any of the other rings that are closer to the planet: as well as being much thicker and wider, it is tilted from the plane of the other rings. It is thus known as the **Phoebe Ring**.

Saturn's rings were first described by astronomer and mathematician **Christiaan Huygens** in 1655. Since then, astronomers have discovered more details about the number of rings in the system and their composition, aided recently by NASA's Cassini–Huygens mission. It was thought that the farthest ring from Saturn — until now the largest known ring in the Solar System — was the **E ring,** which stretches from a distance of 3 R_s (where R_s is the radius of Saturn, equal to 60,330 kilometres) to 8 R_s, and is fed by active geysers on Saturn's icy moon Enceladus.

However, this new ring dwarfs all the others, extending from approximately 128 R_s to 207 R_s with a vertical thickness of 40 R_s.

The Phoebe ring is the largest ring in the Solar system extending millions of kilometers out into space from Saturn and dwarfs all other rings in the Saturnian system yet, is nearly invisible unless photographed in the infrared spectrum.
https://blogdoastronomo.wordpress.com/2009/10/07/astronomos-descobrem-anel-gigante-em-torno-de-saturno/

The extremely sparse nature of the Phoebe ring means that it reflects very little light and is practically invisible, which is why it has previously escaped detection.

Many astronomers believe that the origins of many of Saturn's inner rings are from its satellites. Of course, these are very optically-thin rings and nothing like the spectacular rings we normally associate with Saturn."

The new 'Phoebe Ring' describes the outer boundaries of Phoebe's orbit, and is at an angle of 27°

326

with respect to the other rings. Phoebe travels in the opposite direction to most of the planet's other moons. The ring was detected by thermal emission using the **Spitzer Space Telescope's Multiband Imaging Photometer.**

And in a bonus discovery, the team thinks they may also have solved an astronomical mystery. One hemisphere of Iapetus, the next moon in from Phoebe, is much darker in colour than the other half — an observation that astronomers have never fully explained. The team now believes that the Phoebe Ring could be the missing source of the darker material, as well as being responsible for reddish deposits on Hyperion, another Saturnian moon. However, the team has not yet managed to accurately determine the ring's structure or composition to confirm this hypothesis. **http://www.nature.com/news/2009/091007/full/news.2009.979.html**

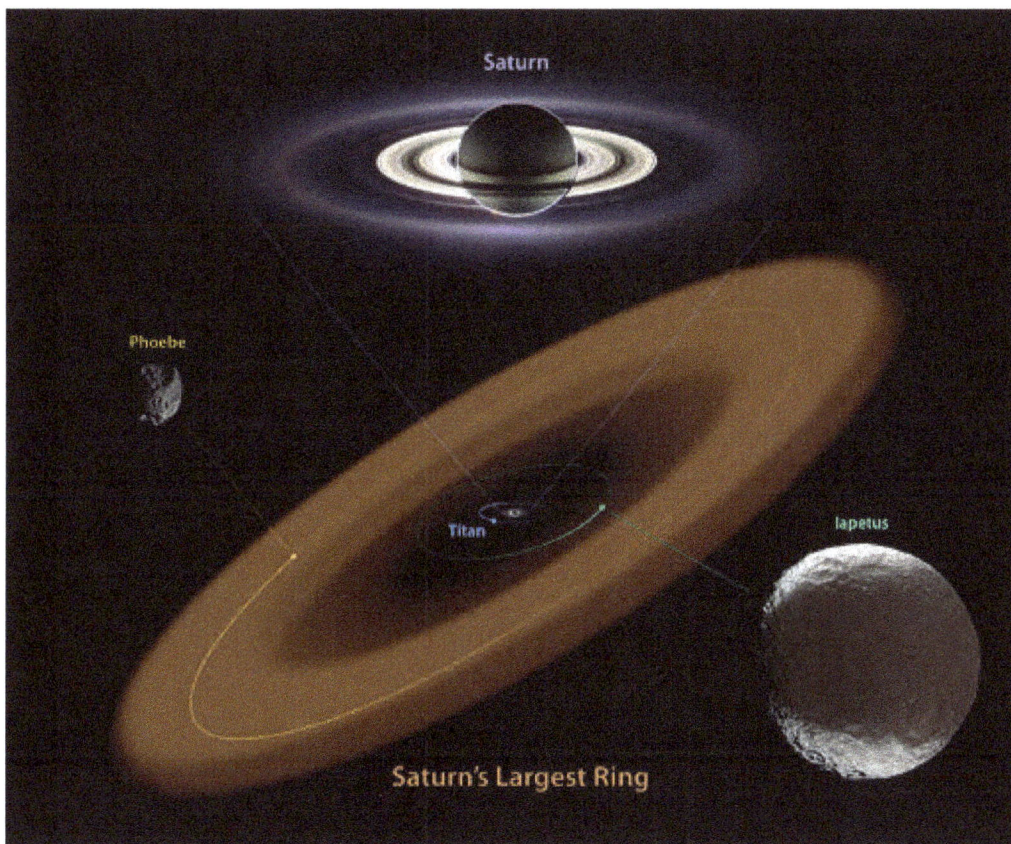

This diagram illustrates the extent of the largest ring around Saturn, discovered by NASA's Spitzer Space Telescope. The ring is huge, and far from the gas planet and the rest of its majestic rings.
https://phys.org/news/2009-10-largest-saturn.html

This explanation while plausible is simply not tenable, even by NASA's self-admission that it's just a hypothesis and it appears to be one more questionable theory of disinformation from NASA that is forcibly foisted upon the public in order to get them to accept their version of reality. **Tom Van Flandern's Exploded Planet Hypothesis (EPH)** is a much better fitting explanation for the high albedos and contrasts on **Hyperion** and **Iapetus** as previously discussed in this section. This explanation forces NASA and astronomers to seriously consider that there is an *"interstellar or intergalactic billiards game" hypothesis (IBGH)* going on with rogue planetary bodies and stable planetary bodies which for the most part is unpredictable and uncontrollable **OR** that planetary rings like Saturn's ring system and even some moons and planets may be the manufactured products of Extraterrestrial Intelligences as posited by **Bergrun**!

Contrary to the position held by astronomers and NASA scientists that the rings are formed by resonance of Saturn's many orbiting satellites and by the blowing of stellar winds scattering dust and debris from off the surfaces of these satellites resulting in the formation of rings and arcs, there stands also, in stark contrast Bergrun's explanation for the proposition that the Saturnian ring system is in fact constructed by **gigantic Extraterrestrial spacecraft**.

NASA in its search for Extraterrestrial technology must have given serious consideration to Bergrun's hypothesis and it may have been one of the main reasons to send a space probe like Cassini specifically in orbit around Saturn and to extend its mission for a few more years to monitor the ET activity around this planet.

The Ringworld of Saturn

Along with this hypothesis of Saturnian ring artificiality created by Extraterrestrial Intelligence must be added the possibility for the existence of a potential "**ringworld system**" around Saturn as proposed by no less than, **Richard C. Hoagland.**

On September 12th and 13th , 2009 at the Static Gallery, China Town in Liverpool, England at the Beyond Knowledge Conference, **Richard Hoagland** gave a lecture to the British audience on the **Secret Space Programs**. http://www.youtube.com/watch?v=NW9r17nEpSc

Hoagland's presentation picks up from where Bergrun left off exploring further the possibilities that part of Saturn's **B-ring** contains artificial structures built into it. His presentation covers other aspects of a **secret black space program** that began with captured WW2 Nazi scientists through **Operation Paperclip** and its connection to the planet Saturn. This secret black space program has been the real space program in the US which has involved other countries like Russia, Britain and Europe, China, Japan and India and not the NASA space program as we have all come to believe, over the last fifty plus years.

This super secret black space program is supported by research work by **Joseph Farrell's** data on the SS Brotherhood of The Bell, the historical research of **Steven Henry** on the secret advanced weaponry and saucer programs of Nazi Germany and the discovery by the British computer hacker, **Gary McKinnon** who found **Pentagon** and **NASA** documents supporting evidence of a massive development and deployment of a secret space program.

However, of particular interest to this subsection of this book concerning Saturn is the hypothesis that Saturn is a gigantic **Torsion Field** device created by a race of giant beings commonly referred to in biblical accounts as the **Nephellum** or in **Sumerian** records as the **Annunaki**. They utilize Saturn whether, as a natural planetary body or as a manufactured planet to generate immense power that scarcely no human can imagine or truly comprehend its ultimate purpose.

Supporting this hypothesis is the raw photographic imagery taken from **Voyager 1 and 2** as well as from NASA's **Cassini space probe**.

Hoagland builds his case with photo images of the rings of Saturn focussing on a small section of the B-ring on the edge of the Cassini Division which reveals structural anomalies that are clearly defined in four lateral bands which cast shadows upon the other rings. These structures rise and fall in size and when compared to the rest of the ring system does not appear to be a natural part of that Saturnian ring system. The extent of these artificial structures cannot be determined at this time as to whether they encircle the entire planet or are merely a small portion within the ring system that is now in a state of ruins or being built. (See photos below).

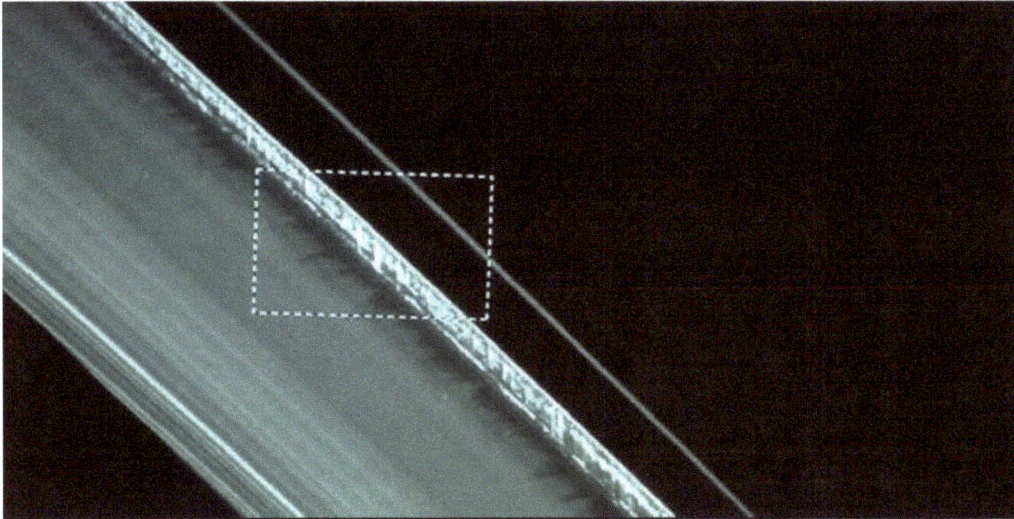

A small section of the B-ring in which artificial structures are located.
http://www.youtube.com/watch?v=NW9r17nEpSc

A close up view of the artificial structures. Note the symmetry between lateral bands and the tall objects casting shadows
http://www.youtube.com/watch?v=NW9r17nEpSc

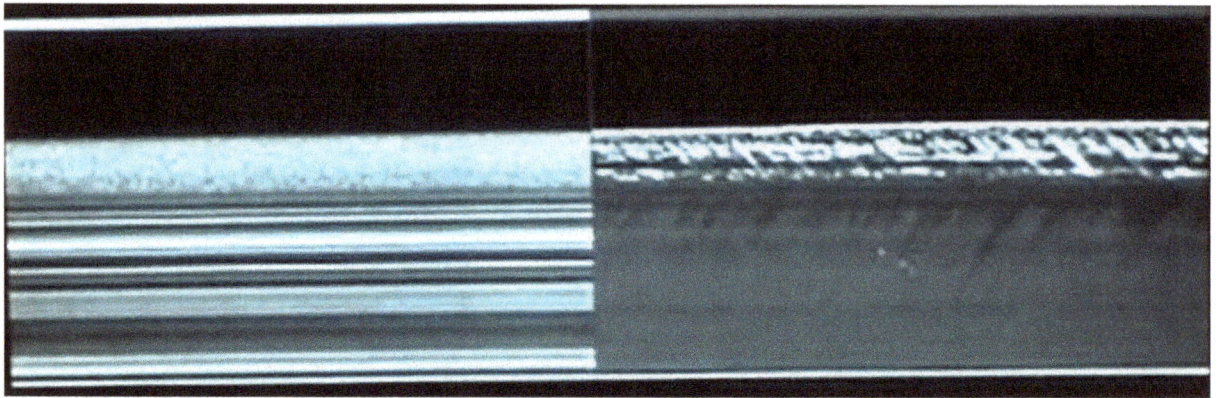

This is the same ring system side by side for comparison to illustrate the section of the ring with artificial structures. Note the ring bands align almost exactly with each other
http://www.youtube.com/watch?v=NW9r17nEpSc

Closer inspection of the rings on the left side in the above photo hints that the entire artificial structure was once covered with some sort of roofing enclosure or casing to enable it to blend with the other ringlets. Note that the enclosure has a mottled appearance that is not clean and crisp with the other rings below it. This encasement may be some type of high tensile fabric.

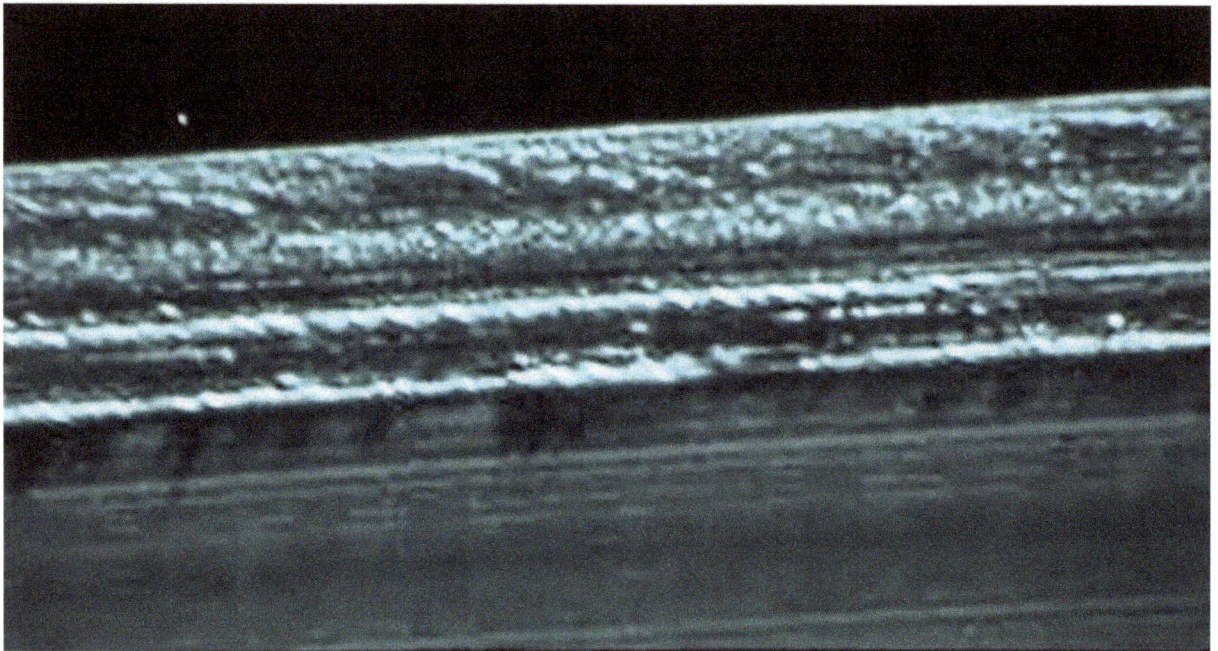

An enlarged view of the ring structure and the shadows it casts across the other rings. Note the clear delineation between light and shadow with the sun shining from above
http://www.youtube.com/watch?v=NW9r17nEpSc

The manufacture of such structures as suggested by Hoagland is based upon the box girder construction similar to how we build bridges on Earth. The actual composite of materials is not really relevant in its construction as there is no gravity or motion vectors exerting upon it out in space where these structures are located. Therefore, this ringworld is not subject to the forces of stress. The only real concern may be the impacts of micrometeorites or larger objects from time to time. As **Hoagland** points out the actual girder construction material could just as well as be made of spaghetti noodles because there are no opposing stress factors to weaken the structure.

So, what is it that we are looking at? Is it a ringworld inhabited by a giant race that created it, a kind of idyllic home away from home for the wealthy corporate elite as portrayed in the recent release of the sci-fi movie; *"Elysium"* or is it vacant and abandoned? Was it designed to be a ringworld habitat or to serve a purpose for something else?

A hypothetical Ringworld habitat orbiting a planet.
Is this an idyllic substitute home-world for the wealthy corporate elite?
http://www.youtube.com/watch?v=NW9r17nEpSc

A possible clue to its purpose according to Hoagland is its position in relationship to Saturn which happens to be at **19.5 radii of Saturn (19.5s)!** This particular number usually triggers a red flag response from Richard Hoagland indicating that **hyperdimensional physics** is at play in the Saturnian system. He posits that the artificial structures in the B-ring is part of a planet-size **Torsion Field Generator (TFG)** designed to generate immense power that may be used to affect travel through space, communications over extreme intergalactic distances, manufacturing of immense machines and buildings, construction of moon-size spacecraft, even time travel!

How such things even possible? Apparently they are! Germany Nazi scientist working on black projects during WWII were not only developing saucer shape craft but, but according to Joseph Farrell and other Ufologists and historians, they were also working on torsion field experiments like **Die Glocke (The Bell)** program which involved levitation, power generation and temporal displacement or time travel. Such experiments were dangerous to human life, particularly if you really don't know the physics behind such research programs, as POWs of Nazi concentration camps who worked on such projects unfortunately found out.

On a larger scale, **Hoagland** sees Saturn as one of many planets in our solar system and perhaps throughout the universe acting as a double tetrahedron hyperdimensional generator as envisioned in the photo image below.

Saturn with a double tetrahedron vortex occupying hyperdimensional space generating a torsion field uniquely symbolized by the numbers "19.5" and "33"
http://www.youtube.com/watch?v=NW9r17nEpSc

The artificial structures on the B-ring are therefore, part of the planet size torsion field generator which is either in disrepair or being constructed or repaired. It is an ancient technology that was originally constructed about 65 million years ago by a **Type 2** or **3 Civilization** that once inhabited this Solar System. Pure speculation, to be sure, but the evidence is slowly mounting that is favouring this hypothesis as part of the original by **Tom Van Flandern.**

The torsion field ring would look something like an atom particle accelerator that is in common use in many countries and universities to study quantum particles and new energy forms. Saturn could actually be an actual generator that harnesses these energy forms for practical usage which we can only guess at this point.

Could Saturn or its moons be hiding other secrets of an advanced Extraterrestrial civilization of giants or other ET species that once inhabited our Solar System? Once again, Hoagland seems to have an answer to that question which may shed more light on the Saturnian system mystery.

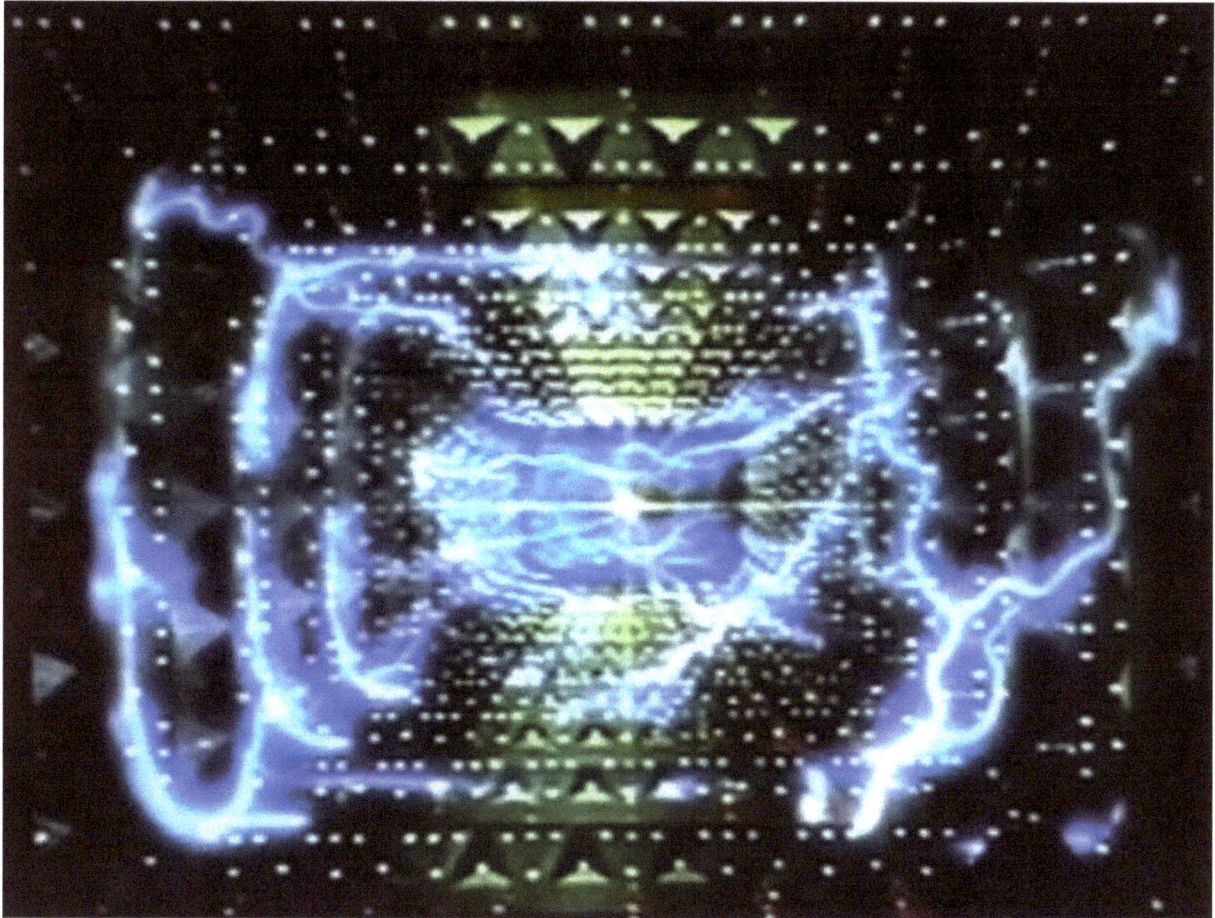

The possible interior of the artificial structures in the B-ring of Saturn generating potentially an immense torsion field of energy about the planet

The **moons of Saturn** are numerous and diverse ranging from tiny moonlets less than 1 kilometer across to the enormous Titan which is larger than the planet Mercury. Saturn has 62 moons with confirmed orbits, 53 of which have names and only 13 of which have diameters larger than 50 kilometers. Saturn has seven moons that are large enough to be ellipsoidal due to having planetary mass, as well as dense rings with complex orbital motions of their own.

In our search for extraterrestrial life in the Solar System we will restrict our search to just the larger moons of Saturn and to a few of its asteroid type satellites which show oddities or anomalies.

CHAPTER 91
THE MOONS OF SATURN

Pan and Atlas

Saturn's moons **Pan** and **Atlas** may have formed in two stages - their cores may be remnants of the breakup of a large icy body early in the solar system's history and their ridges may have formed later, as the cores swept up material from Saturn's rings. The scenario might explain why the ridges appear smooth and the polar regions rough. (See below).

The flying saucer shaped moon of Saturn: Atlas. It is Saturn's most inner moon orbits within the Encke Gap in Saturn's A ring
https://www.newscientist.com/article/dn13014-saturns-flying-saucer-moons-built-of-ring-material/

New observations by the Cassini spacecraft reveal that two of Saturn's small moons look eerily like flying saucers. The moons, which lie within the giant planet's rings, may have come by their strange shape by gradually accumulating ring particles in a ridge around their equators.

The Voyager spacecraft discovered the moons, called Pan and Atlas, in the early 1980s. Pan, which is 33 kilometres wide, orbits Saturn within a gap in the planet's A ring called the **Encke Division** (scroll down for image), while the 39-km-wide Atlas orbits just outside the A ring.

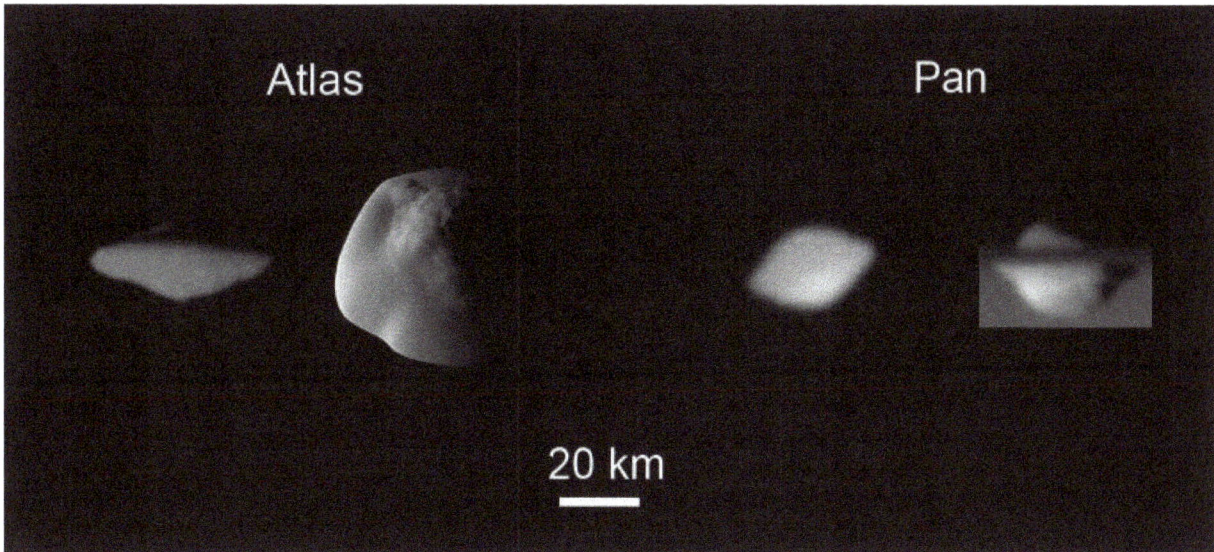

Atlas and Pan are two saucer shaped moons of Saturn
https://www.newscientist.com/article/dn13014-saturns-flying-saucer-moons-built-of-ring-material/

Both moons have a flattened shape, being wider than they are tall. Their appearance is almost like something out of a fantasy or a science fiction movie. But their uncanny resemblance to UFOs only became clear recently, when Cassini imaged them with its powerful cameras.

Hyperion

Hyperion also known as **Saturn VII**, is a moon of Saturn discovered by **William Cranch Bond**, **George Phillips Bond** and **William Lassell** in 1848. It is distinguished by its irregular shape, its chaotic rotation, and its unexplained sponge-like appearance. It was the first non-round moon to be discovered.

Hyperion is one of the largest bodies known to be highly irregularly shaped (non-ellipsoidal, i.e. not in hydrostatic equilibrium) in the Solar System. The only larger moon known to be irregular in shape is Neptune's moon Proteus.

A possible explanation for the irregular shape is that Hyperion is a fragment of a larger body that was broken by a large impact in the distant past. A proto-Hyperion could have been from 350 to 1000 km in diameter.

Like most of Saturn's moons, Hyperion's low density indicates that it is composed largely of water ice with only a small amount of rock. It is thought that Hyperion may be similar to a loosely accreted pile of rubble in its physical composition. However, unlike most of Saturn's moons, Hyperion has a low albedo indicating that it is covered by at least a thin layer of dark material. This may be material from Phoebe (which is much darker) that got past Iapetus. Hyperion is redder than Phoebe and closely matches the color of the dark material on Iapetus.

Hyperion's surface is covered with deep, sharp-edged craters that give it the appearance of a giant sponge. Dark material fills the bottom of each crater. The reddish substance contains long chains of carbon and hydrogen and appears very similar to material found on other Saturnian satellites, most notably Iapetus.

The latest analyses of data obtained by NASA's *Cassini* spacecraft during its flybys of Hyperion in 2005 and 2006 show that about 40 percent of the moon is empty space. It was suggested in July 2007 that this porosity allows craters to remain nearly unchanged over the eons. The new analyses also confirmed that Hyperion is composed mostly of water ice with very little rock.

The **Voyager 2** images and subsequent ground-based photometry indicate that Hyperion's rotation is chaotic, that is, its axis of rotation wobbles so much that its orientation in space is unpredictable. Hyperion is the only moon in the Solar System known to rotate chaotically. It is also the only regular planetary natural satellite in the Solar System known not to be tidally locked.

Hyperion is unique among the large moons in that it is very irregularly shaped, has a fairly eccentric orbit, and is near a much larger moon, Titan.
http://en.wikipedia.org/wiki/Hyperion_%28moon%29

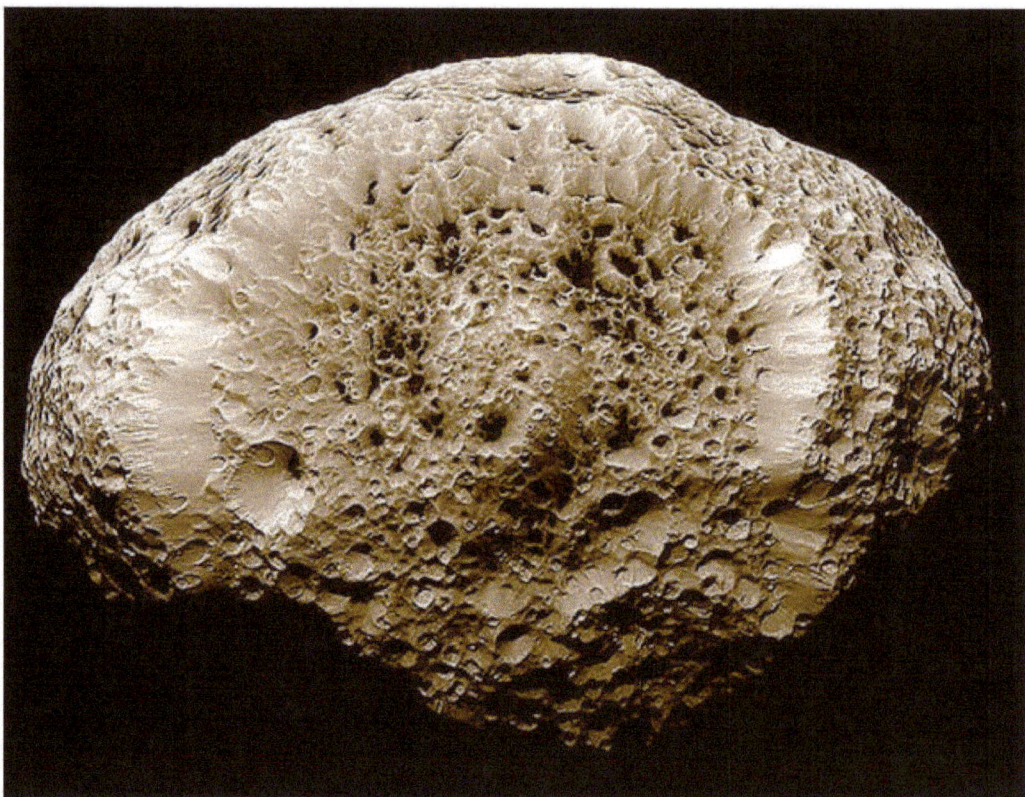

Hyperion is another strange moon of Saturn with its spongy appearance, wobbly rotation and dark reddish surface material trapped in its craters
http://www.sci-news.com/space/science-cassini-saturns-moon-hyperion-02232.html

A report was submitted to **MUFON**, not of a UFO sighting or an alien abduction, but of a green light seen on the surface of one of Saturn's moons, Hyperion. In typical NASA photo lab sanitization of raw photographic data, the original photo has been airbrushed and the 'green light' has been removed. Curious!

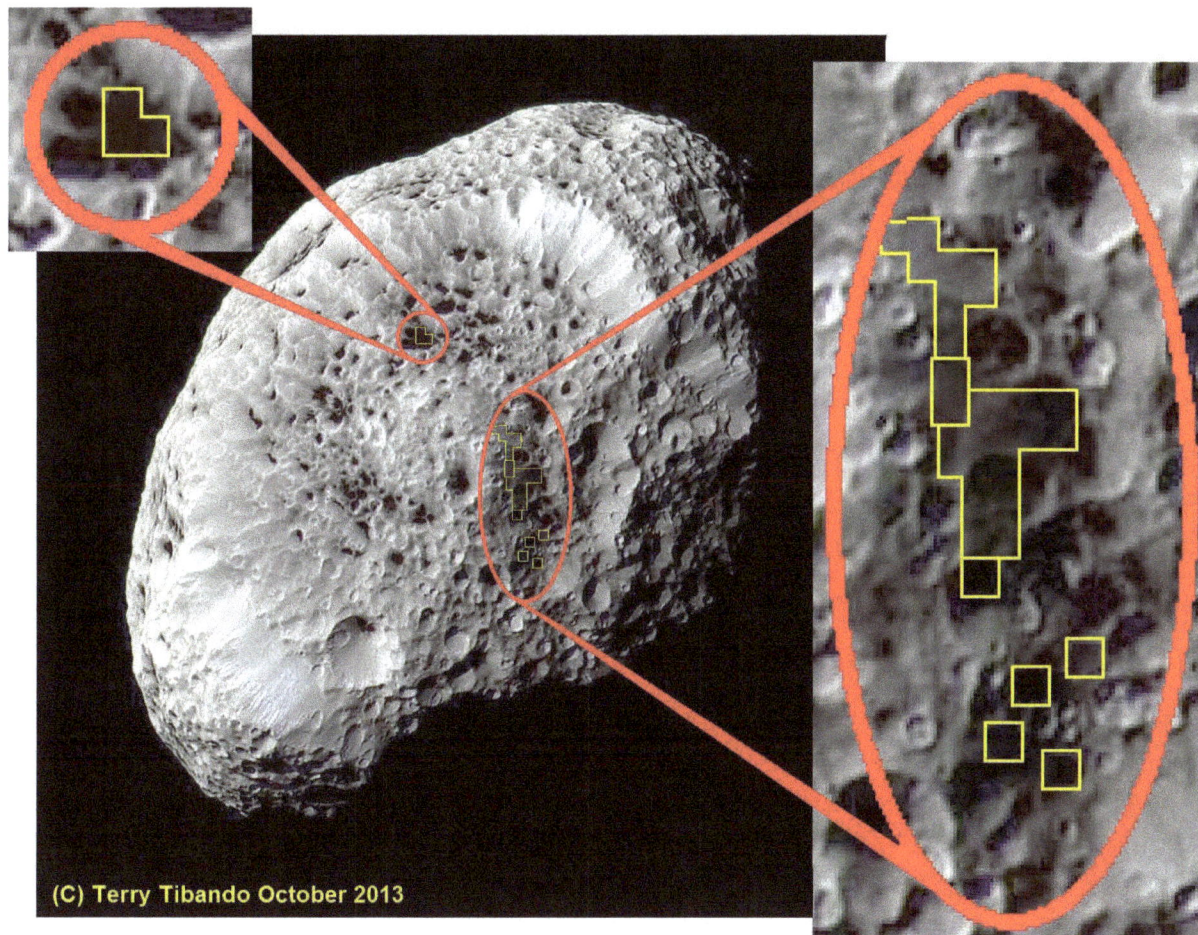

(C) Terry Tibando October 2013

Possible anomalies on Hyperion which may have been airbrushed out of the image. The small red circle and its enlargement is the approximate area described where a green light was originally photographed and then removed by NASA
(c) Terry Tibando

An astute observer in Texas caught the glaring anomaly before it was airbrushed out and filed a report: **MUFON Case # 23389**

"Please forward to someone that is truly interested. Reference NASA's "Feast for the Eyes", Saturn.

The Moon Hyperion has a bright incandescent light, visible on the surface. Place a cross hair on the moon, top right-hand quadrant, approximately 20 degrees, close to the center line.

*There is a bright incandescent green light on the surface. Original picture frame **PIA 07740.jpg** showed the light clearly, if you zoom in approximately 200%.*

The current picture has a circular area where the light is but the coloring has been removed. I have the original on my laptop, with the light, should you be interested. **(Like many researchers, we are all waiting to see this original photo!)** [Bold italics added by author]

Not a UFO but whatever it is, it is not natural. The moon is gray." **Tom Young**
http://lightsinthetexassky.blogspot.ca/2010/05/shine-on-hyperion.html

Titan

Titan, the largest planet-like moon of Saturn is the only natural satellite in the Solar System to have a dense opaque atmosphere and the only object besides the Earth to have stable bodies of surface liquid found on it. **Titan** was discovered in 1655 by the Dutch astronomer Christiaan Huygens, and was the fifth moon to be discovered after **Galileo** discovered the four largest moons of Jupiter in 1610.

Titan is composed primarily of water ice and rocky material and until the Cassini-Huygens space probe arrived in 2004 not much was known. The surface of titan is geologically young with newly discovered hydrocarbon lakes along with mountains and cryovolcanoes, smooth terrain surfaces and few impact craters have been found.

The atmosphere of Titan is largely composed of nitrogen; minor components lead to the formation of methane and ethane clouds and nitrogen-rich organic smog. The climate—including wind and rain—creates surface features similar to those of Earth, such as dunes, rivers, lakes and seas (probably of liquid methane and ethane), and deltas, and is dominated by seasonal weather patterns as on Earth. http://en.wikipedia.org/wiki/Titan_%28moon%29

Titan's hazy atmosphere makes Saturn's largest moon look like a fuzzy orange ball
https://en.wikipedia.org/wiki/Titan_%28moon%29

Whether there is life on Titan at present is an open question and a topic of scientific evaluation and research and no doubt with a lot of debate.

Titan is far colder than Earth, and its surface seems to lack liquid water; factors which have led some scientists to consider life there unlikely. On the other hand, the following points have been

made in favor of Titan's suitability to sustain some form of life:

- Titan appears to have lakes of liquid ethane and/or liquid methane on its surface, as well as rivers and seas, which some scientific models (still tentative and debated) suggest could support non-water-based life.
- It has also been suggested that life may exist in a sub-surface ocean consisting of water and ammonia. Recent data from NASA's Cassini spacecraft have strengthened evidence that Titan likely harbors a layer of liquid water under its ice shell.
- Titan is the only known natural satellite (moon) in the Solar System that is known to have a fully developed atmosphere that consists of more than trace gases. Titan's atmosphere is thick, chemically active, and is known to be rich in organic compounds; this has led to speculation about whether chemical precursors of life may have been generated there.

339

- The atmosphere also contains hydrogen gas, which is cycling through the atmosphere and the surface environment, and which living things comparable to Earth **methanogens** could combine with some of the organic compounds (such as acetylene) to obtain energy.

In June 2010, scientists analyzing data from the Cassini–Huygens mission reported anomalies in the atmosphere near the surface which could be consistent with the presence of methane-producing organisms, but may alternatively be due to non-living chemical or meteorological processes. The Cassini–Huygens mission was not equipped to provide direct evidence for biology or complex organics. http://en.wikipedia.org/wiki/Life_on_Titan

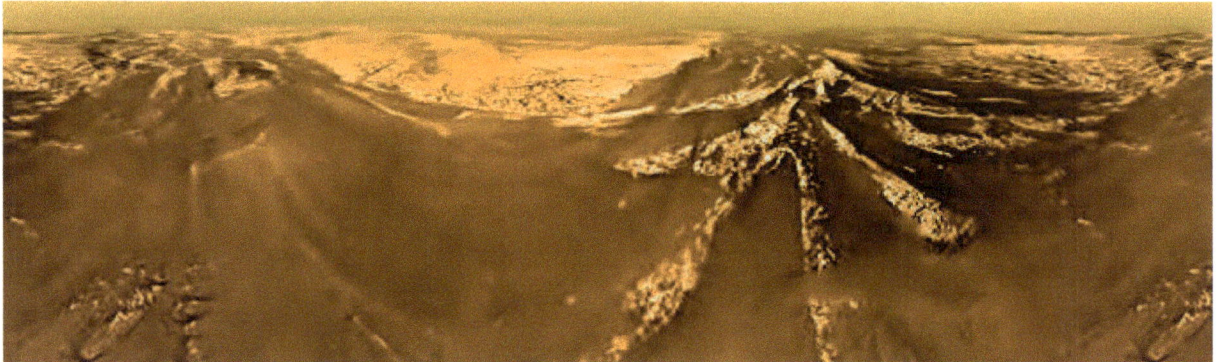

Panoramic view of the surface of Titan as seen by the European Space Agency's Huygens probe as it descended to the surface, the first spacecraft to land on an alien moon.
https://www.nasa.gov/content/ten-years-ago-huygens-probe-lands-on-surface-of-titan

In 2005, astrobiologists **Chris McKay** and **Heather Smith** predicted that if is consuming atmospheric hydrogen in sufficient volume, it will have a measurable effect on the mixing ratio in the troposphere of Titan. The effects predicted included a level of acetylene much lower than otherwise expected, as well as a reduction in the concentration of hydrogen itself.

Evidence consistent with the predictions was reported in June 2010 by **Darrell Strobel** of Johns Hopkins University, who noted an overabundance of molecular hydrogen in the upper atmospheric layers. Near the surface the hydrogen apparently disappears. Another paper released the same month showed very low levels of acetylene on Titan's surface.

**Cassini's radar image of Titan's north polar region where blue coloring
indicates hydrocarbon seas, lakes and tributary networks
filled with liquid ethane, methane and dissolved N$_2$**
https://en.wikipedia.org/wiki/Titan_%28moon%29

Chris McKay agreed with Strobel that presence of life, as suggested in McKay's 2005 article, is a possible explanation for the findings about hydrogen and acetylene; but also cautioned that other explanations are currently more likely: namely the possibility that the results are due to human error, to a meteorological process, or to the presence of some mineral catalyst. He noted that such a catalyst, effective at -178°C (95K), is presently unknown, and would *in itself be a startling discovery*, though less startling than discovery of an extraterrestrial life form.

The June 2010 findings gave rise to considerable media interest, including a report in the British newspaper, the Telegraph, which spoke of clues to the existence of **"primitive aliens".**

In order to assess the likelihood of finding any sort of life on various planets and moons, **Dirk Schulze-Makuch** and other scientists have developed a **Planetary Habitability Index (PHI)** which takes into account factors including characteristics of the surface and atmosphere, availability of energy, solvents and organic compounds. Using this index, based on data available in late 2011, the scientists found that Titan has the highest current habitability rating of any known world, other than Earth.

Titan is presented as a test case for the relation between chemical reactivity and life, in a 2007 report on life's limiting conditions prepared by a committee of scientists under the United States **National Research Council**. The committee, chaired by **John A. Baross**, considered that *"if life is an intrinsic property of chemical reactivity, life should exist on Titan. Indeed, for life not to exist on Titan, we would have to argue that life is not an intrinsic property of the reactivity of carbon-containing molecules under conditions where they are stable..."*
http://en.wikipedia.org/wiki/Life_on_Titan

Richard C. Hoagland in his paper "A Moon With a View" on his Enterprise Mission website states that we must also consider the possibility that *Titan is being "terraformed" by an advanced Extraterrestrial Intelligence.* (We will explode this theory a little later).

During a flyby of Titan on June 27, NASA's Cassini probe photographed a polar vortex or mass of swirling gas in the atmosphere high above the south pole of Titan, hinting that winter may be coming to the huge body's southern reaches. The vortex appears to complete one full rotation in nine hours, while it takes Titan about 16 days to spin once around its axis.

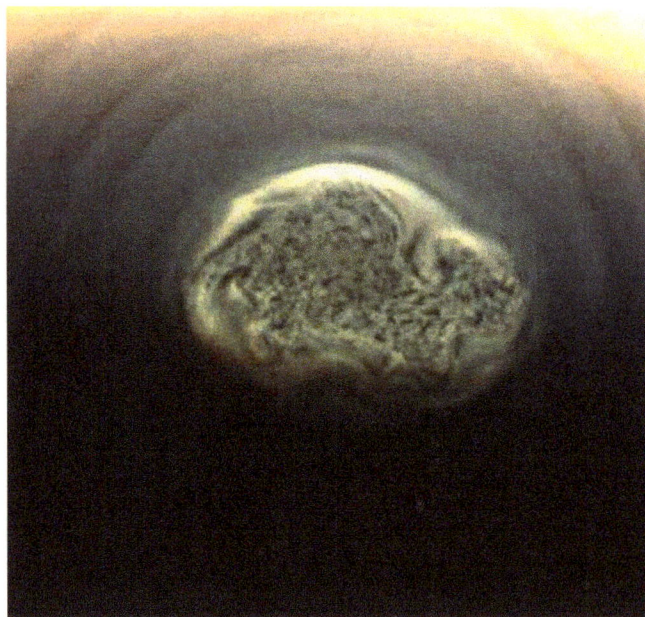

A true color image captured by NASA'S Cassini spacecraft before a distant flyby of Saturn's moon Titan shows a south polar vortex swirling in the moon's atmosphere.
https://en.wikipedia.org/wiki/Climate_of_Titan

Enceladus

When the Voyager 2 spacecraft sped through the Saturnian system more than a quarter of a century ago, it came within 90,000 kilometers of the moon **Enceladus**. Over the course of a few hours, its cameras returned a handful of images that confounded planetary scientists for years. Even by the diverse standards of Saturn's satellites, Enceladus was an outlier. Its icy surface was as white and bright as fresh snow, and whereas the other airless moons were heavily pocked with craters, Enceladus was mantled in places with extensive plains of smooth, uncratered terrain, a clear sign of past internally driven geologic activity. At just over 500 kilometers across, Enceladus seemed far too small to generate much heat on its own. Yet something unusual had clearly happened to this body to erase vast tracts of its cratering record so completely. http://www.scientificamerican.com/article.cfm?id=enceladus-secrets

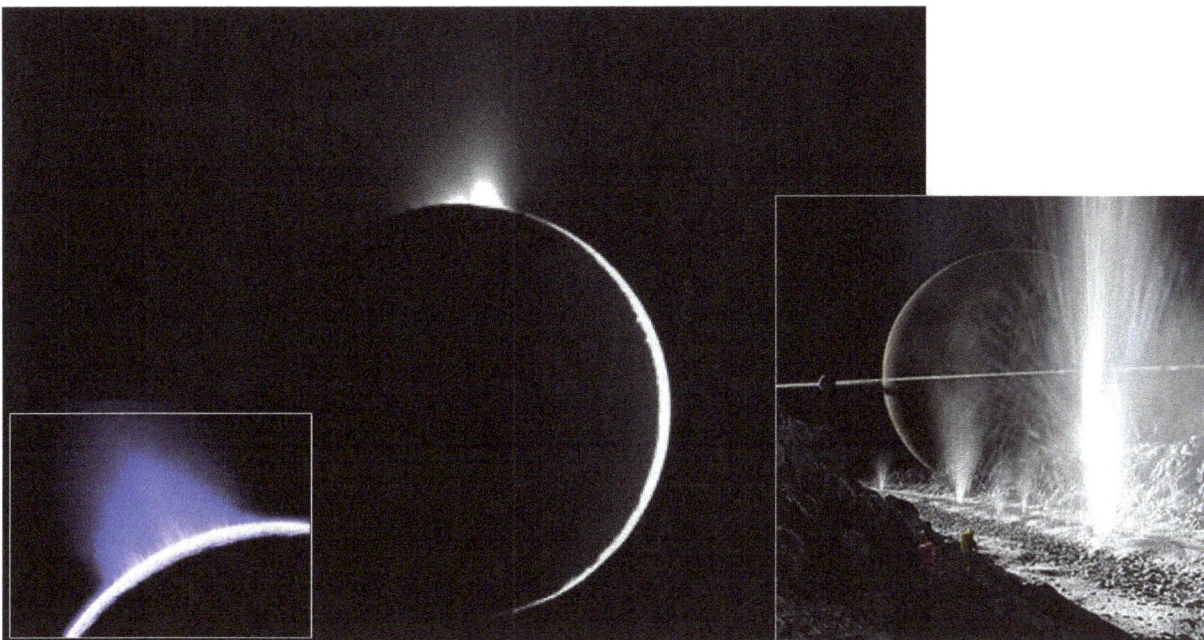

Jets of steam and icy grains erupt from deep fractures in the south polar terrain of Enceladus. This artist's conception includes astronauts for scale.
Cassini took this image (Left) in November 2009 of Enceladus
https://www.nasa.gov/multimedia/imagegallery/image_feature_1510a.html and http://globalnews.ca/news/1189307/the-4-best-places-for-life-in-our-solar-system/ and https://www.scientificamerican.com/slideshow/enceladus-secrets/

Enceladus' surface is a gorgeous white shell of ice, pristine except for a network of fractures near its south pole. These cracks dubbed "tiger stripes" emit fountains of water vapour that instantly turn into icy grains on contact with the chill vacuum of space. Some astrophysicists conclude that the worldlet harbours an ocean of saltwater, which in turn makes it a good candidate as a source for life. But how can a sub-surface sea exist, if the ambient temperature is close to absolute zero (-273 degrees °C) (-460 degrees °F) and the Sun is a distant dot?

The answer, say theorists, lies with a phenomenon called tidal forces. They argue that the gravitational pull exerted by Saturn squeezes Enceladus' innards, causing friction whose heat allows the water to remain in a liquid state.

When Enceladus is closest to Saturn, the plume is at its dimmest, a sign that the fractures are being closed up by a mighty gravitational pull from the giant mother plant, and so relatively little water escapes, according to the new study.

When Enceladus is at its farthest point from Saturn, the plume is several times brighter, suggesting that the fractures open out—rather like an unclenched fist—and more water is disgorged. The evidence comes from 252 infra-red images taken by the great US explorer probe Cassini during its lonely swings around the planet.

They provide "strong evidence that tidal forces do play an important role in controlling Enceladus' plume activity, perhaps by changing the width of the conduits between the surface and various underground reservoirs". Many of the icy grains from Enceladus fall back on its surface, which explains its dazzling white surface. http://phys.org/news/2013-07-gravitational-tide-secret-saturn-weird.html

Mimas

Mimas was first discovered by **William Herschel** in 1789. From that time until Voyager 1 passed by Saturn, it was simply the 7th moon that had been discovered, and a white dot in astronomer's telescope.

Mimas it is 415 km wide which is quite small when compared to our own moon. But it is not actually round, it's more egg shaped, and its exact dimensions are 415.6 × 393.4 × 381.2 km. Mimas has a surface area as large as the country of Spain with a low density, 1.15 g/cm³ indicating that it is composed mostly of water with a small amount of rock.

Mimas's most distinctive feature is a giant impact crater 130 kilometres (81 mi) across, named the Herschel Crater after the discoverer of Mimas. Herschel's diameter is almost a third of Mimas's own diameter; its walls are approximately 5 kilometres (3.1 mi) high, parts of its floor measure 10 kilometres (6.2 mi) deep, and its central peak rises 6 kilometres (3.7 mi) above the crater floor. The impact that made this crater must have nearly shattered Mimas: fractures can be seen on the opposite side of Mimas that may have been created by shock waves from the impact travelling through Mimas's body.

When Voyager 1 passed by Mimas in November, 1980 the first published pictures of its surface showed a strong resemblance to the **"Death Star"** battle station as seen in the popular sc-fi movie "Star Wars: A New Hope" as exclaimed by *Jedi Master*: *Obi-Wan*: *"That's no moon! It's a space station."*

Prophetic fictional words or merely a coincidence as the film was made nearly three years before Herschel Crater was discovered?

344

When seen from certain angles, Mimas's most prominent feature, Herschel Crater resembles the Death Star which is said to be roughly 140 kilometres in diameter. This stems from the fact that Herschel resembles the concave disc of the Death Star's "superlaser". Is Mimas an artificial moon? The possibility cannot be ruled but, it's odd appearance leaves room for speculation and debate.

Mimas is one of many strange moons of Saturn with its "Death Star" Herschel crater
https://en.wikipedia.org/wiki/Mimas_(moon)

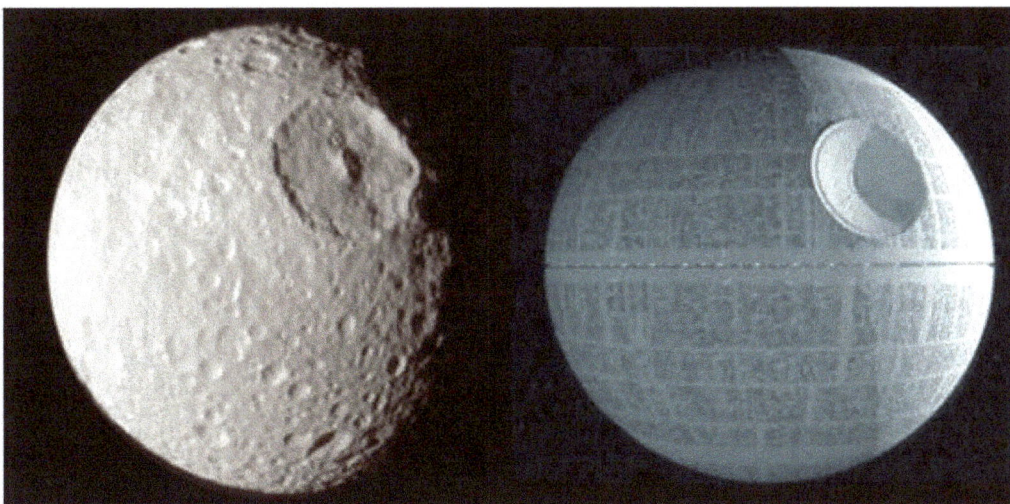

The Saturnian moon Mimas and its remarkable resemblance
to the Death Star Battle Station in the movie Star Wars: A New Hope
https://medium.com/teamindus/the-death-star-moon-mimas-eef4f8b7b237

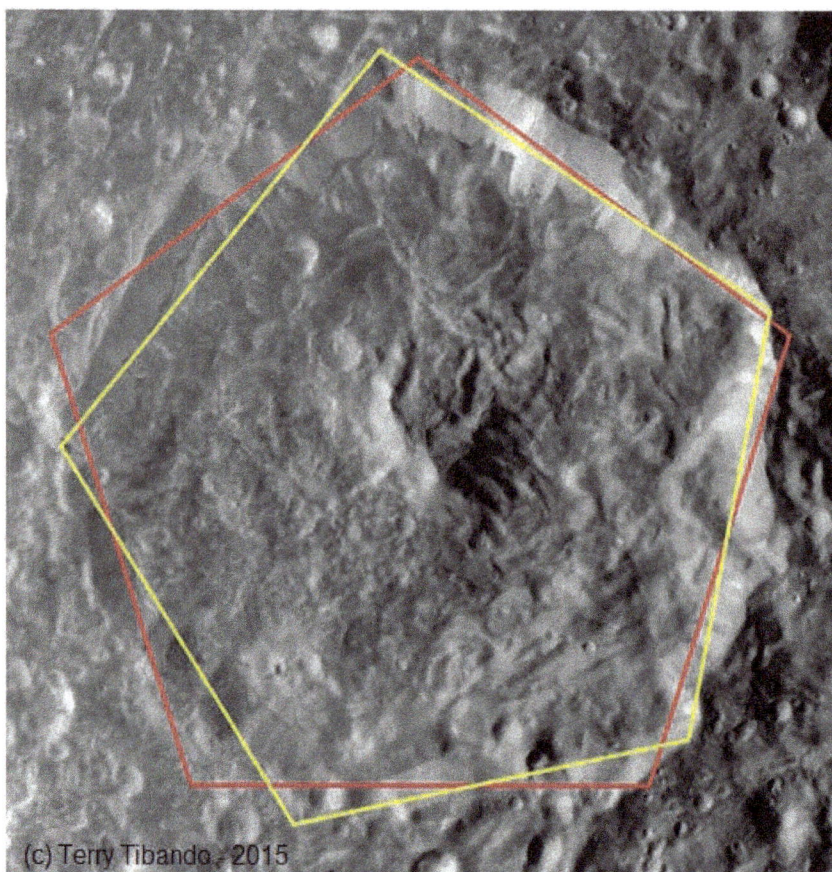

(c) Terry Tibando - 2015

Herschel Crater enlarged. Note the red pentagram indicating crater wall symmetry
and the yellow pentagram overlay in actual alignment to the crater walls
indicating the crater walls are anything but round or circular. This
crater is similar to the hexagon crater on the planetoid, Ceres
(C) Terry Tibando

In 2010, NASA revealed a temperature map of Mimas, using images obtained by *Cassini*. The warmest regions, which are along one edge of Mimas, create a shape similar to the video game character Pac-Man, with **Herschel Crater** assuming the role of an "edible dot" or "power pellet" known from the video game *Pac-Man*. **http://en.wikipedia.org/wiki/Mimas_%28moon%29**

This is indeed strange but, Cassini has discovered that **Tethys** also has the same *Pac-Man* heat signature as Mimas! The cause of which is still unknown to NASA and theorized as being cause by a combination of tidal forces from Saturn and the fact that the large impact crater on each moon may be evidence that each moon contains internally the large meteorite body that once each hit these moons.

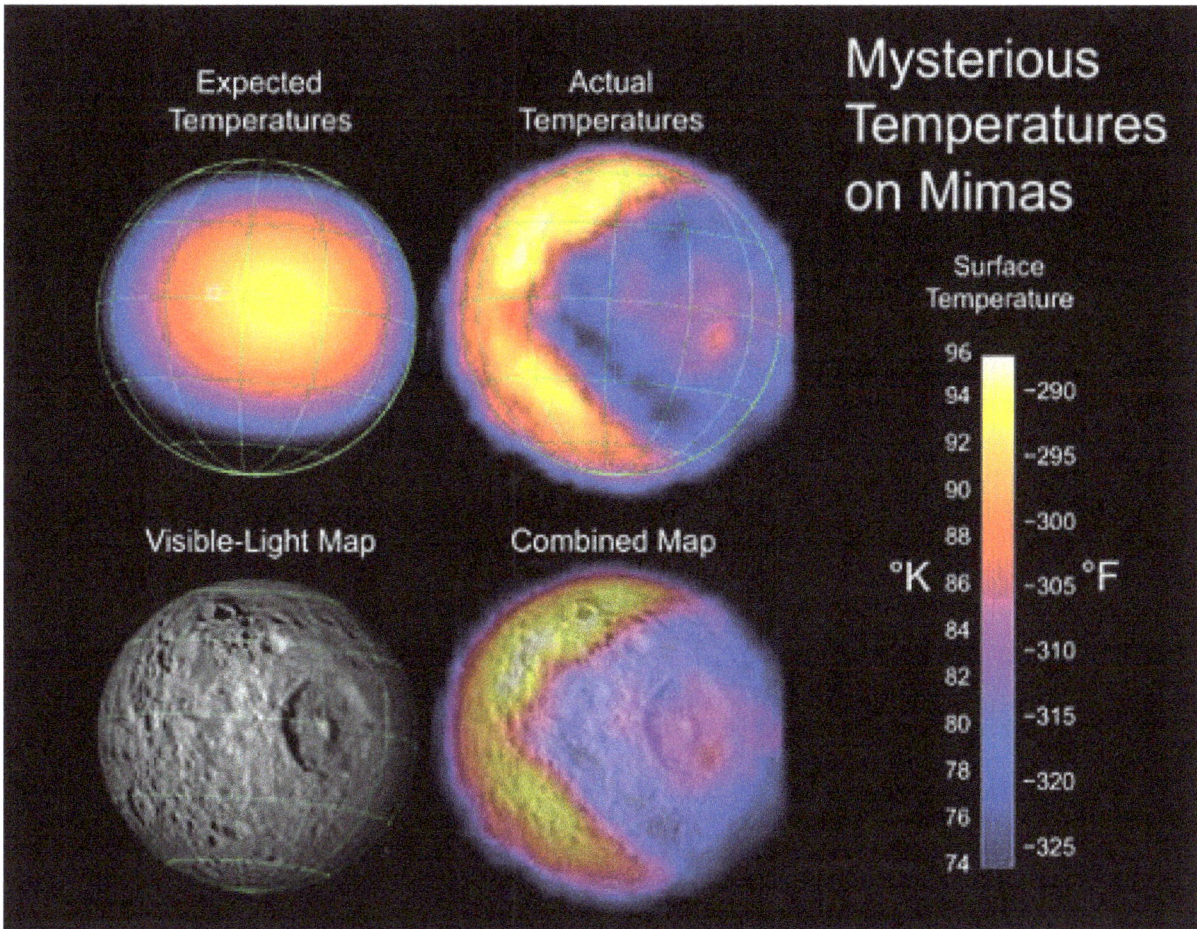

The mysterious temperatures of Mimas as seen through thermal imaging shows the Pac-Man appearance of Mimas because of its heat signature

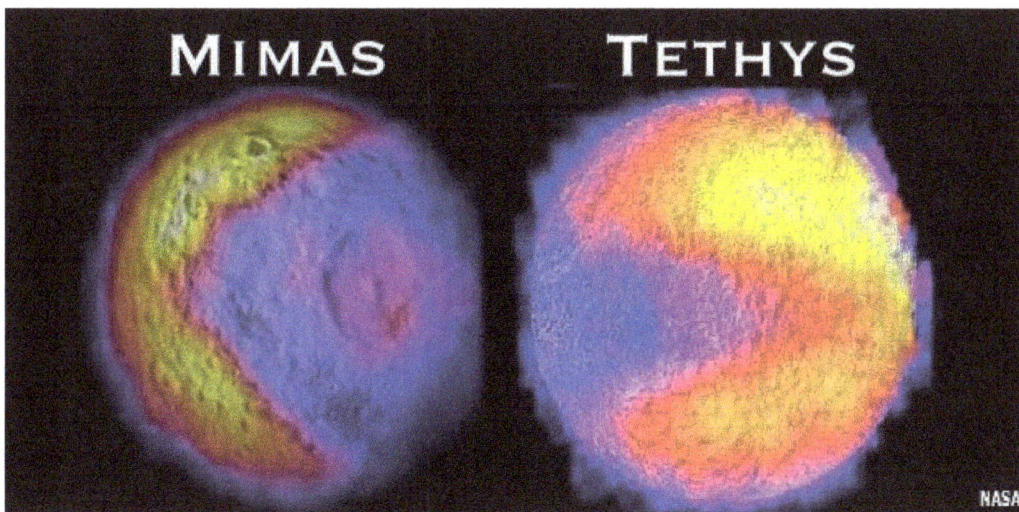

Saturn's two Pac-Man moons: Mimas and Tethys have similar temperature signatures

What are the odds that two moons in the same planetary system would display the same heat signatures as imaged by the Cassini space probe? Is this simply a remarkable coincidence of nature or a product of deliberate artificial design by an advanced Extraterrestrial Intelligence. As strange as Mimas and Tethys look with their large impact craters, the interesting thing is how many of Saturn's moons also have the very same or similar large prominent impact craters.

Tethys

Tethys is believed to be composed almost entirely of water ice. It is one of several such icy worlds in orbit around Saturn, and is very similar to **Dione** and **Rhea** in composition. There may be a small amount of rocky material thrown into the mix, making the moon like a "dirty snowball". The moon has a very low density, and this suggests a much larger concentration of ice than rock. This icy surface is heavily cratered. Cracks and faults can also be seen in the ice. The surface temperature on Tethys averages a chilling -305° F (187° C).

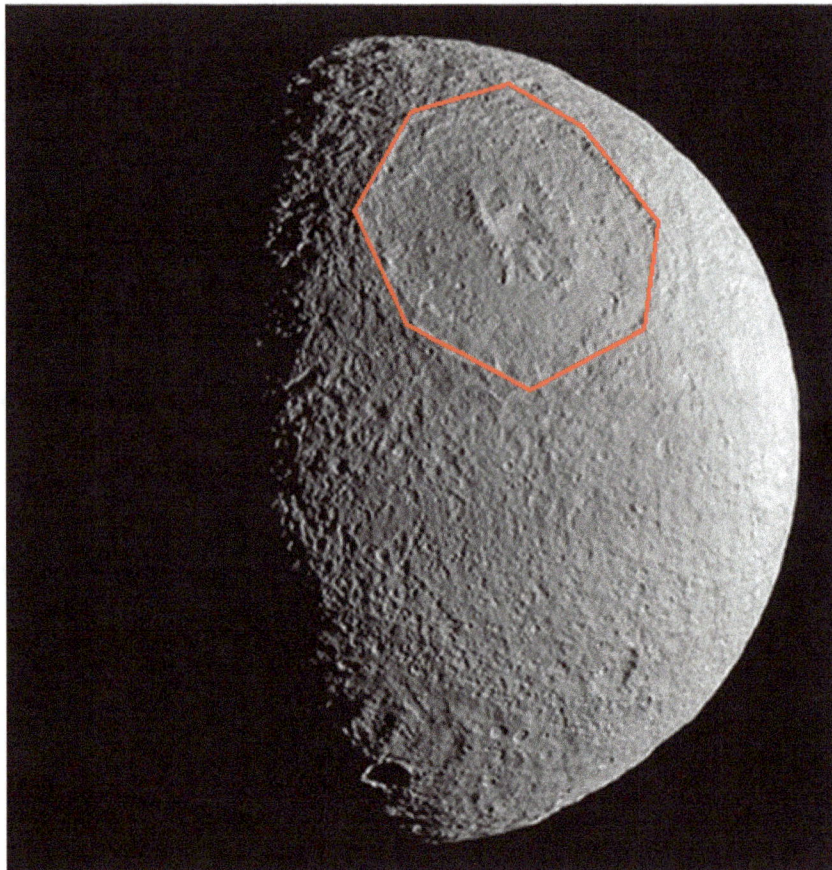

Tethys with it prominent (polygon?) impact crater Odysseus

Tethys is definitely one of the most interesting worlds on our tour, but it does have a few notable surface features. In the western hemisphere can be seen a huge impact crater called **Odysseus** . At 248 miles (400 km) in diameter, this crater is nearly 2/5 the size of the entire moon.

Astronomers are not quite sure why an impact of this size didn't shatter the moon completely. One theory suggests that Tethys may have been in a semi-liquid state at the time of the impact. Odysseus is a very flat crater that conforms to the shape of the moon itself, but it lacks the ringed mountains and central depression that characterize craters on Earth's moon or **Mercury**. The other major feature on Tethys is a huge valley called Ithaca Chasma. This canyon is about 60 miles (96 km) wide and 3 miles (4.8 km) deep, and runs 1,242 miles (2,000 km) long. It runs nearly 3/4 of the way around Tethys. These surface features suggest to astronomers that Tethys was not always frozen solid. Some believe that it may have existed in a liquid state at some time in the past. Tethys has several smaller impact craters, which are believed to have formed more recently. Tethys also contains no albedo features like the ones on Dione and Rhea.
http://www.seasky.org/solar-system/saturn-tethys.html

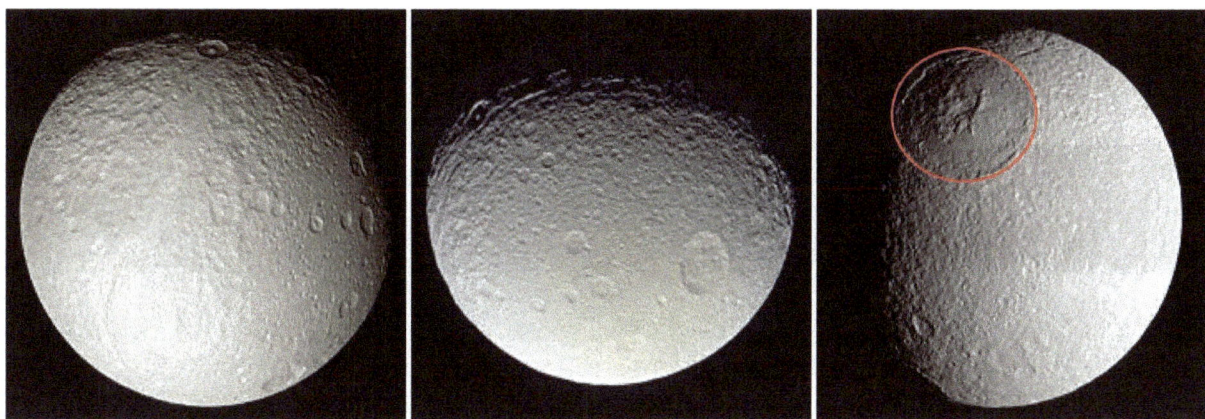

Three views of Tethys with a massive impact crater Odysseus in the north western hemisphere of the moon (right) similar to Dione's and Rhea's primary impact craters
http://www.seasky.org/solar-system/saturn-tethys.html

Dione

Dione is the densest of Saturn's moons with the exception of Titan. It is composed mainly of water ice, but must contain a larger amount of rocky material than Saturn's other ice moons, Tethys and Rhea. Dione is very similar to Rhea in composition, although somewhat smaller. Dione is believed to have a rocky core with less ice coverage than Rhea.

Dione is locked in a synchronous orbit similar to that of Rhea. This causes the same face of the moon to point towards Saturn at all times. Dione has similar albedo features and terrain to that of its close cousin, Rhea. Dione's surface consists of heavily cratered areas, moderate and lightly cratered plains, and bright, wispy features. Most of the heavily cratered areas exist on the trailing hemisphere of the moon. Some of these craters exceed 62 miles (100 km) in diameter, while most of the craters in the plains areas are less than 18 miles (30 km) in diameter. The largest crater is called **Amata** and is 150 miles (241 km) in diameter. ***Like Rhea, these craters lack the high relief features seen on Mercury and the Moon***. Heavy cratering would normally be expected on the leading edge of a tidally locked satellite. Since most of Dione's craters are located on the trailing hemisphere, astronomers believe that Rhea may have once been tidally locked with Saturn in the opposite orientation. Since Dione is relatively small, it would have only taken an impact leaving a 21-mile (35 km) crater to spin the moon around. With many craters on Dione exceeding 21 miles, it is possible that the moon has been spun around more than once throughout its long history. The origin of the bright, wispy streaks is not known. The streaks overlay many of the craters, which indicate that they are newer. They may have been formed by ice eruptions through cracks in the surface. This material may have fallen back to the surface as snow or ash. Dione has no detectable atmosphere. http://www.seasky.org/solar-system/saturn-dione.html

Three views of Dione taken by Cassini spacecraft. Middle image shows white wispy streaks of unknown origin possibly due to ice eruptions on the surface. The image at right shows Amata Crater which is 150 miles across
http://www.seasky.org/solar-system/saturn-dione.html

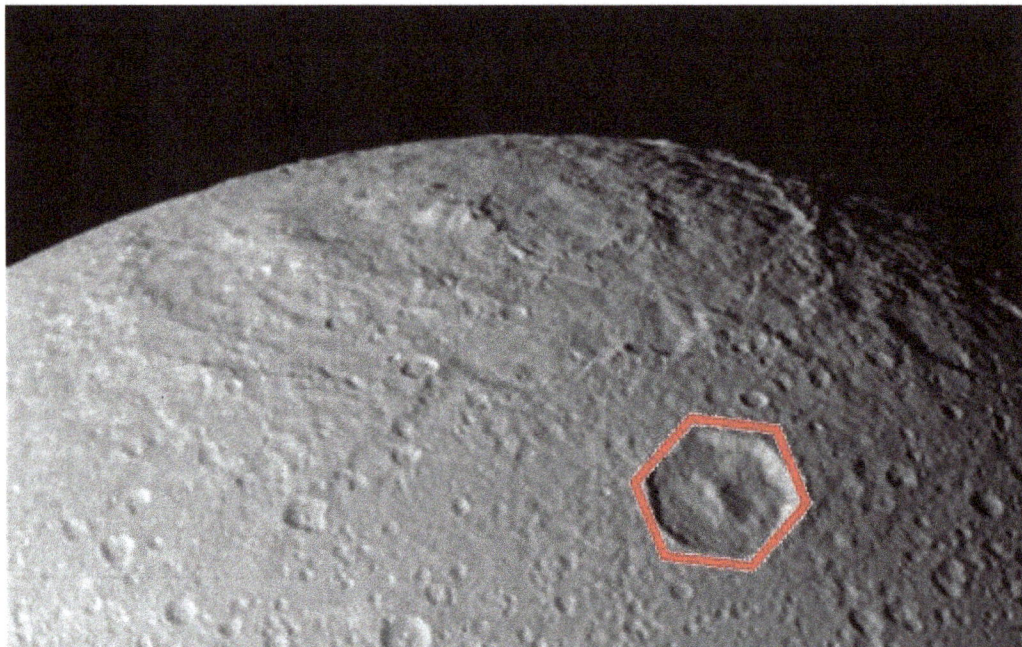

A close-up view of Dione with its large impact crater Amata, note smaller hexagon crater
https://saturn-archive.jpl.nasa.gov/photos/imagedetails/index.cfm?imageId=4433

Rhea

Rhea is an icy body very similar to Dione, although slightly larger in size. It has a very low density, which indicates that it is composed mainly of water ice with rocky material making up less than one third of its total mass. The temperature on Rhea ranges from -281° F (-174° C) in the sunlight to -364° F (-220° C in the shade.

Rhea is one of the most heavily cratered satellites in the Solar System. The surface features of this moon can be divided into two different geological areas based on the density of the craters. The first area contains craters that are larger than 25 miles (40 km) in diameter. The largest of these craters is **Mamaldi**. It has a diameter of 140 miles (225 km). Craters in the second area are smaller than 25 miles in diameter. The difference in these areas indicates that some great event served to resurface parts of this moon at some point in the past. Similar differences in crater size can be found on Mimas and Dione.

Rhea with its large impact craters Mamaldi (left of centre) and beside it, Tirawa (upper right). Note smaller hexagon craters around and inside the larger craters

Like Jupiter's moon Callisto, Rhea's craters lack the high surface features found in craters on

Mercury and the Moon. Like Dione, Rhea's orbit around Saturn is synchronous, meaning that it always keeps the same face toward Saturn. On the trailing hemisphere of Rhea can be found several bright, wispy lines. These linear features are believed to be fractures, and may indicate that Rhea has expanded and contracted due to internal heating and cooling. Similar linear features are found on Dione. Rhea has no detectable atmosphere. http://www.seasky.org/solar-system/saturn-rhea.html

Rhea and Dione appear to be twin-like in many of their features from the white wispy streaks across their surfaces to their very large craters which are nearly the same in size but which are located in hemispheres opposite to each other. Dione's Amata Crater is in the southern latitudes and Rhea's Mamaldi is in the northern latitudes.

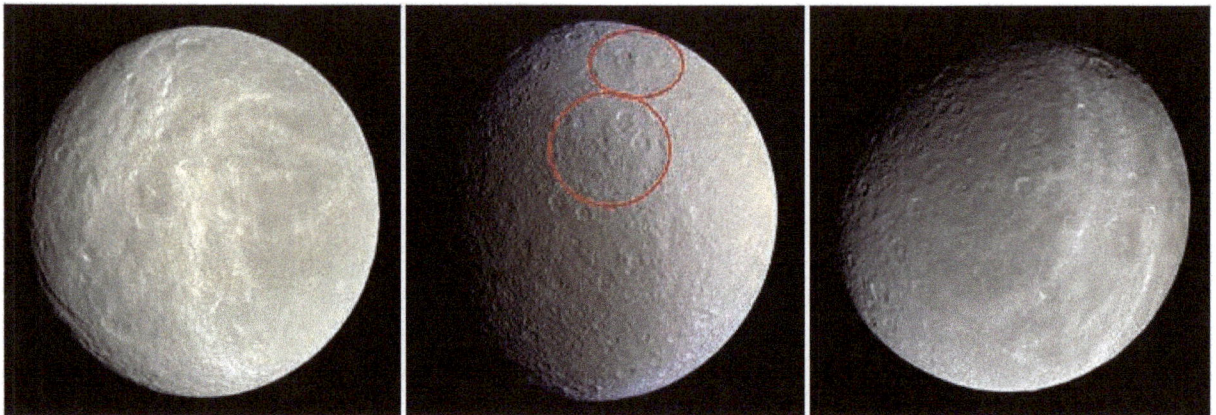

Three views of Rhea, the most heavily cratered moon in the Solar System with two large impact craters (middle), the largest being Mamaldi Note white wispy streaks (left) across the surface of Rhea similar to Dione
http://www.seasky.org/solar-system/saturn-rhea.html

This similarity may be simply coincidence but, the curious fact is that ***there other moons in the Saturnian system with the exact same massive impact craters in the same latitudes or hemispheres and all of these impact craters are approximately the same size as each other!!*** Mimas and Tethys are the cousins each other and Dione and Rhea appear to be twins, in other words, ***we are looking at moons that vary in size but essentially have all the same features of each other.*** All four Saturnian moons appear to be the created from the same "cookie cutter" factory which eliminates the possibility of coincidence and strongly suggests another alternative... the probability of intentional intelligent design! All these moons have the appearance of a "Death Star" quality to them ***as if the placement of the primary craters were deliberately situated in a designated hemispherical location for a particular reason!***

Iapetus - Saturnian Moon of Mystery

If these four moons are unusual with similarity of features then, Iapetus seems to have all the features in spades plus some that the others do not possess, as if Iapetus was the crowning pinnacle in artificial moon creation incorporating all the aspects of the other moons times hundred.

352

NASA's perception of Iapetus is that it's probably the most unusual of Saturn's moons. It has a low density similar to that of Rhea. This indicates that it contains very little rocky material and is composed mainly of water ice. Iapetus is a moon of stark contrasts. Its leading hemisphere is very dark with a slight reddish color, while its trailing hemisphere is very bright. The dark area is one of the darkest terrains in the Solar System, and is darker than asphalt. The reason for this dark matter on the leading side is not known. It could be composed of dark matter swept up from space, or it could be something that flowed out from within the moon. Some astronomers believe that the dark matter could have originated on Saturn's moon, Phoebe. Phoebe has a low albedo. Material could have been blasted into space by impacts, where it was then deposited on the surface of Iapetus. One problem with this theory, however, is that the colour of the material on Phoebe and Iapetus do not match. Another theory states that this dark material may have been formed by eruptions of methane from deep within the moon. The fact that this area is fairly devoid of craters would suggest that this material is replenished regularly. One puzzling aspect of this phenomenon is that the dividing line between the dark and light areas is extremely sharp like a planet size Chinese **Ying Yang icon**. *No one seems to know why*.

The dark, smooth region of Iapetus is known as the Cassini Regio. Very few craters can be found in this region. Astronomers believe that the dark material that covers this region is continually being replenished, thus covering any new craters that may appear. On the opposite side of Iapetus lies the **Roncevaux Terra**. This region is littered with hundreds of craters. The largest of these craters is known as Roland and is 90 miles (144 km) in diameter. The second largest crater in this area, called **Marsilion**, is 84 miles (136 km) in diameter. Iapetus has no detectable atmosphere and no magnetic field. http://www.seasky.org/solar-system/saturn-iapetus.html

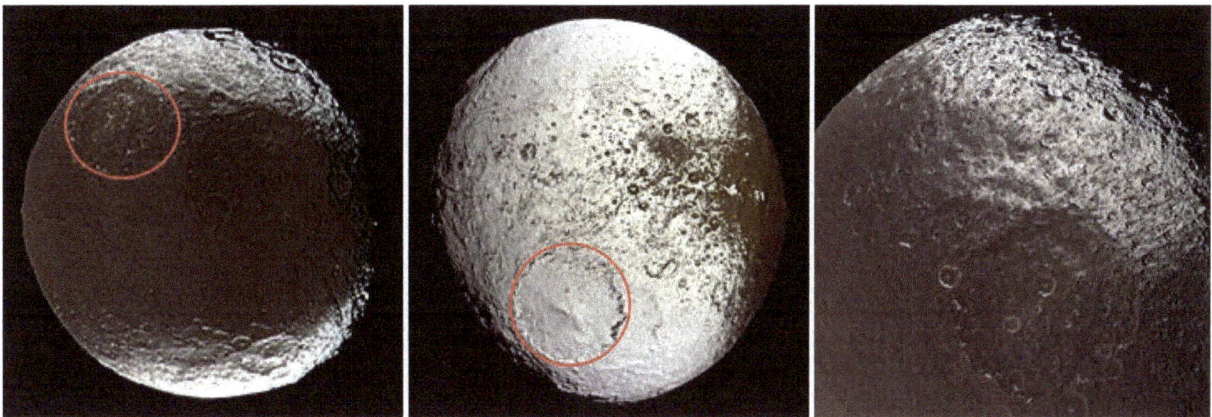

Three views of Iapetus the most unusual moon in the solar System. Note the two large craters, one in the north western hemisphere on the "Dark Side and the other large crater in southern hemisphere on the "Light Side" of Iapetus
http://www.seasky.org/solar-system/saturn-iapetus.html

There is no clear definitive answer for the dark and light material so clearly defined on the surface of Iapetus, except for a few maverick scientists like **Tom Van Flandern** and **Richard C. Hoagland** whose theories and hypotheses have been previously stated**.** Their theories hold as much credence as any other theory and clarify a lot of the mysteries surrounding the Saturnian system as well as the other strange phenomena found throughout the Solar System.

Van Flandern's **Exploded Planet Hypothesis (EPH)** make reasonable sense of the light and dark surface material on Iapetus and why there is a predominance of moons, satellites and planets in the Solar System that are heavily cratered on one hemisphere of their planetary body. Namely, as previously discussed in this section, the **Asteroid Belt** is the remnants of a former planet that exploded approximately 65 million years ago. The explosion of Planet V created solar system-wide devastation scaring many planets and their moons with crater impacts and "planetary soot" debris.

In Van Flandern's reconstructions, there was several successive planets *(not just one planet* - of a significantly more populated former solar system) literally explode (!) and spread their shrapnel far and wide – separated by hundreds of millions of years between explosions – successive "waves" of impacting debris repeatedly collide with all the remaining planets and their satellites, to leave a highly intermittent record of *overlapping* destruction and catastrophe … which is still *on-going*.

A major explosion would send a blast wave through the solar system, blackening exposed, airless surfaces in its path. Most such solar system surfaces are indeed blackened, even for icy satellites. But a few cases have such slow rotation that only a little over half of the moon gets blackened. Saturn's moon Iapetus is one such case, because its rotation period is nearly 80 days long. **Error! Reference source not found.** shows a spacecraft image of Iapetus. One side is icy bright; the other is coal black. The difference in albedo is a factor of five. Gray areas are extrapolations of black areas into regions not yet photographed. As such, they represent a prediction of what will be seen when a future spacecraft (Cassini?) completes this photography.

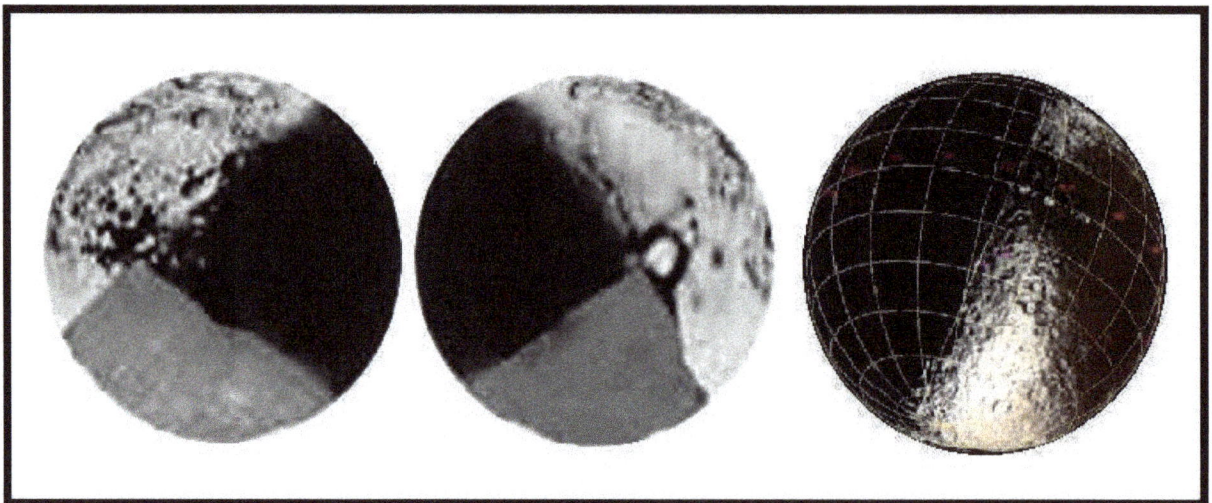

**Van Flandern's "simplified" EPH model for Iapetus (left and middle)
and the actual "gridded" Cassini image of Iapetus (right).**
http://www.enterprisemission.com/moon4.htm

There is a problem with Van Flandern's prediction model for the case of EPH and Iapetus as can be seen in this "gridded" Cassini image from December 8, 2004 (above) as it NOT specifically fulfilled; the distinctly *elliptical* geometry of "Cassini Regio" does NOT conform to his simplified "dark half moon" model (above – left).

This is not a major failure of Van Flandern's model as some may like to argue, as Hoagland's subsequent discovery of completely independent evidence for EPH subsequent to this apparent "EPH failure at Iapetus" is actually due to an oversimplified model for *the interaction of such explosion debris from Planet V with Iapetus itself*... rather than an intrinsic failure of the EPH.

Van Flandern's EPH model for Iapetus is essentially vindicated!

Hoagland's pick up where Van Flandern leaves off and sees planetary and satellite evidence for a **Type Two Civilization** that once existed in the Solar System and Iapetus is a major *"calling card"* of this former highly advance ET civilization.

A closer inspection of Iapetus will reveal just how odd this moon of Saturn really is; the implications of its anomalies seem to indicate that the whole moon is an artificial construct.

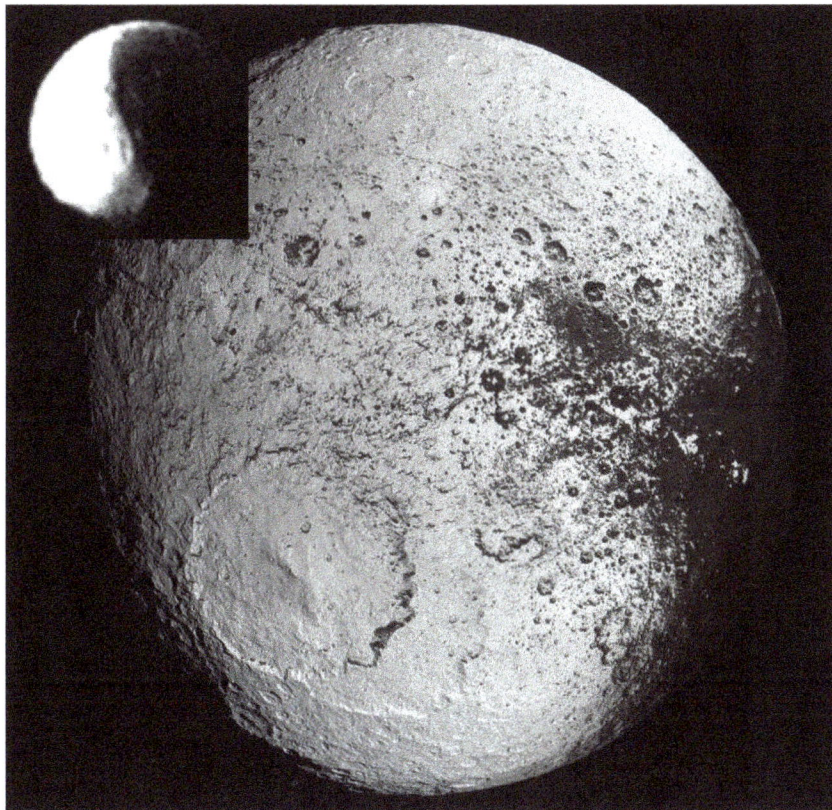

Close up of Iapetus, the Yin/Yang Moon (with insert) showing it high albedo light side and some of its dark charcoal-like material giving it the familiar Chinese iconography
http://islandrepublicofdan.blogspot.ca/2014/03/iapetus-ancient-artificial-moon.html
and https://www.universetoday.com/123045/saturns-yin-yang-moon-iapetus/

Burrowing, rather heavily from Richard Hoagland's Enterprise Mission website and with acknowledgement to him and his research team for their excellent in depth investigation into these planetary phenomena, we continue to follow their lead and to connect the dots on the UFO/ETI phenomenon. **(Author)**

On New Year's Eve, 2004, the Cassini spacecraft got the closest fly-by of Iapetus to date passing within 40,000 miles, revealing even stranger mysteries about this waxing exotic moon in which Arthur C. Clark seemed to know so much about before the Voyager 1 and 2 encounters with it. Cassini confirmed one curious impression left from the Voyager encounters of a quarter century before: in addition to its other unique characteristics, Iapetus does *not* seem to be a perfectly round moon!

Iapetus appeared to be visibly "squashed" by something like 50 miles out of its 900 miles, or about 5%. Iapetus is not round like a real sphere or like any other planet or moon in the Solar System, although Saturn seems to be an odd-ball place for strange moons like Pan and Atlas.

"For solid rocky bodies larger than a few hundred miles across, the relentless force of gravity *always* overcomes the innate tensile strength of such materials, and forces them to assume a spherical geometry. For solid icy bodies (those possessing less tensile strength), the limiting size before a sphere is formed is even smaller." http://www.enterprisemission.com/moon1.htm

Iapetus and the long narrow equatorial ridge girdling the entire moon

According to calculations by both Hoagland and NASA derived from density observations made by both Voyager and Cassini spacecraft, the "specific gravity" of Iapetus is 1.21 slightly more than water which has a "mean density" of 1.0, the Earth has a average density of 3.34 because of

it much denser "silicate rocky core". However, Iapetus which is mostly a solid, icy body measuring 900 miles across, but which only rotates once every 79 days in a synchronous orbit, therefore, any equatorial "centrifugal force" is clearly insignificant. Thus, this cannot be the source of Iapetus' major "out of roundness" according to **Hoagland**.
http://www.enterprisemission.com/moon1.htm

Iapetus' poses a striking, yet bizarre balance of contrast between its "chocolate brown" leading hemisphere and the dazzling white north and south "polar caps.

Stranger still is the long narrow equatorial ridge that completely runs the circumference of the Iapetus appearing much like a gigantic walnut.

Just as baffling is the presence of an arrow-straight, lineal feature, a *12-mile-high* (~60,000 foot!) "wall" which precisely bisects the leading hemisphere, and apparently crosses *the entire width* of this strangely darkened "Cassini Regio" that is over 800 miles in length.

It is Hoagland's opinion that Cassini's discovery of **"the Great Wall of Iapetus"** now forces serious reconsideration of a range of staggering possibilities … that some will most *certainly* find … upsetting: **That, it could really *be* a "wall" … a vast, planet spanning, *artificial* construct!!** *A wall of an extremely ancient construct now in ruins!*
http://www.enterprisemission.com/moon1.htm

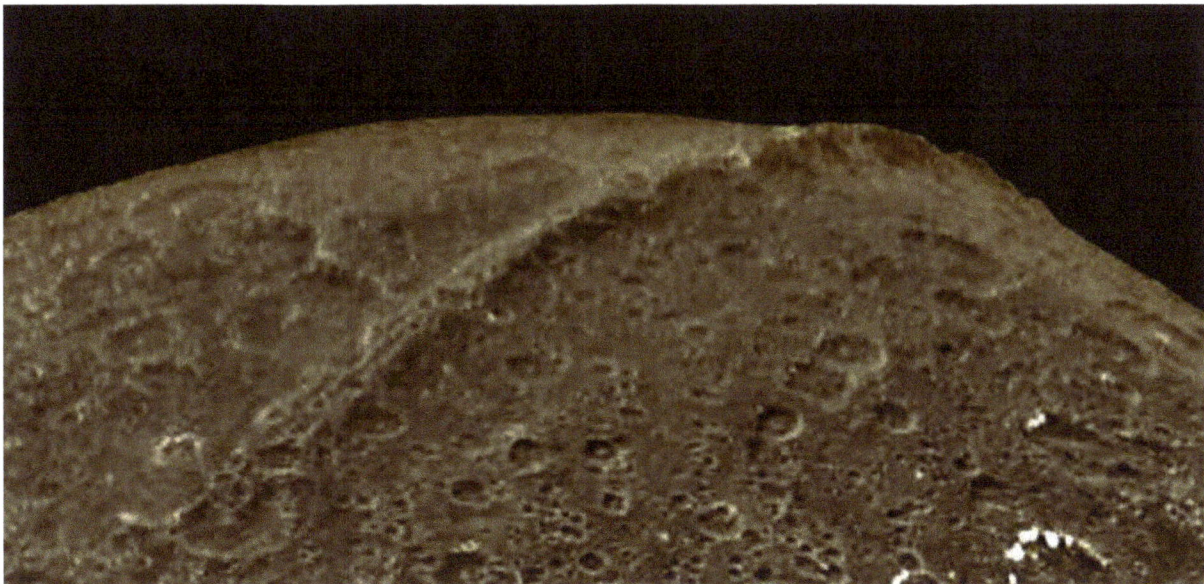

"The Great Wall of Iapetus" is 12 miles high (~60,000 feet!) and about 12 miles wide girdling the equatorial region of the strangest moon in the Solar System
http://www.enterprisemission.com/moon1.htm

This is not the first time that startling new data has prompted scientific consideration of "intelligence" at Saturn.

The extreme albedo range displayed by Iapetus prompted **Donald Goldsmith** and **Tobias Owen** (the NASA discoverer of "the face on Mars!") in 'The Search for Life in the Universe" (1980) to write that Iapetus' brightness variations might be *artificial*:

This unusual moon is the only object in the Solar System which we might seriously regard as an ***alien signpost*** *- a natural object* ***deliberately modified*** *by an advanced civilization to attract our attention* [emphasis added] ....

The photographic revelation discovered by Cassini of unquestionably *the greatest linear feature in the solar system* is of such staggering scientific profoundness as to spur greater urgency for an explanation to find a viable geological model to explain a *sixty thousand-foot-high, sixty thousand-foot-wide, four million-foot-long* "wall" … spanning *an entire planetary hemisphere* … let alone, located in the *precise plane* of its equator!

"Nature doesn't usually create straight lines." is a well-known cliché yet, in apparent contradiction there are *three* such straight lines supposedly created by Nature, all running *parallel*, not only to each other, *but* to the literal equator of the planet.

What could be compared to a veritable "**Maginot Line**" of the geometric complexity and regularity are seen here … certainly not one stretching horizontally, across this one small section of Iapetus, for over *sixty* miles …. http://www.enterprisemission.com/moon1.htm

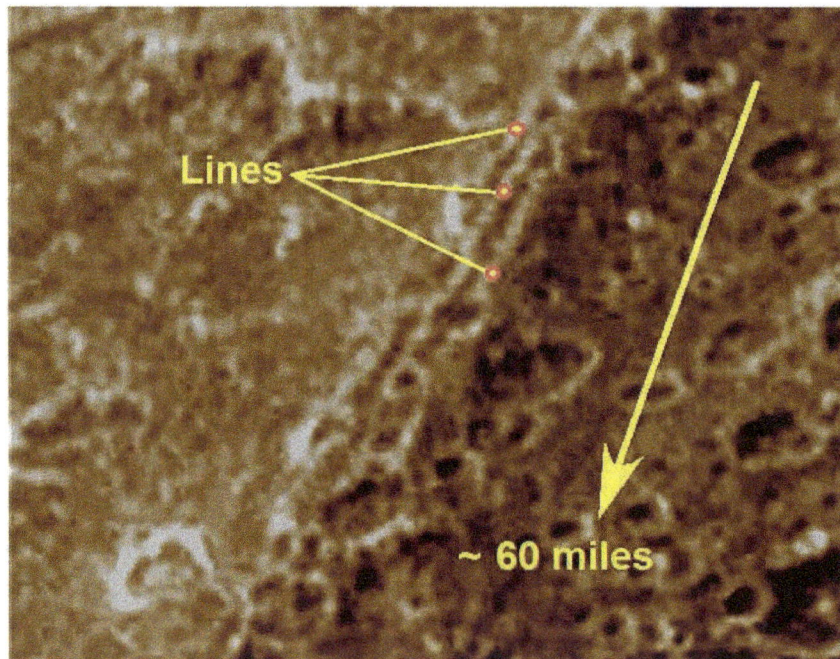

Three parallel lines along the top of the equatorial ridge that stretches for sixty miles
http://www.enterprisemission.com/moon1.htm

This gigantic "Wall" structure boggles the mind when we consider the implications of the engineering construction aspects for its creation. We can only speculate what materials were used, how the construction process actually took place, was construction done by robots or living beings and who were they?

The tallest man-made structure in the world is a skyscraper, the **Burj Dubai Tower (Barj Khalifa)** in the oil rich kingdom of Dubai, United Arab Emirates, standing at an impressive 829.8 m (2,722 ft), that is ***well over a half mile in height!*** (1 mile = 5280 feet = 1.60934 kilometers). Construction began on 21 September 2004, with the exterior of the structure completed on 1 October 2009. The building officially opened on 4 January 2010.

**Earth's tallest structure is the Barj Khalifa in Dubai, UAE
at an incredible 2722 feet in height! That's over half mile in height!!**

Below is a graph of some of the tallest structures ever built on Earth from the **Giza Pyramid** built four thousand years ago in Egypt to the most recent **Barj Khalifa Tower** in Dubai, UAE. To show the comparative proportions of these Earth based buildings to the **Iapetus Equatorial Ridge,** it would require *8 pages of this textbook in height by 12 pages in width!!!*

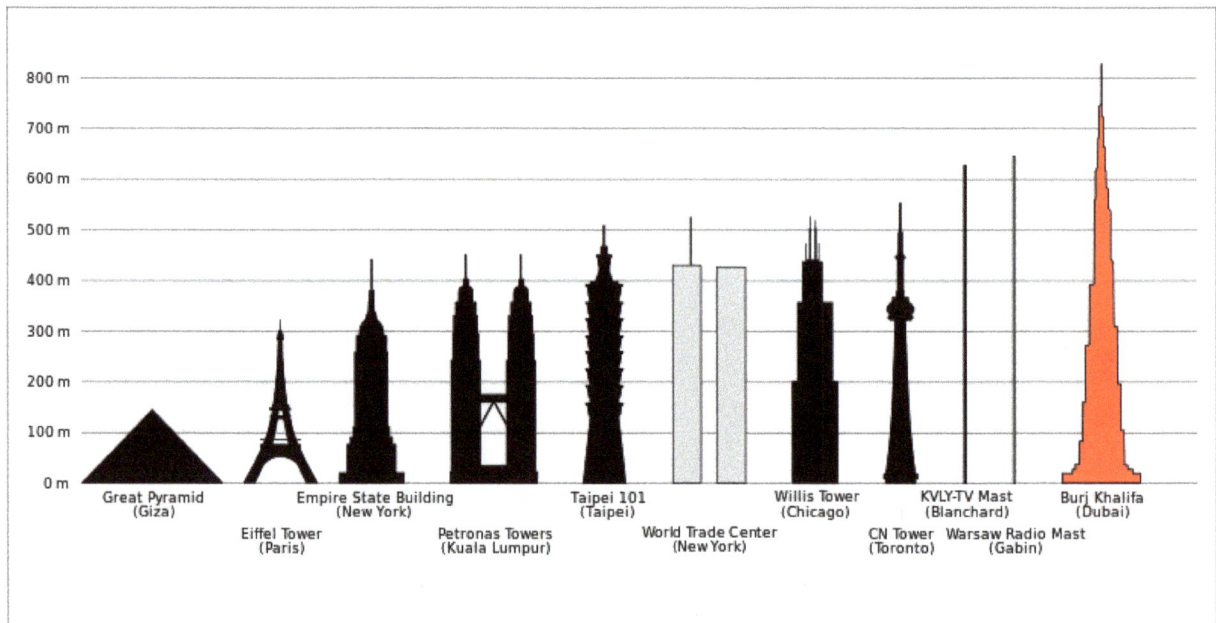

Some of the tallest structures ever built on Earth
http://gadling.com/2011/07/06/exploring-the-worlds-tallest-structure-the-burj-khalifa/

What this "Wall's" function is remains a mystery at this time but, like so many Extraterrestrial artifacts found on the Moon or on Mars or in orbit about planets like Saturn, it is Hoagland's contention that it probably has something to do with Torsion Fields and hyperdimensional physics.

As **Richard Hoagland** has stated earlier in this section with regard to the ringworld of Saturn, the mammoth size of that complex would require giant beings to create it and no doubt those same beings would also have created the equatorial ridge of Iapetus!

Iapetus has a specific gravity 1/40th of that of Earth and a proportional size building like that of the Barj Khalifa would reach a height of 15 miles on Iapetus! In other words the construction of the **Equatorial Ridge** or **"Wall"** is certainly doable and if giant size beings of a **Type Two Civilization** who are highly advanced technologically, they would no doubt implement nanotechnology techniques and materials like nanotubes, carbon fibres, zero-gravity crystalline titanium employing armies of large robotic engineers and billions of **nanobots**. And if this interstellar civilization is extremely advanced they may have little or no need for mechanical construction processes. They may have the ability to utilize **Consciousness Assisted Technology (CAT)** or **Technology Assisted Consciousness (TAC)** in which thought assisted construction or engineering derived from an idea form is created into tangible form with the assistance of thought creation technology).

An example of this type of technology is the classic flying disc or saucer in which witnesses have re-counted an "on board" experience who stated that they saw no seams or rivets to the internal or external structure of the craft. This is a prime example of **CAT** or **TAC** being used to design and construct the spacecraft, even to the point that some spacecraft are alive! They are living sentient entities!! This would explain the reports that damaged spacecraft with "broken or torn" pieces from the craft appear to move toward each other in a process to heal itself or change shape in flight!!!

Now, think of this technology on a grander, megalithic scale in which mammoth-size cities and monuments or even planet-size spacecraft are constructed such as the Saturn moon Iapetus!

Recall what astrophysicist **Carl Sagan** stated that geometry and rectilinear structure on a planet is a sign of intelligence. Now, if we look at Iapetus even closer, we see exactly that, "a striking set of clearly defined, astonishing, repeating, *three-dimensional* rectilinear surface patterns!" These rectilinear structures are "located several hundred miles north of the Wall … near the boundary between the "brown stuff" and the "white stuff" on the leading hemisphere of Iapetus". These are NOT '"square craters" but remarkable, highly ordered evidence of sophisticated, aligned, repeating *architectural* relief!"
http://www.enterprisemission.com/moon1.htm

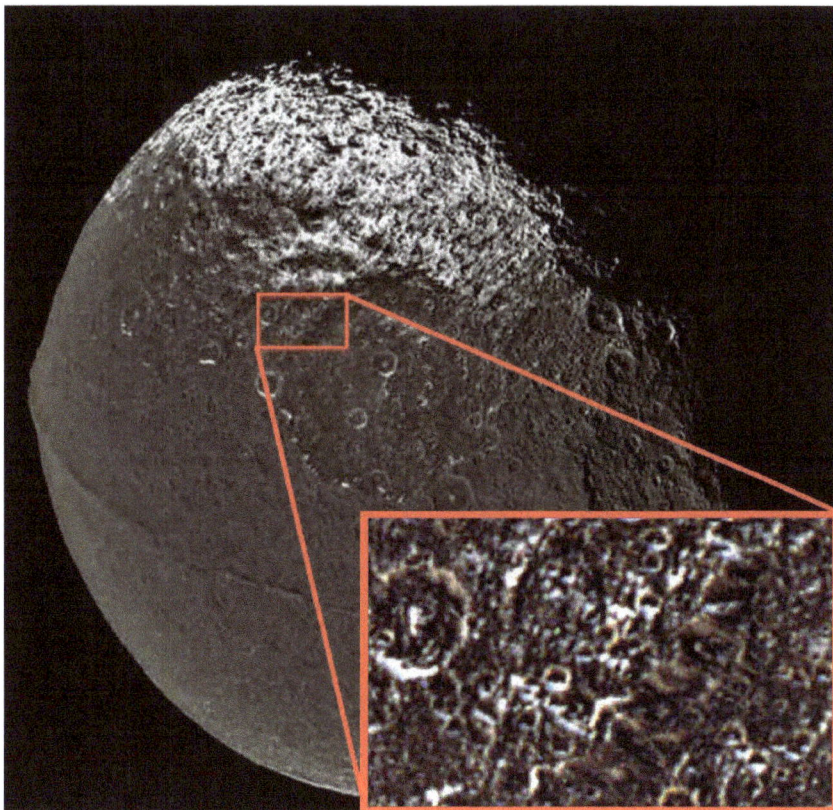

Rectilinear patterns and structures abound on the surface of Iapetus.
These are not "square craters"
http://www.enterprisemission.com/moon1.htm

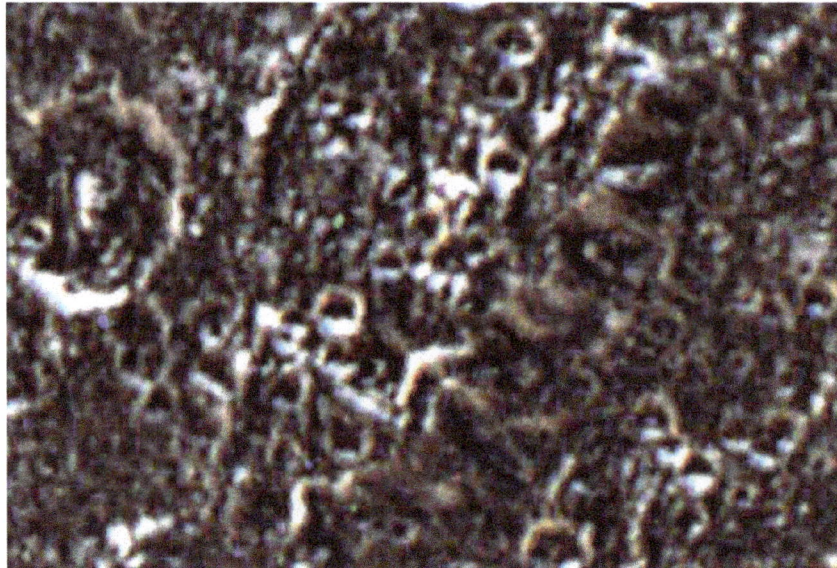

This close up image of Iapetus' surface shows many "box-like" buildings and rooms of a particular cookie-cutter size. Note the "round" crater to the left of the box structures

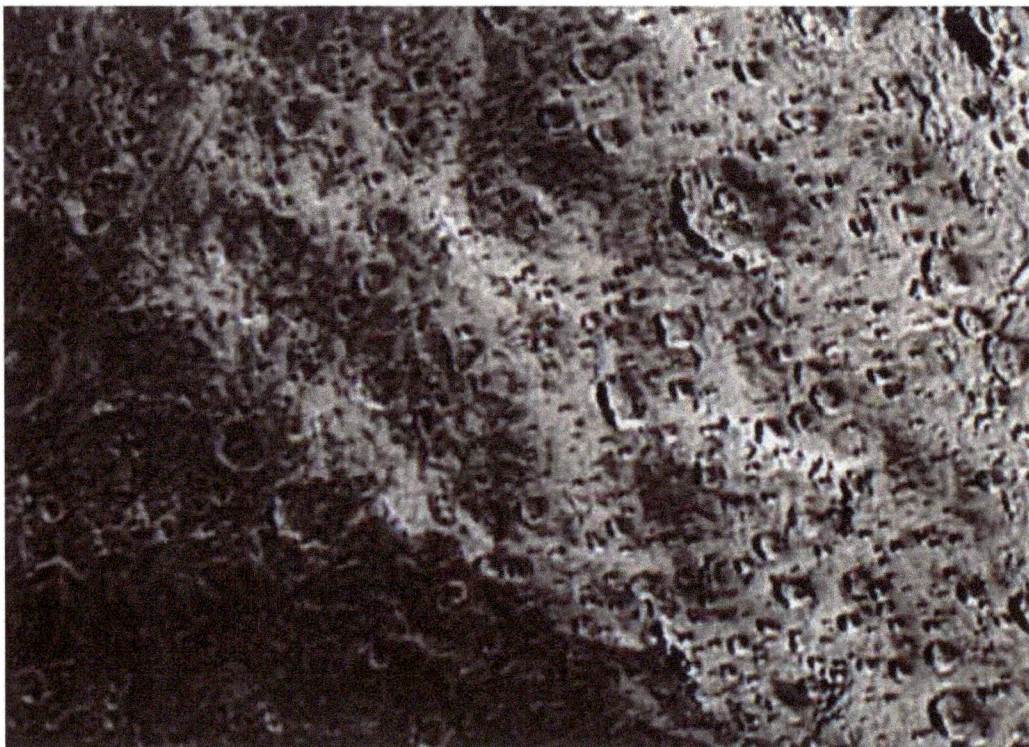

A close-up of the "transitional terrain" revealing more of the 3D "honeycombing" surface features. Note the repeating aligned edges of hundreds of "square" holes, right-angle, uniform-width "walls" which appears to be "mantled" with a heavy "snowfall"

362

The surface of Iapetus is "carpeted" by vast plains of extremely ancient *ruins* that have lain exposed for countless eons as only surviving walls without roofs, now covered by "snow" and what appears as "brown soot or ash debris". There is literally thousands of square miles of **rectilinear ruins** covering both north and south hemispheres of this moon.

Further evidence can be seen from this wide angle Cassini color image (see below), taken of the northern "polar ice" and terminator. The top center (red outlined area) is blown up to reveal an area close to the surface horizon that has repetitive vertical architectural geometry and rectilinear design extending across this entire region, is unmistakable … and totally "unnatural."
http://www.enterprisemission.com/moon1.htm

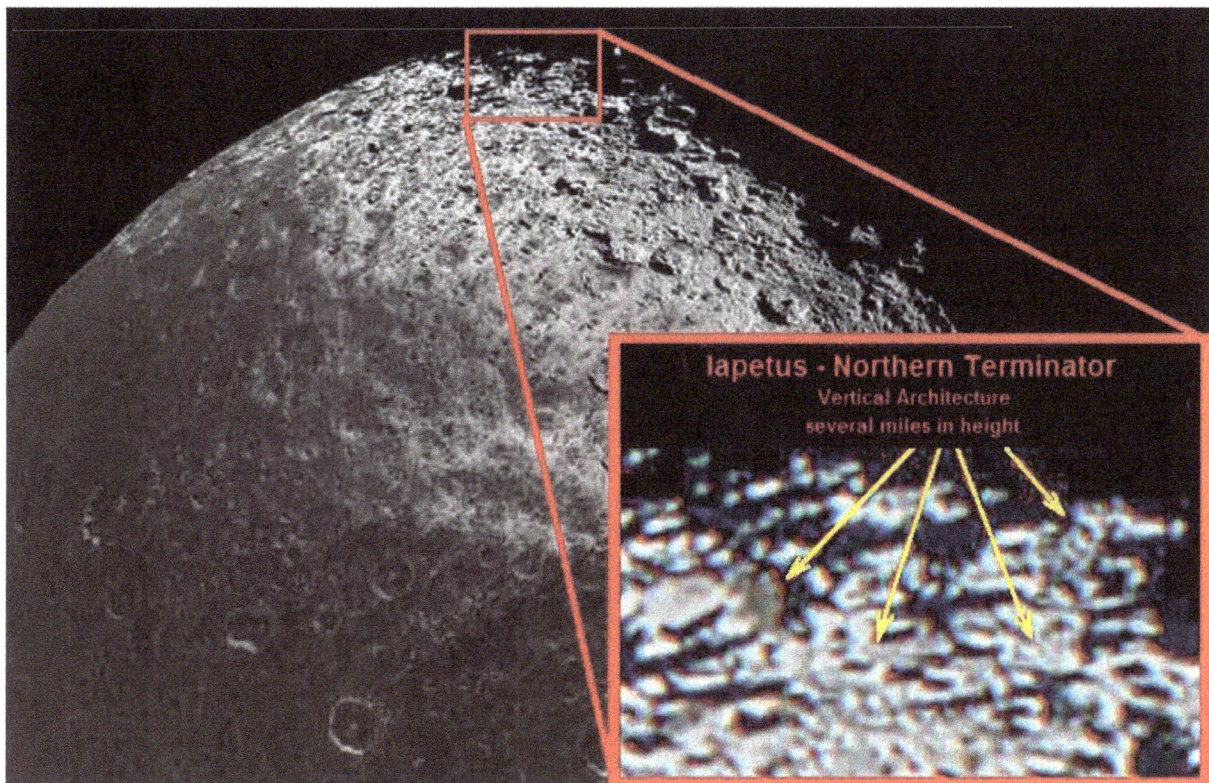

More evidence of rectilinear architecture on the surface of Iapetus
http://www.enterprisemission.com/moon1.htm

Iapetus has other anomalies on its surface that are not natural to the moon like a tower that rises about a mile or so above the surface. (See below). This is similar to some of the other tower structures found upon our own moon, Luna as well as found on Mars and its two moons Phobos and Deimos.

It would appear that these are communication towers strategically located throughout the Solar System enabling one planet or moon to communicate to another planet or moon or possibly with spacecraft traveling across the interstellar distances either within our star system or beyond to other star systems.

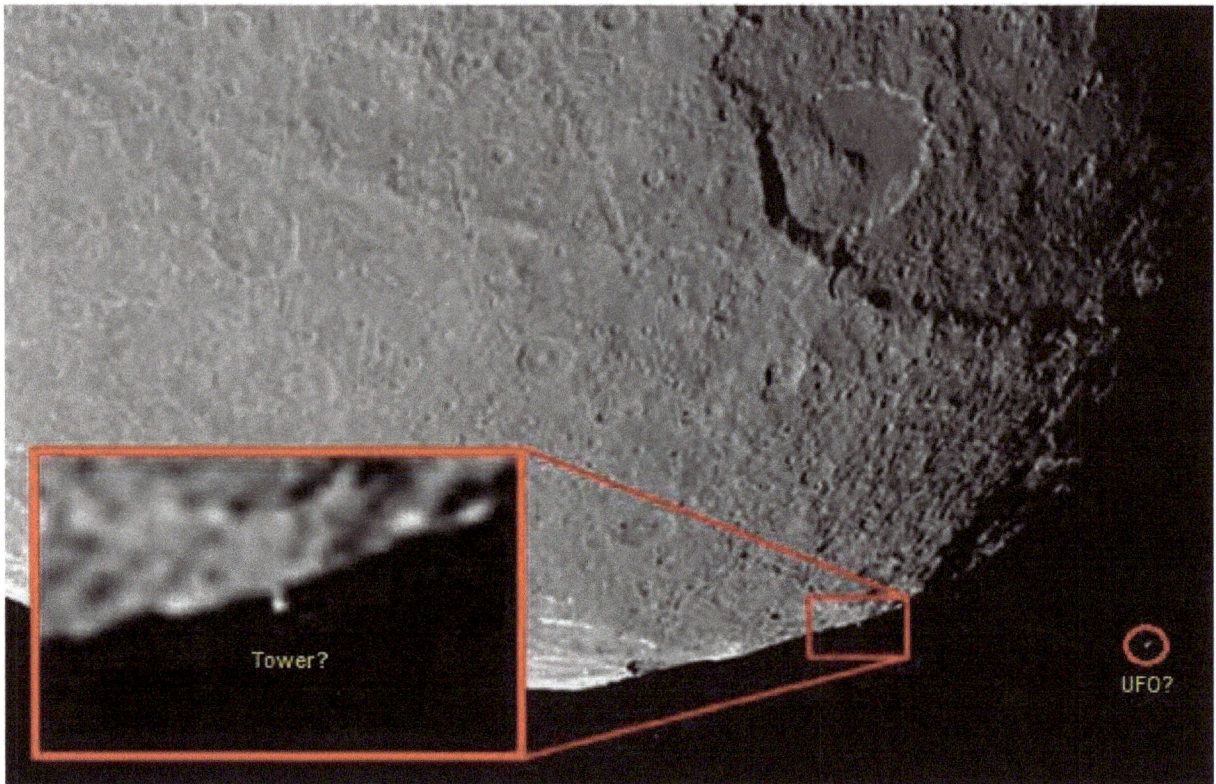

Is this a communication tower in the southern polar area of Iapetus or a rock outcropping?
http://www.enterprisemission.com/moon1.htm

By now, it should be obvious to the reader that Iapetus is not another simple moon in the planetary system of Saturn but a planetary body with a lot of anomalies ranking right beside the Moon and Mars as a place strewn with artifacts of intelligent design.

The Saturnian system is more than just a place of "organic smog" or an atmospheric moon with "liquid methane oceans", it is a system containing unusual ring formations that appear to be constructed and maintained by gigantic planet-size spacecraft plus it is the home of a possible ringworld orbiting Saturn. To this planetary system must now be added strange moons and satellites with Iapetus being the most unusual moon orbiting Saturn,

Iapetus has a near non-existent surface gravity, no atmosphere or winds, no living organisms, has an abundance of real craters, it is an icy cold moon with snow on one hemisphere and dark sooty material on the other hemisphere, yet this moon has extremely ancient ruins that are strewn across thousands of miles amongst the real craters.

How is this possible? These ruins had to have been built at one time upon a breathable surface by an *ancient, extremely advanced extraterrestrial civilization!*

If there was a breathable atmosphere on Iapetus in its ancient past supporting the countless inhabitants that built and occupied these architectural ruins as well as having built the mysterious "Wall", what happen to it, unless, as postulated by Hoagland, the air was never *outside* but, on the *inside?*

364

In other words, what if Iapetus is not a natural satellite at all, but a 900-mile wide spacecraft, an artificial "moon?!" http://www.enterprisemission.com/moon2.htm

Is it possible that Iapetus is a 900 mile wide spaceship? We've already seen and examine photographic evidence of other gargantuan objects in the Solar System. Photos have been take of these planet size objects near the Sun, by our Moon, out in deep space near Mars and Saturn so, it should not come as a surprise to the reader by now that such spacecraft do exist within our solar System. Iapetus seems to be our first real opportunity to get an up close real good look at one of these planet or moon-size spacecraft and Iapetus doesn't disappoint!

Hoagland points out that Iapetus is cover with irregular shaped craters that are not perfectly round as one would expect to find on other planetary bodies like our Moon. In fact most of the craters on Iapetus appear to be polygonal in shape, that are five, six, seven or even eight sided in appearance with an anomalous alignment that is evenly space running parallel to the "Great Equatorial Wall." There are polygon craters and structures that litter the surface of Iapetus!

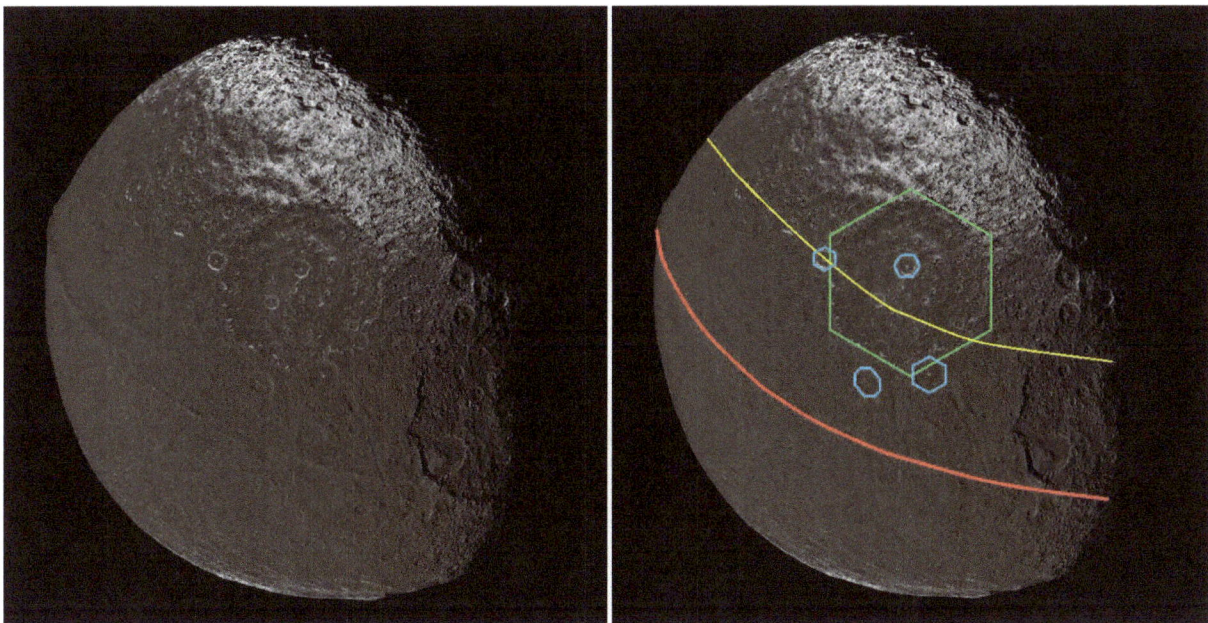

**Crater alignment (yellow line) that is evenly spaced and running parallel to
the equatorial ridge (red line). Note large (green) and small (blue) hexagon craters**
https://www.bibliotecapleyades.net/luna/esp_lunaiapetus02.htm

Now, every astronomer and student of astronomy knows that random same size comets, meteorites and asteroids do not impact a planetary surface over millions or billions of years *in a precise equidistant alignment which just happens to run exactly parallel with the equator of Iapetus*! It just doesn't happen!!

Yet, this is exactly what we find on Iapetus!
A close inspection of one of these hexagon "craters" reveals a major architectural form that is part of an array of equally incredible *geometric* architecture with its uniform-width; "castle-like

walls" and geometric interior which is unlike any actual impact craters imaged anywhere else in the solar system is in fact part of a set of pre-existing structures.

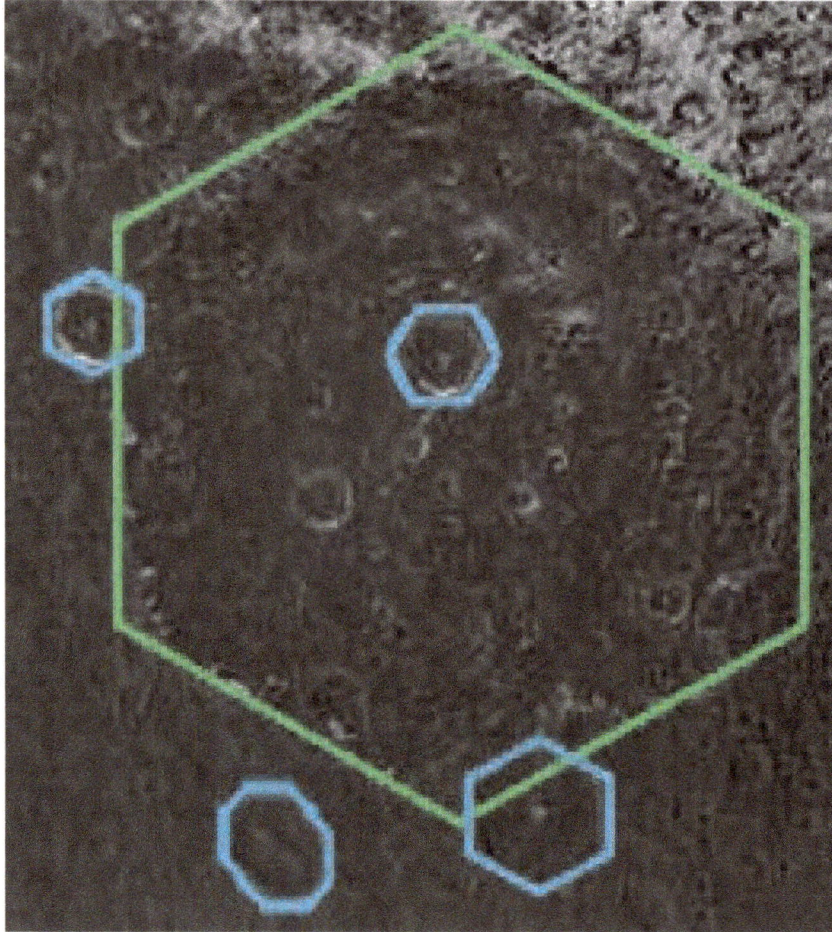

The surface of Iapetus is fractally constructed of various size tetrahedral modular units
https://www.bibliotecapleyades.net/luna/esp_lunaiapetus02.htm

These *"geometrically arranged craters"* are in fact, structurally-defined *surface collapse features* revealing key weak points in the basic sub-modules of this "artificial moon." From and engineering perspective this moon-size spacecraft has been constructed fractally with "multitudes of identical" interconnecting small and large modular units employing the fundamental principal of the "tetrahedral truss", "a replicating **tetrahedral pattern**", the basis of **Buckminster Fuller's** famed "**Geodesic Domes**".

The "latitudes" along which these major sub-units are arranged on Iapetus seems to be determined by the structural strength each one contributes to the integrity of the overall "spherical moon" thus, the hexagon appearance of the craters. The unfortunate aspect is that the major geodesic domes have degraded and collapse over the eons of time from the incessant onslaught of micrometeorites "revealing the geometric nature of the sub-modules themselves" in the form of "deformed hexagons".

366

The largest visible example of this intrinsically *tetrahedral* form is the ~240 mile-diameter basin in the center of the image (above).It too has a basic hexagonal geometry – more supporting evidence for the "two-dimensional, six-sided, sub-module assembly model" for the surface of this entire "moon." As can be seen, smaller collapse features within it (and beside it) are also eroding according to this basic tetrahedral form …. Note also the crater peaks that are smack dab in the centre of the hexagon craters and the elongated ridge in the eight-sided polygon on the bottom left of the above image that runs parallel within the crater. Symmetry is not a natural geological formation but is indicative of artificiality and intelligent design.
http://www.enterprisemission.com/moon2.htm

Buckminster Fuller designed Geodesic domes are light weight and super strong structures that can be constructed on large scales to form villages or cities
http://inhabitat.com/oklahoma-developer-offers-100000-to-anyone-who-will-take-his-geodesic-dome/ and http://geodesicgreenhouse.org/
and https://en.wikipedia.org/wiki/File:Eden_Project_geodesic_domes_panorama.jpg

 A close-up of the vertical, *12-mile drop* -- from the rim to the floor of the major basin (below) - reveals a series of serrated, evenly-spaced "teeth" (similar sized fragments of former structural walls ...) sticking at right angles outward into space from this eroding feature ... creating a series of parallel, ~60,000 foot cliffs. On the basin floor itself, a series of aligned, 90-degree eroded features is also clearly visible (running from lower left to upper right – below) ….

These are all redundant indications that the smaller structural sub-units are, indeed, being fractally eroded backward from the basin center … along each of the six, hundred-mile-long, hexagonal "rim walls." http://www.enterprisemission.com/moon2.htm

By now we must conclude that Iapetus is *not* one of the normal "moons" of Saturn -- but is actually a 900-mile-wide, manufactured, ancient world-sized *spaceship* ... created under 1/40th terrestrial gravity according to a fractally apparent, *"tetrahedral"* pattern! http://www.enterprisemission.com/moon2.htm

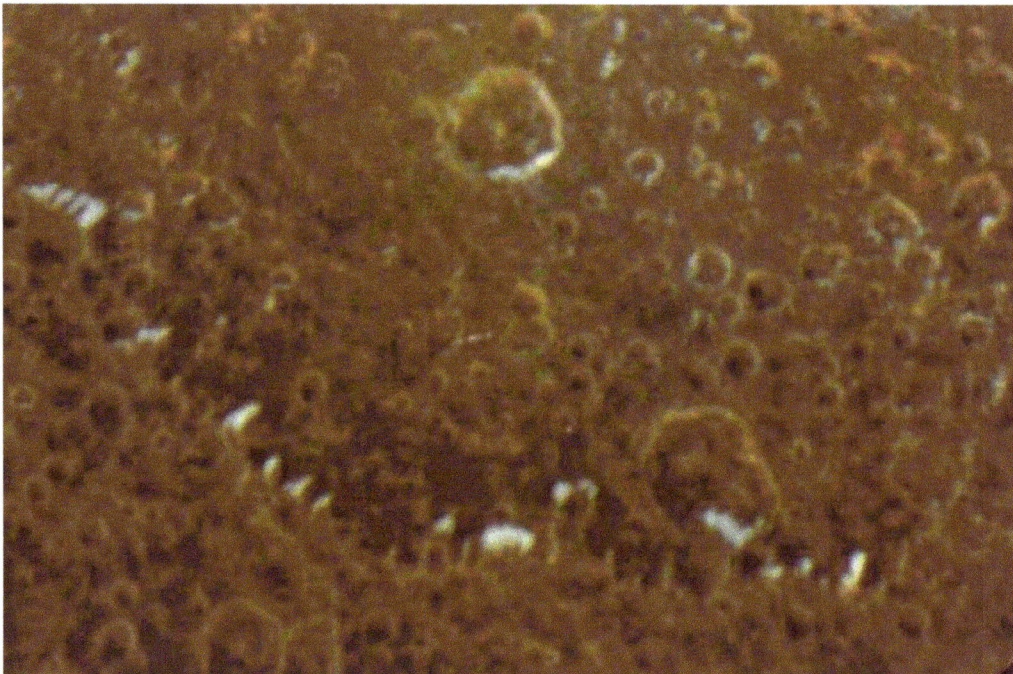

A close-up of the vertical *12-mile drop* from the rim to the floor of the major basin with series of serrated, evenly spaced "teeth" sticking at right angles outward into space from this eroding feature creating a series of parallel 60,000 foot cliffs
https://www.bibliotecapleyades.net/luna/esp_lunaiapetus02.htm

We must come to grip with the realization that we are dealing not with a natural planetary body but, with a planet-sized spacecraft, the supporting evidence of which is found everywhere on and around the surface of Iapetus!

There is at the major impact basin on the right with a "remarkable, rectilinear "waffle" pattern". The explosive formation of that basin to the north would have also totally destroyed any former "surface covering" here as well -- exposing massive inner structural supports.

Close examination reveals a definite "grid" crisscrossing this region … apparently composed of several angled *layers* of overlying structural "rebar" -- but on an incomparably massive scale …. the obvious exposed and layered nature of this "moon" really stands out … as a series of long, linear features which seem to be major structural support in this region for the (now destroyed), overlying "terrain" (below) ….

368

The unique and obviously *layered* nature of Iapetus -- a "shell … within a shell … within a shell" is truly revealed here (below): the "cookie cutter" geometry of this eastern "impact basin" (seen here from a different angle to the previous image) is totally unlike that of any other impact features known ... on any other planet, or satellite across the solar system!

The essentially *vertical*, ~12-mile cliffs (curiously similar to the "12-mile elevation" of Iapetus' "Great Wall"), plunging down *in discrete steps* – as if whatever gargantuan explosion ripped through the surface, exposed successive *shells* of now heavily eroded, *layered Bucky Fuller architecture* – says it all.

These are the shattered, blasted remnants of an ancient, almost incomprehensible science and technology … stark, surviving evidence of a "super-engineering" that once held **together** *an entire world* … and for some reason, *placed it* in the Saturn system.
http://www.enterprisemission.com/moon2.htm

**This major impact basin reveals multiple layers blown away leaving a vertical
12 mile cliff that plunges down in step formation exposing successive shells
of a heavily eroded layered Buckminster Fuller architecture**
https://www.bibliotecapleyades.net/luna/esp_lunaiapetus02.htm

The pièce de rèsistance for the artificiality of Iapetus comes from whole body of Iapetus, itself when viewed 800,000 miles away with the sun shining behind it and "Saturn shine" on the side facing the Cassini camera. (See below). Iapetus has a synchronous orbit around Saturn, always keeping the same side facing in toward Saturn like our own Moon.

We know from a previous discussion in this section, that all major moons of Saturn like Mimas, Tethys, Dione and Rhea have near perfect spherical shapes but in comparison to these moons, something is *strangely bizarre* with Iapetus!

Despite having a larger gravity field and much more mass … and being over 2 million miles from Saturn (so "tides" are totally eliminated) …its overall shape *isn't close to being spherical!*

Look again at the "Saturn shine" image of Iapetus (below left). In the overexposed sunlit portion, the limb of the moon – rather than being round (like Mimas or Dione) – is plainly composed of *a set of sharply slanted planes* (below-right) …. The exact number is difficult to reconstruct (because of the overexposure and the viewing angle), but the outlined areas appear to mark at least six (tetrahedral?) amazingly *flat* "sides" – each measuring *hundreds of miles* in length! http://www.enterprisemission.com/moon2.htm

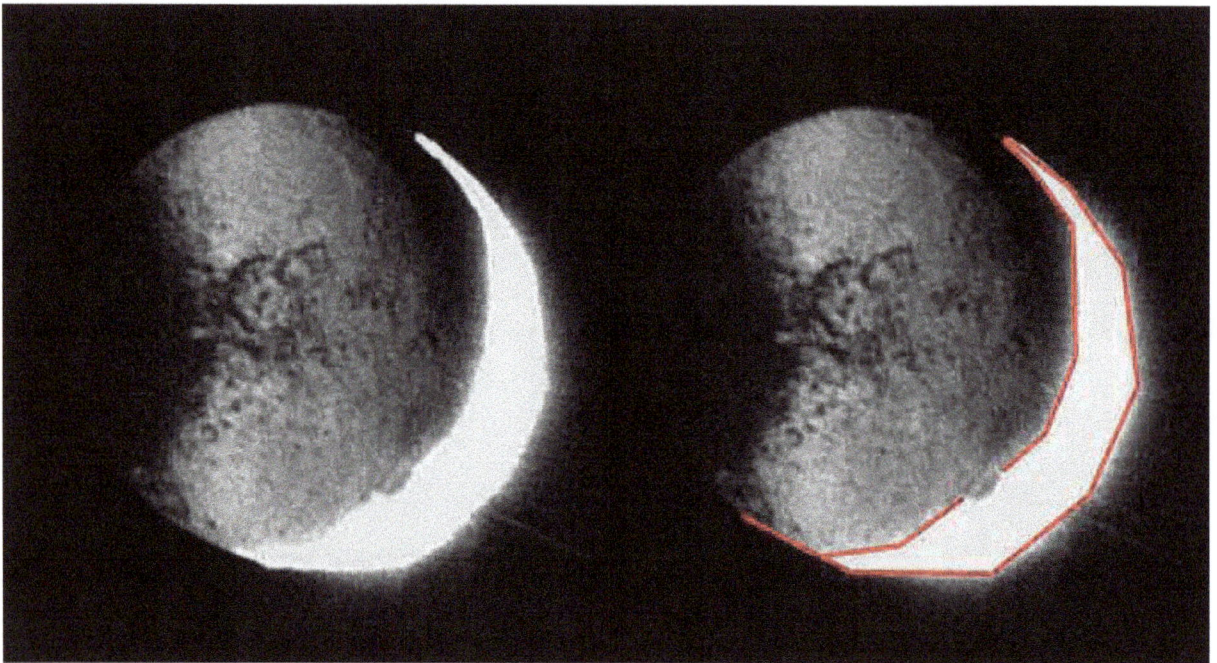

Iapetus in this official NASA press release image (totally overexposed image - left) with sunlit landscape on the right is not spherical like other moons of Saturn but appears to be polygonal in shape (right)

https://www.bibliotecapleyades.net/luna/esp_lunaiapetus02.htm

As Hoagland states, natural planets or satellites do NOT come with sharply defined "straight edges!"

… The image shows mainly the night side of Iapetus; part of the far brighter sunlit side appears at the right and is overexposed due to the long integration time of 180 seconds. Despite this long exposure time, ***almost no blurring due to the spacecraft's motion is apparent*** [emphasis added]. What this means, of course, is that the stark, "impossible" ***straight-edged geometry*** we're seeing on Iapetus' limb … is ***real!***

What are we looking at here? Has someone re-written the laws of the universe as to what a planet should look like, following instead, some other laws of planetary physics and geomorphology?

If we compare this Cassini image of Iapetus with 3D platonic solids, the only regular *polyhedral* ("many sided") geometric forms which will fit perfectly within a *sphere* is an *irregular* polyhedron based on the dodecahedron. In other words, Iapetus planetary structure is not spherical but polyhedral, a dodecahedron in actual fact!

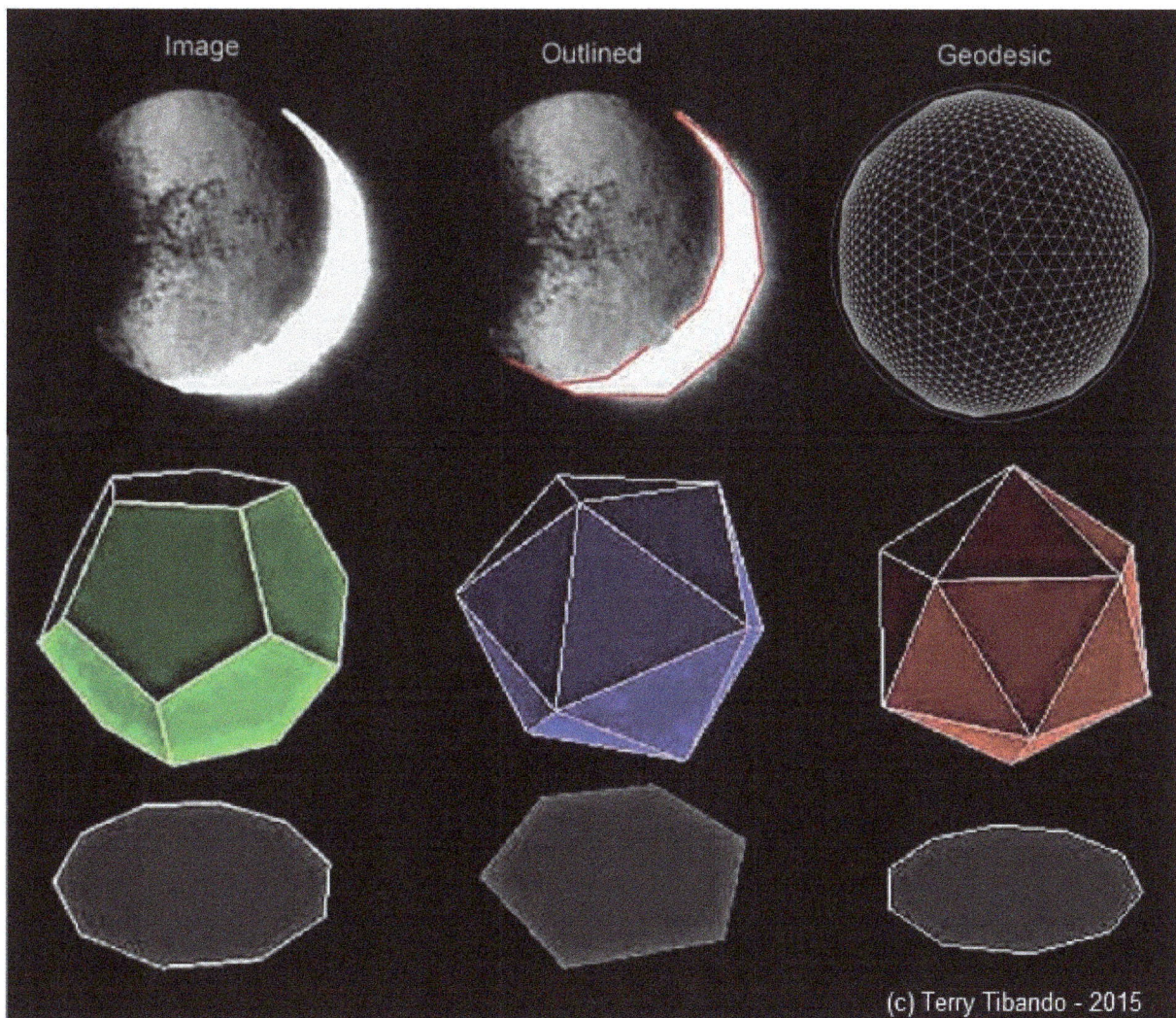

Iapetus is not spherical like other the other moons of Saturn , it exhibits all the features of a polyhedral sphere composed of multiple tetrahedral units

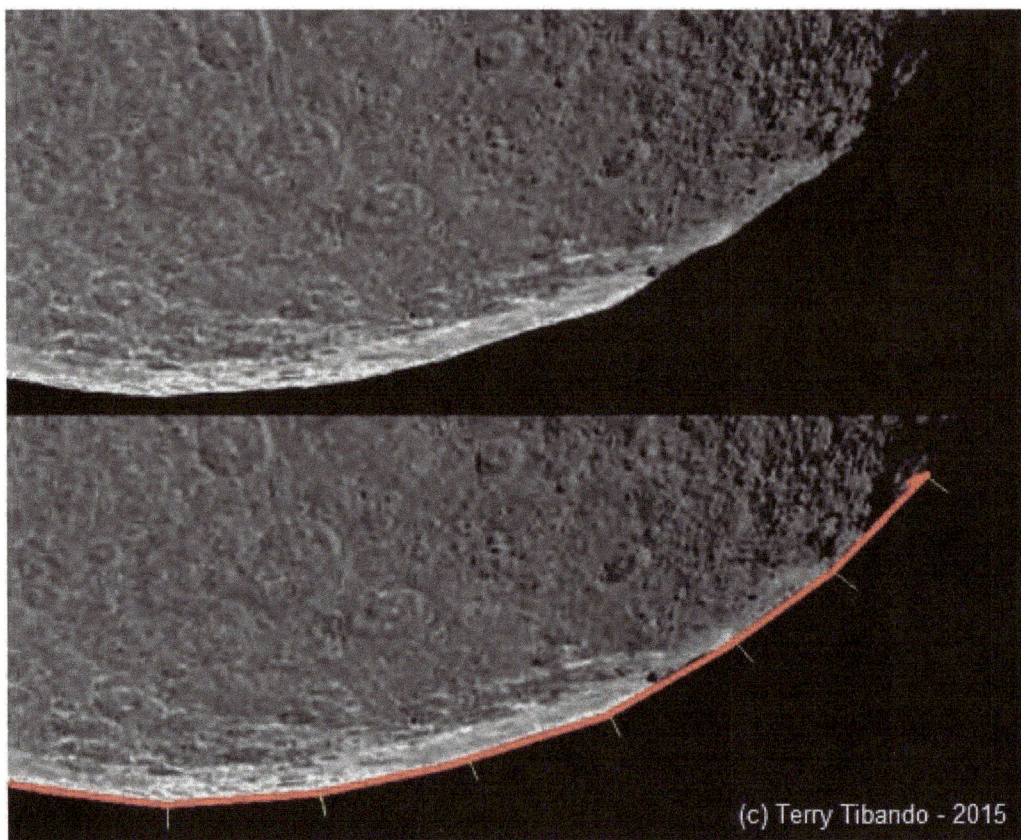

This view of Iapetus shows the impossible geometric planes of the south polar region. Note the equal distance between each plane in this polyhedral moon

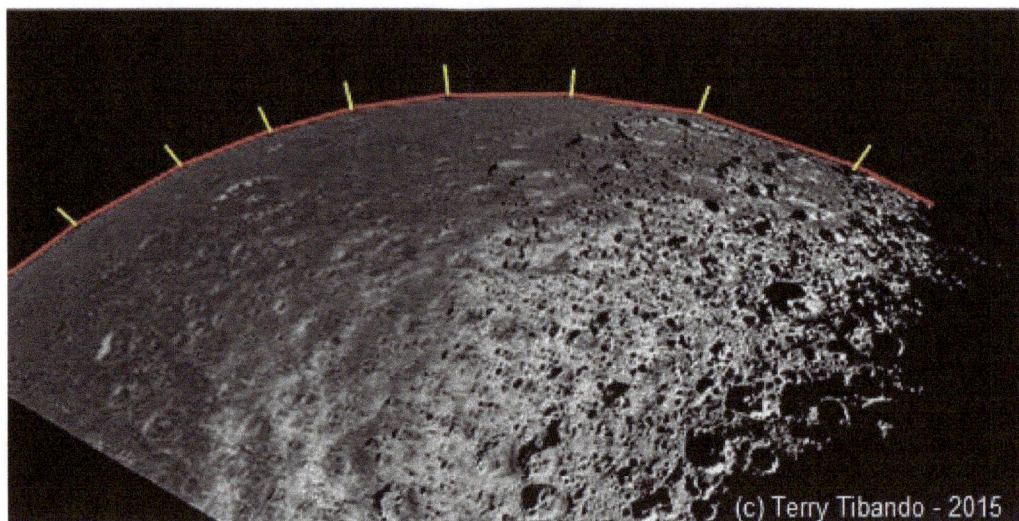

Another view of the Iapetus surface taken as Cassini passed within 80,000 miles revealing a magnified *straight-edged* aspect to the moon's horizon

We are looking at a gargantuan artificial moon-size object *constructed of huge tetrahedral modular units* measuring a hundred of miles in length engineered to be a type of *megalithic Buckminster Fuller (Bucky Ball), polyhedral spherical spacecraft!*

Hoagland points out that there are two great "ring basins" that are 240 degrees apart, one at the east and west ends of the front hemisphere "ellipse that seems to be part of a tetrahedral geometry whose precise placement is unusual. But, it only appears unusual at first appearance and makes perfect sense when understood that the "tetrahedral" location of these two strange "rings" is an imaginary tetrahedron placed inside of a dodecahedron or higher geodesic sphere. Iapetus is in fact an artificial, now anciently eroded, *dodecahedron (or a higher order "geodesic sphere")* – then, suddenly makes perfect sense!

Hoagland says that "if a tetrahedron is placed *inside* a dodecahedron, its vertices can also be precisely placed on select vertices of the dodecahedron itself. This, in turn, generates a "lowest order solution" to the puzzle of why the two "impact basins" marking the Iapetus ellipse would be located on the surface according to a *tetrahedral* geometry.

This means that the basins *aren't* "natural impact basins" after all -- but more ancient, geometric evidence that Iapetus was once a **Platonically-designed world-sized *spaceship 'moon'"***
http://www.enterprisemission.com/moon2.htm

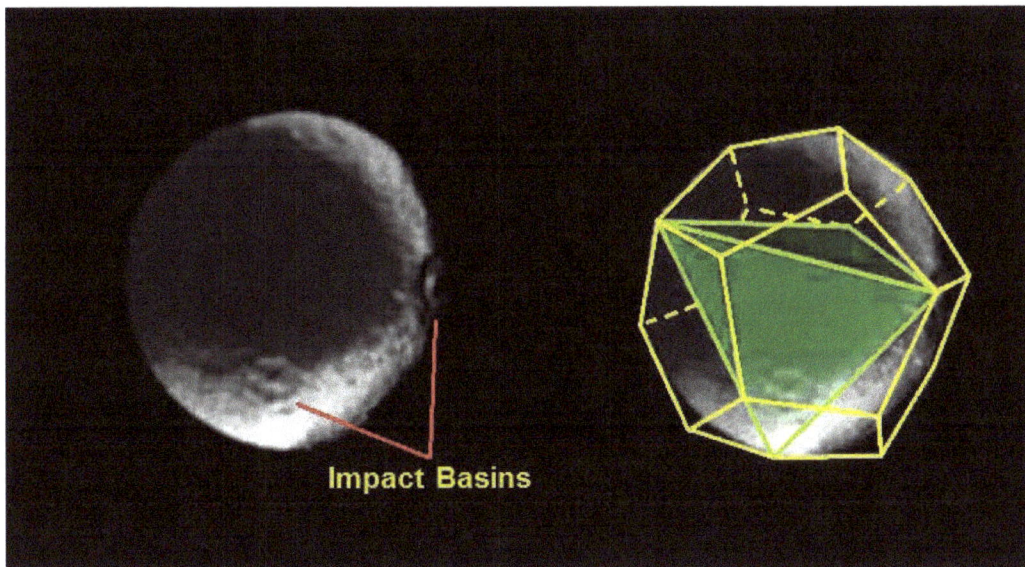

The impact basins are situated in a tetrahedral geometry of a dodecahedron sphere, one more proof of Iapetus as a world-size, spaceship moon
https://www.bibliotecapleyades.net/luna/esp_lunaiapetus02.htm

Perhaps, Iapetus may not be too dissimilar to the Star Wars movie's "Death Star" battle station after all, at least in its basic tetrahedral/ polyhedron design. Certainly the similarities of Iapetus' remarkable equatorial ridge feature bears comparison to the "Death Star's" equatorial trench and the great impact basin in the northern hemisphere is similar to the "super laser". But what must certainly come to mind to the astute reader is the angular planar surfaces of Iapetus that may serve a similar purpose to that of the stealth technology found in such stealth aircraft as **Lockheed's F-117 Nighthawk,** namely radar invisibility!

Iapetus is another Saturnian moon that has more than a passing resemblance to the Star Wars "Death Star battle station, just coincidence or is it something more?

The F-117's pyramidal angular appearance is directly related to this fundamental principle of stealth technology: a shape that *geometrically* redirects by simple *reflection* incoming radar energy *away* from returning to the transmitting antenna! (See below)

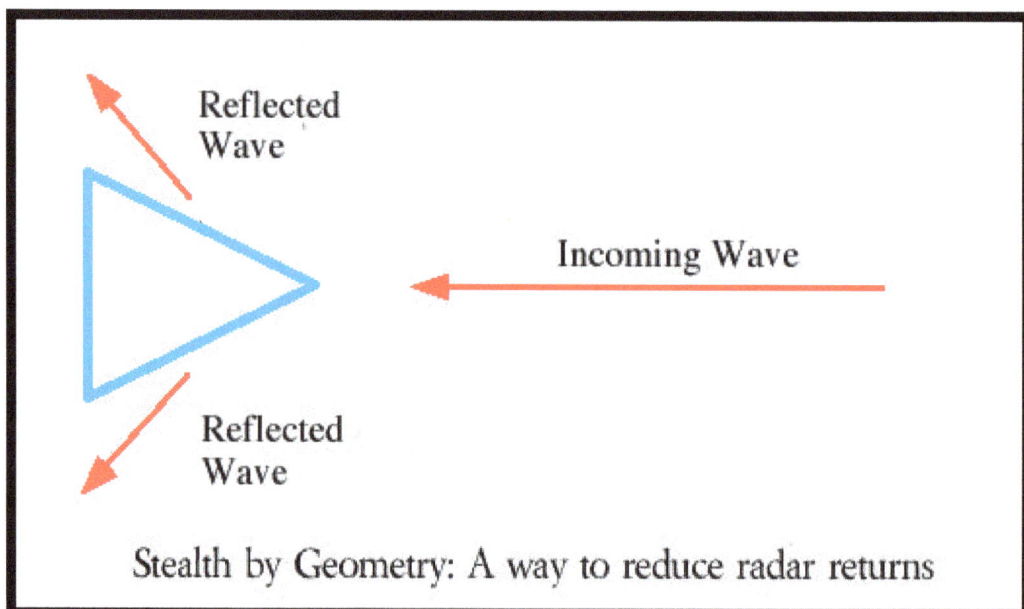

Reflected Wave

Incoming Wave

Reflected Wave

Stealth by Geometry: A way to reduce radar returns

In similar fashion, the platonic geometry of Iapetus is redirecting incoming radar energy *away* from any radar source which gives Iapetus its "anomalously low radar reflectivity"! (See below).

374

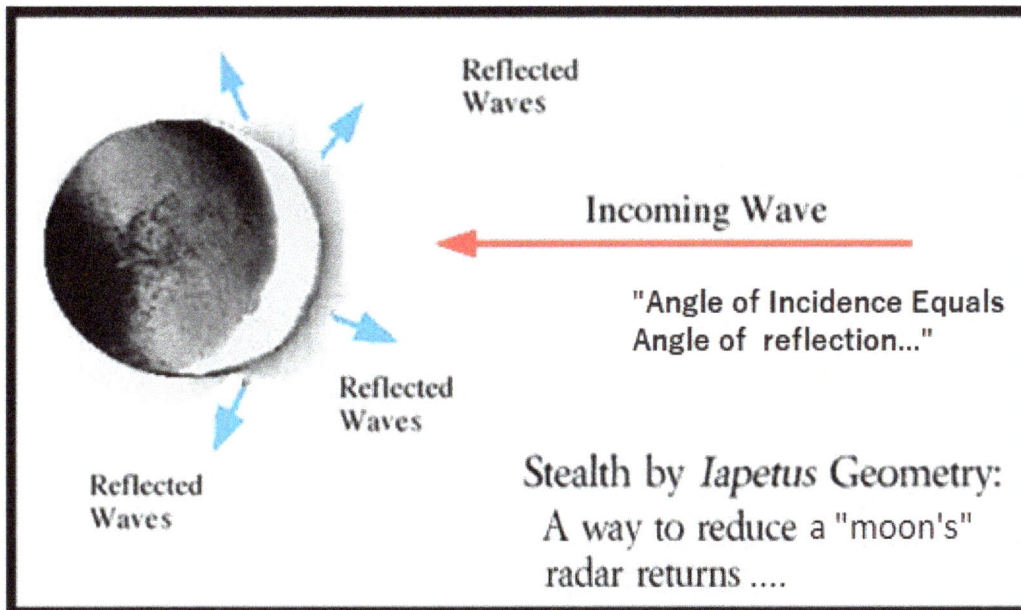

There is also, a major *optical* confirmation that this is what's occurring is provided once again by the new Cassini imaging. **http://www.enterprisemission.com/moon3.htm**

Two images of the "Saturn shine" exposures of Iapetus were taken one day apart on October 21st and 22nd, 2004 which long "diffraction spikes" that emerge in opposite directions in both images. These were caused by oversaturation of "the Cassini CCD camera system is by a brilliant reflection from apparently one small, sunlit region on the trailing side of the 'moon.'"

Now, if you measure the distance of this bright reflection from the sunlit limb, using the "diffraction spikes" as reference -- from one image to the next you'll see that it is *changing* ... the reflection getting significantly closer to the visible "moon's" edge as the orbital viewing *angle* between the spacecraft, Iapetus and the Sun narrows over roughly a 24-hour period.

This progressive movement, relative to the "moon's" edge as the viewing angle changes, is the central hallmark of a *mirror-like* reflection ("specular reflection"), but in this case, not by a normal spherical surface but created by one of those huge, *flat* Iapetus surface areas (simulation, below – right), bouncing the scattered image of the Sun itself, *from thousands of square miles,* directly into the Cassini camera, an enormous flash of sunlight from one of the same *flat* areas seen *profiled* on the limb (below – bottom left) as "a hundred-mile-long *straight edge!*" **http://www.enterprisemission.com/moon3.htm**

We are looking at an artificial moon that is radar and possibly optically invisible!

The implications of this are far reaching like the possibility of being *undetected by an enemy's radar scanning and thus become just another unassuming, uninhabited moon orbiting a planet while actually remaining radar invisible as an actively inhabited world-size spaceship!*

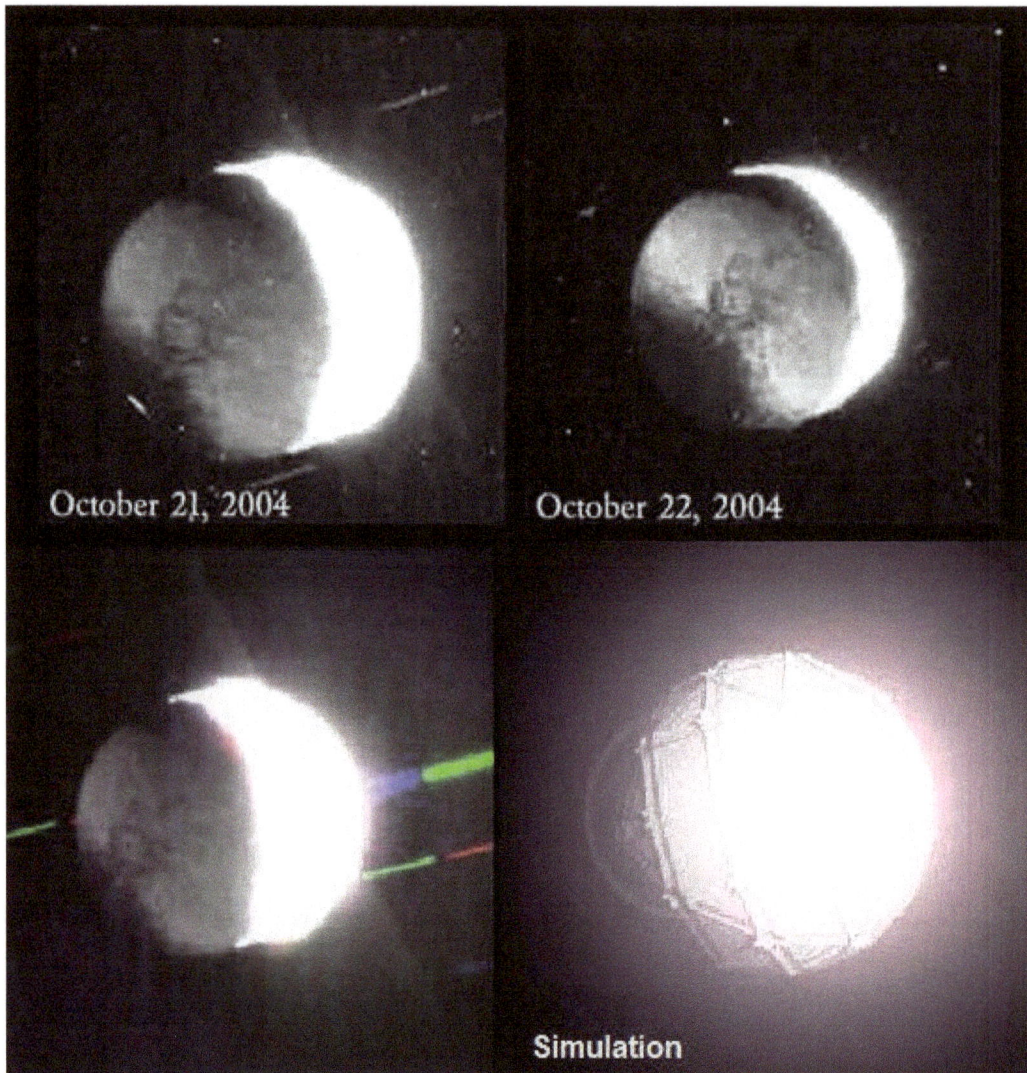

The optical mirror-like reflection of Saturn shine and sunlight is due to Iapetus ' huge flat surface areas causing an enormous flash of sunlight from one of the same flat areas seen profiled on the limb bouncing the scattered image of the Sun itself, from thousands of square miles,
http://www.enterprisemission.com/moon3.htm

The most amazing thing about all this investigative research and analysis comes from NASA's own raw photographic data, in which *none* of this extraordinary information or analysis and its conclusions has received any acknowledgement from "an official NASA press release" despite the availability of the data having been around for years!

The monumental discovery of the Iapetus' surface features by NASA in its space exploration from the last 40 plus years which must rank in prominence with the discoveries of the anomalous structures on Mars; the Face of Mars being the pinnacle of NASA's discoveries, comes not from NASA itself but from public disclosure from independent scientists.

NASA plans to eventually de-commission and replace the very successful Cassini spacecraft in the near future with an even more advanced reconnaissance space probe in the near future and it is hoped at that time that we will get even closer details of Iapetus' extraordinary surface with a distance range 100 times closer. We can only hope!

It should be understood that whenever any private investigator or independent scientist discovers some interesting artifact on any planetary body, you can be assured that some NASA official has already made the same discovery before you found it! The distinct difference is that any researcher or scientist will disclose their findings almost immediately, whereas NASA will keep deathly silent on their discoveries or will spin doctor the release of such information in a controlled manner that emphasizes the least important facts of the discovery while covering up the real prominent aspects of the discovery and off the public radar.

Case in point: There exists "an entire *spaceship world ... trapped in orbit ... around Saturn"*.

How do you cover up a moon-size spaceship that is within plain sight, easily observable to most astronomers and independent investigators without drawing attention to its strangeness of its anomalies?

The answer is, you don't cover up the obvious but, you do emphasis more natural features like albedo, contrast between light and dark materials, impact craters, and provide plenty of scientific rhetoric, speculation and theory and whatever pseudo scientific mumble jumble to keep people side-tracked and off the real trail of investigation. In other words you make the unusual seem ordinary and no real big deal, of no relevance in the scheme of things!

"But, because of the deafening silence coming out of NASA on what it *already* knows but won't release (let alone suspects!), about the glaring anomalies we've now identified about this "moon" ... we are left with only one sad but inevitable conclusion:

NASA, again, has decided to "tough it out" ...to officially say *nothing* -- like it has treated all its *other* discoveries of "extraterrestrial ruins" in the solar system ... over the last 30 or so years ....

Such a policy, of course, is directly due to **"Brookings"** -- the official NASA report of almost 50 years ago (1959) -- which warned the U. S. Government that *any* scientific evidence of extraterrestrial intelligence "could be destabilizing to terrestrial governmental institutions ... if not the future of civilization ...."

The Space Agency, therefore, obviously plans to pretend (certainly, in public ...) -- in full consonance with "Brookings" -- that what exists in these extraordinary Iapetus photographs ... simply isn't there.

Given that the folks at JPL are fully as capable as we are of putting this together, is it possible that this Cassini close-in Iapetus radar return was SO anomalous, that it was correctly analyzed ... for what it was ... within the first few hours of being received at JPL: the radar detection of an entire, artificial moon -- explicitly designed according to **Professor Ufimtsev's** cutting-edge /electromagnetic theories?!

And that soon, "someone" – much higher up in government than JPL (or even NASA) – after seeing this definitive, highly anomalous radar data ... on a place with "hundred-mile-long-edges!" quietly issued a "gag order" on this entire "Iapetus intelligence experiment" ... in consonance with "Brookings?!

(And then some folks wonder why I've been saying for a long, long time ... *we need new leadership at NASA* ....") http://www.enterprisemission.com/moon4.htm

Iapetus is certainly different. Though the third largest satellite in the Saturnian system after Titan) at slightly over 900 miles across, **Iapetus *orbits significantly inclined to the rest of Saturn's moons at some ~15 degrees and over 2 million miles (~ 60 Saturn radii) away***. All the other "regular" moons orbit in the plane of Saturn's equator, along with the trillions of particles making up the rings. This is the natural balance found in any interstellar system of planets and it is true in the Saturnian system with all its moons, except for Iapetus.

Iapetus' orbit is nearly circular with a very high inclination orbit and when we look at all the above factors we wonder if this moon's formation is through "random chance" is just coincident or something more.

Once again, the sharp reader will notice from the preceding references, that Iapetus currently orbits slightly less than 60 radii away from Saturn (59.091 radii, to be exact), the difference is due to orbital drift, since it was originally parked in Saturn's orbit, due to Saturnian/sun tides or other forces.

But, it is the reader's attention to the that *"ideal"* Iapetus distance from Saturn just *"happens"* to also be ***base 60 -- another tetrahedral number*** -- suddenly appearing in the first Sumerian civilization on Earth some ~ 6000 years ago .... A number that fits ***perfectly*** into the redundant, equally mysterious tetrahedral placement of the two major "ring basins" we've previously discussed on Iapetus ... 120/240 degrees apart.

And, the *same* number that is also redundantly communicated via Iapetus' own baffling Platonic ***shape!*** http://www.enterprisemission.com/moon4.htm

Richard Hoagland goes out on a limb to speculate about the ***"purpose of Iapetus creation"*** which is based upon his observations and in depth analysis of Iapetus.

"If you take the inclination of Iapetus' orbit (~ 15 degrees) and multiply by its distance in Saturn radii (60), the result is the ***average*** of the current Cassini "triaxial measurements" of Iapetus' diameter ~ 900 miles!

The measured Iapetus' diameter!!

(The slight discrepancies between the "ideal numbers," and those currently observed, can easily be explained as slight changes – occurring over a literally geological period of time – in the evolving ***orbital elements*** of Iapetus ... again, due to external solar system forces.)

All these numbers – Iapetus' size, distance from Saturn, and orbital inclination -- are "independent variables." Meaning – *none* of them are automatically *interrelated*, or mandated by *any* current theory of satellite formation. Yet, for some reason, they have all come together in Iapetus … this one bizarre "moon" … orbiting Saturn. This simply makes no sense, and the odds of it happening coincidentally – especially, resulting in the actual diameter of Iapetus expressed in *miles!* -- are (really!) "astronomical"— Unless this was *designed!*

Now, if Iapetus was *created* (literally, "from the ground up") -- to memorialize "something" extremely valuable and historically *important* -- both its size and orbital elements could easily have been precisely *engineered* … as recurring aspects of the same "tetrahedral message" embodied in *other* aspects of this satellite … a much vaster variant on our previously discovered (and deciphered) tetrahedral **"Message of Cydonia."**

Curiously, if you again take that orbital inclination of Iapetus in degrees (~ 15), and divide it into the ~ 60 Saturn radii of its orbit, the result is 4 … the number of *the very planet* where we found our first extraterrestrial "tetrahedral" design.

In other words, is Iapetus actually a *time capsule* -- with its own haunting, multi-leveled Message? A message ultimately ending with (below):

"Mars is where our solar-system-wide civilization was centered … and where its shattering demise began …?" http://www.enterprisemission.com/moon4.htm

What Mars may have looked like 50 – 70 million years ago
http://www.enterprisemission.com/moon4.htm

Regardless of the details, the point of such an elaborate, redundantly encoded communication – if it is "communication" -- could *only* be to signal the presence of *vital information on/in Iapetus* regarding its strange presence in the Saturn system … to whoever discovered this unique "moon" when they finally developed (redeveloped?) the technology able to reach Saturn once again!

But, a signal that would only be successful *if* those reaching it this time understood the crucial, ancient "code key" of **hyperdimensional physics: base 60.** Which, among other essentials, includes the non-arbitrary, elemental reason for a "360-degree circle" … and the size of the British mile … based on **sexagesimal "tetrahedral geometry "** itself.

Ok, you now see where I'm going ….

In the "spaceship moon model" for Iapetus -- leaving aside for the moment the non-trivial *reasons* for building such a stupendous "craft" -- there are only two possible points of origin:

1) A "vehicle" from somewhere, far beyond the solar system … some kind of "interstellar ark"– which came to this system a long, long time ago … and ended up at Saturn, or
2) a spaceship "moon" built within this solar system, for equally obscure reasons … which also ended up at Saturn -- but with a visible signature, the baffling "light/dark dichotomy" -- which would flag it across the entire system and future millennia as "anomalous" … for the returning *descendents* of whoever originally left it circling eternally in orbit ….

What was it **Goldsmith** and **Owen** said, back in 1980 …?

 *"… the only object in the Solar System which we might seriously regard as an alien signpost – a natural object **deliberately modified** by an advanced civilization to **attract our attention".** [Emphasis added]

We'll consider in some detail the "interstellar option" later on. But for now, what about the "solar system explanation": if Iapetus *is* truly a derelict "spaceship moon" -- *who* built it, and *where* could they have come from … in *this* system?

In keeping with years of previous research published here on **Enterprise,** would it surprise anyone to learn that the answer to those questions could lead us all the way back across the solar system … from the distant, icy realm of Saturn … back to—*Mars?*

What if **(we're speculating, remember?)** … the inconceivable, increasingly confirmed catastrophe which overtook **Planet V** – the last major solar system planet to explode in the Van Flandern model, and the one *Mars used to orbit as a moon* – was **anticipated**?! What if the real science of **planetology** was able then – utilizing a sophisticated **hyperdimensional** model – to be aware of a developing instability in Planet V's core … centuries or even *millennia* before it was destined to explode? http://www.enterprisemission.com/moon4.htm

Faced with such an overwhelming but certain, planet-wide catastrophe -- what would *that* incredibly advanced society (judging by the awesome scale of the ruins still present on its surface) have done?

If it was discovered that *nothing* could be accomplished technologically to prevent such a catastrophic core explosion, the only reasonable alternative would have been a mass migration of the Martian population (or, a reasonable fraction thereof …) to *another* planet. And one definitely well away from the inner solar system ... as the effects of the explosion would be felt even ***billions*** of miles away ….

And that would have called for either an interstellar migration to ***another*** solar system, or the terraforming of another planet … in the outer reaches of ***this*** one. Or— the creation of a totally new "planet!"

So, was Iapetus part of all three options …?

In other words, was Iapetus a specifically designed **"interstellar ark" (as proposed by this author in this current text!)** -- created on a crash basis in the Saturn system (see below) to transport a significant population from a doomed Mars (if not from other worlds in the ***entire imperiled inner solar system!)*** … to the ***stars***?

Is that what the "number 4" – indelibly encoded in the very ***orbit*** of Iapetus – is telling us ... across the millions of miles and the literally ***millions*** of years since that inconceivable explosion: that it all goes back to … ***Mars!?***

Did mass interstellar migration with such a large population turn out to be impossible, in time … so Iapetus was built as the "replacement planet" in *this* system -- for some small percentage of the teeming populations of those soon to be destroyed inner system worlds (the Earth, the Moon, Mars …) whose peoples would literally have no place else to go …?

Was this **"ark"** (Noah anyone …?) then left in orbit around ***Saturn*** … because – like in the interstellar scenario above -- the Saturn system was an abundant source of raw materials for its construction, if not the vital resources needed to sustain a long-term biosphere for those who would be **"saved?"** Was Saturn also chosen because it was far enough from the impending cataclysm to insure survival … yet still close enough to the warm center of the solar system to allow ***the next phase*** in this extraordinary Plan to be initiated--

The increasingly fascinating enigma of *Titan* -- as a literal "Saturn system terraforming project?" A valiant attempt to recreate *a whole new world* for the rescued populations of the soon-to-be-destroyed entire inner solar system … but on an even grander scale …?
http://www.enterprisemission.com/moon4.htm

Now, readers with good memories will remember that in Part 3 we discussed ferro-electric coatings ("paint") -- like "carbonyl iron ferrite" -- applied to the F-117 to make it "stealthy." And, in the February, 2005 issue of **Scientific American**, in an article called "Nanotubes and the Clean Room," by **Gary Stix**, the following comment, regarding an unwanted side effect of current nanotube manufacturing techniques, was casually made—
" … nanotubes, purchased from bulk suppliers, are a form of high-tech soot *that contains a residue that averages 5 percent iron,* a contaminant whose very mention can produce involuntary tremors in managers of multi-million-dollar clean rooms. The Nantero team devoted

much of its early development [of the new nanotube chip] to devising a complex filtration process to reduce the amount of iron to the parts-per-billion-level..." [emphasis added]

Given that the natural moons of Saturn formed in a region of the solar system **essentially *devoid of iron*** -- the ultra-frigid realm of the outer solar system, where the condensation of *ices* from the Saturn nebula far more readily occurred – the presence of copious amounts of free iron **on the surface** of Iapetus (but **missing** on the other Saturnian satellites) presents another serious impediment to Iapetus' formation and evolution as "just another moon"

Unless—

1)   what Burris and her colleagues are seeing in the VIMS data of Iapetus is an iron contaminant of *the nanotube manufacturing process* used in Iapetus' original construction; or

2)   the anomalous iron is part of a ferro-electric coating, ***deliberately applied*** to Iapetus (along with its intrinsic geometric shape – below) -- to make it … ***stealthy!***

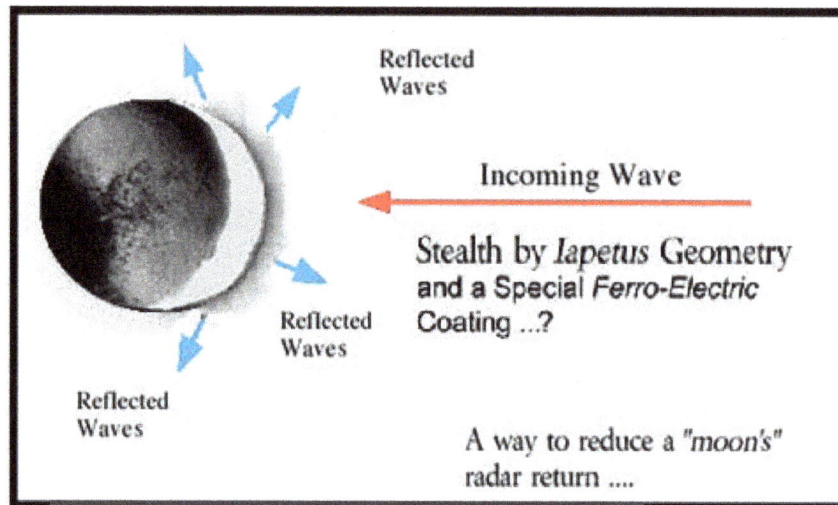

Reflected Waves

Incoming Wave

Stealth by *Iapetus* Geometry and a Special *Ferro-Electric* Coating ...?

Reflected Waves

Reflected Waves

A way to reduce a *"moon's"* radar return ....

http://www.enterprisemission.com/moon5.htm

The implications of a *deliberately* stealthy Iapetus – constructed with both the precise geometry required for reflecting radar waves away, as well as a dead black, iron-based coating for absorbing the remaining visible and radio electromagnetic energy – are definitely *non-trivial* in this context! Including—

The possibility that Iapetus was not a "rescue ark" at all, but in actuality—

A world-size *warship!*

This, of course, would immediately raise the very troubling possibility that the demise of Planet V -- and the environmental destruction of the entire inner solar system as a consequence, including Mars itself -- was *not* a naturally-occurring catastrophe at all ….
http://www.enterprisemission.com/moon4.htm

**If Iapetus is not a "natural" moon of Saturn, is it possible that it is
a "Death Star " warship or possibly an "Ark of Salvation"?**
http://www.enterprisemission.com/moon5.htm

Is this, indeed, the origins of that chilling and peculiar mythos … which has, for millennia, equated "Mars" – where our ancestors in this scenario ultimately came from -- with a bloodthirsty "god of war" ….?

If any of this is true ... it could now make this striking comparison (below) *far more* than just a metaphor ...

Either prospect is extraordinary – Iapetus as "ark" … or Iapetus as "Death Star" -- and opens up major new avenues for using Cassini's onboard radar during the next fly-by to determine critical dielectric parameters regarding the true composition of Iapetus' surface -- which could ultimately allow scientific determination of its possible *artificial* origins … if not which "design hypothesis" is true ...

Mainstream defense research into **"buckytube"** dielectric and electromagnetic properties – and their application to current stealth technology -- leave no doubt that, if Iapetus is truly an artificial shell structure, composed of trillions of manufactured carbon nanotubes underneath its remaining covering of ice, their "anomalous" (compared to natural absorption models) *radar signature* from Cassini should ultimately tell us.

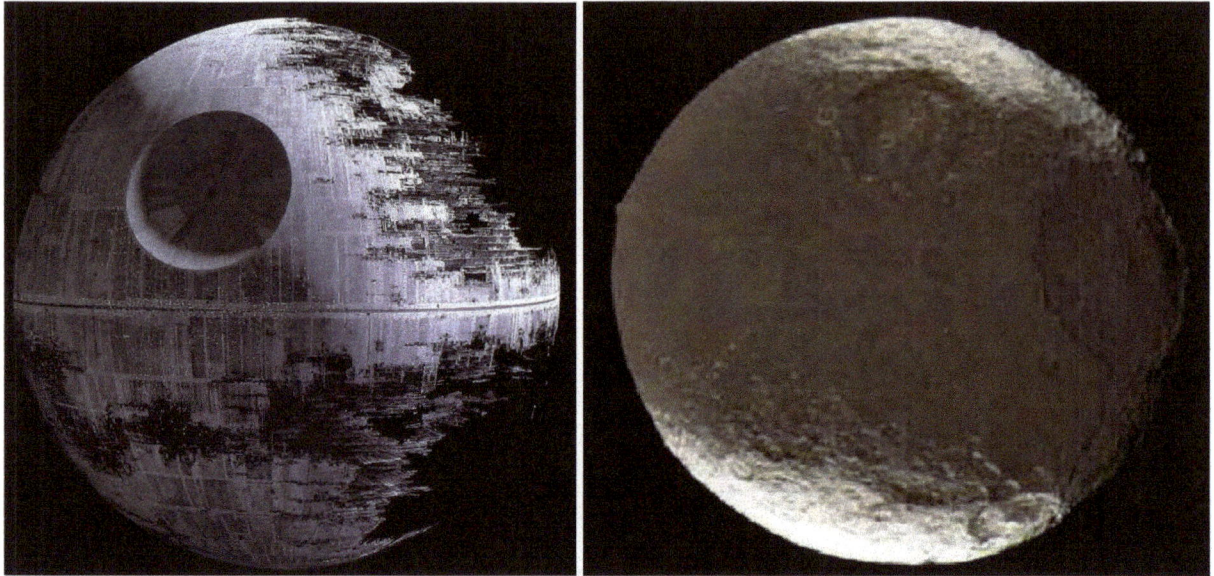

Iapetus as a possible "Death Star" moon-size warship
http://www.enterprisemission.com/moon5.htm

Maybe – given JPL's obvious reticence to releasing the results of its *existing* Cassini Iapetus' radar observations -- they already have ...

Which brings us to the other curious observation that JPL's Buratti talked about: the presence in the Cassini Iapetus spectra of "organics." Where – in our artificial "moon" scenario -- would those *organic* molecules originate?

Unfortunately, looking at the global Iapetus images taken by Cassini last December – with the enormous impact scars still etched across its surface – the answer now is all too obvious:

The organic component of the exposed "dark ellipse' – and thus, by inference, the rest of the ice-covered blackened surface of Iapetus … still hiding underneath its frozen layers of ancient inside air – could be direct clues to *the incineration, explosive decompression and subsequent "cold trapping" on the surface … of its former organic biosphere inside!*
The magnitude of such a potential cosmic cataclysm boggles the imagination.
The thought of countless beings – along with their entire rich interior ecology -- destroyed in one hellish moment … by the inferno of the *deliberate* impact of one (or more) asteroid-sized objects into Iapetus … releasing *a hundred million megatons or more* … is almost unimaginable.

But, if true, the lasting signatures of this world-shattering catastrophe -- the *"intimate mixture of water ice, amorphous carbon, and a nitrogen-rich organic compound …"* – would indeed be spread for all eternity across the surface of such a shattered "world"--

Creating an immortal "winking" epitaph across the solar system … down through the countless millions of ensuing years … written in the mysterious "dark ellipse" that now forever scars the surface of this frozen tomb -- if not the psyche of the few who may have managed to escape … to

384

start new lives in the dim pre-history of our own world ....
http://www.enterprisemission.com/moon4.htm

(C) 2005 The Enterprise Mission

**The Iapetus "Death Star" battle station "Ark of Salvation" receives a "killing blow"
either from an asteroid impact or from battling another interstellar civilization**

http://www.enterprisemission.com/moon5.htm

**Author's Rant: Before we leave Iapetus, I must relate an experience in remote viewing
back in July 1996 in Crestone, Colorado while attending Dr. Steven Greer's week long
"Ambassadors to the Universe" training seminar. It was at that time in the San Luis Valley
that I experienced many remote view sessions of ETIs and ET craft that were vectored to
our location in that valley.**

**Dr. Greer had stated, just before our session of meditation into a higher state of
consciousness (a prerequisite to remote viewing and the psychic powers), that there were
many ET bases on Earth, on the Moon, on Mars and on one of the satellites or moons of
Saturn in which our group of 32 CSETI members could remote view and call upon the ETI
at those bases to come to our field site.**

**I already had some very successful remote sessions in Crestone and decided to call the ETI
on one of the moons of Saturn and as I journeyed to that unknown moon in my mind, I
perceived that its surface structure was constructed of many geometric shapes and forms
that were primarily triangular and pentagonal in shape!**

**I was surprised to find that the moon which had both light and dark areas over it's surface
was essentially a gigantic, polygonal, planet-size spacecraft!!**

It also had a very tall tower that I believe was tens of miles in size, to be exact I thought that the tower was telescopic, extending 70 miles in height and it could be retracted back into the interior of the moon-size spaceship!!! This immense miles-high tower was possibly the communications tower for this behemoth spacecraft allowing it to communicate across the Solar System and across the interstellar vastness of the galaxy. This was before the Cassini-Huygens space probe had even photographed the surface of Iapetus in 2004!!

What I understood back in 1996 from this remote view observation based upon the simple information given by Dr. Greer and my perception of this "moon" (*I was not aware that it was called Iapetus at the time, but as a former astronomy student, I should have known that, no excuses!*) is that this gigantic spaceship was still "ACTIVELY OCCUPIED"!!!

This remote viewing session was long before it was known that this moon, Iapetus was even considered or understood to be a world-size spacecraft by Richard C. Hoagland as he only wrote about it in 2005 ("A Moon With a View")!

In some way, my remote view of Iapetus is another confirmation to obtain accurate information from remote viewing as well as the subtle confirmation of Hoagland's in-depth research and investigation into this Saturnian moon, Iapetus.

It appears that Richard Hoagland and fellow research partner Mike Bara have struck major pay dirt once again, with Iapetus and have fleshed out a possible pre-history of mankind ancestors, who had developed a **Type Two Civilization** within our Solar System. These ancestors of humanity had either originated from Mars or had come to Mars having originated possibly from another distant star system and made Mars the planet "show-place" in our Solar System, as the *"Administrative/Cultural Center"* for their civilization.

Then, something happened about 65 million years ago or thereabouts and borrowing from a scene from the movie "**Forbidden Planet**" the "poor Krill civilization" or in our case, the "poor Martian civilization" came to an abrupt end that seems to collapse their civilization almost overnight. Was it an immediate or gradual collapse, was it a natural calamity, a manmade natural disaster or an interstellar war with another star civilization remains unknown yet, there is evidence to support all these possible scenarios.

The evidence, regardless of the demise of this Type Two Civilization strongly indicates that the Earth became a new home for some fragment of this civilization; the other fragments of Martian civilization may have fled to some other colonies in the neighbouring constellations. One can only speculate until some tangible proof becomes available for research and analysis.

CHAPTER 91

URANUS ("GOD OF THE SKY")

Uranus is the seventh planet from the Sun. It has the third-largest planetary radius and fourth-largest planetary mass in the Solar System. Uranus is similar in composition to Neptune, and both are of different chemical composition than the larger gas giants Jupiter and Saturn. For this reason, astronomers sometimes place them in a separate category called **"ice giants".** Uranus's atmosphere, although similar to Jupiter's and Saturn's in its primary composition of hydrogen and helium, contains more "ices" such as water, ammonia, and methane, along with traces of hydrocarbons. It is the coldest planetary atmosphere in the Solar System, with a minimum temperature of 49 K (−224.2 °C), and has a complex, layered cloud structure, with water thought to make up the lowest clouds, and methane the uppermost layer of clouds. In contrast, the

The third largest planet in the solar system, Uranus is a featureless ice-cold planet
https://en.wikipedia.org/wiki/Uranus

The only planet derived from a figure from Greek mythology, Ouranos, god of the sky rather than Roman mythology like the other planets. Like the other giant planets, Uranus has a ring system, a magnetosphere, and numerous moons. The Uranian system has a unique configuration among those of the planets because ***its axis of rotation is tilted sideways***, nearly into the plane of the elliptic as it revolves around the Sun. Its north and south poles, therefore, lie where most other planets have their equators. In 1986, images from ***Voyager 2*** showed Uranus as a virtually featureless planet in visible light without the cloud bands or storms associated with the other giants. Terrestrial observers have seen signs of seasonal change and increased weather activity in recent years as Uranus approached its equinox. The wind speeds on Uranus can reach 250 meters per second (900 km/h, 560 mph). **http://en.wikipedia.org/wiki/Uranus**

The complexity of the rings of Uranus with its many moons and Uranus tilted 90 degrees with its north and south poles to the equatorial plane of the ecliptic
https://en.wikipedia.org/wiki/Uranus

Uranus has rings similar to Saturn that are composed of extremely dark particles, which vary in size from micrometers to a fraction of a meter. Thirteen distinct rings are presently known, the brightest being the ε ring. All except two rings of Uranus are extremely narrow – they are usually a few kilometres wide. The rings are probably quite young; the dynamics considerations indicate that they did not form with Uranus. The matter in the rings may once have been part of a moon (or moons) that was shattered by high-speed impacts. From numerous pieces of debris that formed as a result of those impacts, only a few particles survived in a limited number of stable zones corresponding to present rings.

In all respects, Uranus looked like a dynamically dead planet when Voyager 2 flew by in 1986 during the height of Uranus's southern summer and could not observe the northern hemisphere. At the beginning of the 21st century, when the northern polar region came into view, the **Hubble Space Telescope (HST)** and **Keck telescope** initially observed neither a collar nor a polar cap in the northern hemisphere. So Uranus appeared to be asymmetric: bright near the south pole and uniformly dark in the region north of the southern collar. In 2007, when Uranus passed its equinox, the southern collar almost disappeared, whereas a faint northern collar emerged near 45° of latitude.

Recent observation also discovered that cloud features on Uranus have a lot in common with those on Neptune. For example, the dark spots common on Neptune had never been observed on Uranus before 2006, when the first such feature dubbed **Uranus Dark Spot** was imaged. The speculation is that Uranus is becoming more Neptune-like during its equinoctial season.

The **Uranus Dark Spot (UDS)** is located at the latitude of about 28 ± 1° and measured approximately 2° (1300 km) in latitude and 5° (2700 km) in longitude. The feature moves in the prograde direction relative to the planet with an average speed of 43.1 ± 0.1 m/s, which is almost 20 m/s faster than the speed of clouds at the same latitude.

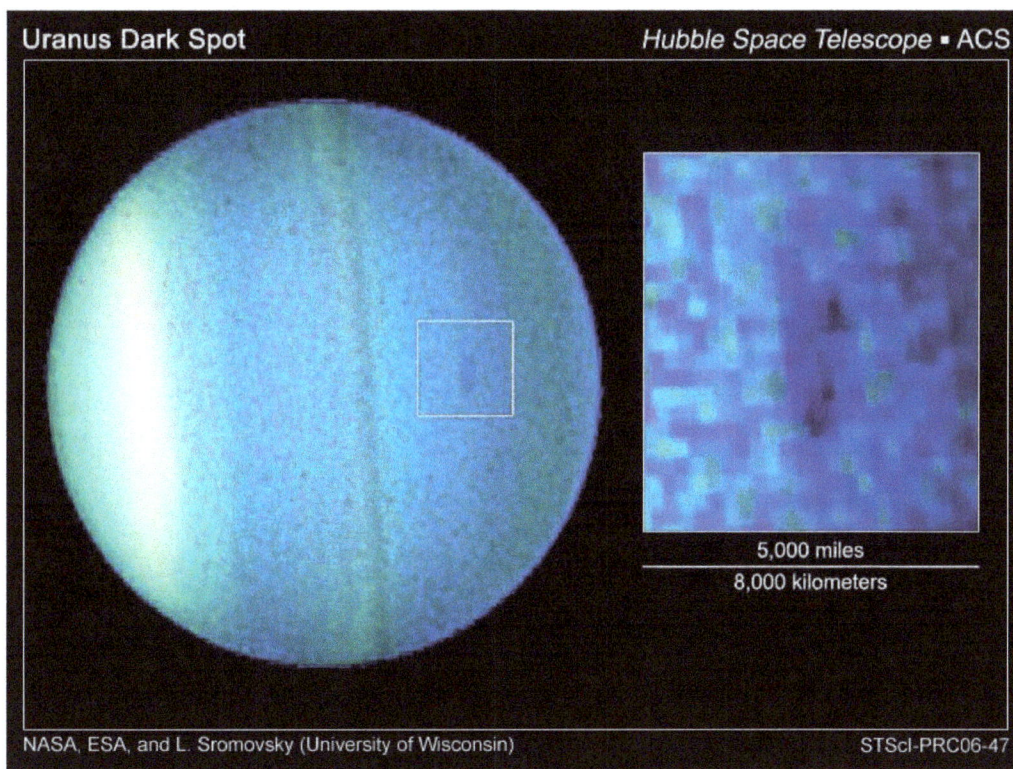

The Dark Spot of Uranus
https://en.wikipedia.org/wiki/Uranus

A part from the unusual **Dark Spot** on Uranus, if we are searching for additional anomalies like UFOs and Extraterrestrials around or on Uranus, other than the natural ones thus far discovered, there does not appear to be any reports that have come to light at this present time to confirm this aspect.

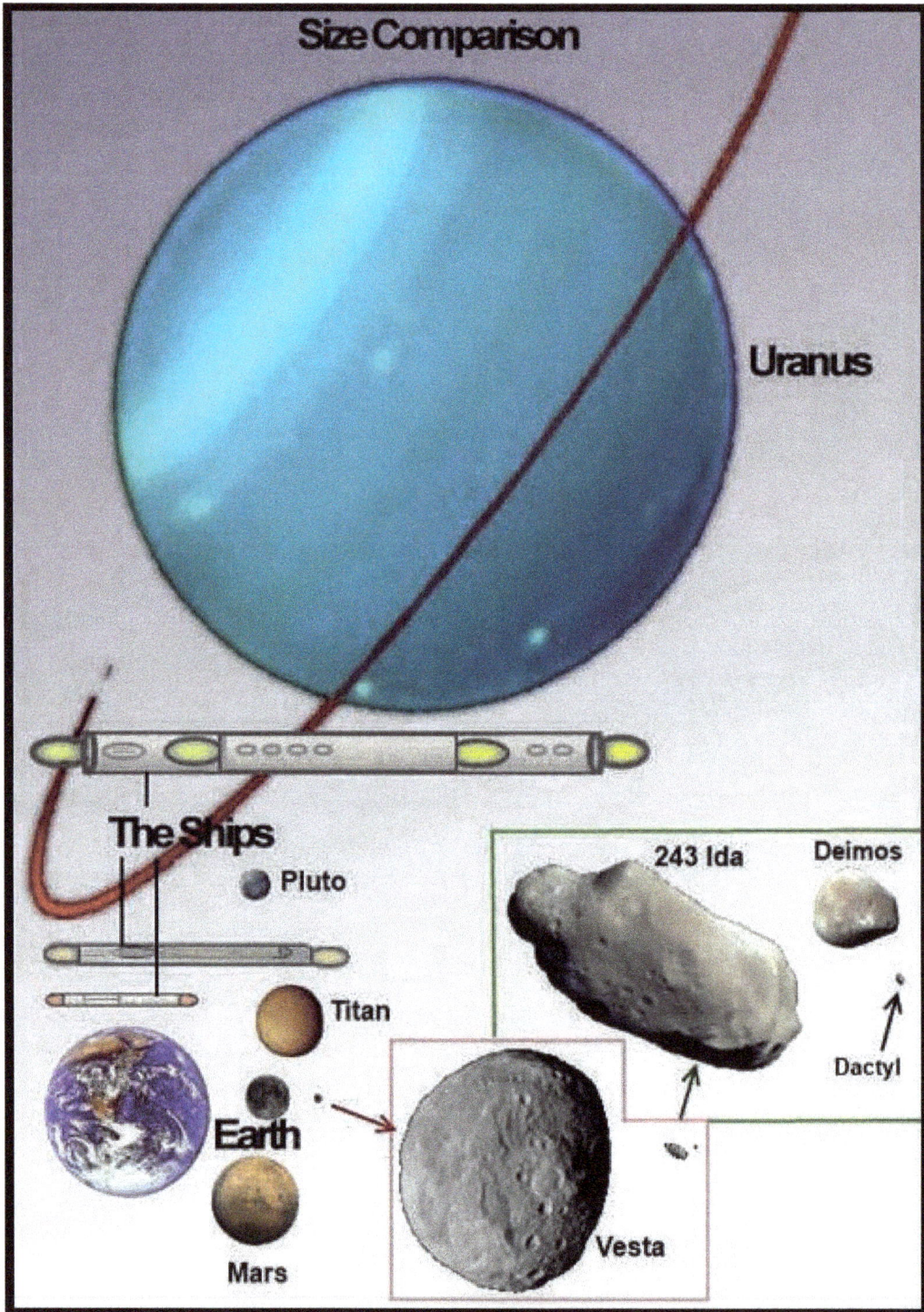

The Ringmaker ships of Saturn and their size comparison with Earth, Moon and Uranus

The only thing that this author has come across with regard to UFOs in connection with Uranus is the size of the **Ringmaker Spaceships** of Saturn, which in the chart above show the comparison size of the Ringmaker Spacecraft length to that of Uranus' diameter. This topic has been previously discussed earlier in this section.

Moons of Uranus

Uranus, the seventh planet of the Solar System, has 27 known moons, all of which are named after characters from the works of **William Shakespeare** and **Alexander Pope**. **William Herschel** discovered the first two moons, **Titania** and **Oberon**, in 1787, and the other three ellipsoidal moons were discovered in 1851 by **William Lassell** (**Ariel** and **Umbriel**) and in 1948 by **Gerard Kuiper** (**Miranda**). These five have planetary mass, and so would be considered (dwarf) planets if they were in direct orbit about the Sun. The remaining moons were discovered after 1985, either during the *Voyager 2* flyby mission or with the aid of advanced Earth-based telescopes.

Uranian moons are divided into three groups: thirteen inner moons, five major moons, and nine irregular moons. The inner moons are small dark bodies that share common properties and origins with the planet's rings. The five major moons are massive enough to have achieved hydrostatic equilibrium, and four of them show signs of internally driven processes such as canyon formation and volcanism on their surfaces. The largest of these five, Titania, is 1,578 km in diameter and the eighth largest moon in the Solar System, and about 20 times less massive than Earth's Moon. Uranus's irregular moons have elliptical and strongly inclined (mostly retrograde) orbits at great distances from the planet.
http://en.wikipedia.org/wiki/Moons_of_Uranus

Miranda

Miranda looks like a moon that was once pulverized into large chunks from the massive impact of a large comet or asteroid that thoroughly destroyed the moon and yet, somehow over eons of time coalesced back into a sphere due to the strong gravitational forces from the large chunks of moon being attracted to each other. There are chevron and square continent size pieces that have been *jig-saw fitted* back into the surface of Miranda as well as large gouges or canyons that run across the surface of the moon, all evidence of something catastrophic in Miranda's past.

So far the only close-up images of Miranda are from the *Voyager 2* probe, which made observations of the moon during its Uranus flyby in January 1986. During the flyby, the southern hemisphere of the moon was pointed towards the Sun so only that part was studied. Miranda shows more evidence of past geologic activity than any of the other Uranian satellites.

Miranda's surface may be mostly water ice, with the low-density body also probably containing silicate rock and organic compounds in its interior.

Miranda's surface has patchwork regions of broken terrain indicating intense geological activity in the moon's past and is crisscrossed by huge canyons. Large 'racetrack'-like grooved structures, called **coronae**, may have formed via extensional processes at the tops of **diapirs**, or upwellings of warm ice. The ridges probably represent extensional tilt blocks. The canyons probably represent graben formed by extensional faulting. Other features may be due to **cryovolcanic eruptions** *of icy magma*. The diapirs may have changed the density distribution within the moon, which could have caused Miranda to reorient itself, similar to a process believed to have occurred at Saturn's geologically active moon Enceladus.

An earlier theory, proposed shortly after the Voyager 2 flyby, was that a previous incarnation of Miranda was *"torn apart and then thrown back together, again"* perhaps being shattered by a massive impact, with the fragments reassembling and denser ones subsequently sinking to produce the current strange pattern. http://en.wikipedia.org/wiki/Miranda_%28moon%29

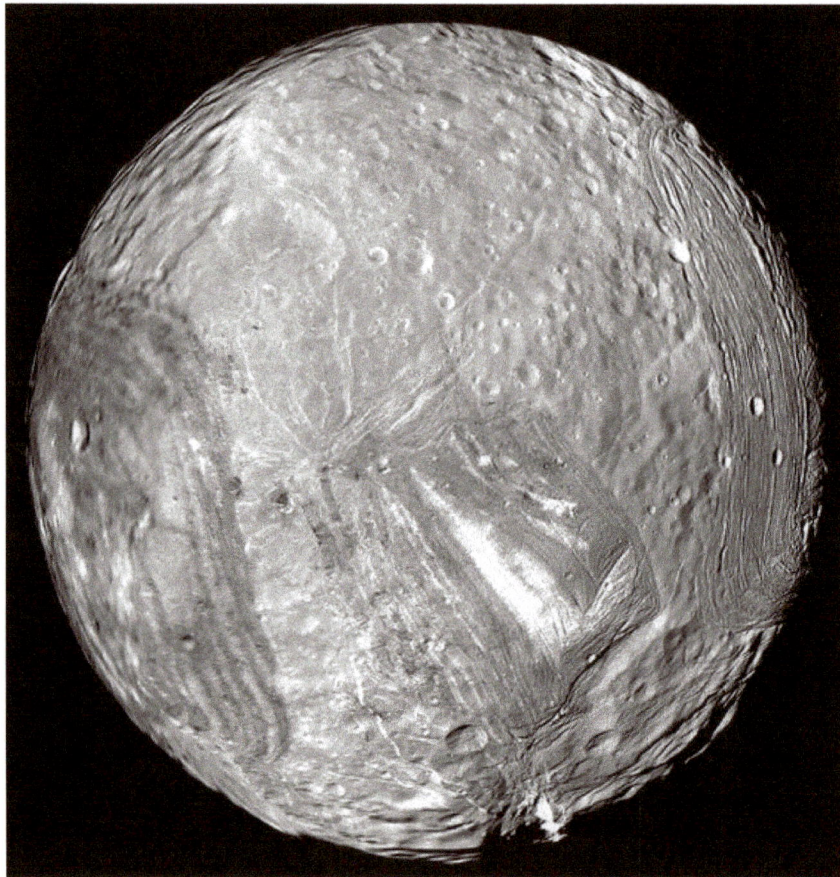

Miranda shows massive scarring all over its surface due to a large meteorite impact in its distant past resulting in a continent-size chevron to square land masses and deep canyon gouges. Miraculously, through gravitational forces, the remnant land masses coalesced back into a sphere
https://en.wikipedia.org/wiki/Miranda_(moon)

Miranda recalls to mind, the exploded planet hypothesis (EPH) model formulated by **Tom Van Flandern** that caused Planet V to become the Asteroid Belt between Mars and Jupiter yet, whatever hit Miranda did not create another asteroid belt or a ring of debris circling Uranus as would be expected. The impact of devastation was enough to blow out large chunks of lunar mass but, internal pressure of Miranda may have been low in its original formation during the nebula stage of the Solar System while it orbited Uranus and thus, the debris scatter was minimal and confined to a small area allowing for the moon debris to coalesce back into a globe because of gravitational and tidal forces of Uranus and its other moons.

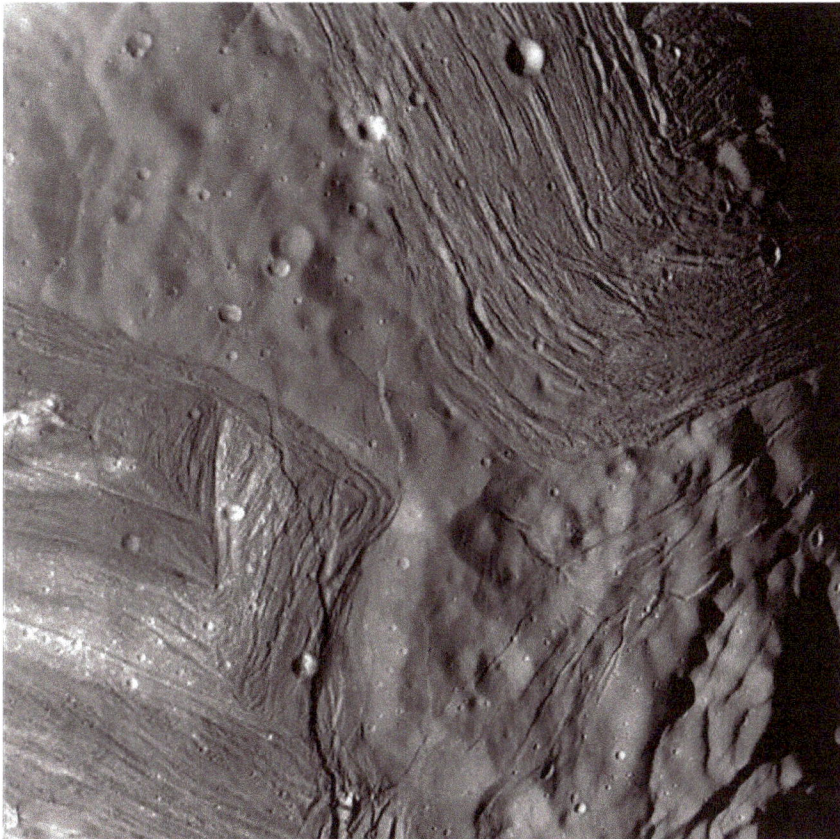

Close-up of Miranda's continent-size chevron and the angularity of other surface terrain

http://www.seasky.org/solar-system/uranus-miranda.html

There is a unique ***geometric phenomenon*** present on Miranda highly indicative of the fundamental physics underlying all these solar system changes and anomalies that we have been closely examining, a process that directly supports the ***HD model***.

Acquired by ***Voyager 2***, in January 1986, the startling images of a bright, obviously geometric, "***L- shaped*" *formation*** on ***Miranda*** (see images above and below) is very unambiguous and completely without any theoretical explanation in conventional geological models for the moon's formation, or subsequent evolution. Look closely, and study the two sides of the image (below) carefully.

What is the likelihood of seeing a set of similar angles, in such nearby proximity and relationship with such straight-lined geometric perfection, if this were only a "*natural formation*" (in the mainstream sense)? Even the slightly widened nature of the two higher, smaller triangles is not unexpected: since the shape of the underlying geometric "stress patterns" is being projected onto a spherical surface. The geometry is patently obvious, even without a theoretical basis in place -- with the major clue to its origin being that the largest

393

observable *"triangle"* on *Miranda* is or once was perfectly equilateral.
http://www.bibliotecapleyades.net/esp_dayaftertomorrow3.htm

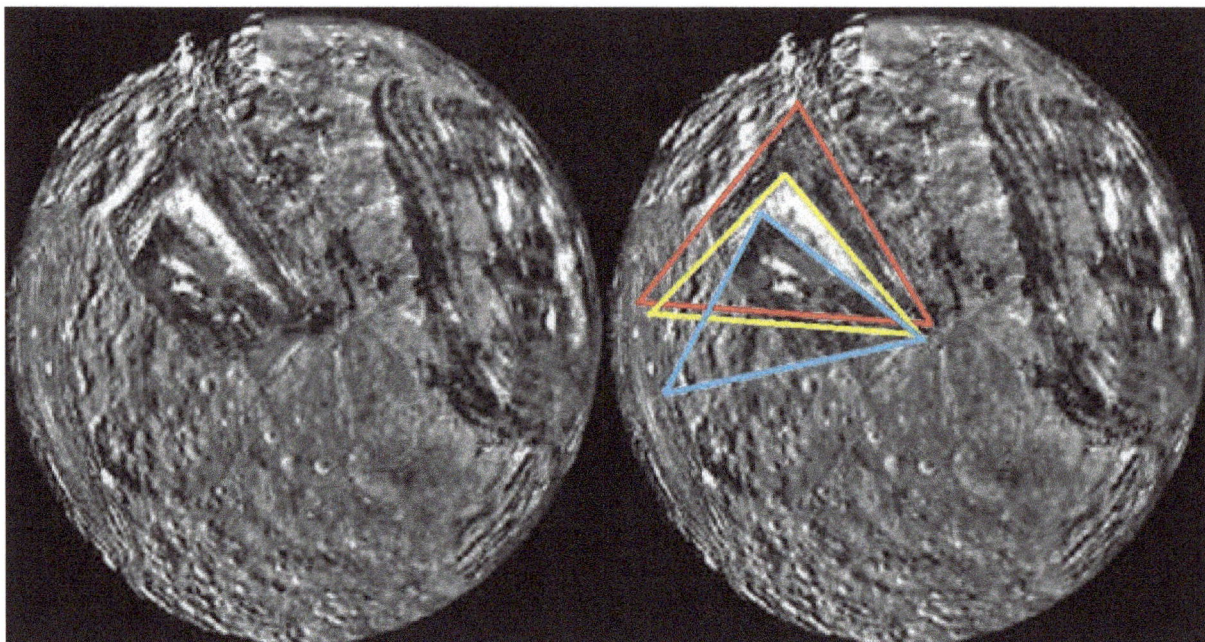

Voyager image of Uranian moon Miranda (L) and underlying three-fold triangular geometry (R). (NASA (L) with additions by Wilcock (R), 2004.)
http://www.bibliotecapleyades.net/esp_dayaftertomorrow3.htm

A more complete overview of the scenario that accounts for this unique satellite geometry has been presented in **David Wilcock**'s *"Divine Cosmos"*

After *Miranda* formed, in the dust and gaseous nebula which orbited *Uranus* in the forming solar system, there was a subsequent, apparent physical expansion of *Miranda* … shortly (geologically speaking) after its own formation. The process appears to have been shaped by *internal "geometric forces" -- resonant internal energy patterns still unacknowledged (let alone explained) by any mainstream planetary models.*

When this process was occurring, most of the surface of the now icy *Miranda* was composed of a high percentage of liquid water (!) for a time -- ideal conditions for a *fluid-like HD energy* to express itself as *"formative geometry"* in our dimension. In that early era, the normally-invisible internal geometric resonances that we have proposed elsewhere in this Report -- as shaping surface other features on other planets and their moons -- were apparently able to leave their unmistakable geometric signature on the rapidly-cooling, icy surface layers of this "ice ball" moon … for Voyager to find.

Below is another set of geometric triangular formations developed by the author which follow the natural geological features of rifts, valleys, and grooves along the large "L" shape chevron. Note that the triangles are all nearly equilateral triangles seen in descending order from large to gradually smaller size triangles. Here again, is another possible example of geological features confirming the internal geometric resonances of Miranda's HD energy.

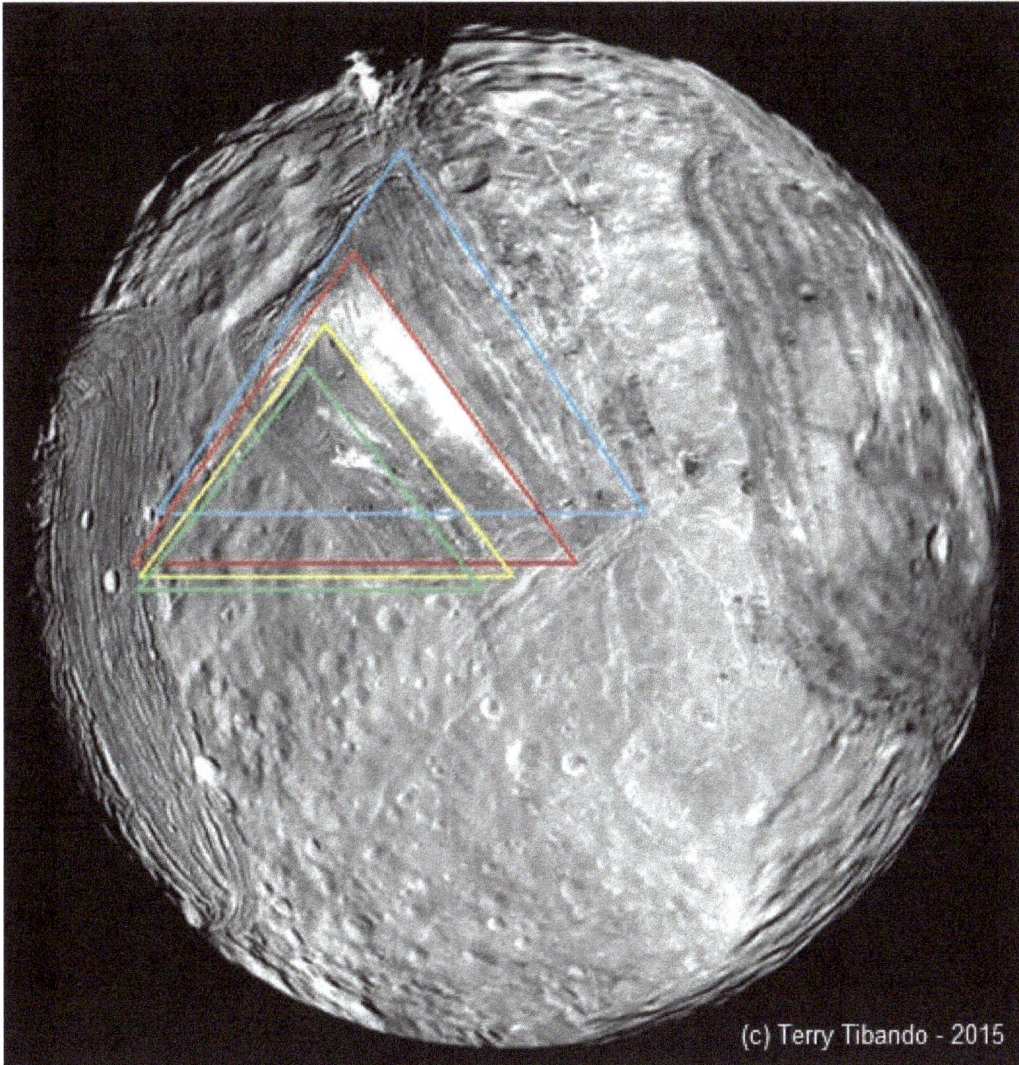

Triangular geometry found in the large chevron surface feature of Miranda
(c) Terry Tibando

Oberon

Oberon, also designated **Uranus IV**, is the outermost major moon of the planet Uranus. It is the second largest and second most massive of the Uranian moons, and the ninth most massive moon in the Solar System. Discovered by William Herschel in 1787, Oberon is named after the mythical king of the fairies who appears as a character in Shakespeare's *A Midsummer Night's Dream*. Its orbit lies partially outside Uranus's magnetosphere.

It is likely that Oberon formed from the accretion disk that surrounded Uranus just after the planet's formation. The moon consists of approximately equal amounts of ice and rock and is probably differentiated into a rocky core and an icy mantle. A layer of liquid water may be present at the boundary between the mantle and the core. The surface of **Oberon**, which is dark and slightly red in color, appears to have been primarily shaped by asteroid and comet impacts. It is covered by numerous impact craters reaching 210 km in diameter. Oberon possesses a system of **chasmata** (graben or scarps) formed during crustal extension as a result of the expansion of its interior during its early evolution.

The Uranian system has been studied up close only once: the spacecraft **Voyager 2** took several images of Oberon in January 1986, allowing 40% of the moon's surface to be mapped.

Following **Richard Hoagland**'s lead and looking with an impassionate eye for anomalous features on Oberon, one is immediate struck with some polygonal crater features that seem to leap off this moon's surface. There are craters that are hexagonal in shape instead of the circular features one would come to expect from a natural planetary body. There is symmetry in the unusual positioning of craters around the dark crater, Hamlet which is just below center of the equatorial region of the moon.

To its upper left is the smaller **Othello crater.** Above the limb at lower left rises an 11 km high mountain, probably the central peak of another crater or possibly it is a **cryovolcanoe. Mommur Chasma** runs along the terminator at upper right.

The symmetry that catches the eye almost right away is that there are six craters nearly equidistant from each other around Hamlet crater forming a hexagon pattern. Closer inspection of the other craters reveals that some of them are also hexagon in shape and not the traditional circular "punch-bowl" shape as seen on so many moons and satellites throughout the Solar System. In fact, **Hamlet Crater** is also a six-sided crater and one of these smaller hexagon craters just right of Hamlet and above the equator intersects almost dead centre with one of the apexes of the larger hexagon arrangement of craters.

Normally, we would dismiss this type of hexagonal anomaly as a natural occurrence from meteoric bombardment of the moon's surface, a sort of one in a million kind of probability but, given that so many other moons in the **Solar System** have this type configuration as we have already discovered and because nature traditionally abhors straight lines and geometrical symmetry, it is highly unlikely that these are just simple natural occurring craters, as they seem to reflect artificiality and deliberate design.

If one looks closely to the terminator Oberon, you will see craters of similar size spaced almost equidistant from each other running the length of the terminator which again, which is also unusual feature.

There is also a subtle hemispherical resemblance to Saturn's moon Iapetus! When looking at the overall circumference or limb of Oberon, there does appear to be straight lines or planes which have been highlighted with a large red circle to show this polygonal aspect of Oberon.

Author's Rant: Admittedly, the reader may feel that I am stretching my perception a bit to justify the hypothesis of another moon size spacecraft in our Solar System and given the

limited number of raw photographic data available of Oberon, my perception of this moon may be in question but, I'll let the reader judge for themselves, what they see.

Oberon is not a round spherical moon but possibly another dodecahedron shaped moon with less severe meteoric weathering like **Iapetus** and thus, is somewhat more pristine in its appearance. (See images below and examine carefully!)

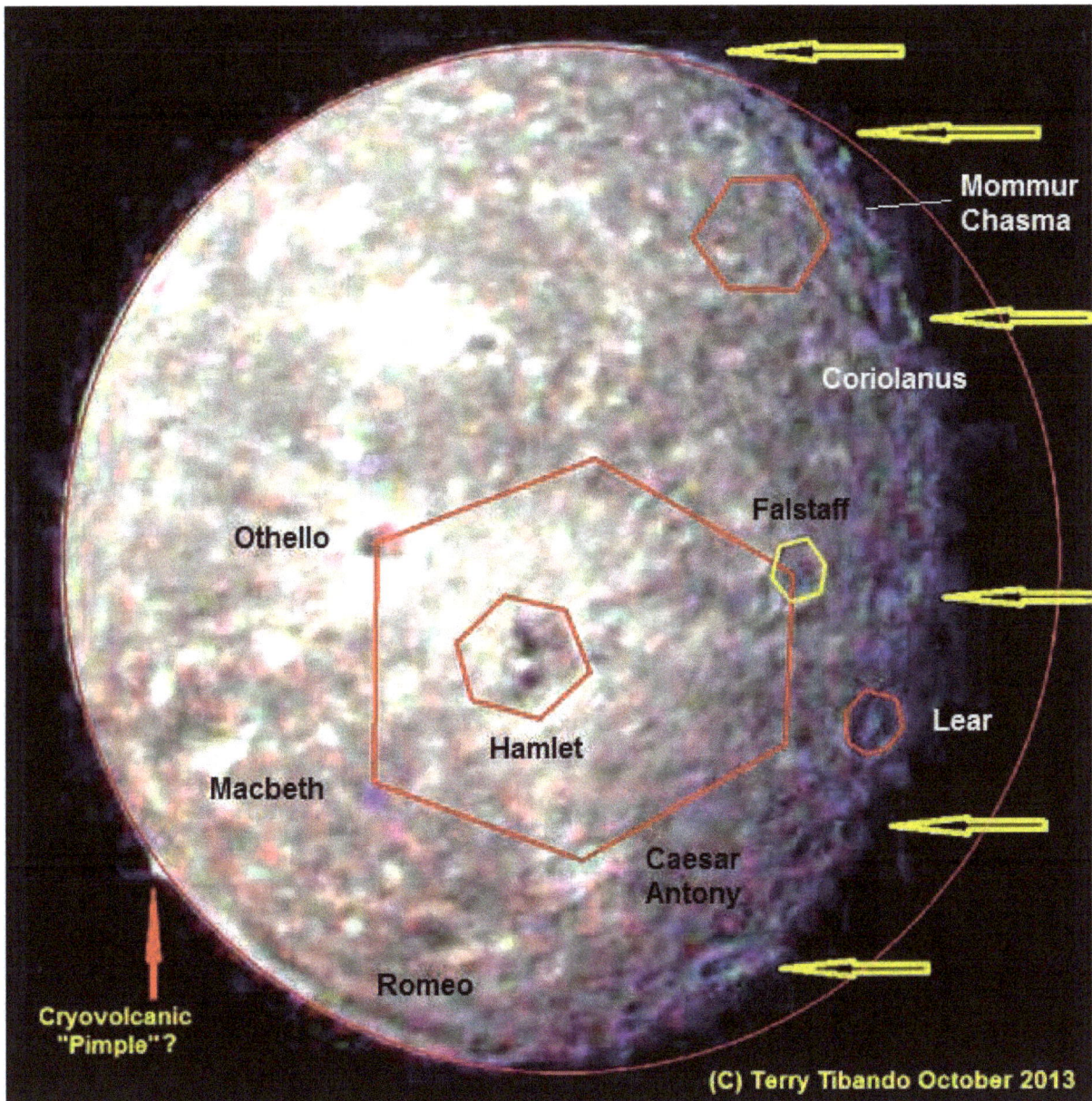

**Oberon with hexagon craters in a hexagonal formation around
Hamlet Crater slightly left of center colour enhanced**
(c) Terry Tibando

CHAPTER 92

NEPTUNE ("GOD OF THE SEA")

Neptune is the eighth and farthest planet from the Sun in the Solar System. It is the fourth-largest planet by diameter and the third largest by mass. Among the gaseous planets in the solar system, Neptune is the densest. Neptune is 17 times the mass of Earth and is slightly more massive than its near-twin Uranus, which is 15 times the mass of Earth but not as dense.[c] On average, Neptune orbits the Sun at a distance of 30.1 AU, approximately 30 times the Earth–Sun distance. Named for the Roman god of the sea.

The Great Dark Spot of Neptune located at the familiar 19.5° degrees latitude in the southern hemisphere by Voyager 2 in 1989, before it literally vanished but, a new dark spot has appeared in the northern hemisphere at 19.5° degrees latitude!
https://cosmosmagazine.com/space/hubble-confirms-new-dark-spot-on-neptune

Neptune was the first planet found by mathematical prediction rather than by empirical observation. Unexpected changes in the orbit of Uranus led **Alexis Bouvard** to deduce that its orbit was subject to gravitational perturbation by an unknown planet. Neptune was subsequently observed on 23 September 1846 by **Johann Galle** within a degree of the position predicted by **Urbain Le Verrier.** http://en.wikipedia.org/wiki/Neptune

Neptune is similar in composition to Uranus, and both have compositions which differ from those of the larger gas giants, Jupiter, and Saturn. Neptune's atmosphere, while similar to Jupiter's and Saturn's in that it is composed primarily of hydrogen and helium, along with traces of hydrocarbons and possibly nitrogen, contains a higher proportion of "ices" such as water, ammonia, and methane. Astronomers sometimes categorise Uranus and Neptune as "ice giants" in order to emphasize these distinctions. The interior of Neptune, like that of Uranus, is primarily composed of ices and rock.

In contrast to the hazy, relatively featureless atmosphere of Uranus, Neptune's atmosphere is notable for its active and visible weather patterns. For example, at the time of the 1989 Voyager 2 flyby, the planet's southern hemisphere possessed a **Great Dark Spot** comparable to the Great Red Spot on Jupiter. http://en.wikipedia.org/wiki/Neptune

But, by June 1994, Neptune's great Dark Spot has mysteriously disappeared. Then in April 1995, it or a similar Dark Spot re-appeared in the ***northern hemisphere at 19.5 degrees*** accompanied by many bright, high-altitude clouds, all to the surprise of NASA! It was a mirror image of the Great Dark spot first imaged by Voyager 2.

Neptune, it seems can undergo extraordinary dynamics where the weather can change within a matter of weeks.

NASA says that "Energy from the *Sun* drives the Earth's weather system. However, the mechanism on *Neptune* must be very different because the planet ***radiates 2 times more energy than it receives from the distant, dim Sun.*** (Emphasis added).

Hoagland and Wilcock point out that two years after these official descriptions, *NASA* wrote of "a looming mystery":

When the planetary probe *Voyager* visited *Neptune* in 1989, it detected the ***Great Dark Spot***, a pulsating feature nearly the size of the Earth itself. Two years ago, ***Hubble observations*** showed the spot had disappeared, and that another smaller spot had emerged. But instead of growing to a large-scale storm like the *Great Dark Spot*, the ***new spot appears to be trapped*** at a fixed latitude and may be declining in intensity, said **Lawrence Sromovsky**, a senior scientist.

What, exactly, would "*trap*" the new spot at a fixed latitude, exactly the same number of degrees above the equator as the previous spot was below the equator!? This is easily explainable in the **Hyperdimensional Model**, as an 180-degree phase shift in the ***simplest resonant*** (***tetrahedral***) ***pattern*** underlying *Neptune*'s internal fluid dynamics… which forces the precise positioning of "the **Great Dark Spot**" -- the vortex resonantly ***shifting from 19.5 degrees south latitude… to 19.5 degrees north***. http://www.bibliotecapleyades.net/esp_dayaftertomorrow3.htm

If you're thinking that this "**HD phase shift**" of *Neptune* is somehow correlated with the shift of vortex activity away from *Jupiter*'s equatorial regions, towards the polar regions… and the 58.6% slowdown of cloud rotations at *Saturn*'s equatorial region… with the surprise emergence of x-ray emissions along *Saturn*'s equator, rather than the poles as *NASA* would expect… and the disappearance of the so-called "*spoke*" formations in **Saturn's rings**… then, like the fictional character of the Matrix, **Neo**… you are to be congratulated. You are beginning to see with new "*eyes*" and not through the eyes of someone else. You are beginning to see "*the real world*" beyond the confines of a three-dimensional box, beyond the "**Matrix, the world that has been pulled over your eyes**"!

You have *seen* that *it is not the spoon that bends but, that there is no spoon*"… Well done!

It gets better. By 1996, less than a year after this "*hyperdimensional Neptunian pole shift*," **Dr. Lawrence Sromovsky** noticed an increase in *Neptune*'s *overall brightness,* which continued dramatically surging upwards through 2002 (Figure 43). Though the false-color photo speaks far louder than the statistics, the fact is that in only six short years, blue light became 3.2% brighter on *Neptune*, red light 5.6% brighter… and near-infrared light intensified by a whopping 40%. Even more surprisingly, *some areas of latitude became fully 100 percent brighter!*
http://www.bibliotecapleyades.net/esp_dayaftertomorrow3.htm

NASA explains away, how these unprecedented, planet-wide changes in brightness are a "*simple seasonal variation model*" related to *Neptune*'s tilt angle to the Sun… (…yawn…). Yeah, right!

"April 22, 2002, Madison, WI: *Hubble Space Telescope* (*HST*) imaging observations in August 2002 show that *Neptune*'s brightness has increased significantly since 1996… and now appears to be consistent with a simple seasonal variation model… Comparing August 2002 observations to similar observations in 1996, the authors found that *Neptune*'s reflectivity averaged across the planet's face (disk-averaged) increased by 3.2% at 467 nm (blue), 5.6% at 673 nm (red), and 40% in the 850 nm–1000 nm band (near infrared). These changes result from even larger brightness increases in restricted latitude bands, reaching 100% in some cases. The reason for the increases may be "**seasonal forcing**", which is the seasonal variation in local solar heating."

NASA article makes it all seem good and simple, like "*a neat little light show brought on by very ordinary meteorology*"…

The weather on Neptune, the eighth planet from the sun, is an enigma, to begin with. The mechanism that drives its near-supersonic winds and giant storms have yet to be discerned.

Scientists themselves have stated: "*…Some of the wildest, weirdest weather in the solar system… a planet whose blustery weather -- monster storms and equatorial winds of 900 miles per hour - bewilders scientists…*"

On *Earth*, weather is driven by energy from the sun as it heats the atmosphere and oceans. On *Neptune*, the sun is 900 times dimmer and scientists have yet to understand how *Neptune*'s *weather-generating machinery* can be so efficient. "It's an efficient weather machine compared to *Earth*," said **Sromovsky**. "It seems to run on almost no energy."…

Atmospheric brightness has increased on Neptune since 1996-2002
http://www.bibliotecapleyades.net/esp_dayaftertomorrow3.htm

Sromovsky said that compared to the look provided by the *Voyager* spacecraft, *Neptune* is a different place: *"The character of **Neptune** is different from what it was at the time of **Voyager**. The planet seems stable, yet different."* (Emphasis added)
http://www.bibliotecapleyades.net/esp_dayaftertomorrow3.htm

Springtime has come to the Solar System courtesy of the internal geometric resonances of the Sun's hyperdimensional (HD) energy as well as the HD energy emanating from the galactic centre of the Milky Way galaxy!

Triton: Backward Moon of Neptune

Neptune has 13 known moons, though most are small and orbit closer to Neptune than its rings. (Yes, Neptune has rings like Saturn but, they are fainter and less dense in their formation).

Triton is the largest moon of the planet Neptune, discovered on October 10, 1846, by English astronomer William Lassell. It is the only large moon in the Solar System with a retrograde orbit, which is an orbit in the opposite direction to its planet's rotation. At 2,700 kilometres (1,700 mi) in diameter, it is the seventh-largest moon in the Solar System. Because of its retrograde orbit and composition similar to Pluto's, Triton is thought to have been captured from the Kuiper belt. Triton has a surface of mostly frozen nitrogen, a mostly water ice crust, an icy mantle and a substantial core of rock and metal.

401

Triton is one of the few moons in the Solar System known to be geologically active. As a consequence, its surface is relatively young, with a complex geological history revealed in intricate and mysterious cryovolcanic and tectonic terrains. Part of its crust is dotted with geysers thought to erupt nitrogen.

The sublimation of surface ices creates a **thin, hazy atmosphere** on Triton made mostly of Nitrogen. The atmosphere creates wind streaks on the surface, and unknown processes pump unusual plumes of gas and particles into the atmosphere.
http://en.wikipedia.org/wiki/Triton_%28moon%29

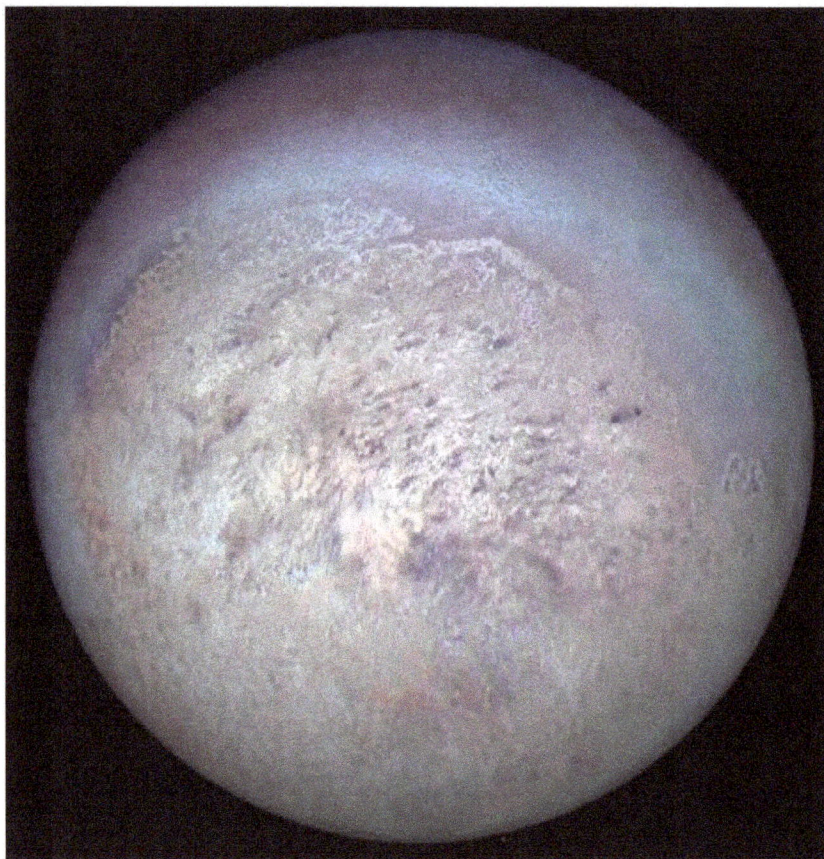

Triton's pinkish, south polar cap composed of nitrogen and methane ice and streaked by dust deposits left by nitrogen gas geysers. The darker region above it includes Triton's "cantaloupe terrain" also, cryovolcanic and tectonic features. The fewness of craters is evidence of extensive geologic activity.
http://solarviews.com/cap/nep/triton5.htm

Neptune's major moon, *Triton*, also experienced great changes… in this case, a "very large" 5% temperature increase between 1989 and 1998. According to MIT researchers, this is comparable to the Earth's atmosphere globally heating up by 22 degrees Fahrenheit… in only 9 years! It is believed that *Triton*'s atmospheric pressure also has "at least doubled… since the time of the Voyager [1989] encounter."

It is curious that each of these components we have discovered throughout the solar system, like the warming trend on *Triton*, are so often discussed as single events, or perhaps with "one or two others" included... in a few, rare instances.

NASA does give us all the hard evidence we would ever need to make a case... but they, or the media reporting on their discoveries, simply **never** put this evidence together under one roof. Thus, the data continues to quietly slip out into the open undetected, while the gaping silence of yawns from the public perpetually haunts the prospects of any newly proposed space mission ever getting off the ground.

If NASA had a truly open policy toward public disclosure of all its discoveries they would find that the public would get behind them in support by 1000% for any planned space exploration mission! Public interested and funding would literally "skyrocket" (pun intended) with the interest generated by NASA's discoveries and how they would affect our world, hopefully toward an ever advancing civilization.

Alas! NASA is in complete denial and suppression of truth with every discovery that it comes across in its exploration of our Solar System. The "puppet masters" behind NASA are in control and they have no plans to relinquish their grip on the reigns of our future on this planet!

We know that the *Earth* is also undergoing major changes, as we will explore in unprecedented detail in *Part 4* of this Report.

In the meantime, the only unexplored territory still remaining is the distant **Pluto** -- the icy planet on a long, elliptical orbit at the far outer reaches of our solar system, recently downgraded to a "*planetesimal*" **status** as perceived in the eyes of most planetologists'. If *Pluto* can be shown to have any changes at all, then we most certainly are dealing with a solar-system-wide effect. Case closed.

Pluto does not disappoint.

CHAPTER 93

PLUTO ("GOD OF THE UNDERWORLD")

Pluto, minor-planet designation **134340 Pluto**, is the largest object in the **Kuiper belt** with a diameter 2368 km (±20 km) or (1471.4 mi.) (±12.4 mi) and the tenth-most-massive body observed directly orbiting the Sun. It is the second-most-massive known dwarf planet, after **Eris**. Like other Kuiper-belt objects, Pluto is composed primarily of rock and ice and is relatively small, approximately one-sixth the mass of the Earth's Moon and one-third its volume. It has an eccentric and highly inclined orbit that takes it from 30 to 49 AU (4.4–7.4 billion km) from the Sun. This causes Pluto to periodically come closer to the Sun than Neptune. As of 2011, it is 32.1 AU from the Sun. http://space-facts.com/pluto/

Pluto, the "wondering" dwarf planet" was recently photographed for the first time by the New Horizon Probe on July 14, 2015. It can at times be closer to the Sun when it crosses over the orbit of Neptune or it can become the outermost planet in the Solar System. Note the heart shape south polar region on an almost craterless surface
https://fsicafascinante.blogspot.ca/2016_01_01_archive.html

Discovered in 1930, Pluto was originally classified as the ninth planet from the Sun. However, its status as a major planet fell into question following further study of it and the outer Solar System over the ensuing 75 years. Starting in 1977 with the discovery of minor planet **2060 Charon**, numerous icy objects similar to Pluto with eccentric orbits were found. The most notable of these was the scattered disc object Eris—discovered in 2005, which is 27% more massive than Pluto.

Pluto's Mountainous South Polar Region

35 miles

Pluto reveals for the first time a terrain of ice, rock, dark valleys, sharp jagged mountains, deep chasms, and almost no crater scarring at all over its surface
https://www.nasa.gov/feature/possible-ice-volcano-on-pluto-has-the-wright-stuff

The understanding that Pluto is only one of several large icy bodies in the outer Solar System prompted the **International Astronomical Union (IAU)** to formally define what it means to be a "planet" in 2006. This definition excluded Pluto and reclassified it as a member of the new ***"dwarf planet"*** category (and specifically as a ***plutoid***). A number of scientists hold that Pluto should have remained classified as a planet and that other dwarf planets should be added to the roster of planets along with Pluto.

With the recent flyby of the **New Horizon probe** on July 13-14, 2015 the designation for Pluto as a dwarf planet may change again as NASA scientists and astronomers gather and assess the new photographic and instrument data of the planet from the space probe's flyby.

Is There Life on Pluto?

According to NASA life is most probable within the ***"Goldilocks Zone"*** or the **Zone of Habitability** which is between Venus and Mars which includes the Earth, However, with the flybys of both **Voyager 1 and 2** that have left the Solar System and the **Cassini** and **Juno** space probes to Saturn and Jupiter to photograph and do scientific measurements of those planets and their many moons, it now seem very probable that extraterrestrial life may be present under the thick ice moon surfaces of **Enceladus** of Saturn and **Europa** of Jupiter.

If indeed, with near future space probes to be launched by NASA to these moons of the giant gas planets that will orbit satellites and land probes to their surfaces, do discover life forms (no doubt they will be acclaimed, initially as microbes with latter announcements of larger, deeper creatures, similar to our own Earth crustaceans and tubeworms around volcanic vents on these moons ocean floors.

Should this be the future discovery of extraterrestrial life besides our own in the Solar System, this will be a nice, safe, self-contain and self-confined acknowledgement that we are not alone in the universe or at least in our Solar System.

Safe, self-contained and self-confined small microbes or simple life forms are much easier to deal with than having to be challenged with another sentient intelligence life form that may be capable of also leaving its planet and travelling through space. No doubt this type of discovery will necessitate an expansion of the Zone of Habitability to include some of the moons of Jupiter and Saturn and given that Titan of Neptune has liquid methane lakes on its surface, which may also contain life forms, this moon too, may also, fall into the Zone of Habitability!

Once again, we need to revisit the statement of **Baha'u'llah** who said back in the mid-1800s, quite poignantly and prophetically: ***"Know thou that every fix star has its planets and every planet has its creatures whose number no man can compute."***

It seems science (chiefly, NASA) is beginning to come around, although somewhat slowly, to acknowledging Baha'u'llah's insightful utterance on science.

This brings us to ask, is there any life on Pluto?

Pluto has an atmosphere that has minimal methane but is rich in nitrogen gas with a hot rocky core. Given what we now know about NASA's search for life on other planets and their satellites as well as UFO researchers' investigations into other intelligent life in the Solar System and no doubt through the universe, we may yet, find that Pluto is also inhabited with life forms. With the potential for finding life on Pluto, this will mean that the Zone of Habitability must be expanded to include Pluto and any other far distant icy worlds and their moons.

Pluto like some of the inner planets and their moons has the possibly of a life-supporting subsurface ocean as identified by Japanese in the area known as **Sputnik Planitia.** This Texas size ice cover area has as a liquid water ocean beneath it that's insulated from the otherwise frozen conditions of a dwarf world. "Pluto's ocean is capped and insulated by gas hydrates", according to **Shunichi Kamata**, Associate Professor at Hokkaido University. Scientists think that there could be an insulating layer of gas (probably methane) that could be keeping an ocean in liquid form. https://www.forbes.com/sites/jamiecartereurope/2019/05/22/is-there-alien-life-on-pluto-liquid-ocean-the-size-of-texas-could-change-where-we-look-for-life/#35fbcef926b3

This is the highest-resolution color departure shot of Pluto's receding crescent from NASA's New Horizons space probe
NASA/Johns Hopkins University Applied Physics Laboratory/Southwest Research Institute
https://www.forbes.com/sites/jamiecartereurope/2019/05/22/is-there-alien-life-on-pluto-liquid-ocean-the-size-of-texas-could-change-where-we-look-for-life/#35fbcef926b3

Rather intriguingly, NASA did photograph some unusual forms on the surface on the ice-covered **Sputnik Planitia** area that seem to look like giant "slug or snail" life forms sliding over the icy surface of Pluto! The images, published by NASA, show oddly-shaped objects in pathways "similar in form to the slime trails left by earth snails on concrete. In one image, (see below) an object appears to be approaching a fork in the path.

NASA explanation is that these bumps are frozen iceberg on top of convection cracks formed when gas bubbles up to the surface from the core cracking the icy surface, According to NASA, the objects are "dirty" icebergs traveling across an icy nitrogen plain known as the Sputnik Planum. Some of the icebergs are as big an office building. (See photos below).

However, it doesn't explain why the nitrogen icebergs are directly on the cracked areas or why there are so many such nitrogen icebergs on these convection cracks!
https://www.newser.com/story/218930/no-those-arent-snails-on-pluto.html

It should also be understood that life on other planets may not follow the same rules of evolution as is found on Earth, but rather are conditioned by the planet's environment, atmosphere, gravity, its position in orbit near to its parent star, etc. These are conditions that need to be seriously considered in any planetary investigation.

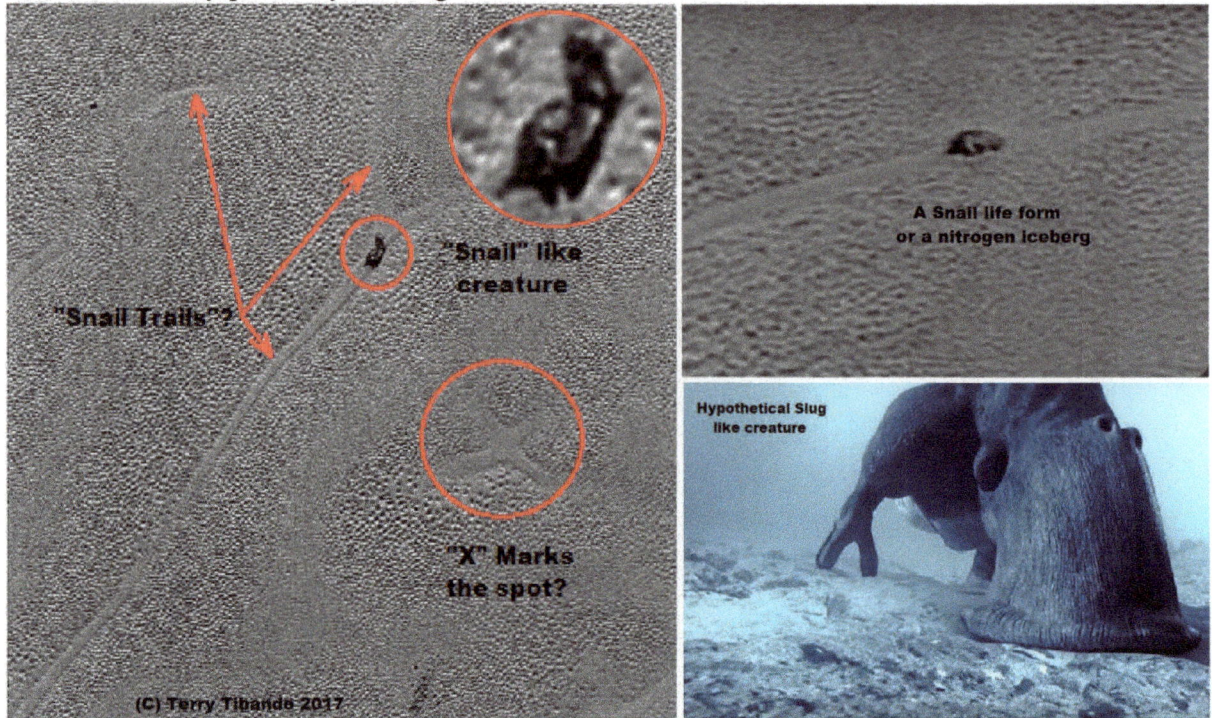

Is there giant snail or slug life forms on Pluto or merely geological anomalies of nitrogen icebergs on convection cracks within the ice surface of Sputnik Planitia?
Google images

The Moons of Pluto

Pluto has five known moons: **Charon** (the largest, with a diameter just over half that of Pluto), **Nix, Hydra, Kerberos,** and **Styx.** Pluto and Charon are sometimes described as a binary system because the barycenter of their orbits does not lie within either body. However, the IAU has yet to formalize a definition for binary dwarf planets, and as such Charon is officially classified as a moon of Pluto.

On July 14, 2015, the Plutonian system was visited for the first time by the **New Horizons space probe** which performed a flyby during which it made detailed measurements and images of the **plutoid** and its moons. It revealed a heart shaped south polar region of ice and virginal mountains. Its surface structure showed very little cratering scars unlike the inner iron core planets Mercury, Venus, Earth and its Moon and Mars and some of the moons of the gaseous outer planets. NASA scientists were, to say the least "over the moon" with the images of Pluto and its moon Charon!

False colour enhancements to show surface detail on Pluto and its largest moon Charon

https://www.nasa.gov/audience/forstudents/k-4/stories/nasa-knows/what-is-pluto-k4.html

Pluto's origin and identity had long puzzled astronomers. One early hypothesis was that Pluto was an escaped moon of Neptune, knocked out of orbit by its largest current moon, Triton. This notion has been heavily criticized because Pluto never comes near Neptune in its orbit.

Pluto's true place in the Solar System began to reveal itself only in 1992 when astronomers began to find small icy objects beyond Neptune that were similar to Pluto not only in orbit but also in size and composition. This trans-Neptunian population is believed to be the source of many short-period comets. Astronomers now believe Pluto to be the largest member of the Kuiper belt, a somewhat stable ring of objects located between 30 and 50 AU from the Sun. Like other **Kuiper-belt objects (KBOs),** Pluto shares features with comets; for example, the solar wind is gradually blowing Pluto's surface into space, in the manner of a comet. If Pluto were placed as near to the Sun as Earth, it would develop a tail, as comets do.

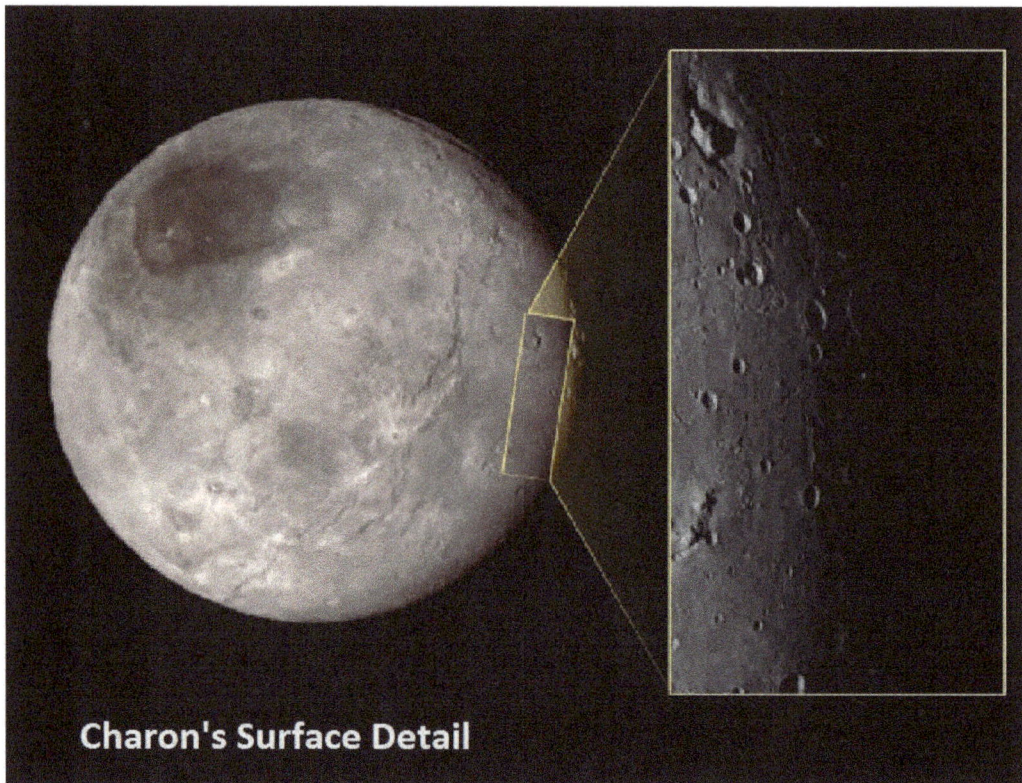

Charon's Surface Detail

Charon, Pluto's largest moon shows deep canyons or chasms, many small craters, mountains, and wide plain areas
https://www.nasa.gov/feature/pluto-and-charon-new-horizons-dynamic-duo

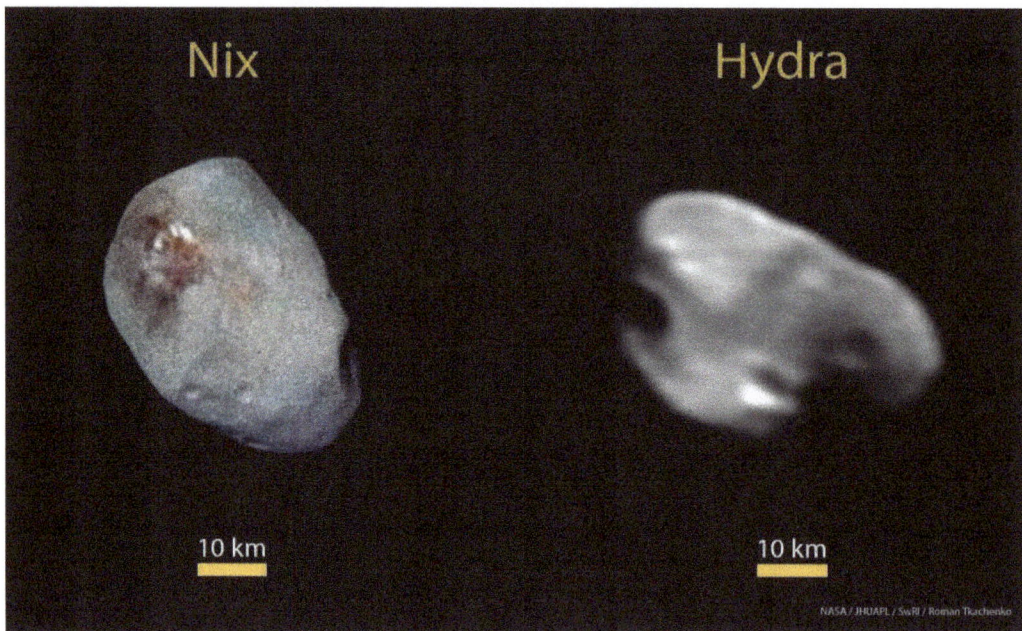

Nix

Hydra

10 km

10 km

NASA / JHUAPL / SwRI / Roman Tkachenko

Pluto's smaller moons Nix and Hydra appear to be more like captured Kuiper Belt Objects (asteroids) found beyond Pluto
http://www.plutorules.com/page-33-kbo.html

**Of the five orbiting moons of Pluto, only Charon has been photographed
with any clarity or definition of its surface detail and shape**

https://phys.org/news/2015-07-pluto-moon-hydra.html

Though Pluto is the largest of the Kuiper belt objects discovered so far, Neptune's moon Triton, which is slightly larger than Pluto, is similar to it both geologically and atmospherically and is believed to be a captured Kuiper belt object. Eris (see below) is about the same size as Pluto (though more massive) but is not strictly considered a member of the Kuiper belt population. Rather, it is considered a member of a linked population called the *scattered disc.*

A large number of **Kuiper belt objects**, like Pluto, possess a 2:3 orbital resonance with **Neptune KBOs** with this orbital resonance are called "plutons", after Pluto.

Like other members of the Kuiper belt, Pluto is thought to be a residual **planetesimal**; a component of the original protoplanetary disc around the Sun that failed to fully coalesce into a full-fledged planet. Most astronomers agree that Pluto owes its current position to a sudden migration undergone by Neptune early in the Solar System's formation. As Neptune migrated outward, it approached the objects in the proto-Kuiper belt, setting one in orbit around itself (Triton), locking others into resonances, and knocking others into chaotic orbits. The objects in the scattered disc, a dynamically unstable region overlapping the Kuiper belt, are believed to have been placed in their current positions by interactions with Neptune's migrating resonances.

A computer model created in 2004 by **Alessandro Morbidelli** of the Observatoire de la Côte d'Azur in Nice suggested that the migration of Neptune into the Kuiper belt may have been triggered by the formation of a 1:2 resonance between Jupiter and Saturn, which created a gravitational push that propelled both Uranus and Neptune into higher orbits and caused them to switch places, ultimately doubling Neptune's distance from the Sun. The resultant expulsion of objects from the proto-Kuiper belt could also explain the Late Heavy Bombardment 600 million years after the Solar System's formation and the origin of Jupiter's Trojan asteroids. It is possible that Pluto had a near-circular orbit about 33 AU from the Sun before Neptune's migration perturbed it into a resonant capture. The Nice model requires that there were about a thousand Pluto-sized bodies in the original planetesimal disk; these may have included the early Triton and Eris. http://en.wikipedia.org/wiki/Pluto#cite_note-Levison2007-134

Throughout this section in our of search for Extraterrestrial life in the Solar System and in the universe, we have look at NASA's and ESA's raw photographic data of the planets and their moons and satellites and we have not been disappointed in our exploration for signs of life.

The signs of Extraterrestrial life appear to be everywhere particularly from the Earth to Mars and out toward Jupiter, Saturn, Uranus and Neptune and along the way we have had our pioneering guides: Richard C. Hoagland, Mike Bara, Tom Van Flandern, Norman Bergrun, David Wilcock and many others who have pointed out mysterious anomalies for our consideration and inspection. We have found through review of their logical arguments that life has occurred or shown evidence of its self within our Solar System and that it is extremely ancient.

Along the way, we have also been directed to examine the curious science of hyperdimensional physics, a science that seems to have been lost in this modern day and age of space exploration and in all the sciences of physics, chemistry, biology, and mathematics, etc. Within the ranting of these above mention pioneers in science, who many in the hallowed hall of scientific academia have condemned politely as mavericks, while others have labeled blasphemers and charlatans of pseudoscience, there is no denying that these people have reawakened what was lost from our memories of that knowledge that we once held in high esteem in hope for a brighter future which over the millennia seems to have been deliberately cover-up by the men of power, control, and greed.

Hyperdimensional physics (HD) is a true reality and a true science that appears again and again on every planet, moon and even on the Sun in which we have looked at while we searched vigorously for signs of intelligent life, elsewhere in the universe. HD is the calling-card of the ETI, it is their clarion message trumpeted to us, their descendants to re-awaken from our beds of heedlessness and re-investigate who we really are from the signs that they have been scattered across our Solar System on every planet and moon and that the science of hyperdimensional physics is the key to unlocking immense power to run our world as well as to propel us across the vast oceans of the universe in the twinkling of an eye.

In this last recognized planet in our star system, Pluto, we find that hyperdimensional physics reaches even out to this distant planet with a force equal to none in its impact and confirmation of its reality. There are changes occurring not only on Earth in a major way that is being felt by all upon this planet but, those same changes are occurring everywhere in our Solar System and it dare be said, throughout the entire galaxy nay, even in the whole universe! **Hoagland** and **Wilcock** suggest that it is the result of a *Hyperdimensional Springtime!*

It is a Regenesis of our Solar System and the Universe as a whole or as some of the more spiritually enlightened would say: *"It is a Spiritual Springtime and the Birth of a New Day"!!!* Baha'i World Faith (Selected Writings of Bahá'u'lláh and Abdu'l-Bahá; 1943; National Spiritual Assembly of the United States; Baha'i Publishing Trust; Wilmette, USA

Hoagland and Wilcock in their paper: "Interplanetary Day After Tomorrow, an Enterprise Mission Hyperdimensional Report" explains these unprecedented changes that are been discovered and photographed throughout our solar System, particularly on Pluto.

Before we discuss the likelihood of any real changes going on with **Pluto**, we have to begin by taking note of something important. Most conventional **NASA** explanations for these changes, as we have seen, revolve around the notion (pun intended!) that the planet or satellite's angular tilt (obliquity) relative to the Sun, is by far the most likely cause of any observable changes. In the case of **Pluto**, the 248-year elliptical orbit it traces around the *Sun* actually brings it closer to the **Sun** than **Neptune** at certain times… which, incidentally, just occurred between 1979 and 1999… and much farther away from the **Sun** at other times.

Obviously, we would naturally assume that if a planet moves closer to the **Sun**, it is exposed to more heat than if it moves farther away from the **Sun**. Simple, right? If you're heating your house with a single fireplace, you're not going to hang out in the kitchen if the fire is in the living room. So, where is **Pluto** now?

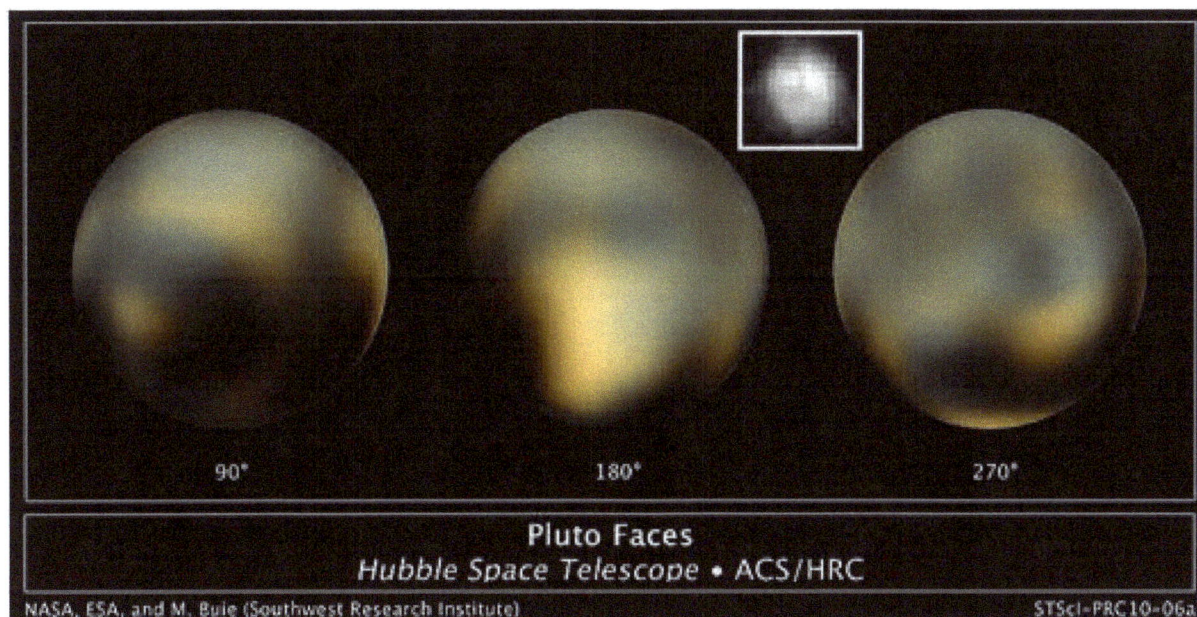

Pluto Faces
Hubble Space Telescope • ACS/HRC

NASA, ESA, and M. Buie (Southwest Research Institute)   STScI-PRC10-06a

The best image of Pluto by the Hubble Space Telescope (actual image, upper right).
The colour images below it are computer-synthesized maps
made from the raw Hubble images
http://www.tested.com/science/space/461876-tested-explains-why-we-have-no-good-images-pluto/

Above is a **Hubble Space Telescope** image of **Pluto**, taken several years ago. Because **Pluto** was slightly less than 3 billion miles from Earth (and still, at the time, inside **Neptune**'s orbit) when the image was acquired, even with **Hubble**'s superb resolution, the size of each "pixel" on the surface of the tiny planet was more than 100 miles across!

At that enormous distance, the strength (and thus the heating effect) of sunlight reaching **Pluto**'s surface was 800 times less than sunlight at **Earth**'s distance … and getting smaller with each passing hour!

This is because -- crucially -- since 1989, **Pluto** has been moving away from the **Sun** in its highly elliptical, 248-year orbit. As you probably already guessed, 1989 was exactly halfway in the middle of the 1979-1999 period when **Pluto** was inside **Neptune**'s orbit.
http://www.bibliotecapleyades.net/esp_dayaftertomorrow3.htm

Despite this drifting into the nether regions, where we would logically expect it to get colder and colder, something phenomenal is going on… something that utterly, totally puts the capstone on our *Hyperdimensional Model*.

Pluto's *temperature is increasing*. Its *atmospheric pressure is increasing*. And not just a little… **A lot!**

No, scratch that… by a *truly tremendous amount*!!

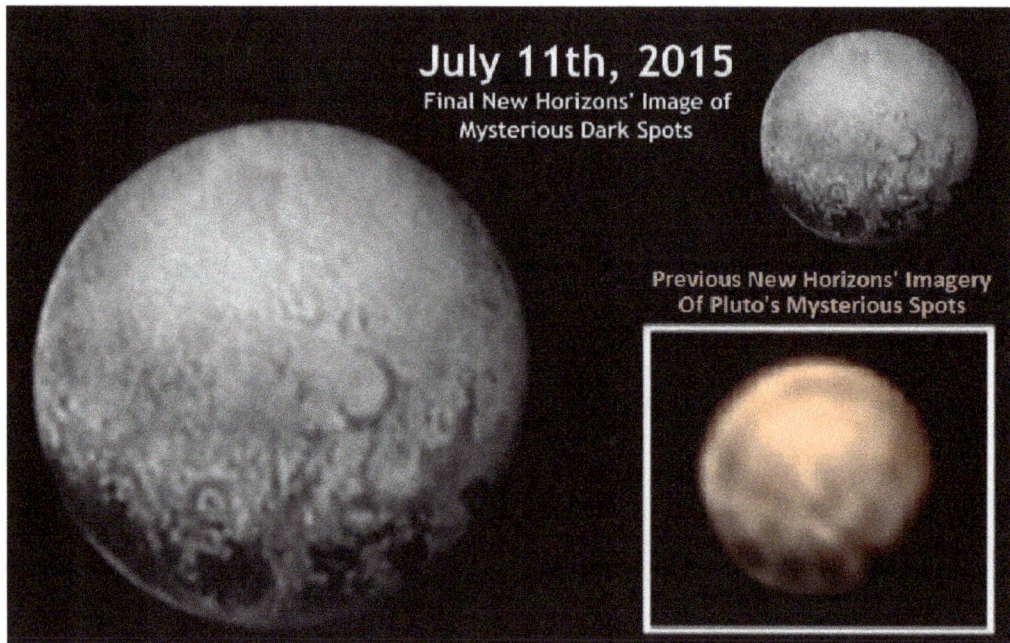

Another view of the New Horizon probe's flyby approach of Pluto reveals greater Surface detail never before seen by any space-based Earth-orbiting telescope
https://www.nasa.gov/image-feature/new-horizons-last-portrait-of-pluto-s-puzzling-spots
and http://www.bbc.com/news/science-environment-33421208

Yet, this is all occurring even as **Space.com** admits, *"Logic suggests the planet might cool as it receives less sunlight each day."* Indeed.

A groundbreaking Pluto study, recently led by **Dr. James L. Elliot**, took advantage of a rare event, similar to the one we previously discussed with Saturn's satellite, Titan. Pluto passed in front of a star in 2002, and this allowed **Dr. Elliot** and his associates to determine whether Pluto's structure and composition had stayed the same, as previously observed in 1989, or if it had changed in some way.

To their obvious surprise, they discovered that the ***atmospheric pressure of Pluto*** has ***increased... by a full three hundred percent... (!)*** ... between 1989 and 2002! This has also caused a noticeable rise in Pluto's surface temperatures. Again, this is attributed by mainstream planetologists to... you guessed it... ***"seasonal change."***

Remember just a bit earlier when we were discussing Neptune's moon, Triton... how its global warming compares to the Earth becoming globally 22 degrees Fahrenheit hotter -- in only nine years? According to **Dr. Elliot**,

*"The changes observed in **Pluto**'s atmosphere are much more severe [than the "global warming" seen on **Triton**]. The changes observed on **Triton** were subtle. **Pluto**'s changes are not subtle... We just don't know what is causing these effects."*
Indeed, elsewhere **Dr. Elliot** says that the idea of *"seasonal changes"* being responsible for such a *"severe"* increase is *"counterintuitive,"* because, by orbiting farther from the **Sun**, it would be expected that *Pluto*'s temperatures would naturally fall... not rise!

Hence, **Dr. Elliot** and his other ***NASA team members*** acknowledge this unexpected *"global warming"* of Pluto, but they also say that this warming trend is "likely not connected with that of the *Earth*" since the "Sun's output is much too steady." Furthermore, "some longer term change, analogous to long-term climatic changes on Earth" could be responsible. Without identifying precisely what this *longer-term change* could be, they tread very closely to ***suggesting some single, unified cause***... like what we present here in our ***hyperdimensional model***.

Furthermore, not only has Pluto's atmospheric pressure increased, but it is also showing signs of weather... for the first time, as was reported in a **Space.com** article:

Meanwhile, the new studies reveal what appear to be the first signs of weather on Pluto, small fluctuations of air density and temperature. Sicardy's team figures the changes, seen as spikes in the data, are caused "either by strong winds between the lit and dark hemispheres of the planet or by convection near the surface of Pluto."

Scientists have long suspected that pressure difference in the tenuous atmosphere, created by stark temperature differences from the day side to the night side, would fuel brisk breezes. The researchers did not attempt to estimate the strength of Pluto's apparent winds. Pluto gives up its secrets slower than any planet.

Given that even NASA seems to be vaguely aware that these distant, totally unexplainable changes in Pluto's environment are, somehow, analogous to equally inexplicable *"global warming"* occurring here on Earth… by saying that the *"global warming"* of the *Earth and of Pluto* are *"likely not connected"*… in the closing section of this Report we shall, therefore, turn our attention back towards the Earth, where all of these changes truly matter most .

These are seasonal changes occurring all around us… a *"Hyperdimensional Spring"* that is blooming throughout our *entire solar system*.

CHAPTER 94

DWARF PLANETS OF THE KUIPER BELT, THE OORT CLOUD, EXOPLANETS, AND BEYOND

As we tour ever outwards from the Solar System in our search for extraterrestrial life, we eventually come to the Kuiper Belt of small planetesimal objects or dwarf planets, asteroids and comet-like objects. It is the last stop before we leave our star system and enter into deep interstellar space.

The term dwarf planet was adopted in 2006 by the **International Astronomical Union (IAU),** the organization that assigns designations to celestial bodies. With continually improving technologies and more Sun-orbiting bodies of all sizes being discovered all the time, how to designate them is a matter of great scientific debate. The only thing certain is that not all scientists agree.

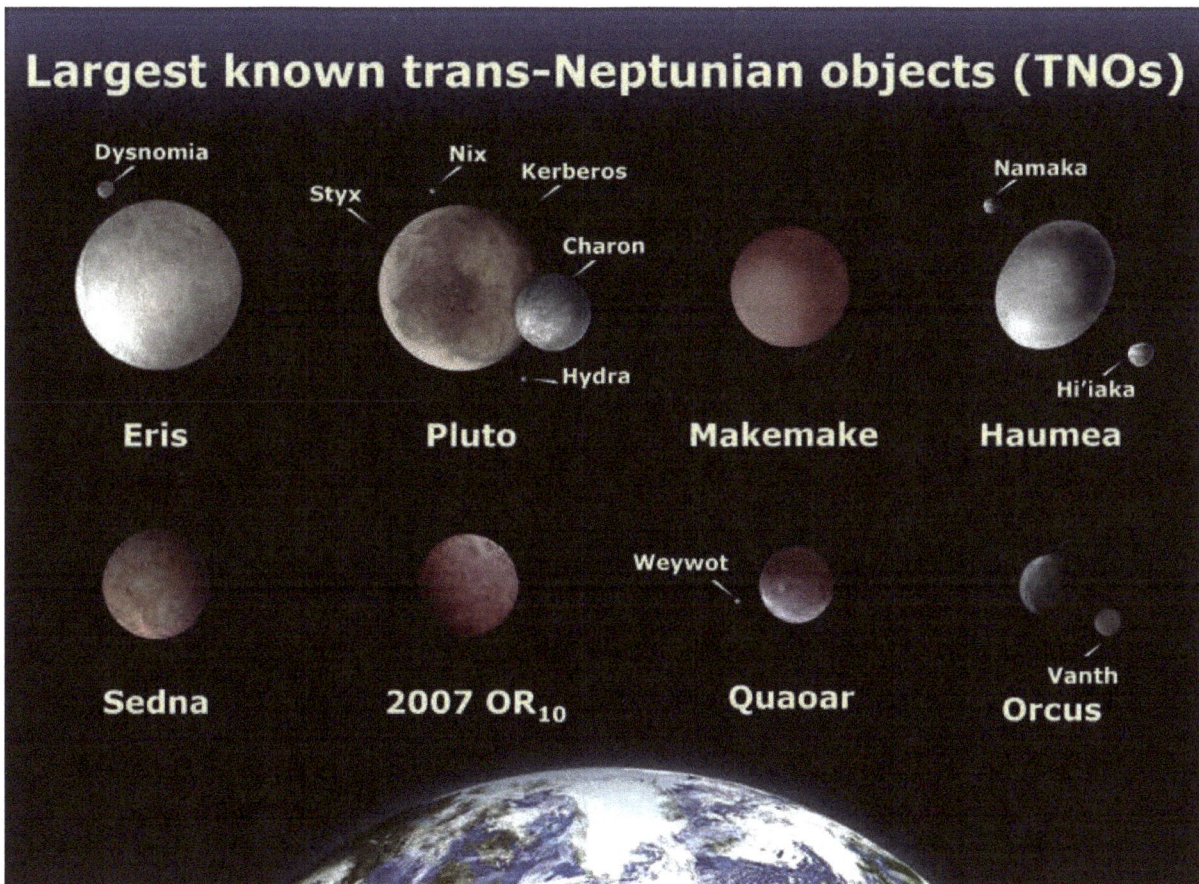

Artistic comparison of Eris, Pluto, Makemake, Haumea, Sedna, 2007 OR$_{10}$, Quaoar, Orcus, and Earth.
http://lists.physicswiki.net/index.php/Planet

The IAU currently recognizes five dwarf planets—Pluto, **Eris, Ceres, Makemake,** and **Haumea**. It is suspected that 40 more known solar system objects meet current the criteria for dwarf planets. When the entire Kuiper belt region is explored, there may be 200 more. If objects beyond the Kuiper belt are considered, there could be 2,000. Stay tuned – the definition may change again as we continue to better understand our Solar System.

The upper and lower size and mass limits of dwarf planets have not been specific by the IAU.

Ceres is the dwarf planet closest to the Sun. It resides in the asteroid belt and orbits the Sun every 4.6 years. Eris is farthest from the Sun, located in the distant icy region called the scattered disk. It takes 557 years to orbit the Sun. The three remaining dwarfs are in the Kuiper belt, the massive region beyond Neptune that consists mainly of leftovers from the solar system's formation.

An asteroid belt object such as Ceres is expected to be made of mainly rock and metal, while bodies from the Kuiper belt and beyond are composed largely of frozen ices of methane, ammonia, and water.

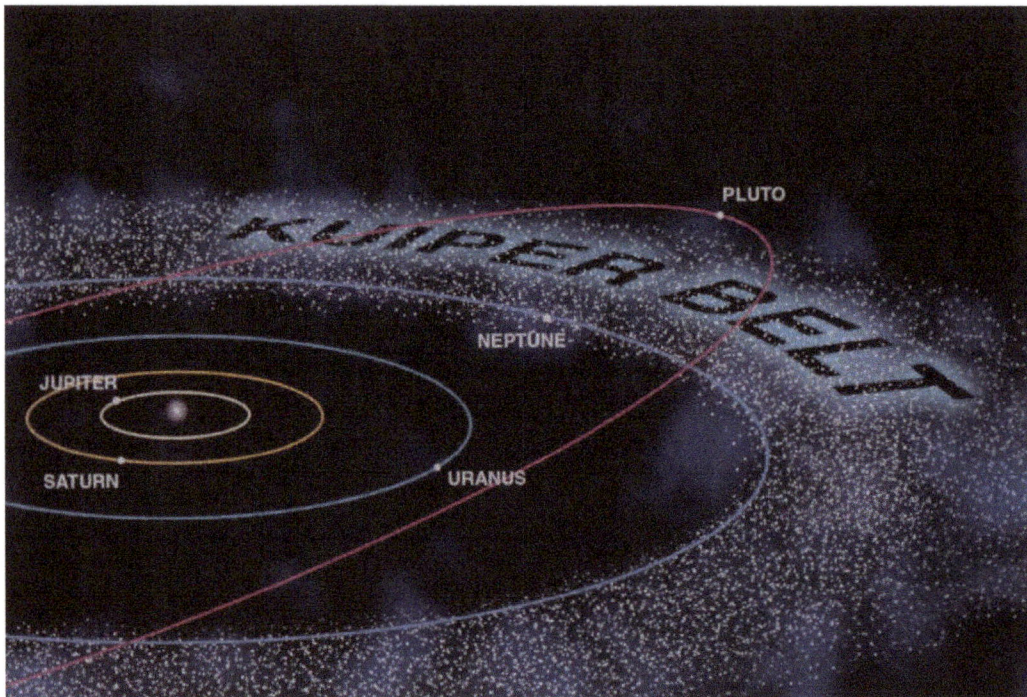

Three dwarf planets orbit in the Kuiper belt the vast region between beyond Neptune from the Earth to the Sun, or 93 million miles).
https://pics-about-space.com/solar-system-diagram-kuiper-belt?p=1

Voyager I and II Spacecraft

As of this writing, we stated earlier in this section that the space probes Voyager 1 was near to leaving the solar system completely as they were near the outer fringes of the **Heliosphere** and **Kuiper Belt**. NASA has officially confirmed to the public Voyager 1 has entered interstellar space leaving our Solar System for good on August 25, 2012, making it Earth's first extraterrestrial spacecraft.

Voyager 1 and 2 each carried recorded messages on 12-inch gold-plated copper disk phonograph record containing sounds and images selected to portray the diversity of life and culture on Earth. The contents of the record were selected for NASA by a committee chaired by Carl Sagan of Cornell University, et al. Dr. Sagan and his associates assembled 115 images and a variety of natural sounds, such as those made by surf, wind and thunder, birds, whales, and other animals. To this, they added musical selections from different cultures and eras, and spoken greetings from Earth-people in fifty-five languages, and printed messages from **President Carter** and **U.N. Secretary General Waldheim**.

Voyager 1 has left the Solar System and entered into interstellar space becoming Earth's first interstellar Extraterrestrial (ET) spacecraft
https://www.space.com/22778-voyager-1-spacecraft-interstellar-space-photo-timeline.html

Each record is encased in a protective aluminum jacket, together with a cartridge and a needle. Instructions, in symbolic language, explain the origin of the spacecraft and indicate how the record is to be played. The 115 images are encoded in analog form. The remainder of the record is in audio, designed to be played at 16-2/3 revolutions per minute. It contains the spoken greetings, beginning with **Akkadian**, which was spoken in **Sumer** about six thousand years ago, and ending with **Wu**, a modern Chinese dialect. Following the section on the sounds of Earth, there is an eclectic 90-minute selection of music, including both Eastern and Western classics and a variety of ethnic music. Once the Voyager spacecraft leave the solar system (by 1990, both will be beyond the orbit of Pluto), they will find themselves in empty space. It will be *forty thousand years* before they make a close approach to any other planetary system.

The Voyager carries a 12-inch gold-plated copper disk phonograph record containing sounds and images selected to portray the diversity of life and culture on Earth
http://imgur.com/gallery/jtZ2I

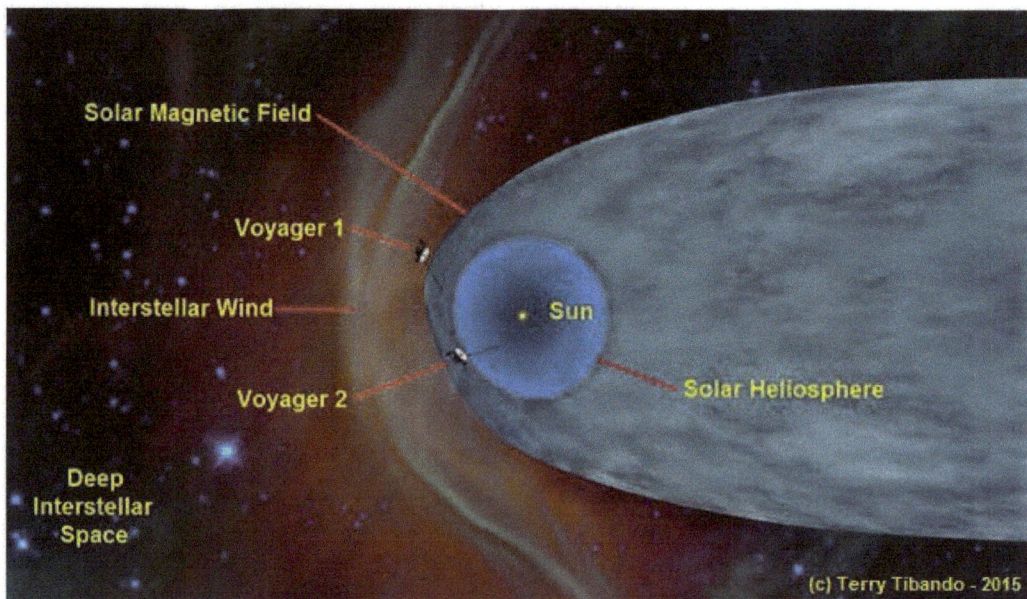

Voyager 1 (top) has sailed beyond our solar bubble into interstellar space, the space between stars. Voyager 2 (bottom) is still exploring the outer layer of the solar bubble
https://www.jpl.nasa.gov/interstellarvoyager/ (c) Terry Tibando

As Carl Sagan has noted, *"The spacecraft will be encountered and the record played only if there are advanced space-faring civilizations in interstellar space. But the launching of this bottle into the cosmic ocean says something very hopeful about life on this planet."*

CHAPTER 95

EXOPLANETS AND EXTRATERRESTRIAL LIFE
IN THE UNIVERSE

Exoplanets are planets that exist outside our solar system, orbiting other stars which are extremely distant our star system. Even the most powerful telescopes like NASA's **Kepler, Spitzer, Webb Space Telescope** (soon to be launched in 2021), and ESA's **Hipparcus**, etc. can't image anything as small as a planet outside our solar system. Even within our own solar system, Neptune and Pluto are blurry coloured balls of light when observed from Earth's orbit.

Searching for exoplanets outside our solar system are essentially invisible. However, planets orbiting around other star systems can be detected in indirect ways by how they affect their stars in measurable ways. Generally, astronomers have resorted to indirect methods to detect extrasolar planets. Here is a list of different indirect methods that have proven successful: **Radial velocity, Transit photometry, Relativistic beaming, Ellipsoidal variations, Pulsar timing, Variable star timing, Transit timing, Transit duration variation, Eclipsing binary minima timing, Gravitational microlensing, Direct imaging, Polarimetry, Astrometry, X-ray eclipse, Disc kinematics** and there are methods being developed all the time.

The two most widely used methods are **transits** (the blinking method) and **Doppler shifting** (the wobble method).

Transit Method

When a planet orbits its star, the planet will sometimes cross between it and the Earth. This crossing is called a **transit**, and when it happens, the planet blocks a bit of the star's light. It may be well under one percent of the light, but that's enough for special telescopes to measure. If that star is blinking in a regular, cyclical pattern, that tells astronomers there's a planet circling it – as well as the size (width) of the planet and how big its orbit is.

Wobble Method

In the wobble method, astronomers rely on the fact that just as stars tug on their planets to keep them in orbit, planets also tug on their stars. So, as a planet circles, its star will wobble back and forth very slightly. This wobble is usually too small to see in an image, but it does show up as a wiggle in the spectrum, or color, of the star. Again, astronomers look for a pattern to that wiggle, which tells them how massive a planet is and how far away it orbits.
https://en.wikipedia.org/wiki/Methods_of_detecting_exoplanets

Radial Velocity

A star with a planet will move in its own small orbit in response to the planet's gravity. This leads to variations in the speed with which the star moves toward or away from Earth, i.e. the variations are in the **radial velocity** also known as **Doppler spectroscopy**) of the star with respect to Earth. The radial velocity can be deduced from the displacement in the parent star's spectral lines due to the Doppler effect. The radial-velocity method measures these variations in order to confirm the presence of the planet using the binary mass function.

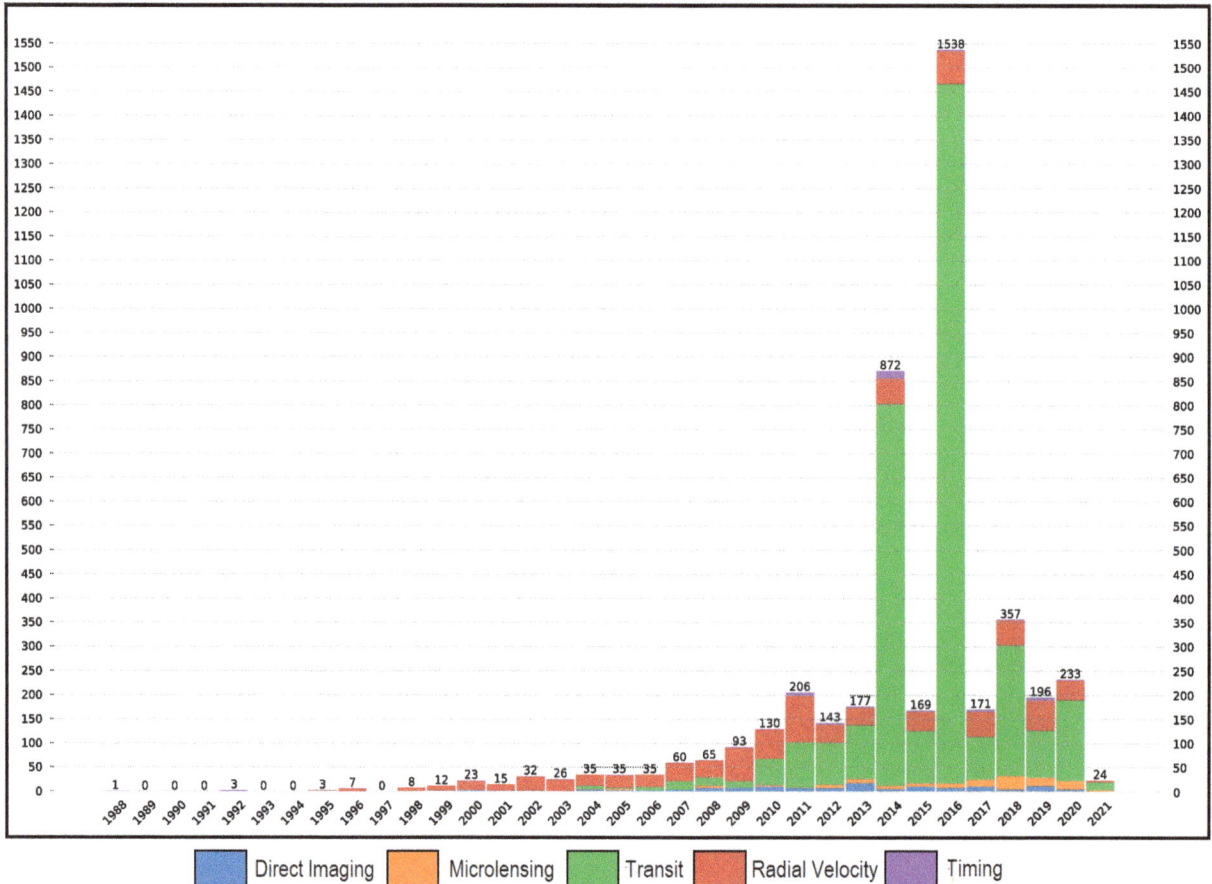

Number of extrasolar planet discoveries per year through 2020, with colors indicating method of detection
https://upload.wikimedia.org/wikipedia/commons/f/fd/Confirmed_exoplanets_by_methods_EPE.svg

A simulated silhouette of Jupiter (and 2 of its moons) transiting our Sun, as seen from another star system

Transit Photometry

The **Photometric** method can determine the planet's radius. If a planet crosses (transits) in front of its parent star's disk, then the observed visual brightness of the star drops by a small amount, depending on the relative sizes of the star and the planet. For example, in the case of HD 209458, the star dims by 1.7%. When compared to an Earth -size planet transiting a Sun-like star it produces a dimming of only 80 parts per million (0.008 percent), so is enough to determine the planet's radius. https://en.wikipedia.org/wiki/Methods_of_detecting_exoplanets

Reflection and Emission Modulations

Short-period planets in close orbits around their stars will undergo **reflected light variations** because, like the Moon, they will go through phases from full to new and back again. In addition, as these planets receive a lot of starlight, it heats them, making **thermal emissions** potentially detectable. Since telescopes cannot resolve the planet from the star, they see only the combined light, and the brightness of the host star seems to change over each orbit in a periodic manner.

Relativistic Beaming

A separate novel method to detect exoplanets from light variations uses **relativistic beaming** of the observed flux from the star due to its motion. It is also known as **Doppler beaming** or **Doppler boosting**. As the planet tugs the star with its gravitation, the density of photons and therefore the apparent brightness of the star changes from observer's viewpoint.

Planets transit in front of their star blocking a bit of its sunlight enabling astronomers to measure the width of a planet and its orbit © Terry Tibando

Ellipsoidal Variations

Massive planets can cause slight tidal distortions to their host stars. When a star has a slightly **ellipsoidal shape**, its apparent brightness varies, depending if the oblate part of the star is facing the observer's viewpoint. Like with the relativistic beaming method, it helps to determine the minimum mass of the planet, and its sensitivity depends on the planet's orbital inclination. In addition, the planet distorts the shape of the star more if it has a low semi-major axis to stellar radius ratio and the density of the star is low. This makes this method suitable for finding planets around stars that have left the main sequence.
https://en.wikipedia.org/wiki/Methods_of_detecting_exoplanets

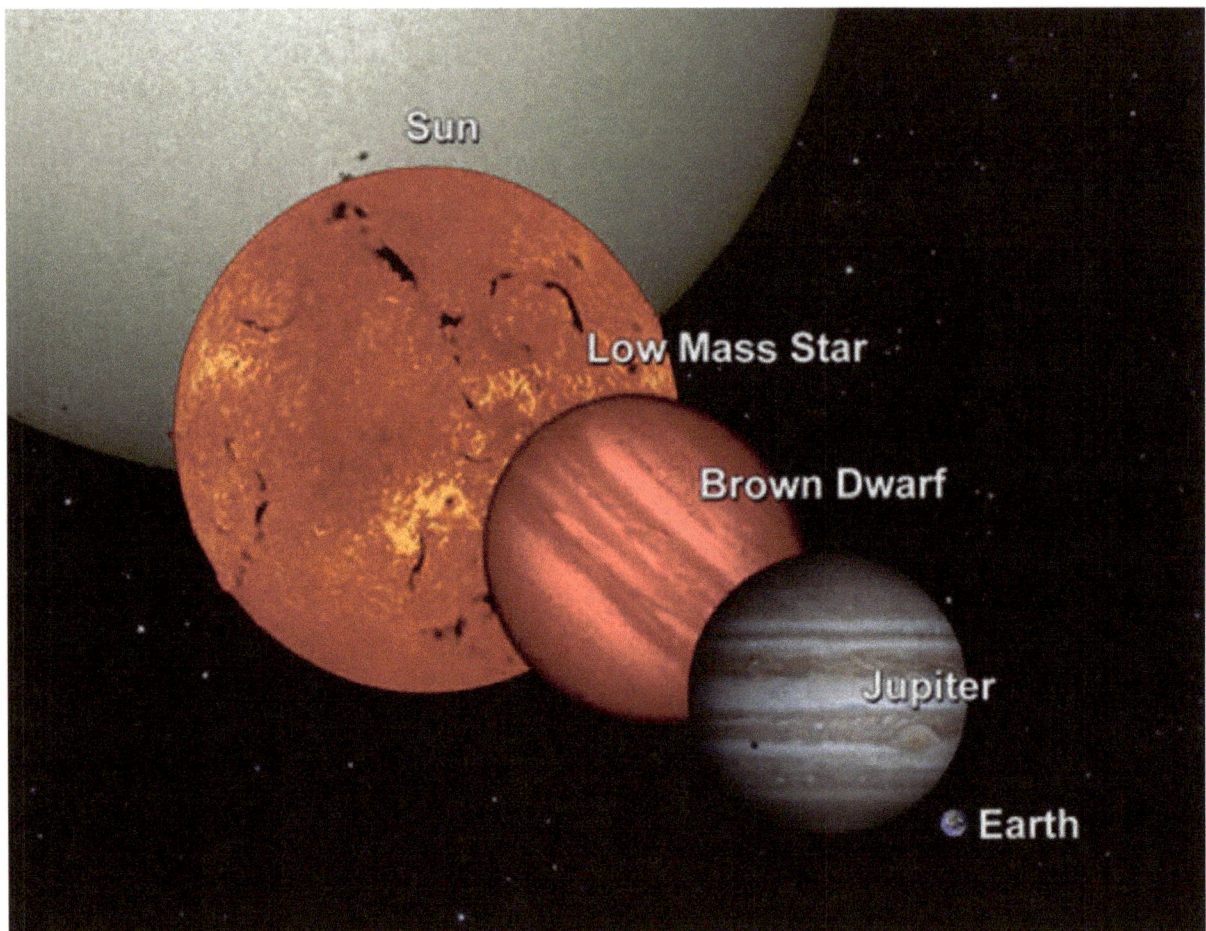

This image shows the relative sizes of brown dwarfs and large planets.
https://en.wikipedia.org/wiki/Methods_of_detecting_exoplanet

Pulsar Timing

A **pulsar** is a **neutron star**: the small, ultradense remnant of a star that has exploded as a **supernova**. Pulsars emit radio waves extremely regularly as they rotate. Because the intrinsic rotation of a pulsar is so regular, slight anomalies in the timing of its observed radio pulses can be used to track the pulsar's motion. Like an ordinary star, a pulsar will move in its own small orbit if it has a planet. Calculations based on pulse-timing observations can then reveal the parameters of that orbit.

Variable Star Timing

Like pulsars, some other types of **pulsating variable stars** are regular enough that **radial velocity** could be determined purely photometrically from the **Doppler shift** of the pulsation frequency, without needing **spectroscopy.** The first success with this method came in 2007, when V391 Pegasi b was discovered around a pulsating subdwarf star.

Transit Timing

The **transit timing** variation method considers whether transits occur with strict periodicity, or if there is a variation. When multiple transiting planets are detected, they can often be confirmed with the transit timing variation method. This is useful in planetary systems far from the Sun, where radial velocity methods cannot detect them due to the low signal-to-noise ratio. If a planet has been detected by the transit method, then variations in the timing of the transit provide an extremely sensitive method of detecting additional non-transiting planets in the system with masses comparable to Earth's. It is easier to detect transit-timing variations if planets have relatively close orbits, and when at least one of the planets is more massive, causing the orbital period of a less massive planet to be more perturbed.

Transit Duration Variation

"**Duration variation**" refers to changes in how long the transit takes. Duration variations may be caused by an **exomoon**, **apsidal precession** for eccentric planets due to another planet in the same system, or general relativity. When a **circumbinary planet** is found through the transit method, it can be easily confirmed with the transit duration variation method.

Eclipsing Binary Minima Timing

When a **binary star system** is aligned such that – from the Earth's point of view – the stars pass in front of each other in their orbits, the system is called an **"eclipsing binary" star system**. The time of minimum light, when the star with the brighter surface is at least partially obscured by the disc of the other star, is called the primary eclipse, and approximately half an orbit later, the secondary eclipse occurs when the brighter surface area star obscures some portion of the other star. These times of minimum light, or central eclipses, constitute a time stamp on the system, much like the pulses from a pulsar (except that rather than a flash, they are a dip in brightness). If there is a planet in **circumbinary orbit** around the binary stars, the stars will be offset around a binary-planet center of mass. As the stars in the binary are displaced back and forth by the planet, the times of the eclipse minima will vary. The periodicity of this offset may be the most reliable way to detect **extrasolar planets** around close binary systems. With this method, planets are more easily detectable if they are more massive, orbit relatively closely around the system, and if the stars have low masses.
https://en.wikipedia.org/wiki/Methods_of_detecting_exoplanets

Gravitational Microlensing

Gravitational microlensing occurs when the gravitational field of a star acts like a lens, magnifying the light of a distant background star. This effect occurs only when the two stars are almost exactly aligned. Lensing events are brief, lasting for weeks or days, as the two stars and Earth are all moving relative to each other. More than a thousand such events have been observed over the past ten years. If the foreground lensing star has a planet, then that planet's own gravitational field can make a detectable contribution to the lensing effect.

Direct Imaging

Planets are extremely faint light sources compared to stars, and what little light comes from them tends to be lost in the glare from their parent star. So, in general, it is very difficult to detect and resolve them directly from their host star. Planets orbiting far enough from stars to be resolved reflect very little starlight, so planets are detected through their **thermal emission** instead. It is easier to obtain images when the star system is relatively near to the Sun, and when the planet is especially large (considerably larger than Jupiter), widely separated from its parent star, and hot so that it emits intense infrared radiation; images have then been made in the infrared, where the planet is brighter than it is at visible wavelengths. **Coronagraphs** are used to block light from the star, while leaving the planet visible. **Direct imaging** of an Earth-like **exoplanet** requires extreme **optothermal stability.**

Other emerging exoplanet detection methods are **Polarimetry, Astrometry, X-ray eclipse, Disc kinematics, Flare and variability echo detection, Transit imaging, Magnetospheric radio emissions, Auroral radio emissions, Optical interferometry,** and **Modified interferometry.**
https://en.wikipedia.org/wiki/Methods_of_detecting_exoplanets

Since the first confirmation of an exoplanet orbiting a Sun-like star in 1995, more than **4,375 exoplanets** have been discovered and are considered "confirmed." However, there are **5856 "candidate" exoplanet** detections that require further observations in order to say for sure whether or not the exoplanet is real.

Remarkably, the first exoplanets were just discovered about two decades ago. We live in an extraordinary time where in the span of a single generation, the centuries-old question "Are there planets orbiting other stars?" has been answered with a resounding "Yes!"

There are now **3247 planetary systems** that have been discovered; it is a reasonable expectation that when humanity becomes a true interstellar travelling civilization, most likely these star systems will be the first ones to be explored for extraterrestrial life.
https://exoplanets.nasa.gov/faq/6/how-many-exoplanets-are-there/

Massachusetts Institute of Technology - MIT own physics professor **Sara Seager** looks for possible chemical combinations that could signal the presence of alien life. She and her biochemistry colleagues first focused on the six main elements associated with life on Earth: carbon, nitrogen, oxygen, phosphorous, sulfur and hydrogen. The **James Webb Space Telescope,** to be launched in 2021, could get the first glimpses of these mix of gases (oxygen, carbon dioxide, methane) in the atmospheres of Earth-sized exoplanets which would be strong indications of possible life.

MIT astronomer Sara Seager, is a McArthur Genius Fellow and leader in the scientific race to find another Earth in the near future.

A statistical estimate based on data from NASA's **Kepler Space Telescope** revealed that there are more planets than stars in our galaxy. That means there are more than a trillion planets in our galaxy alone, many of them in Earth's size range. https://exoplanets.nasa.gov/discovery/how-we-find-and-characterize/

Here, it can be seen that astronomers and scientists expect that the chances of life in the universe will increase if the planets are similar to our Earth and are within the **Zone of Habitability**. It is believed that this Zone will increase in range as humans start to explore the other planets in our Solar System such as Mars and some of the Jovian, Saturnian and outer planet moons. Using our own solar system model of habitability, it can be used as a measure tool for possible life in other star systems.

Future telescopes might even pick up signs of **photosynthesis** – the transformation of light into chemical energy by plants – or even gases or molecules suggesting the presence of animal life. Intelligent, technological life might create atmospheric pollution, as it does on our planet, also detectable from afar. Of course, the best we might be able to manage is an estimate of probability. Still, an exoplanet with, say, a 95 percent probability of life would be a game changer of historic proportions. https://exoplanets.nasa.gov/search-for-life/can-we-find-life/

The search for extraterrestrial life and in particular, intelligent life may be found within our own neighborhood, possibly beneath the Martian surface or in the dark, subsurface oceans of

427

Jupiter's moon, **Europa**. It may be the dream of the ages come true, and when we fortuitously, eavesdrop on the communications of extraterrestrial civilizations. We might even capture evidence of **"technosignatures,"** (traces of technology).

Once again, however, the Baha'I writings on life in the universe must be restated as this is the new path to be explored: *"Knowst thou that every fix (stable) star has its own planets and every planet its own creatures whose number no man can compute!"* – **Baha'u'llah**

What Kind of Life May be Found on Exoplanets?

In Volume Four of this series "A Citizen's Disclosure on UFOs and ETI" in Chapters 79 and 80, and in this volume under Chapters 81, 82, and 83, it has been clearly demonstrated that extraterrestrial life exists as revealed by raw photographs from the various NASA Mars rovers travelling over the Martian terrain.

Life may turn up beneath the Martian surface as drilling and scooping soil samples taken by the Mars rovers which in turn will analyze for material elements and microorganisms that may be present under the surface.

There is now proof that liquid water exists besides ice water as evident from seepage and run-off from hills and crater walls which is a major ingredient for life as we know and understand it on Earth. Perhaps, it will be discovered in the near future that there may be life on Jupiter's moons, **Europa** and **Ganymede,** as well as that of the Saturn satellite **Enceladus**, beneath in its dark subsurface oceans.

Venturing out beyond the confines of our solar system on our virtual tour for the search of extraterrestrial life, astronomers have found a nearby **"super-Earth" exoplanet** that may be capable of supporting life as we know it.

An international group of astronomers discovered the planet using NASA's Transiting Exoplanet Survey Satellite (TESS) earlier this year in the constellation Hydra, about 31 light-years from Earth, according to a statement by NASA.

The exoplanet, named, is believed to be around twice the size of Earth and harbor six times Earth's mass. Located in the outer edge of its host star's **"habitable zone,"** scientists believe that this super-Earth could have water on its surface.

But GJ 357 b's possibly habitable neighbor planet soon stole the show. Further observations showed that GJ 357 d orbits its star every 55.7 days at a distance of around a fifth of Earth's distance from the sun, and could have Earth-like conditions, according to a statement from Cornell University. https://www.space.com/super-earth-exoplanet-gj-357d-may-support-life.html

"We built the first models of what this new world could be like," **Jack Madden**, doctoral candidate at Cornell and co-author of the study, said in the statement. "Just knowing that liquid water can exist on the surface of this planet motivates scientists to find ways of detecting signs of life."

Humanity is on the verge of discovering alien life, high-ranking NASA scientists say.

The other planet in the system, **GJ 357 c**, is at least 3.4 times more massive than Earth and orbits the star every 9.1 days. GJ 357 c probably has a surface temperature around 260 Fahrenheit (127 C), NASA officials said.

An illustration of Planet GJ357 orbiting around its dwarf sun.
(Image credit: Jack Madden/Cornell University) and
https://www.space.com/super-earth-exoplanet-gj-357d-may-support-life.html

"I think we're going to have strong indications of life beyond Earth within a decade, and I think we're going to have definitive evidence within 20 to 30 years," NASA chief scientist **Ellen Stofan** said during a panel discussion that focused on the space agency's efforts to search for habitable worlds and alien life.

Farther afield, observations by NASA's **Kepler space telescope** suggest that nearly every star in the sky hosts planets — and many of these worlds may be habitable. Indeed, Kepler's work has shown that rocky worlds like Earth and Mars are probably more common throughout the galaxy than gas giants such as Saturn and Jupiter. https://www.space.com/29041-alien-life-evidence-by-2025-nasa.html

Paul Hertz, director of NASA's Astrophysics Division said, "We can see water in the interstellar clouds from which planetary systems and stellar systems form. We can see water in the disks of debris that are going to become planetary systems around other stars, and we can even see comets being dissipated in other solar systems as [their] star evaporates them."

will scan the starlight that passes through the air of super-Earths, which are more massive than our own planet but significantly less so than gaseous worlds such as Uranus and Neptune. This method, called transit spectroscopy, will likely not work for potentially habitable Earth-size worlds.

Searching for **biosignature gases** on small, rocky exoplanets will instead probably require direct imaging of these worlds, using a **"coronagraph"** to block out the overwhelming glare of their parent stars, Hertz added. https://www.space.com/29041-alien-life-evidence-by-2025-nasa.html

By examining the charts of Habitable Zone Planets above and below, it is easy to see why astronomers are looking for Earth type planets because their proximity to their sun or star may provide one of the requirements for life to evolve on the planet.

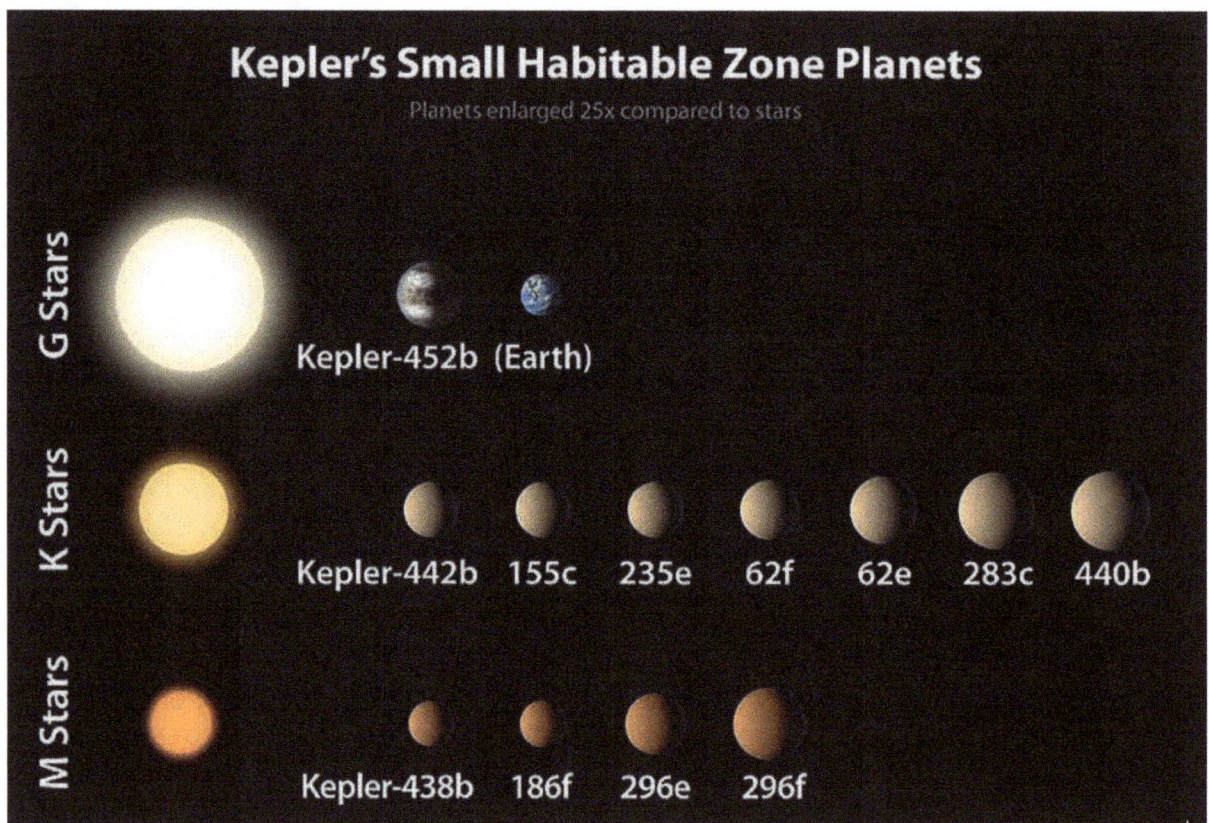

Comparison of small planets found by *Kepler* in the habitable zone of their host stars.
https://en.wikipedia.org/wiki/List_of_potentially_habitable_exoplanets

Artist's impression of Kepler-442b compared to Earth

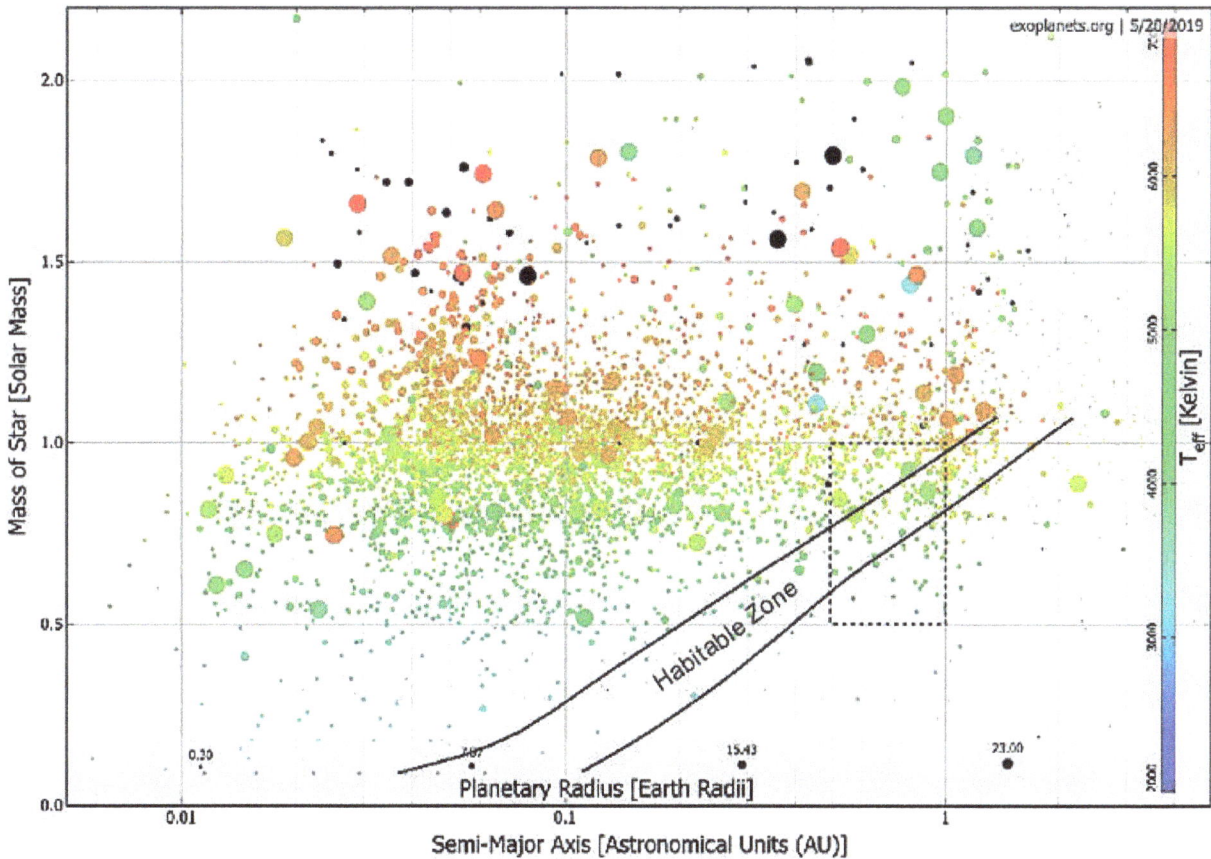

Star-planet distances and mass of the host star of roughly 4500 exoplanets and exoplanet candidates. The temperatures of the stars are indicated with symbol colors; planetary radii are encoded in the symbol sizes.

431

Conservative Sample of Potentially Habitable Exoplanets

This is a list of the exoplanets that are more likely to have a rocky composition and maintain surface liquid water (*i.e.* 0.5 < Planet Radius ≤ 1.5 Earth radii or 0.1 < Planet Minimum Mass ≤ 5 Earth masses).

Name	Type	Mass (M_E)	Radius (R_E)	Flux (S_E)	T_{eq} (K)	Period (days)	Distance (ly)	ESI
001. Teegarden's Star b	M-Warm Terran	≥1.05	—	1.15	264	4.9	12	0.95
002. TOI-700 d (N)	M-Warm Terran	—	1.14	0.87	246	37.4	101	0.93
003. K2-72 e	M-Warm Terran	—	1.29	1.11	261	24.2	217	0.90
004. TRAPPIST-1 d	M-Warm Subterran	0.41	0.77	1.14	263	4.0	41	0.90
005. Kepler-1649 c (N)	M-Warm Terran	—	1.06	0.75	237	19.5	301	0.90
006. Proxima Cen b	M-Warm Terran	≥1.27	—	0.70	228	11.2	4.2	0.87
007. GJ 1061 d (N)	M-Warm Terran	≥1.64	—	0.69	218	13.0	12	0.86
008. GJ 1061 c (N)	M-Warm Terran	≥1.74	—	1.45	275	6.7	12	0.86
009. Ross 128 b	M-Warm Terran	≥1.40	—	1.48	280	9.9	11	0.86
010. GJ 273 b	M-Warm Terran	≥2.89	—	1.06	258	18.6	12	0.85
011. TRAPPIST-1 e	M-Warm Terran	0.62	0.92	0.66	230	6.1	41	0.85
012. Kepler-442 b	K-Warm Terran	—	1.35	0.70	233	112.3	1193	0.84
013. Wolf 1061 c	M-Warm Terran	≥3.41	—	1.30	271	17.9	14	0.80
014. GJ 667 C c	M-Warm Terran	≥3.81	—	0.88	247	28.1	24	0.80
015. GJ 667 C f	M-Warm Terran	≥2.54	—	0.56	221	39.0	24	0.77
016. Kepler-1229 b	M-Warm Terran	—	1.40	0.49	213	86.8	865	0.73
017. TRAPPIST-1 f	M-Warm Terran	0.68	1.04	0.38	200	9.2	41	0.68
018. Kepler-62 f	K-Warm Terran	—	1.41	0.41	204	267.3	981	0.68
019. Teegarden's Star c	M-Warm Terran	≥1.11	—	0.37	199	11.4	12	0.68
020. Kepler-186 f	M-Warm Terran	—	1.17	0.29	188	129.9	579	0.61
021. GJ 667 C e	M-Warm Terran	≥2.54	—	0.30	189	62.2	24	0.60
022. tau Cet f	G-Warm Terran	≥3.93	—	0.32	190	636.1	12	0.58
023. TRAPPIST-1 g	M-Warm Terran	1.34	1.13	0.26	181	12.4	41	0.58
024. GJ 682 b	M-Warm Terran	≥4.40	—	0.31	190	17.5	16	0.57

http://phl.upr.edu/projects/habitable-exoplanets-catalog

Optimistic Sample of Potentially Habitable Exoplanets

This is a list of the exoplanets that are less likely to have a rocky composition or maintain surface liquid water (*i.e.* 1.5 < Planet Radius ≤ 2.5 Earth radii or 5 < Planet Minimum Mass ≤ 10 Earth masses).

Name	Type	Mass (M_E)	Radius (R_E)	Flux (S_E)	T_{eq} (K)	Period (days)	Distance (ly)	ESI
001. Kepler-452 b	G-Warm Superterran	–	1.63	1.11	261	384.8	1799	0.83
002. Kepler-62 e	K-Warm Superterran	–	1.61	1.15	264	122.4	981	0.83
003. Kepler-1652 b	M-Warm Superterran	–	1.60	0.84	244	38.1	822	0.83
004. Kepler-1544 b	K-Warm Superterran	–	1.78	0.90	248	168.8	1092	0.80
005. Kepler-296 e	M-Warm Superterran	–	1.52	1.50	276	34.1	737	0.80
006. Kepler-283 c	K-Warm Superterran	–	1.82	0.90	248	92.7	1526	0.79
007. K2-296 b	M-Warm Superterran	–	1.87	1.15	264	28.2	519	0.78
008. Kepler-1410 b	K-Warm Superterran	–	1.78	1.34	274	60.9	1196	0.78
009. K2-3 d	M-Warm Superterran	2.80	1.65	1.50	282	44.6	144	0.78
010. Kepler-1638 b	G-Warm Superterran	–	1.87	1.39	276	259.3	4973	0.76
011. Kepler-296 f	M-Warm Superterran	–	1.80	0.66	225	63.3	737	0.75
012. Kepler-440 b	K-Warm Superterran	–	1.91	1.44	273	101.1	981	0.75
013. Kepler-705 b	M-Warm Superterran	–	2.11	0.83	243	56.1	903	0.74
014. Kepler-1653 b	K-Warm Superterran	–	2.17	1.04	258	140.3	2461	0.74
015. GJ 832 c	M-Warm Superterran	≥5.40	–	0.99	253	35.7	16	0.74
016. Kepler-1606 b	G-Warm Superterran	–	2.07	1.41	277	196.4	2710	0.73
017. Kepler-1090 b	G-Warm Superterran	–	2.25	1.20	267	198.7	2800	0.72
018. Kepler-61 b	K-Warm Superterran	–	2.15	1.39	273	59.9	1092	0.72
019. K2-18 b	M-Warm Superterran	8.92	2.37	1.08	257	32.9	124	0.71
020. Kepler-443 b	K-Warm Superterran	–	2.35	0.89	247	177.7	2615	0.71
021. Kepler-1701 b (N)	K-Warm Superterran	–	2.22	1.37	275	169.1	1904	0.71
022. Kepler-22 b	G-Warm Superterran	–	2.38	1.10	261	289.9	635	0.71
023. LHS 1140 b	M-Warm Superterran	6.98	1.73	0.50	214	24.7	49	0.70
024. Kepler-1552 b	K-Warm Superterran	–	2.47	1.10	261	184.8	2507	0.70
025. K2-9 b	M-Warm Superterran	–	2.25	1.45	279	18.4	270	0.70
026. Kepler-1540 b	K-Warm Superterran	–	2.49	0.92	250	125.4	799	0.70
027. GJ 180 c	M-Warm Superterran	≥6.40	–	0.78	239	24.3	39	0.70
028. Kepler-1632 b	F-Warm Superterran	–	2.47	1.27	270	448.3	2337	0.69
029. Kepler-298 d	K-Warm Superterran	–	2.50	1.29	271	77.5	1689	0.68
030. GJ 163 c	M-Warm Superterran	≥6.80	–	1.41	277	25.6	49	0.67
031. HD 40307 g	K-Warm Superterran	≥7.09	–	0.67	226	197.8	42	0.66
032. K2-288 B b	M-Warm Superterran	–	1.91	0.44	207	31.4	214	0.65
033. GJ 3293 d	M-Warm Superterran	≥7.60	–	0.59	223	48.1	66	0.63
034. GJ 229 A c (N)	M-Warm Superterran	≥7.27	–	0.53	216	122.0	19	0.62
035. Kepler-174 d	K-Warm Superterran	–	2.19	0.43	206	247.4	1254	0.61
036. GJ 357 d	M-Warm Superterran	≥6.10	–	0.38	200	55.7	31	0.58

http://phl.upr.edu/projects/habitable-exoplanets-catalog

The Search for Exo-Earths

The search for life on distant exoplanets requires the right elements to occur in the right perimeters for life to evolve. Those perimeters are water, an atmosphere, a magnetic field, a

433

planetary core dynamo, and tectonic plates; all of which must be situated in the optimum region of space from its host star, the **Habitable Zone**. We need to explore these perimeters further.

The dreams of life beyond Earth pervade literature, TV shows and drive Hollywood blockbusters – but the truth of life beyond the Earth continues to evade us. However, scientists believe that within a few decades or sooner, we will have definite proof that life exists on other planets, either within our Solar System or on an exoplanet within fifty lightyears distance from our star.

But before we look at what we think are the various ingredients needed to contribute to the recipe of a perfect world, we need to look at the history of the search for life elsewhere in the universe.

Giordano Bruno, a Dominican Friar and the forefather astrobiology who was brought before the **Roman Inquisition** as a heretic for propagating a non-Earth centric universe. He saw an infinite universe with infinite number of suns and an infinite number of planets which was counter to the Church's point of view that the Earth was held as the centre of the universe. He was, for his lack of recanting his position, burnt at the stake, but now, his point of view has been repeatedly validated by scientists.

When we view Mars through a telescope, we see polar caps and it has an angle of inclination similar to the Earth permitting Mars to go through seasonal changes like the Earth. Spectroscopic analysis of Mars' atmosphere shows that it a lack of oxygen or water and the **"canali"** long thought to be actual canals built by an intelligence for the irrigation of the planet was merely an optical illusion.

However, we have demonstrated in this book and the volume preceding it, that there may very well exist a type of irrigation canals of glass tubes for the flow of water or an extensive subterranean transport system. This will of course need to be verified by Mars rovers or rover-helicopters.

The first signs of life will probably be the discovery of microorganisms or bacteria which is a nice safe way of saying that NASA has discovered life on another planet as oppose to stating that small mammals or reptiles have been found on Mars.

The worst case scenario would be to discover that another intelligence exists in the Solar System besides humans because then, NASA and many scientists, military, intelligence, and government officials will enter the picture and try to spin-control the narrative in a way that perhaps, portrays this other intelligence as potentially hostile or worst! We are, however, leaping ahead of ourselves on this matter.

It is now almost two decades since the discovery of the first planet orbiting another sun-like star. That planet, **51 Pegasi b**, was about as alien and hostile to life as it is possible to imagine. A behemoth – comparable in size and mass to Jupiter – it spins around its host star at a distance approximately 1/20th that between the Earth and the sun.

The temperature of 51 Pegasi b's cloud tops is likely in excess of 1,000C, so hot that clouds of metal and silicate probably float among the hydrogen that most likely makes up the bulk of its atmosphere. A fascinating planet, yes, and one that revolutionised our understanding of planet formation and evolution. But Earth-like, it is not.

As each year passes, we come closer to finding the first planets that truly resemble Earth – detecting ever smaller planets, orbiting at more temperate positions in their host system. It seems almost certain that the first truly terrestrial planets will be discovered in the next decade, and the search for life upon them will begin in earnest. https://theconversation.com/exo-earths-and-the-search-for-life-elsewhere-a-brief-history-33096

The observations required to detect clear and incontrovertible evidence of life on one of these exo-Earths will be hugely time-consuming and costly. A recent paper suggests that even the incredible **James Webb Space Telescope**, scheduled for launch in 2018, will be unable to perform the requisite observations.

Because of this great difficulty in performing the required observations, the selection of the most promising target will be key. Our solar system illustrates this need remarkably nicely.

In the inner reaches lurk three Earth-like planets – Venus, Earth and Mars. From a distance, all three could be considered to fulfil our requirements, rocky worlds of around the same size. How would we chose which to target?

The terrestrial planets (l to r): Venus, Earth and Mars.
All images from NASA (Venus taken by Mariner 10, Earth from Apollo 17 and Mars from the Hubble Space Telescope)

Despite their superficial similarities the three planets are vastly different. Venus is far too hot, the victim of a runaway greenhouse effect with a surface that would melt lead. Mars is too cold, too arid and with too thin an atmosphere. Only one, Earth, is just right and is covered with abundant life.

Based on this comparison, when new exoplanets are detected, the discussion of the most Earth-like usually boils down to one thing – the likely surface temperature of the planet. The reason is simple: water.

Despite their superficial similarities the three planets are vastly different. Venus is far too hot, the victim of a runaway greenhouse effect with a surface that would melt lead. Mars is too cold, too arid and with too thin an atmosphere. Only one, Earth, is just right and is covered with abundant life.

Based on this comparison, when new exoplanets are detected, the discussion of the most Earth-like usually boils down to one thing – the likely surface temperature of the planet. The reason is simple: water.

Artist's impression of the habitable zone for the solar system (top) and the planetary system around the nearby star Gliese 581. sits in the middle of the 'Goldilocks zone' making it a likely candidate for life ESO
https://theconversation.com/exo-earths-and-the-search-for-life-elsewhere-a-brief-history-33096

For this reason, astronomers arrived at the concept of the **"habitable zone"**, or **"Goldilocks Zone"**, around a star. Move a planet too close to the star and it will be so hot that liquid water will not be able to exist on its surface (as in Venus). Move it too far away and any water will freeze out (just like Mars, or the moons of the outer planets). Between these extremes lays a broad region in which an Earth-like planet will be neither too warm nor too cold – the **"habitable zone"**. https://theconversation.com/exo-earths-and-the-search-for-life-elsewhere-a-brief-history-33096

Because there will be an increasing growing list of potential Earth-like exoplanets, there needs to be a tentative recipe for a habitable planet. We need to consider some of the many factors that come together to make one planet a more promising location for life like ours to develop and thrive. With this precept in mind, we need to know what ingredients are in that recipe to produce life on a planet.

We imagine that in the birth of a solar system of a star and its planets come about by starting off as a **proto-solar system** of gases and debris that rotate around a central core , the **proto-star** and gradually over a very long period of time gases and debris coalesce into **proto-planets** that orbit the star and spin on their axis. Eventually, gases stabilize about the planet to form atmospheres and cores of planets develop a magnet field keeping everything attracted to the planetary surface whether it is purely gases like what we find on Jupiter, Saturn, Uranus and Neptune or they become rocky planets like Mercury, Venus, Earth and Mars.

This was the way scientist viewed the creation of a solar system, but recent evidence now suggests that planet formations may be quite violent where hit and/or run collisions by meteorites, asteroids and comets from within and outside the proto-solar system smashed into larger bodies destined to become planets.

Such evidence can be determined from Mercury, which made have been twice its current size or the Earth-Moon system where collisions stripped the mantle and the crust of the Earth to leave a cloud of debris to accrete to form the Moon.

When we look at Mars, half of its hemisphere shows a dichotomy of one side being highland and one side being lowlands from impact that almost shattered the planet. These types of impacts may explain Uranus' peculiar spin tipped 90 degrees to the rest of the Solar System as may explain Venus' slow spin.

Now, realize that this is all a hypothesis based on scientific assumptions and theories and no actual proof can determine this to be fact at this time until we observe such a planetary impact by major sizeable body.

We need to recall that **Tom Van Flandern** hypothesized that such collisions could be due to a number of reasons, all of which are viable as an explanation, such as his **Exploded Planet Hypothesis** as explained early in this book.

Our Moon is an oddity within our solar system far larger, with respect to its host planet, the Earth. Most satellites/moons within our solar system have a ratio size of **0.025%** or smaller of its parent planet, but the **Moon/Luna** is **0.25 %** the size of the Earth, that is by far, larger than any other satellite in comparison to its host planet! Astronomers think that this is due to a giant **impact collision** in Earth distant past as a **proto-planet**.

Pluto and its largest moon, **Charon** is the other exception rivalling our moon's scale with its moon about a sixth the size of its host.

Following the current line of scientific astronomical and cosmological thinking, it can be reasoned that not all impacts are bad, unless they are an **extinction event** wiping out all life on the planet. https://theconversation.com/impacts-extinctions-and-climate-in-the-search-for-life-elsewhere-34791

It is believed from the best models of planet formation that given our position relative to the Sun, the Earth should be dry and arid, yet we have water, both frozen as ice and in liquid water. Literally, water cover over 70% of our planet which is made up of ocean and scientists feel that the bulk of Earth's water was delivered from asteroids and comets.

The problem is actually exacerbated by the collision that formed the moon. That giant impact occurred after the proto-Earth had differentiated – with the heaviest elements (such as iron and nickel) settling to our planet's core. This means that the mantle and crust of the Earth, stripped off by the collision, would also have contained most of Earth's water at the time.

Since volatile impacts are a chance occurrence and some planets are poorly set up from the point of view of the delivery of volatiles to any terrestrial worlds therein.

On the other hand, studies of the formation and evolution of the "hot Jupiters" – planets like Jupiter orbiting far closer to their hosts than Mercury orbits the sun – suggest that the inward migration of such planets could drag with them vast amounts of volatiles.

In those models, so much water is delivered to the inner reaches of those systems that any Earth-like planets that form are water worlds – drenched in oceans hundreds of kilometres deep.

While such worlds might well be teeming with life, it is unlikely that it would be easy to detect. Indeed, without continents, the oceans could be almost completely lifeless, with the only source of nutrients being volcanoes on the ocean floor.

If life on such water worlds did exist, it might be so deeply buried in the ocean that any sign of it would be extremely challenging to detect, particularly from a distance measured in tens or hundreds of light-years. As such, ocean worlds would most likely be poor targets for the initial stages of the search for life elsewhere. https://theconversation.com/impacts-extinctions-and-climate-in-the-search-for-life-elsewhere-34791

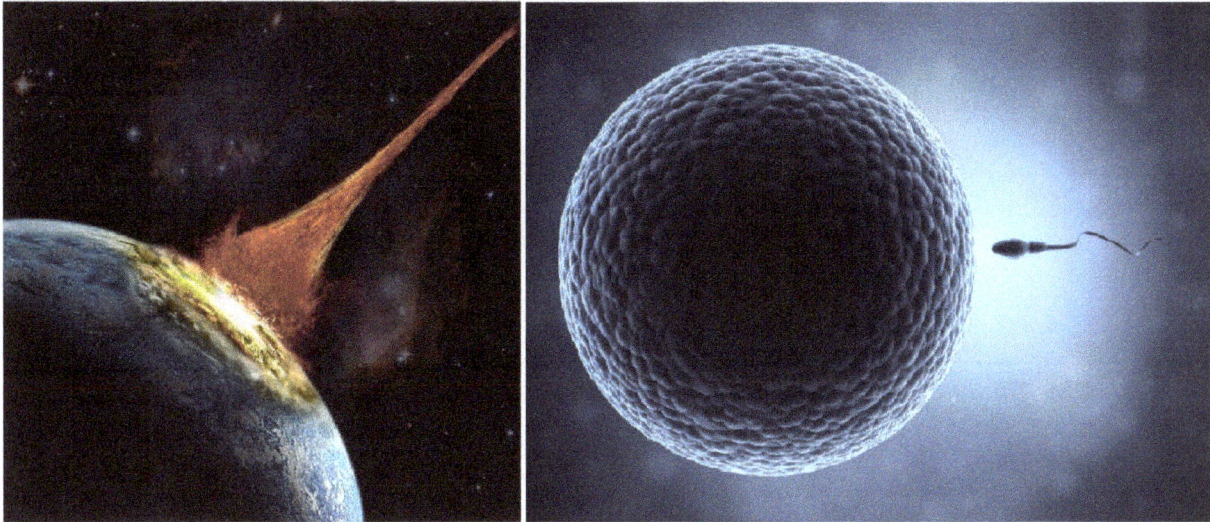

Is the creation of life on a planet delivered by a comet or asteroid the same as to the creation of life within the womb? Some scientists think so!
Google Images

Most of the objects that make up the tail of this accretion – grains of dust, lumps of ice, and pieces of rock – smash into our atmosphere and ablate harmlessly many kilometres above the ground, visible only as shooting stars.

Larger impacts do, however, continue to occur – as illustrated on February 15, 2013, in the Russian city of Chelyabinsk. On that day, with no warning, a small near-Earth asteroid detonated in the atmosphere, and outshone the noon-day sun.

Though the object itself was relatively small, around 20m in diameter, it exploded with sufficient force to shatter windows many kilometres away, damaging more than 7,000 buildings. Amazingly, nobody was killed – but the impact served as a stark reminder of the dangers posed by rocks from space. https://theconversation.com/impacts-extinctions-and-climate-in-the-search-for-life-elsewhere-34791 and
https://www.youtube.com/watch?v=dpmXvJrs7iU&t=1s

Does Jupiter size planets have an effect on habitable planets within a star system? It seems the debate goes either way from protection toward other inner smaller habitable planets or creating the potential for comets and asteroids to be hurdled toward inner planets. Admittedly, scientists say the model shows that Jupiter size planets may be more foe than friend. Greater modelling is needed by more powerful and faster computers as currently the modelling is time consuming.

Without Jupiter in our star system, the Earth probably would experience more impacts than it would were Jupiter not present. More remarkably, the situation would be far worse were Jupiter was smaller than its current mass. By dropping Jupiter's mass to one fifth its current value would lead to a dramatic increase in the rate at which both comets and asteroids plough into the Earth.

Comet Shoemaker–Levy 9 (formally designated **D/1993 F2**) was a comet that broke apart in July 1992 and collided with Jupiter in July 1994, providing the first direct observation of an extraterrestrial collision of Solar System objects. This generated a large amount of coverage in the popular media, and the comet was closely observed by astronomers worldwide. The collision provided new information about Jupiter and highlighted its possible role in reducing space debris in the inner Solar System.

The comet was discovered by astronomers **Carolyn** and **Eugene M. Shoemaker** and **David Levy** in 1993. **Shoemaker–Levy 9 (SL9)** had been captured by Jupiter and was orbiting the planet at the time. It was located on the night of March 24 in a photograph taken with the 46 cm (18 in) Schmidt telescope at the **Palomar Observatory** in California. It was the first active comet observed to be orbiting a planet and had probably been captured by Jupiter around 20–30 years earlier.

Calculations showed that its unusual fragmented form was due to a previous closer approach to Jupiter in July 1992. At that time, the orbit of Shoemaker–Levy 9 passed within **Jupiter's Roche limit**, and Jupiter's tidal forces had acted to pull apart the comet. The comet was later observed as a series of fragments ranging up to 2 km (1.2 mi) in diameter. These fragments collided with Jupiter's southern hemisphere between July 16 and 22, 1994 at a speed of approximately 60 km/s (37 mi/s) (Jupiter's escape velocity) or 216,000 km/h (134,000 mph). The prominent scars from the impacts were more easily visible than the **Great Red Spot** and persisted for many months.

It would appear that Jupiter in this particular astronomical event was a friend to the inner planets and only time will tell if it is also a foe to the Earth or other inner planets.

Shoemaker–Levy 9, disrupted comet on a collision course
(total of 21 fragments, taken on July 1994)

Brown spots mark impact sites on Jupiter's southern hemisphere
https://en.wikipedia.org/wiki/Comet_Shoemaker%E2%80%93Levy_9

From the point of view of impacts alone, we need to tread a careful middle ground. Too few impacts early on, and any Earth-like planets may be too dry for life to take hold and thrive. Too many impacts, and life will be extinguished, and never become observable.

So perhaps our ideal system would be something like our own – neither too many nor too few impacts – but a rate that is just right, delivering much needed volatiles, but allowing time for life to spread and prosper between occasional resets (such as that, 65 million years ago, which paved the way for mammalian life to take over from our former reptilian overlords).

But the other planets in the solar system don't just affect the impact rate experienced by the Earth – they also play a more subtle ongoing role, sculpting the climate of the planet on which we live.

The climate of an exo-Earth will clearly play a role in determining its suitability as a host for life. A planet cannot be too hot, or too cold, or it would automatically be excluded from the **Habitable Zone**. **https://theconversation.com/impacts-extinctions-and-climate-in-the-search-for-life-elsewhere-34791**

For the past few million years, Earth's climate has oscillated from ice age to interglacial period and back again. Our current climate sits at the cusp of glaciation, and very small variations in the amount of radiation Earth's poles receive from the sun over the course of a year can have a surprisingly large climatic effect.

As the Earth orbits around the sun, it is continually tweaked by the gravitational influence of the other planets which affects the eccentricity and inclination of our orbit to change with time. Sometimes the orbit is more circular, sometimes a bit more egg-shaped. Sometimes it is more tilted, sometimes less.

Beyond these changes to our orbit, the Earth's spin axis and orbit both precess over time – just like a wobbling spinning top.

Taken together, these variations are known as the **Milankovitch cycles**. In combination, they result in small changes in the total amount of energy received from the sun at the Earth's poles and are thought to drive the current glacial-interglacial cycles.

For other planets, orbiting other stars, the Milankovitch cycles would be different. The variations experienced on Earth are the result of the influence of all the other objects in our solar system – all pulling and nudging our planet. If the architecture of the solar system was different, so too would be these oscillations.

Depending on the precise architecture of a planetary system, the oscillations that drive periodic climate change would change. They might have a greater or lesser amplitude and may happen faster or more slowly. The climate of the planet in question would respond in kind. https://theconversation.com/impacts-extinctions-and-climate-in-the-search-for-life-elsewhere-34791 and https://www.youtube.com/watch?v=zcs3THdFPpo

In the search for life in the universe, a planet to develop life on it must be of the right size for in a solar system of planets, size matters, at least in this current age of astronomy and space exploration.

In a proto solar system, planets cannibalize disk material around a nascent star. Small pieces of dust collide and grow, devouring their neighbours and as this continues consuming all things around them they grow ever larger and accelerates.

If a planet swells to more than around ten times the mass of the Earth before the nuclear furnace within its host ignites, it will be able to accrete vast amounts of gas. It will grow to become a gas giant, like Jupiter and Saturn in our own solar system. Such planets would definitely not be good places to look for life.

At the other end of the spectrum, an object that is too small and of too low a mass (such as Mercury, or Earth's moon) won't have enough gravity to hold an atmosphere. https://theconversation.com/what-makes-one-earth-like-planet-more-habitable-than-another-33479

The more massive a planet, the more massive its atmosphere it can acquire and maintain. This is important because the mass of a planet's atmosphere will directly influence its climate. The

442

location of the "habitable zone" around a star will therefore be a function of the mass of the planet in question.

A more massive planet, with a more massive atmosphere, will likely have a stronger greenhouse effect. Such a planet would most likely be habitable at distances that would result in smaller planets icing over.

Similarly, a smaller planet, with a thinner atmosphere, will likely be habitable at distances at which the oceans of a more massive world would boil.

Young planets are continually bombarded by material and radiation that is flung out by the host star which to some degree is vital for life. Without this energy that the Earth receives from the Sun, there would be no light and our planet would be a frozen ice-ball. With light however, is the curse comes solar winds blowing from the star which can tear away a planet's atmosphere, stripping it to nothingness!

Fortunately, the Earth is protected from such solar wind hostilities by a magnetic field. As the solar wind strikes Earth's magnetic field, it is deflected around, shielding the planet within. Only the strongest solar storms, and the most energetic particles, can penetrate that shielding.

Cascades of charged particles that penetrate the Earth's magnetic shield pour down to the poles, causing the magnificent **Aurora Boraelis** and **Australis.** https://theconversation.com/what-makes-one-earth-like-planet-more-habitable-than-another-33479

The Aurora Borealis, imaged here from northern Canada, is the result of charged particles flung outward by the Sun smashing into Earth's atmosphere.
Wikimedia/Kshitijr96, CCBY

If a planet had no **magnetic field**, the situation would be grave. Our nearest planetary neighbour, Mars, lacks a strong magnetic field and most likely has been unshielded since the early youth of the solar system. That lack of shielding has caused Mars's once thick atmosphere to have been whittled away to almost nothing.

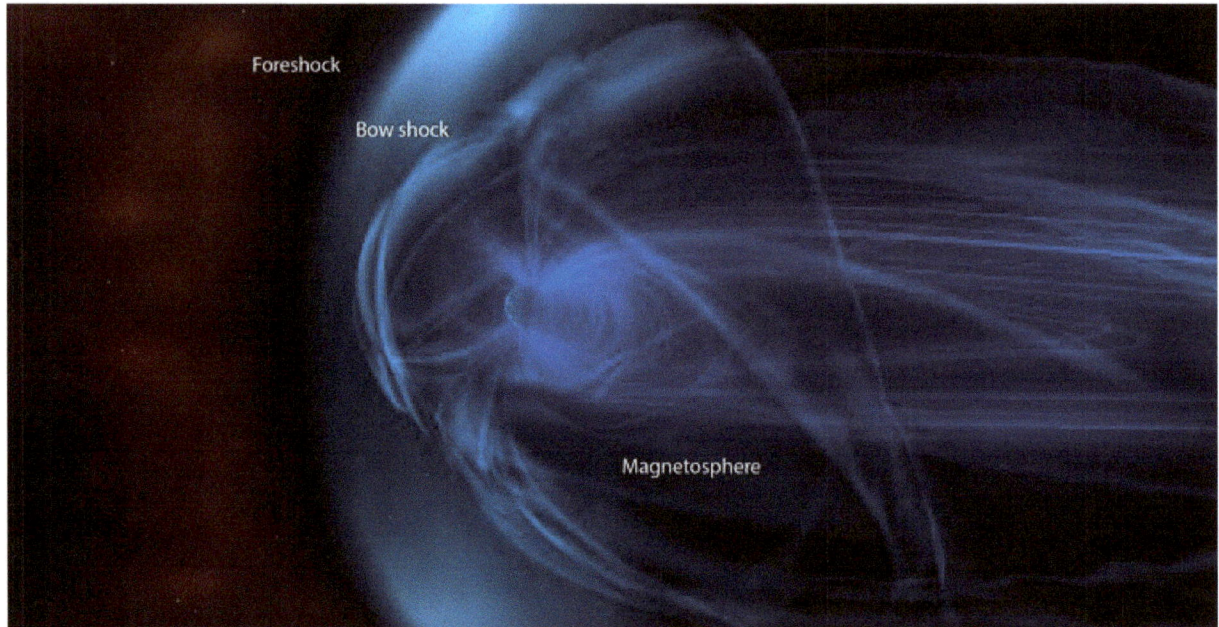

The Earth's magnetosphere protects our atmosphere from the erosive power of the Solar wind. But without convection in Earth's core, that field would not exist. NASA/GSFC
https://theconversation.com/what-makes-one-earth-like-planet-more-habitable-than-another-33479

The interior of the Earth is very hot, primarily as a result of the decay of radioactive elements trapped in the interior. Our planet slowly sheds that heat to space through its crust, with the result that the surface layers of our planet's interior are cooler than those near the

The presence of a magnetic field is particularly important during a planet's youth – when its host star is still young and energetically shedding material into space. As stars age they remain capable of slowly stripping planets of their gaseous shrouds. It is clear, therefore, that for a planet to be considered a promising target in the search for life, it must possess a strong magnetic shield to protect its atmosphere.

The Earth's magnetic field is driven by a **colossal dynamo**, deep in the Earth's outer core. It is thought that this dynamo is driven by **convection currents** in that region of our planet's interior. But therein lies a problem – how can you maintain convection inside a planet over periods of billions of years?

The interior of the Earth is very hot, primarily as a result of the decay of radioactive elements trapped in the interior. Our planet slowly sheds that heat to space through its crust, with the result that the surface layers of our planet's interior are cooler than those near the core.

In order to setup and maintain convection, you need a large temperature difference between two locations. So the convection currents in the Earth's mantle are the result of the fact that the part of the mantle just beneath the crust is far cooler than that next to the core.

But here's the strange thing. If the Earth didn't have plate tectonics, the upper layers of the mantle would not be able to cool anywhere near as efficiently. With a warmer upper mantle, the temperature difference between that region and the Earth's deep interior would be smaller, and convection would eventually cease.

And without that convection, the Earth's dynamo would die, and its magnetic field would weaken and disappear.

It is thought that this is precisely what happened to Mars. At first, the red planet may well have been sufficiently hot inside for a certain degree of **tectonic activity** to occur. Certainly, the crust of Mars bears evidence of an ancient **magnetic field**, frozen into the rocks.

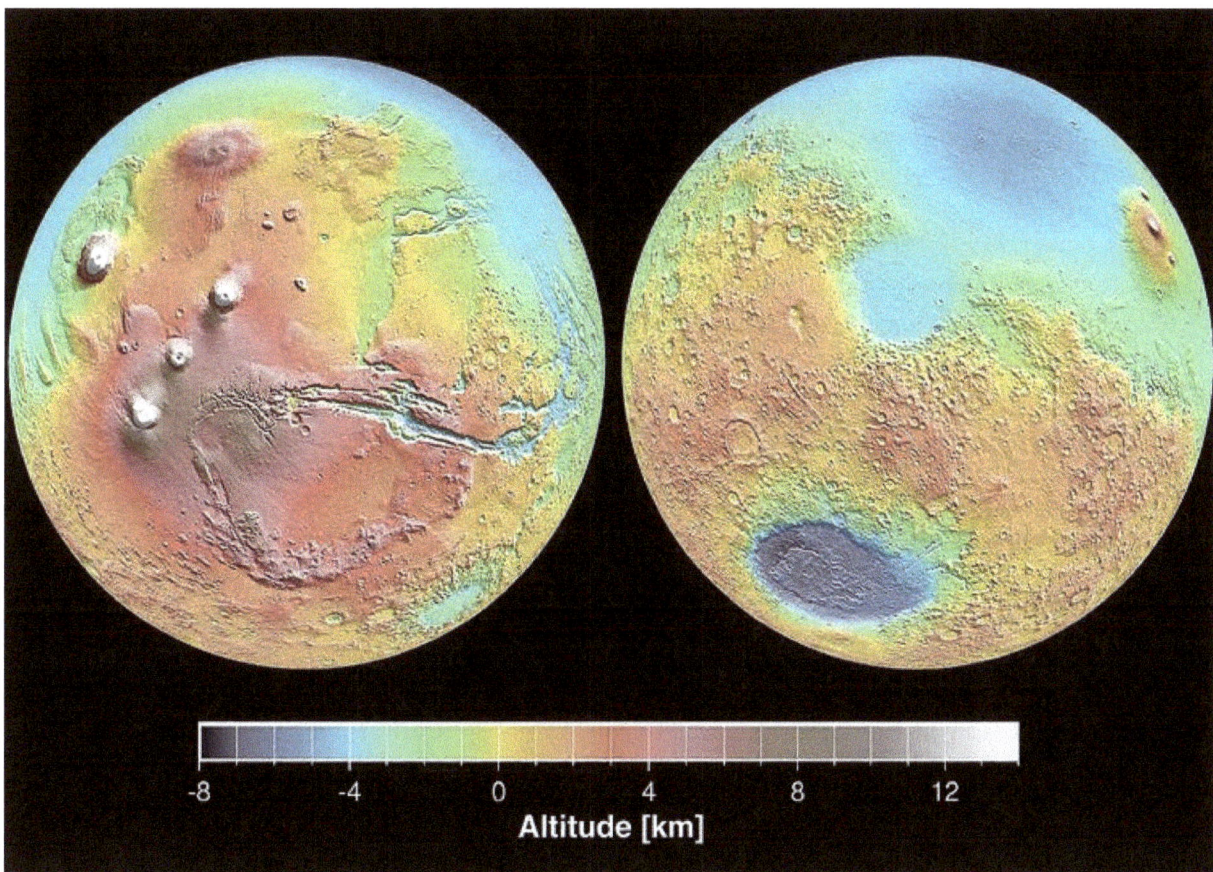

The topography of Mars, based on observations taken by the Mars Orbiter Laser Altimeter. Despite its famous volcanoes, Mars does not exhibit plate tectonics, which may be tied to the loss of its atmosphere over the past four billion years. NASA/JPL
https://theconversation.com/what-makes-one-earth-like-planet-more-habitable-than-another-33479

But Mars is smaller than Earth and so lost its interior heat to space much more quickly. Eventually, it cooled sufficiently that it could not maintain convection, and its magnetic field died down.

At that point, the solar wind began to strip the atmosphere away, while at the same time, chemical processes on the surface were pulling the atmosphere down into the rocks. Without tectonics to recycle the rocks, and to drive the magnetic field, Mars slowly became the world we see today.

In addition to maintaining our planet's magnetic field, plate tectonics are important for many other reasons, particularly when it comes to the creation of the atmosphere we breath today.

Tectonic activity is clearly important for life on Earth and so would certainly be a key criterion in the search for life elsewhere. But some researchers suggest that the story may be even more complicated than we once thought.

Over the years, a number of studies have suggested that without water, the Earth wouldn't have **plate tectonics** – a theory that remains heavily debated.

The general idea is that water acts as a lubricant – either between colliding plates (helping them overcome the friction between them), or within Earth's mantle (increasing its fluidity, helping it to move, and easing the convection currents therein).

Without water, a planet might have to be somewhat more massive than the Earth to support plate tectonics. Had Earth formed much drier, the motion of its plates might have seized long ago, causing convection in the mantle to die, and our atmosphere to be been purged, just like that of Mars.

So, with this theory, we've come full circle. According to the latest understanding in astronomy,

Life itself needs water, but for a planet to keep its atmosphere, it needs a magnetic field. To maintain a magnetic field, a planet needs a dynamo. To have a dynamo, it needs plate tectonics and to keep plate tectonics, it needs water. https://theconversation.com/what-makes-one-earth-like-planet-more-habitable-than-another-33479

Did Astronomers Find An Alien Megastructure Around Another Star?

In 2015, scientists noticed unusual fluctuations in the light from a star named **KIC 8462852**. This otherwise-normal F-type star, which is slightly larger and hotter than Earth's sun, sits about 1,480 light-years from Earth, in the constellation **Cygnus**.

A mysterious star whose repeated bouts of darkening might be due to **"alien megastructures,"** according to some researchers' conjectures, may now have more than a dozen counterparts that display similarly mystifying behavior, a new study finds.

Further research into all of these stars might help solve the puzzle of their bewildering flickering, the study's author said.

When the researchers analyzed data from NASA's **Kepler space telescope**, astronomer **Tabetha "Tabby" Boyajian,** then at Yale University, and her colleagues found dozens of odd instances of KIC 8462852 dimming by up to 22%, with such dips lasting anywhere from a few days to a week. These events did not appear to follow any pattern and seemed far too substantial to be caused by planets or dust crossing the star's face.

These analyses of KIC 8462852 — now nicknamed **"Boyajian's star"** (formerly **Tabby's star**) after its discoverer — raised the possibility that astronomers had detected signs of intelligent alien life. Specifically, researchers have suggested that the star is surrounded by a **Dyson sphere**, a hypothetical megastructure that is built around a star to capture as much of its light as possible. Mathematician and physicist **Freeman Dyson** suggested that such megastructures could help power an advanced civilization. (Science fiction often depicts Dyson spheres as solid shells around stars, but the megastructures also could be globular swarms of giant solar panels.)

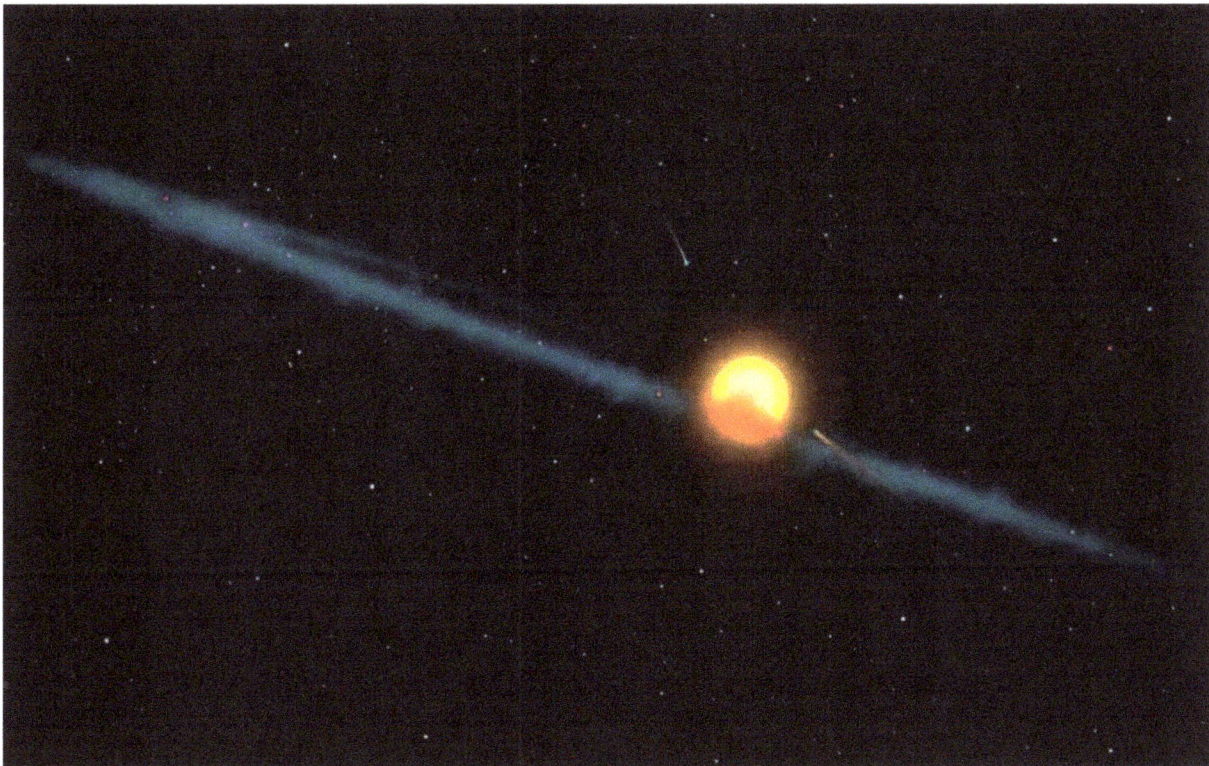

This artist's illustration depicts a hypothetical dust ring orbiting KIC 8462852, also known as Boyajian's Star or Tabby's Star. (Image credit: NASA/JPL-Caltech)
https://www.space.com/alien-megastructure-mysteriously-dimming-stars.html

The megastructure hypothesis is near the bottom of most astronomers' lists these days when it comes to Boyajian's star, however; further analyses have pointed to more prosaic explanations, such as clouds of dust or comet fragments. Still, scientists have not yet nailed down the precise cause of the odd dimming. The answer remains elusive in part because Boyajian's star seemed unique; there were no known counterparts to provide additional clues that might help researchers solve this cosmic mystery.

A star is surrounded by a Dyson sphere, a hypothetical megastructure that is built around a star to capture as much of its light as possible.
https://earthsky.org/space/tabbys-star-more-weirdness

Now, study author **Edward Schmidt**, an astrophysicist at the University of Nebraska-Lincoln, suggests that he may have discovered more than a dozen stars like Boyajian's star.

Schmidt looked for counterparts of Boyajian's star using software that searched for analogous dimming events from about 14 million objects with varying brightness monitored in the **Northern Sky Variable Survey** from April 1999 to March 2000. He then followed up on promising candidates by examining their long-term behavior, using data from the **All-Sky Automated Survey for Supernovae**, ruling out sources whose dimming could be caused by conventional explanations such as an eclipsing companion star or some intrinsic variability in brightness.

Schmidt identified 21 stars that showed possibly unusual dimming. These fell into two distinct categories: 15 were "slow dippers" that dimmed at rates similar to **Boyajian's star**, and six were "rapid dippers" that showed even more extreme variability in their dimming rates.

"The thing that surprised me the most were these stars that had so many dips, the ones I called 'rapid dippers,'" Schmidt told Space.com. "I expected more occasional dips like Boyajian's star."

Further analysis using data from the **European Space Agency's Gaia space observatory** found that these potential dippers tended to be either conventional **"main-sequence" stars** with about the same mass as the sun or **red giant stars** with about twice the sun's mass. The slow and rapid

dippers are seen in both groups, which may suggest that they represent varying degrees of the same mechanism, Schmidt said.

It now appears that the universe is beginning to reveal its secrets as we explore planets in our solar system and into other interstellar systems using radio telescopes, orbiting space telescopes, voyaging space probes, and interplanetary rovers.

We are discovering that not only are there the important elemental ingredients to provide and sustain life on most of the planets, but that there is life on or within stars themselves. In fact, the whole universe appears to be alive, conscious and aware with clusters of planets, stars, and galaxies functioning like a gargantuan mega-brain. Within this universe that is gigaparsecs or perhaps, tetra-parsecs in size are galactic clouds and strands connecting to each other in a simulacrum of galactic neurons and synapses. The universe is infinite in size and beyond our ability to explore it, no matter how fast a spacecraft is able to speed through the universe.

In the vastness of an infinite universe with all its mysteries, its gigantic proportions and its extremely fine subtleties from the **microverse** of the atomic system of protons and its electron to even smaller constituents of quarks and charms, etc. to the **macroverse** of the universe in which we have virtually explored, life is prevalent at all levels because the universe is alive, it is a sentient being and we humans and all life within it are an integral part of it!

As overwhelming as this concept may be to us and how insignificant we may feel by comparison, **Baha'u'llah**, the Prophet and Founder of the Baha'i Faith and the most recent Manifestation of God for this day and age has stated *"Dost thou reckon thyself only a puny form When within thee the universe is folded?"* Baha'u'llah; Seven Valleys and Four Valleys; 1942 and 1952; Baha'I Publishing Trust; Wilmette, Illinois, USA Library of Congress C.C. 53-12275

For men of science such as astronomers, who expand the frontiers of knowledge for the betterment and progress of all mankind, **Baha'u'llah** states:

"O people of God! Righteous men of learning who dedicate themselves to the guidance of others and are freed and well guarded from the promptings of a base and covetous nature are, in the sight of Him Who is the Desire of the world, stars of the heaven of true knowledge…They are indeed fountains of soft-flowing water, stars that shine resplendent, fruits of the blessed Tree, exponents of celestial power, and oceans of heavenly wisdom."

This is the current scientific thinking for an exoplanet to be a viable world which can develop and evolve. The problem with this type of thinking is that there are missing factors in the requirements for life to appear on planet and a presumption that life must conform to basic anthropocentric thinking, especially, how life occurred and evolved on Earth.

No doubt, the science of astronomy will progress and evolve beyond its current understanding. Perhaps, a few pioneering astronomers and scientists will need to step outside the traditional paradigm of thinking to look for new resources of knowledge that few have considered for the answers they seek.

One physicist, **Vahid Houston Ranjbar** asserts that there are some Baha'i writings which may have relevant insights into the question of life in the universe.

The *"Tablet of the Universe"* written by **Abdu'l-Bahá,** the son of **Baha'u'llah,** the Prophet and founder of the **Baha'i Faith** has been translated into English in a provisional form. It was originally written as the *"Lawh-i-Aflákiyyih"* and is primarily a piece on spirituality and therefore, not an exposition of physics.

It is too longer to include here in full, but the reader may see full English translation of this tablet at Baha'i Library Online at **https://bahai-library.com/abdulbaha_lawh_aflakiyyih**

Abdu'l-Baha states in the Tablet :

"These are spiritual truths relating to the spiritual world. In like manner, from these spiritual realities infer truths about the material world. For physical things are signs and imprints of spiritual things;".

As Vahid Ranjbar points out that, "perhaps something can yet still be deduced about the physical universe from this tablet. It is my opinion that the words of Baha'u'llah and Abdu'l-Baha's exposition represent a profound source of knowledge. Humanity has only barely touched the surface of this great ocean. This article represents my flawed initial attempt to try and correlate what relates to my training as a physicist with my understanding of this Tablet. In fact, we should from the outset understand that the meanings hid in this Tablet will only reveal themselves fully through passage of time".

What follows are *selected extracts* from the Tablet which struck me and my attempt to relate them to current understandings of physics.

Tablet of the Universe Comments by Vahid Houston Ranjbar

3rd paragraph

"Divine and all-encompassing Wisdom hath ordained that motion be an inseparable concomitant of existence, whether inherently or accidentally, spiritually or materially. This movement must be governed by some check or rein, some regulator or director, otherwise order will be disrupted and the spheres and bodies will fall from the heavens. For this reason God brought into being a universal attractive force between these bodies to hold sway over them and govern them, a force deriving from the firm ties, the mighty correspondence and affinity that exist between the realities of these limitless worlds. By the operation of this attractive force those holy and resplendent suns, with their luminous worlds, satellites and planets, circling and orbiting in their heavens, at once exerted attraction and were subject to

it, induced motion and were themselves moved, began orbiting and set into orbit other bodies, shone forth and caused others to shine."

Non-existence of Absolute Rest

The first sentence is an idea has reiterated in many other tablets, that absolute rest is impossible. This is a well understood consequence of Heisenberg's uncertainty principle: the fact that absolute zero Kelvin is impossible to reach.

Gravitational Attraction as Entanglement?

"The third sentence, although seemingly trivial, when carefully parsed actually might contain some very profound insights. He relates that the universal attractive force is derived "*from the firm ties, the mighty correspondence and affinity that exist between the realities.*" One might understand firm ties, correspondence and affinity between realities to represent the nature of quantum entanglement. At its heart quantum entanglement ensures a correlation or correspondence between entangled particles over arbitrary distances. On this point I would be very curious to have experts in the original language and culture weigh in. The question is, how would an individual from Persia in the 19th and early 20th century describe entanglement with the language at his disposal? Furthermore, how would he go about explaining this to an audience of individuals scientifically illiterate by the standards of today?" https://bahai-library.com/abdulbaha_lawh_aflakiyyih

Studies of Hawking radiation from Black holes and information preservation has lent support to the idea of space-time being product of quantum entanglement
https://bahai-library.com/abdulbaha_lawh_aflakiyyih

$$S = \frac{c^3 k A}{4 \hbar G}$$

Entropy Equation for a Black Hole

"Most are probably aware from general relativity, that gravity is currently understood as being a result of space-time distortions created by mass-energy. One of the most exciting developments of recent theoretical physics has been the conjecture that space-time is actually a product of quantum entanglement. This idea is supported by the consideration of information preservation in black holes as well as work done in condensed-matter physics and quantum information theory. This has now blown-up into an area of study which has attracted the brightest minds of our day." An excellent introduction to this new idea can be found at
https://www.nature.com/news/the-quantum-source-of-space-time-1.18797

Baby stars are forming near the eastern rim of the cosmic cloud Perseus, in this infrared image from NASA's Spitzer Space Telescope.

Stellar Formation

The fourth sentence is interesting here due to the fact that he links the universal attractive force to the process of sun's shining forth. From a physics point of view gravitation of course is what drives stellar formation and causes suns to shine, a fact which was not known until the process of nuclear fusion was well understood."

Paragraphs 5 and 6

"Know thou that the expressions of the creative hand of God throughout His limitless worlds are themselves limitless. Limitations are a characteristic of the finite, and restriction is a quality of existent things, not of the reality of existence.

This being the case, how can one, without proof or testimony, conceive of creation being bound by limits? Gaze with penetrating vision into this new cycle. Hast thou seen any matter in which God is bounded by limits which He cannot overstep? Nay, by the excellence of His glory! On the contrary, His tokens have encompassed all things and are sanctified and exalted beyond computation in the world of creation."

Cosmos or Universe is infinite

"This theme of limitlessness is repeated several times in the Tablet as well as in other writings of both Abdu'l-Baha and Baha'u'llah. It would seem that even if we discover a bound to our current universe the existence of multi-verses would needs exist. Further he says a very interesting thing; *"restriction is a quality of existent things, not the reality of existence"*. This makes me think of something like the relationship between the quantum wave equation and a particle. The quantum wave in theory permeates all space and thus can be thought of as unlimited while its expression as a particle is limited in space. I have a sense that there is much more to this idea though that relates back to **Plato's** ideal forms, which underlie the true nature of existence."

Paragraph 7 and 8

"These are spiritual truths relating to the spiritual world. In like manner, from these spiritual realities infer truths about the material world. For physical things are signs and imprints of spiritual things; every lower thing is an image and counterpart of a higher thing. Nay, earthly and heavenly, material and spiritual, accidental and essential, particular and universal, structure and foundation, appearance and reality and the essence of all things, both inward and outward — all of these are connected one with another and are interrelated in such a manner that you will find that drops are patterned after seas, and that atoms are structured after suns in proportion to their capacities and potentialities. For particulars in relation to what is below them are universals, and what are great universals in the sight of those whose eyes are veiled are in fact particulars in relation to the realities and beings which are superior to them. Universal and particular are in reality incidental and relative considerations. The mercy of thy Lord, verily, encompasseth all things!

Know then that the all-embracing framework that governs existence includes within its compass every existent being — particular or universal — whether outwardly or inwardly, secretly or openly. Just as particulars are infinite in number, so also universals, on the material plane, and the great realities of the universe are without number and beyond computation."

Scale Invariance or Self Similarity

What **Abdu'l-Baha** is describing I believe is the principle of **scale invariance** and the related idea of **self similarity** which underly the notion of oceans being like drops. In physics this is well recognized and actively studied. In fact, **Fractal mathematics** has been applied as (a) powerful tool to model and (to) study this phenomena. For more in-depth discussion of this phenomena see self similarity at https://en.wikipedia.org/wiki/Self-similarity or scale invariance at https://en.wikipedia.org/wiki/Scale_invariance.

However, what is perhaps beginning to be understood is how it might apply to the hierarchy of physical laws, moving from particular laws to increasingly universal laws. An example of this would be the discovery of the particular physical laws such as **Ampere's law** for moving charges and **Faraday's Law** for **magnetic flux**. Then several years later **Maxwell** and **Einstein** with relativity, unified them in such a way that they could be expressed as a single field represented by a single **four-dimensional tensor**. Then later **quantum field theory** demonstrated that this field arose from the application of a **rotational phase symmetry** of the **quantum field**.

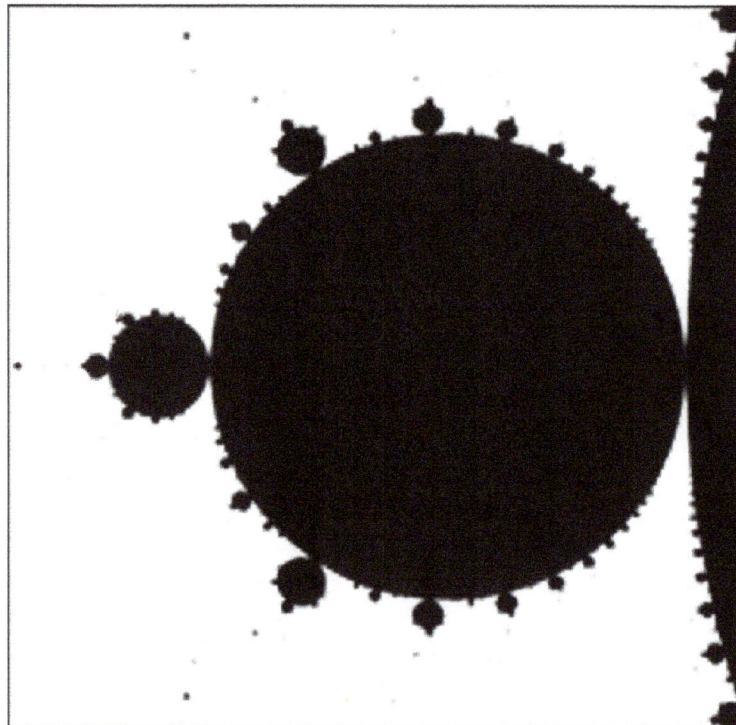

Zooming in on the Mandelbrot Set
https://bahai-library.com/abdulbaha_lawh_aflakiyyih

"There is also an old idea, that our universal constants like the speed of light or **Planck's Constant** actually might have changed over the deep time of the universe and are properties given by some as yet not understood underlying geometry of space-time (e.g. see "New varying speed of light theories" at http://cds.cern.ch/record/618057/files/0305457.pdf). **Paul Dirac**, one of the developers of **Quantum Field Theory**, was a proponent of such ideas."
https://en.wikipedia.org/wiki/Self-similarity
https://en.wikipedia.org/wiki/Scale_invariance

"Whether or not the physical constants have varied over the life of our particular universe is one question which may or may not be true. However personally I am more inclined towards the idea that the value of our constants are probably related to each other and a product of some sort of geometry or an overarching structure inherent in reality. This seems an antidote to arguments about the fine tuning of the **universe's constants**. Some may promote fine tuning as evidence of divine design however, I have an instinctual distaste for this viewpoint since I think the divine hand is much more subtle and works through the universal laws of nature and not like a watch maker. As **Baha'u'llah** says in the **Tablet of Wisdom**, *"Nature in its essence is the embodiment of My Name, the Maker, the Creator"*.

P.A.M. Dirac at the blackboard
https://bahai-library.com/abdulbaha_lawh_aflakiyyih

What is not yet appreciated and seems new is the idea that this progression from particular laws to more general laws might be an infinite process and could never terminate into some final theory of everything. For example, I could imagine the day may come where we understand all the major forces as manifestations of single field in a so-called **unified field theory**, however, there still might be principles which govern how the underlying field and its constants are created and sustained which will remain unanswered. Thus, one could imagine a process of infinite progression towards greater and greater universal laws which **Abdu'l-Baha** seems to be referring to.

Paragraph 20

"Know then that, as hath been clearly handed down in the accounts of old, these great orbits and circuits fall within subtle, fluid, clear, liquid, undulating and vibrating bodies, and that the heavens are a restrained wave because a void is impossible and inconceivable."

Electron Orbitals

This sentence describes the great orbits in a manner which reminds me of how the **electron orbitals** are understood. Electron orbitals are described mathematically as **spherical harmonics** or a kind of standing wave bound by the attraction of the force generated by the electric field of the atom. Now, **Abdu'l-Baha** appears to be describing the physical heavens and not the **atomic orbitals** or the spiritual realm, though given the nature and process of the Tablet, he could be also, be speaking of any number of **grades of existence**.

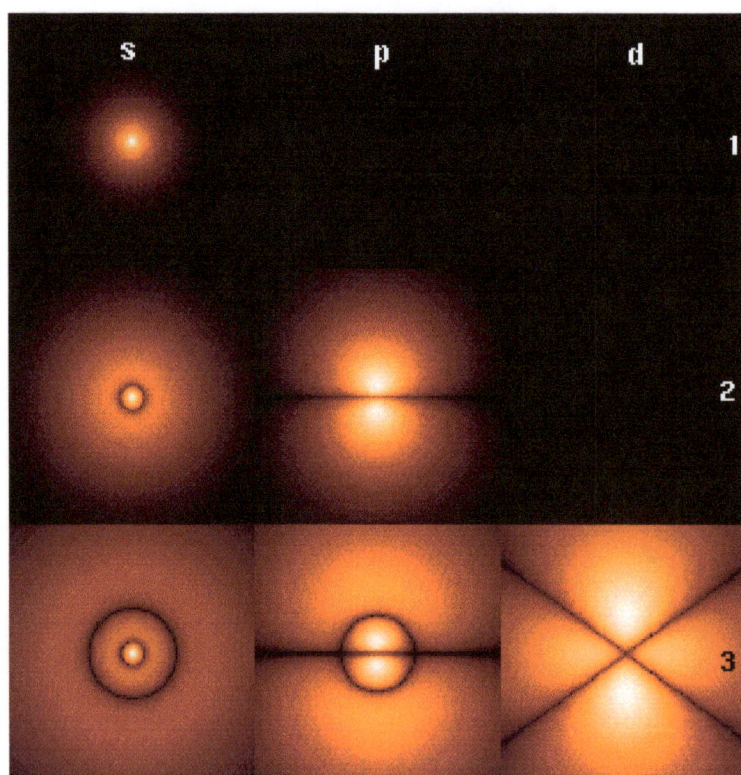

False-colour density images of some hydrogen-like atomic orbitals
(f orbitals and higher are not shown)

However, if we proceed taking it at face value, both **quantum field theory** and **general relativity** are **field theories**, in that the forces arise from the effects of the **continuum**. As a result, both sustain the propagation of waves. More recently the prediction of **gravitational waves** given by **general relativity** has been demonstrated in the ground-breaking **LIGO experiment**. Thus, strictly speaking the heavens or space-time could be understood as a 'restrained wave' held by the **mass-energy** of the universe and acted on by **dark energy**.

Non-existence of a void

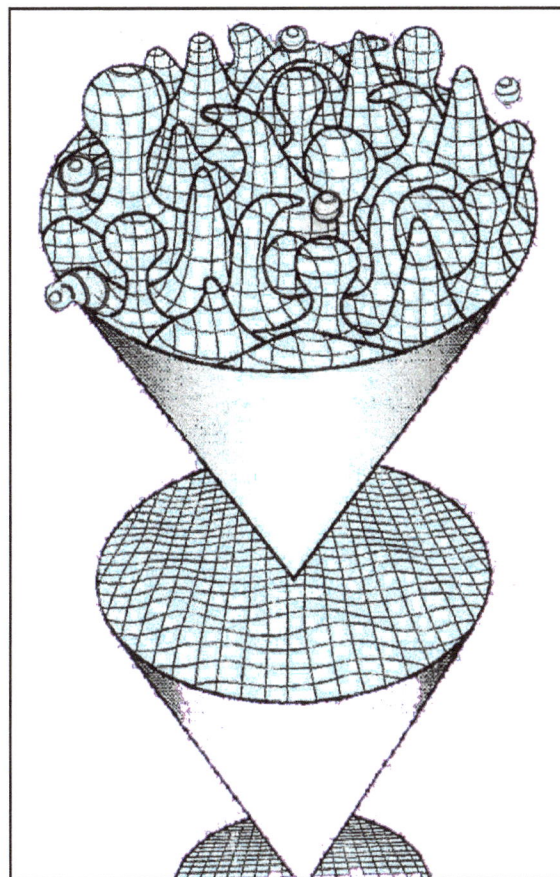

A graphic representation of Wheeler's calculations of what quantum reality may look like at the Planck length. By Jarrokam (Own work)
https://bahai-library.com/abdulbaha_lawh_aflakiyyih

This idea is repeated several places in the tablet and I would claim it is supported by current **quantum field theory** for which empty space is not really empty, underlying it is a field from which particles pop in and out of existence. It is bubbly or foamy thing. Actually, the vacuum can never reach absolute zero energy. In all space there exists what is called 'zero-point' energy (ZPE). This again is a direct consequence of **quantum mechanics**. Incidentally it has also been speculated that the zero-point energy is somehow related to dark energy which drives the accelerating expansion of the universe. However, **zero-point energy** is orders of magnitude too large to account for the observed **cosmological constant**. This actually represents one of the big unanswered questions of physics. https://bahai-library.com/abdulbaha_lawh_aflakiyyi

Some Baha'i Quotes About UFOs, Abductions, and Life on Other Planets

Thou hast, moreover, asked Me concerning the nature of the celestial spheres. To comprehend their nature, it would be necessary to inquire into the meaning of the allusions that have been made in the Books of old to the celestial spheres and the heavens, and to discover the character of their relationship to this physical world, and the influence which they exert upon it. Every heart is filled with wonder at so bewildering a theme, and every mind is perplexed by its mystery. God, alone, can fathom its import. The learned men, that have fixed at several thousand years the life of this earth, have failed, throughout the long period of their observation, to consider either the number or the age of the other planets. Consider, moreover, the manifold divergencies that have resulted from the theories propounded by these men. Know thou that every fixed star hath its own planets, and every planet its own creatures, whose number no man can compute. Bahá'u'lláh, Gleanings from the Writings of Bahá'u'lláh, p. 162-163

As to your question whether the power of **Bahá'u'lláh** extends over our solar system and to higher worlds: while the Revelation of Bahá'u'lláh, it should be noted, is primarily for this planet, yet the spirit animating it is all-embracing, and the scope therefore cannot be restricted or defined. Shoghi Effendi, Lights of Guidance, p. 481

There is no record in history, or in the teachings, of a Prophet similar in Station to Bahá'u'lláh, having lived 500,000 years ago. There will, however, be one similar to Him in greatness after the lapse of 500,000 years, but we cannot say definitely that His Revelation will be inter-planetary in scope. We can only say that such a thing may be possible. What Bahá'u'lláh means by His appearance in 'other worlds' He has not defined, as we could not visualize them in our present state, hence He was indefinite, and we cannot say whether He meant other planets or not. Shoghi Effendi, Lights of Guidance, p. 472

Mr. Garcia refers to a book entitled Abduction, written by a Harvard psychiatrist, Dr. John Mack, which posits that ... alien beings of a vastly superior intelligence, who are possibly from other planets or dimensions, or from our own distant future, have been conducting genetic experiments on unwilling humans to produce a crossbreed between humans and aliens (reportedly to repopulate our planet with a more peaceful species after we destroy ourselves).

Mr. Garcia comments on the popularity of this topic and mentions that a number of Bahá'ís claim to have had experiences similar to those associated with abduction by aliens. He requests guidance concerning how to respond to such matters.

The Bahá'í Teachings do not deal specifically with the subjects of alien abduction and genetic engineering. The following extracts concerning unidentified flying objects might be of assistance to Mr. Garcia:

There is nothing in the Teachings about spaceships; and the Guardian does not feel this is a subject on which he can offer the friends any advice whatsoever. Indeed, to be frank, he is so busy with the work of the Cause that he seldom has time to devote much thought to speculation of this nature, however fascinating it may be. **From a letter dated 15 February 1957 written on behalf of Shoghi Effendi to several believers https://bahai-library.com/abdulbaha_lawh_aflakiyyih** and **https://bahaiquotes.com/subject/ufos**

As you rightly state, Bahá'u'lláh affirms that every fixed star has its planets, and every planet its own creatures. The House of Justice states however, that it has not discovered anything in the Bahá'í Writings which would indicate the degree of progress such creatures may have attained. Obviously, as creatures of earth have managed to construct space probes and send them into outer space, it can be believed that creatures on other planets may have succeeded in doing likewise.

Regarding the attitude Bahá'ís should take toward unidentified flying objects, the House of Justice points out that they fall in the category of subjects open to scientific investigation, and as such, may be of interest to some, but not necessarily to everyone. In any case, Bahá'ís have a fundamental obligation at this stage of the development of the earth's people, that is, the responsibility of spreading the unifying Message of Bahá'u'lláh. **(From a letter dated 11 January 1982 written on behalf of the Universal House of Justice to an individual believer). See The World Order of Bahá'u'lláh: Selected Letters (Wilmette: Bahá'í Publishing Trust, 1991), pp. 170-186** and **The Universal House of Justice, 1996 Aug 06, Sabeans, UFOs, Alien Abduction and Genetic Engineering**

The earth has its inhabitants, the water and the air contain many living beings and all the elements have their nature spirits, then how is it possible to conceive that these stupendous stellar bodies are not inhabited? Verily, they are peopled, but let it be known that the dwellers accord with the elements of their respective spheres. These living beings do not have states of consciousness like unto those who live on the surface of this globe: the power of adaptation and environment moulds their bodies and states of consciousness, just as our bodies and minds are suited to our planet.

For example, we have birds that live in the air, those that live on the earth and those that live in the sea. The sea birds are adapted to their elements, likewise the birds which soar in the air, and those which hover about the earth's surface. Many animals living on the land have their counterparts in the sea. The domestic horse has his counterpart in the seahorse which is half horse and half fish.

The components of the sun differ from those of this earth, for there are certain light and life-giving elements radiating from the sun. Exactly the same elements may exist in two bodies, but in varying quantities. For instance, there is fire and air in water, but the allotted measure is small in proportion.

They have discovered that there is a great quantity of radium in the sun; the same element is found on the earth, but in a much smaller degree. Beings who inhabit those distant luminous bodies are attuned to the elements that have gone into the composition of their respective spheres. **The Universal House of Justice, 1992 June 08, Gaia Concept, Nature, p. 11 https://bahaiquotes.com/subject/ufos**

As humanity develops and expands its space exploration in its search for other life within our Solar System and out toward the exoplanets in other star systems, it is hoped that our spiritual maturity will evolve effectively to match our technical ability to reach the stars.

After all, science and religion are really the same fruit on the tree of knowledge, they are the refreshing waters of knowledge that flows from two founts, but whose source is from the ocean of all knowledge, the divine Creator of all life within the universe. It would, therefore, be only right that we partake from this divine source of knowledge in our search for extraterrestrial life.

As we move out into the vastness of space among the stars, we will find life everywhere, as life is a natural and a universal common theme, not relegated to one small corner of the galaxy in the infinite universe.

We will find that the markers for life in all its diversity will be found by the very stars in which its planets orbit, that being a stable star, a non-prototype star or a dying star nearing a state of becoming a **supernova**, but will be found within a healthy, vibrant, life -giving star system!

"Know thou that every fix (stable) star has its own planets and every planet its own creatures whose number no man can compute"! - Baha'u'llah

BIBLIOGRAPHY, WEBLIOGRAPHY AND VIDEOGRAPHY

The following list includes all books and major journals, newspapers and web based material, including other reference ebooks and materials such as web links and video links found on the internet in researching this book. Not all chapters use reference material and therefore, these chapter are not listed.

Bibliography listed refers to books marked in RED,
Webliography refers to websites marked in BLUE,
Videography refer to video websites marked in GREEN,
News Service websites are marked in LIGHT BLUE, and
Newspaper websites, magazines and professional papers are marked in PURPLE.

It does not include specific government documents, archival repositories, or various journals or other web based material.

CHAPTER 81
THE MONUMENTS OF MARS - A TYPE TWO CIVILIZATION IN RUINS

http://en.wikipedia.org/wiki/Cydonia_%28region_of_Mars%29

http://www.bibliotecapleyades.net/marte/esp_marte_41.htm

https://www.youtube.com/watch?v=JPJRpBWuGbY

https://www.youtube.com/watch?v=0DTfPW9hHb8

http://en.wikipedia.org/wiki/Tom_Van_Flandern

http://www.bibliotecapleyades.net/marte/esp_marte_41.htm

http://www.tmgnow.com/repository/mars/mars_jordan.html

The Martian Enigmas, A Closer Look by Mark J. Carlotto; 1991; published by North Atlantic Books; ISBN 1-55643-092-2

http://www.enterprisemission.com/catbox.htm

http://en.wikipedia.org/wiki/Arcology

http://www.greatdreams.com/geology.htm

http://www.aulis.com/mars.htm

http://en.wikipedia.org/wiki/E_%28mathematical_constant%29

http://www.enterprisemission.com/ken2b.html

http://www.enterprisemission.com/message.htm

http://www.enterprisemission.com/Path-sphinx.html

http://science1.nasa.gov/science-news/science-at-nasa/2001/ast24may_1/

http://www.bcvideo.com/bmars.html

http://herotwins.hypermart.net/Crowned/CrownedFace.htm

http://metaresearch.org/home/viewpoint/archive/010313GlassyTubes/Meta-in-News010313.asp

http://www.viewzone.com/marsobject.html

Insights into Martian water reservoirs from analyses of Martian meteorite QUE94201 by Laurie A. Leshin; December 1, 1999; published in Geophysical Research Letters Volume 27, Issue 14, pages 2017–2020, 15 July 2000

http://palermoproject.com/lowell2004/legacy.htm

http://palermoproject.com/lowell2004/legacy2.htm

http://palermoproject.com/lowell2004/legacy6.htm

http://www.mactonnies.com/imperative42.html

http://www.marsanomalyresearch.com/evidence-reports/2001/029/huge_nozzle.htm

http://www.gods-and-monsters.com/cynocephalus.html

https://www.youtube.com/watch?v=urdtDcua5ik

"Avian Formation on a South-Facing Slope along the Northwest Rim of the Argyre Basin" by Michael A. Dale, George J. Haas, James S. Miller, William R. Saunders, A. J. Cole, Joseph M. Friedlander and Susan Orosz; 6/10/2011; Journal of Scientific Exploration, Vol. 25, No. 3

http://www.marsanomalyresearch.com/evidence-reports/2009/167/parrotopia.htm

https://www.youtube.com/watch?v=RxbCtBjkW98

CHAPTER 82
THE EVIDENCE IS CONCLUSIVE - MARTIANS DO EXIST!!!

http://marsrover,nasa.gov/gallery/panoramas/spirit/2005.html

462

http://en.wikipedia.org/wiki/Aerial_photographic_and_satellite_image_interpretation

http://www.marsanomalyresearch.com/evidence-reports/2006/102/mars-humanoid-skull.htm

https://www.youtube.com/watch?v=RacHZpYXhU8

Dark Mission: The Secret History of NASA by Richard C. Hoagland and Mike Bara; 2007; a Feral House Book; Los Angeles, CA; ISBN: 978-1-932595-26-0

http://archives.weirdload.com/nasa-shame.html

CHAPTER 83
ARE MARTIANS WATCHING OUR MARS SATELLITES AND ROVERS?

http://www.nasa.gov/mission_pages/msl/multimedia/pia16453.html#.VO7PDuH7OVp

UFOs and the National Security State, the Cover-up Exposed 173 – 1991 by Richard M. Dolan; 2009; Published by Keyhole Publishing Company; USBN 978-0-9677995-1-3

http://www.paranoiamagazine.com/2013/01/alternative-3-end-game-of-the-new-world-order/

https://www.youtube.com/watch?v=jSDBl0FMX0s

https://www.youtube.com/watch?v=gmNFzBVKqyE

http://breakawaycivilization.com/

http://archives.weirdload.com/nasa-shame.html

CHAPTER 82
PHOBOS AND DEIMOS

4http://en.wikipedia.org/wiki/Moons_of_Mars

"Mars' Moon Phobos May Be The Death Star"; published by Martin J. Clemens; 26 March, 2013

http://martinjclemens.com/mars-moon-phobos-may-be-the-death-star/

http://www.bibliotecapleyades.net/marte/marte_phobos05.htm

http://www.enterprisemission.com/Phobos.html

https://www.youtube.com/watch?v=j4LArNlspjg

Mars' Moon Phobos May Be The Death Star"; published by Martin J. Clemens; 26 March, 2013

http://www.paranormalpeopleonline.com/mars-moon-phobos-may-be-the-death-star/

Casey Kazan, "European Space Agency: Mars Moon Phobos 'Artificial'"

http://realityzone-realityzone.blogspot.ca/2010/06/european-space-agency-mars-moon-phobos.html

http://www.enterprisemission.com/Phobos.html

http://www.youtube.com/watch?v=oaiSfn8jlxY

http://www.mactonnies.com/imperative17.html

The Monuments of Mars: A City on the Edge of Forever by Richard C. Hoagland; 1987; published by North Atlantic Books; ISBN 1-55643-118-X

http://www.enterprisemission.com/Phobos2.html

http://www.mactonnies.com/imperative17.html

http://www.dailymail.co.uk/sciencetech/article-1204254/Has-mystery-Mars-Monolith-solved.html

https://www.youtube.com/watch?v=Pnt5WKo6SCY

https://www.youtube.com/watch?v=uvkr9ZXnnGs

http://www.oneism.org/orig_right_wit.php

http://thehiddenrecords.com/mars.htm

CHAPTER 85
THE ASTEROID BELT - DID OUR SOLAR SYSTEM EXPERIENCE AN ANCIENT INTERSTELLAR WAR?

http://www.nasca.org.uk/Ancient_Nuc__War/ancient_nuc__war.html

The 12th Planet by Zecharia Sitchin; 1976; published by Avon Books, New York, New York; ISBN0-380-39362-X

http://en.wikipedia.org/wiki/Valles_Marineris

http://www.abovetopsecret.com/forum/thread518421/pg1

http://www.nasca.org.uk/Ancient_Nuc__War/ancient_nuc__war.html

Carl Sagan, "Cosmos" (Random House, 1980)

http://www.enterprisemission.com/Rosetta/Rosetta-analysis-test.htm

http://www.metaresearch.org/solar%20system/eph/eph2000.asp

http://science.nationalgeographic.com/science/prehistoric-world/dinosaur-extinction/

CHAPTER 86
VESTA AND CERES, THE DWARF PLANETS IN THE ASTROID BELT

http://en.wikipedia.org/wiki/4_Vesta

http://www.nasa.gov/mission_pages/dawn/ceresvesta/#.VRjOUOH7OVo

http://hagablog.co.uk/demos/enceladus/volcanism/

https://www.youtube.com/watch?v=jSDBl0FMX0s

http://dawn.jpl.nasa.gov/feature_stories/Bright_Spot_Ceres_Dimmer_Companion.asp

CHAPTER 87
JUPITER ("KING OF THE GODS")

http://en.wikipedia.org/wiki/2010_%28film%29

http://planetfacts.org/mass-and-density-of-jupiter/

https://en.wikipedia.org/wiki/Atmosphere_of_Jupiter

http://www.enterprisemission.com/europa2000.htm

http://en.wikipedia.org/wiki/Europa_%28moon%29

https://en.wikipedia.org/wiki/Ganymede_%28moon%29

https://en.wikipedia.org/wiki/Callisto_%28moon%29

http://slushpup62.blogspot.ca/2010/01/nasa-pics-show-ufos-orbiting-jupiters.html

CHAPTER 88
SATURN ("GOD OF PLENTY, WEALTH, AGRICULTURE AND FEASTING")

http://en.wikipedia.org/wiki/Saturn

http://www.dailygalaxy.com/my_weblog/2013/04/the-vortex-of-saturns-colossal-hexagon-storm-.html

http://www.dailygalaxy.com/my_weblog/2012/11/saturns-hexagon-one-of-the-most-bizarre-things-seen-in-the-solar-system.html

http://www.universetoday.com/24029/atmosphere-of-saturn/#ixzz2alFWBoht

http://www.abc.net.au/science/space/planets/saturn.htm

http://en.wikipedia.org/wiki/Rings_of_Saturn

http://spacespin.org/article.php?story=cassini_new_details_saturn_rings

CHAPTER 89
"PREPARE YOUR MINDS FOR A WHOLE NEW SCALE OF PHYSICAL SCIENTIFIC VALUES, GENTLEMEN!"

http://en.wikipedia.org/wiki/Forbidden_Planet

http://voyager.jpl.nasa.gov/

Ringmakers of Saturn© Norman R. Bergrun; 1986 by The Pentland Press Ltd Kippielaw by Haddington East Lothian EH41 4PY Scotland; Library of Congress Catalogue Card Number 86-81530; ISBN 0 946270 33 3

http://www.ringmakersofsaturn.com/About%20Bergrun.htm

The Seven Valleys and the four Valleys by Bah'u'llah; translated by Ali-Kuli Khan and Marieh Gail in 1945; Baha'i Publishing Trust; Wilmette, Illinois

The Hidden Words of Bahá'u'lláh by Bahá'u'lláh; translated by Shoghi Effendi; 1970; Baha'i PublishingTrust; Wilmette, Illinois

CHAPTER 90
GIGANTIC ANOMALIES ORBITING SATURN

Ringmakers of Saturn© Norman R. Bergrun; 1986 by The Pentland Press Ltd Kippielaw by Haddington East Lothian EH41 4PY Scotland; Library of Congress Catalogue Card Number 86-81530; ISBN 0 946270 33 3

http://saturn.jpl.nasa.gov/mission/introduction/

https://www.youtube.com/watch?time_continue=257&v=ibT4SFNcGcY

http://saturn.jplnasa.gov/news/press-release-details.cfm?newswsID=589

http://www.space.com/5800-partial-rings-discovered-saturn.html

http://www.nature.com/news/2009/091007/full/news.2009.979.html

http://www.youtube.com/watch?v=NW9r17nEpSc

CHAPTER 91
THE MOONS OF SATURN

http://en.wikipedia.org/wiki/Hyperion_%28moon%29

http://lightsinthetexassky.blogspot.ca/2010/05/shine-on-hyperion.html

http://en.wikipedia.org/wiki/Titan_%28moon%29

http://en.wikipedia.org/wiki/Life_on_Titan

http://www.scientificamerican.com/article.cfm?id=enceladus-secrets

http://phys.org/news/2013-07-gravitational-tide-secret-saturn-weird.html

http://en.wikipedia.org/wiki/Mimas_%28moon%29

http://www.seasky.org/solar-system/saturn-tethys.html

http://www.seasky.org/solar-system/saturn-dione.html

http://www.seasky.org/solar-system/saturn-rhea.html

http://www.seasky.org/solar-system/saturn-iapetus.html

http://www.enterprisemission.com/moon1.htm

http://www.enterprisemission.com/moon2.htm

http://www.enterprisemission.com/moon3.htm

http://www.enterprisemission.com/moon4.htm

CHAPTER 91
URANUS ("GOD OF THE SKY")

http://en.wikipedia.org/wiki/Uranus

http://en.wikipedia.org/wiki/Moons_of_Uranus

http://en.wikipedia.org/wiki/Miranda_%28moon%29

http://www.bibliotecapleyades.net/esp_dayaftertomorrow3.htm

CHAPTER 92
NEPTUNE ("GOD OF THE SEA")

http://en.wikipedia.org/wiki/Neptune

http://www.bibliotecapleyades.net/esp_dayaftertomorrow3.htm

http://en.wikipedia.org/wiki/Triton_%28moon%29

CHAPTER 93
PLUTO ("GOD OF THE UNDERWORLD")

http://space-facts.com/pluto/

https://www.forbes.com/sites/jamiecartereurope/2019/05/22/is-there-alien-life-on-pluto-liquid-ocean-the-size-of-texas-could-change-where-we-look-for-life/#35fbcef926b3

https://www.newser.com/story/218930/no-those-arent-snails-on-pluto.html

http://en.wikipedia.org/wiki/Pluto#cite_note-Levison2007-134

Baha'i World Faith (Selected Writings of Bahá'u'lláh and Abdu'l-Bahá); 1943; National Spiritual Assembly of the United States; Baha'I Publishing Trust; Wilmette, USA

http://www.bibliotecapleyades.net/esp_dayaftertomorrow3.htm

CHAPTER 94
DWARF PLANETS OF THE KUIPER BELT, THE OORT CLOUD, EXOPLANETS, AND BEYOND

No References for This Chapter

CHAPTER 95
EXOPLANETS AND EXTRATERRESTRIAL LIFE IN THE UNIVERSE

https://en.wikipedia.org/wiki/Methods_of_detecting_exoplanets

https://exoplanets.nasa.gov/faq/6/how-many-exoplanets-are-there/

https://exoplanets.nasa.gov/discovery/how-we-find-and-characterize/

https://exoplanets.nasa.gov/search-for-life/can-we-find-life/

https://www.space.com/super-earth-exoplanet-gj-357d-may-support-life.html

https://www.space.com/29041-alien-life-evidence-by-2025-nasa.html

https://theconversation.com/exo-earths-and-the-search-for-life-elsewhere-a-brief-history-33096

https://theconversation.com/impacts-extinctions-and-climate-in-the-search-for-life-elsewhere-34791
https://www.youtube.com/watch?v=dpmXyJrs7iU&t=1s

https://www.youtube.com/watch?v=zcs3THdFPpo

https://theconversation.com/what-makes-one-earth-like-planet-more-habitable-than-another-33479

Baha'u'llah; Seven Valleys and Four Valleys; 1942 and 1952; Baha'I Publishing Trust; Wilmette, Illinois, USA Library of Congress C.C. 53-12275

https://bahai-library.com/abdulbaha_lawh_aflakiyyih

https://www.nature.com/news/the-quantum-source-of-space-time-1.18797

https://en.wikipedia.org/wiki/Self-similarity

https://en.wikipedia.org/wiki/Scale_invariance

https://en.wikipedia.org/wiki/Self-similarity

https://en.wikipedia.org/wiki/Scale_invariance

Bahá'u'lláh, Gleanings from the Writings of Bahá'u'lláh, p. 162-163

Shoghi Effendi, Lights of Guidance, p. 481

Shoghi Effendi, Lights of Guidance, p. 472

From a letter dated 15 February 1957 written on behalf of Shoghi Effendi to several believers

https://bahai-library.com/abdulbaha_lawh_aflakiyyih

https://bahaiquotes.com/subject/ufos11

From a letter dated 11 January 1982 written on behalf of the Universal House of Justice to an individual believer). See

The World Order of Bahá'u'llah: Selected Letters (Wilmette: Bahá'í Publishing Trust, 1991), pp. 170-186. And

T he Universal House of Justice, 1996 Aug 06, Sabeans, UFOs, Alien Abduction and Genetic Engineering

Universal House of Justice, 1992 June 08, Gaia Concept, Nature, p.11

https://bahaiquotes.com/subject/ufos

INDEX (VOLUME FIVE)

D

H

I

About the Author

Terry Tibando's background experience and understanding of this phenomenon spans 65 years of personal UFO sightings and ET contact that began at the age of five years. This childhood experience initiated a lifetime of many other-worldly sightings and encounters into a mysterious universe of Unidentified Flying Objects, Extraterrestrial Intelligence and the paranormal. As an experiencer, researcher, and investigator in Ufology he brings a unique and refreshing perspective on this subject based on a world view.

While attending Victoria High School in the mid sixties, Terry began attending UFO lectures meeting such people as Dr. Edward Edwards, a linguist from the University of Victoria and a fellow member of APRO (Aerial Phenomenon Research Organization and also Daniel Fry from New Mexico, USA, well known contactee and UFO author.

Terry attended the University of Victoria majoring in astronomy, physics, math and other sciences. During those university years other alien craft were sighted near his family's home in Victoria leading Terry to theorized that a possible undersea ET base existed off the coast of Vancouver Island which may account for the numerous UFO sightings seen over the Island.

He was a former member of APRO and its Canadian sister organization CAPRO during the sixties. His investigative research culminated back in the summer of 1996 when he met with Dr. Steven M. Greer during a one week "Ambassadors to the Universe" training seminar. They soon discovered that they shared similar UFO/ETI experiences during their early life.

Terry was a speaker at the Bellingham UFO Group (BUFOG) UFO seminar in 1996, and as a panel speaker along with Peter Davenport from NUFORC and Sharon Filip, alien abduction researcher.

He has talked on the Grimerica blog talk radio and been interviewed on the Discovery Channel during their "Alien Week" series in 1997 which had two ET spacecraft show up during the TV interview; he has been interviewed on BCTV News and appeared briefly in Dr. Greer's successful documentary movie "Sirius" and was a major financial contributor to the current documentary "Unacknowledged"!

He was instrumental in coordinating, hosting and emceeing the first Disclosure Project event on UFOs and ETS in Canada as a part of Dr. Greer's Disclosure Witness Tour held at Simon Fraser University in Vancouver on September 9, 2001, which included guest speakers Dr. Steven Greer, Dr. Carol Rosin and Dr, Alfred Webre.

For the last 27 years, Terry has been the field coordinator of CSETI Vancouver leading teams of people on field expeditions to successfully establish contact and communications with extraterrestrial intelligences visiting the Earth. Currently, he is finishing the remaining four volumes in this series in preparation for publishing and printing.